AF832174

Information Systems Engineering and Management

Volume 35

Series Editor

Álvaro Rocha, ISEG, University of Lisbon, Lisbon, Portugal

Editorial Board

Abdelkader Hameurlain, Université Toulouse III Paul Sabatier, Toulouse, France

Ali Idri, ENSIAS, Mohammed V University, Rabat, Morocco

Ashok Vaseashta, International Clean Water Institute, Manassas, VA, USA

Ashwani Kumar Dubey, Amity University, Noida, India

Carlos Montenegro, Francisco José de Caldas District University, Bogota, Colombia

Claude Laporte, University of Quebec, Québec, QC, Canada

Fernando Moreira, Portucalense University, Berlin, Germany

Francisco Peñalvo, University of Salamanca, Salamanca, Spain

Gintautas Dzemyda, Vilnius University, Vilnius, Lithuania

Jezreel Mejia-Miranda, CIMAT - Center for Mathematical Research, Zacatecas, Mexico

Jon Hall, The Open University, Milton Keynes, UK

Mário Piattini, University of Castilla-La Mancha, Albacete, Spain

Maristela Holanda, University of Brasilia, Brasilia, Brazil

Mincong Tang, Beijing Jaiotong University, Beijing, China

Mirjana Ivanovíc, Department of Mathematics and Informatics, University of Novi Sad, Novi Sad, Serbia

Mirna Muñoz, CIMAT Center for Mathematical Research, Progreso, Mexico

Rajeev Kanth, University of Turku, Turku, Finland

Sajid Anwar, Institute of Management Sciences, Peshawar, Pakistan

Tutut Herawan, Faculty of Computer Science and Information Technology, University of Malaya, Kuala Lumpur, Malaysia

Valentina Colla, TeCIP Institute, Scuola Superiore Sant'Anna, Pisa, Italy

Vladan Devedzic, University of Belgrade, Belgrade, Serbia

The book series "Information Systems Engineering and Management" (ISEM) publishes innovative and original works in the various areas of planning, development, implementation, and management of information systems and technologies by enterprises, citizens, and society for the improvement of the socio-economic environment.

The series is multidisciplinary, focusing on technological, organizational, and social domains of information systems engineering and management. Manuscripts published in this book series focus on relevant problems and research in the planning, analysis, design, implementation, exploration, and management of all types of information systems and technologies. The series contains monographs, lecture notes, edited volumes, pedagogical and technical books as well as proceedings volumes.

Some topics/keywords to be considered in the ISEM book series are, but not limited to: Information Systems Planning; Information Systems Development; Exploration of Information Systems; Management of Information Systems; Blockchain Technology; Cloud Computing; Artificial Intelligence (AI) and Machine Learning; Big Data Analytics; Multimedia Systems; Computer Networks, Mobility and Pervasive Systems; IT Security, Ethics and Privacy; Cybersecurity; Digital Platforms and Services; Requirements Engineering; Software Engineering; Process and Knowledge Engineering; Security and Privacy Engineering, Autonomous Robotics; Human-Computer Interaction; Marketing and Information; Tourism and Information; Finance and Value; Decisions and Risk; Innovation and Projects; Strategy and People.

Indexed by Google Scholar. All books published in the series are submitted for consideration in the Web of Science.

For book or proceedings proposals please contact Alvaro Rocha (amrrocha@gmail.com).

Andreia de bem Machado · Maria Jose Sousa ·
Andrea Brambilla · Antonio Pesqueira ·
Alvaro Rocha
Editors

Environmental, Social, Governance and Digital Transformation in Organizations

Springer

Editors
Andreia de bem Machado
Universidade Federal de Santa
Catarina (UFSC)
Florianópolis, Brazil

Andrea Brambilla
Dipartimento ABC
Politecnico di Milano
Milano, Italy

Alvaro Rocha
University of Lisbon
Lisboa, Portugal

Maria Jose Sousa
Iscte - University Institute of Lisbon
Lisboa, Portugal

Antonio Pesqueira
Iscte - University Institute of Lisbon
Lisboa, Lezíria do Tejo, Portugal

ISSN 3004-958X ISSN 3004-9598 (electronic)
Information Systems Engineering and Management
ISBN 978-3-031-86078-2 ISBN 978-3-031-86079-9 (eBook)
https://doi.org/10.1007/978-3-031-86079-9

© The Editor(s) (if applicable) and The Author(s), under exclusive license to Springer Nature Switzerland AG 2025, corrected publication 2025

This work is subject to copyright. All rights are solely and exclusively licensed by the Publisher, whether the whole or part of the material is concerned, specifically the rights of translation, reprinting, reuse of illustrations, recitation, broadcasting, reproduction on microfilms or in any other physical way, and transmission or information storage and retrieval, electronic adaptation, computer software, or by similar or dissimilar methodology now known or hereafter developed.
The use of general descriptive names, registered names, trademarks, service marks, etc. in this publication does not imply, even in the absence of a specific statement, that such names are exempt from the relevant protective laws and regulations and therefore free for general use.
The publisher, the authors and the editors are safe to assume that the advice and information in this book are believed to be true and accurate at the date of publication. Neither the publisher nor the authors or the editors give a warranty, expressed or implied, with respect to the material contained herein or for any errors or omissions that may have been made. The publisher remains neutral with regard to jurisdictional claims in published maps and institutional affiliations.

This Springer imprint is published by the registered company Springer Nature Switzerland AG
The registered company address is: Gewerbestrasse 11, 6330 Cham, Switzerland

If disposing of this product, please recycle the paper.

Foreword

Dear Reader,

Our time is marked by two major transformations—Sustainability and Digitalization.

Sustainability encompasses an immense set of ideas, policies and actions that seek to answer a question that Humanity had never faced: the awareness of its own limits and the smallness of its planet in the face of its population growth and its model of economic development. Perhaps this realization was more vivid when we behold the photograph of the Earth as seen from space. A lonely, blue, marine, and misty sphere, the only place where we live, in the immensity of the Universe around us.

In the meantime, technology has created the electronics of information, and the Network has covered the whole world. The most varied phenomena have come to have a digital description and have become "information". Processing this type of knowledge allowed us to obtain a greater volume and quality of new data, and to collect more and more, continuing and increasing the process, until Artificial Intelligence.

Environmental, Social, and Governance (ESG), as a performance accountability model, extended to the dimensions of risks and opportunities in relation to the Environment and the Social Environment, also includes, and in a very innovative way, the dimension of the Organization itself and its performance in the way it ensures its own resilience and incorporates the human requirements of ethics, dignity, diversity, equity, justice, safety, predictability, building the future, health and well-being, in short, how the Organization enables individual happiness while fulfilling its Purpose.

Thus, the challenge of producing knowledge about these phenomena was posed, to which the Academy responds with research and reflection.

The book that you hold in your hands, *Environmental, Social, Governance and Digital Transformation in Organizations*, explores the intersection of Sustainability, ESG report and digital transformation.

Authored by experts in the field, coming from all over the world, this is a comprehensive guide explores the complexities of implementing ESG frameworks within organizations and elucidates the relationship between digitalization and sustainable business practices.

As the world runs forward into the technological advancement and the global interconnection and interdependence, the imperative for organizations to prioritize ESG principles has never been more urging.

From "understanding the historical context of ESG in companies" to "leveraging technology for sustainable development", this book provides invaluable insights and practical strategies for the complexities of the modern business environment. Also, whether it's "environmental management in the digital age" or "the role of ESG in risk management", each chapter offers crucial insights and some amazing strategies for organizations who wants to acquire corporate responsibility.

As an ESG professional deeply committed to sustainability programs, involved in initiatives that aims to take in progress this development of sustainability, such as the ISO 26000 Working Group on Social Responsibility and the United Nations Global Compact, I find this book to be an amazing contribution for the responsible business practices.

It is my sincere hope that this reading will inspire and empower all the organizations to embrace ESG principles, driving positive change and shaping a future that is not only prosperous but also equitable and sustainable for all.

Together, let's tile the way for a future that prioritizes prosperity, equality, and sustainability for everyone, leaving no one behind.

Lisbon, Portugal Mário Parra da Silva

Introduction

In the realm of Environmental, Social, and Governance (ESG) programs, a significant evolution is observed from ethical or value-based initiatives to mainstream approaches. This transition is underscored by the increasing importance of best-practice ratings in evaluating the performance of ESG programs, as highlighted by Blank, Sgambati, and Truelson (2016). Furthermore, the role of ESG frameworks in enhancing compatibility in technology-based education, particularly for special groups, is evident in the work of Pavlis-Korres and Barriocanal (2013). Additionally, the development of toolkits for supporting export-oriented ESG projects, as noted by Perskaya et al. (2023), reflects the growing practical application of ESG frameworks in international development, education, and research. However, the literature reveals a notable gap in the integration of ESG programs with digital transformation strategies, particularly in a global context, indicating a need for further research in this area.

The discourse on digital transformation reveals its characterization as an ongoing process within incumbent firms, driven by factors such as agility, business model transformation, and cultural shifts through the adoption of new digital technologies (Warner and Wäger 2019). The impact of digital transformation on Small and Medium Enterprises (SMEs) is particularly notable, as it enhances capabilities and processes but requires external support and consistent investment in Information and Communication Technology (ICT) (Ulaş 2019). Zaki (2019) further elucidates the transformative effect of digitalization in service industries, highlighting opportunities for creating and capturing value through new business models and customer experiences. Despite these insights, the literature exhibits a lack of longitudinal studies assessing the long-term impacts of digital transformation and its intersection with ESG programs.

The topics covered in this book answer the following questions: How will the digital transformation take place in organizations? What is the relationship between ESG and digital transformation? What is the relationship between ESG and the COVID-19 pandemic in the digital transformation of organizations? How can organizations promote actions to implement ESG through digital transformation? This paper will present some situations for improving, enhancing and qualifying ESG actions through digital transformation. Thus, the questions have been answered with the 22 chapters described below.

Chapter "ESG and Digital Transformation in Organizations" aims to map out, in the light of a systematic review, how digital transformation can contribute to the implementation of ESG. This research answered the following questions: (1) What is digital transformation? (2) What is ESG? (3) How can digital transformation contribute to ESG implementation? How can digital transformation contribute to ESG implementation? The results of the study made it possible to organize an agenda for proposing social transformation in companies through the ESG pillars.

Chapter "Digital Transformation in ESG Programs: Understanding Environmental, Social, and Governance Factors" explores the dynamic convergence of these two critical domains: Digital transformation in ESG programs—Understanding environmental, social and governance factors

Chapter "Does Corporate Governance Impact ESG Performance Reporting in Healthcare?" aims to analyze the factors that influence the reporting of Environmental, Social and Governance (ESG) performance by companies in the health sector in European Union (EU) countries.

Chapter "Environment, Social, and Governance Business: Challenges and Opportunities" presents an in-depth analysis of environmental, social and governance (ESG) criteria in the business context, outlining the intricate balance between challenges and emerging opportunities.

Chapter "AI and the Next Generation: Protecting Childhood in the Digital Age", meanwhile, explores the multifaceted impact of AI on children, examining both the potential advantages and the inherent risks.

Chapter "Political Governance and State Modernization: Bridging AI Research with e-Democracy in Portugal" analyzes and compares the implementation of NEPAP (NATO Environmental Policy and Action Plan) and NGAP (NATO Governance and Accountability Policy) in the three branches of the Portuguese Armed Forces.

Chapter "ESG and Stakeholder Engagement: Guidelines for Sustainable Organizational Development" addresses the integration of multiple and plural actors through the dimensions of stakeholder equity and sustainable organizational development (SOD).

Chapter "Strategic Leadership and Value Creation: Evaluating the Impact of ESG Program Implementations in the Life Sciences Industry from Real World Evidence and Market Access Perspectives" explains how real-world evidence and market access (RWEMA) connects commercial teams, research and development (R&D), and external stakeholders.

Chapter "Governance and Transparency: Ethical Leadership in the Information Age", goal is to conduct a comparative/quantitative analysis of the models used by organizations to assure ethical leadership and what's their impact on organizational behaviour and performance.

Chapter "Diffusion of Sustainable Business Ecosystems by Services Businesses", study is to investigate how business service can contribute to the diffusion of a sustainable business ecosystem.

Chapter "Smart Cities, People at the Center: A Collaborative Mapping with AI for a Sustainable Future" explores how collaborative mapping with Generative Artificial Intelligence (Gen-AI), centered on citizens for resilient and sustainable development, is vital in this environment and has emerged as an innovative technique to improve sustainable performance analysis. In order to fill the knowledge gap on the effects of this technology on the reliability and transparency of data relating to the sustainability of territories, this study investigates the role that Gen-AI plays in the use of collaborative mapping with citizens.

The ESG chapter "The Positive Impact of ESG in the Business World" explains how strong ESG performance is better positioned for long-term success, also making organizations more competitive.

Chapter "Opportunities and Challenges of Sustainability Reports for Corporate Governance and Transparency in Portugal", study provides an overview about the 100 biggest companies of the district of Leiria, to evaluate how prepared they are to delivery their sustainability report next year.

Chapter "Assessing the Digital Competencies of Graduates for Sustainable Development: Bridging the Gap Between Academia and Industry" study aims to assess organizations' perceptions of new graduates' digital skills and compare them to the students' own perceptions. The study results can be used to redesign academic programs to better contribute to the promotion of SDGs and ESG.

Chapter "Sustainable Healthcare Investments: ESG Strategies for Hospital Infrastructure Renovation" aims to study the healthcare sector and adaptability to ESG (Environmental, Social, Governance) challenges.

Chapter "Strengthening Organizational Resilience in Response to Technostress Challenges", explores the relationship between technostress and countermeasures, with the aim of improving resilience and sustainability. ESG-based strategies, including social responsibility programs, education, awareness initiatives, detoxification, ethical leadership, risk programs and green IT are proposed to mitigate technological stress.

Chapter "The Relationship Between Technostress and Environment, Social, and Governance (ESG)", study examines the link between technostress and ESG (Environmental, Social, and Governance) in businesses. It discusses technostress's impact on business sustainability and potential countermeasures.

Chapter "Examining Successful ESG Strategies: A Collection of Case Studies from Various Sectors in Oman", presents a collection of case studies highlighting successful environmental, social, and governance (ESG) strategies implemented across various sectors in Oman.

Chapter "Toward a Sustainable Digital Future: A Comprehensive Study on Digital Transformation and Environmental Sustainability" explains the gap between digital transformation and environmental sustainability. Through a meticulous review of existing literature, the study will expose the various disruptions that digital transformation has triggered in this domain. Conclusions will culminate in the development of a robust framework, outlining transformations in four key areas: urban sustainability, waste management, pollution control and sustainable production.

Chapter "Implementing Digital ESG Strategies in Waste Management Companies", explores integrating digital technologies with ESG initiatives within waste management companies, highlighting how digital transformation can enhance sustainability practices.

Chapter "ESG in an industry 4.0 Context: A Promising Encounter" aims to explore the relationships between industry 4.0 and ESG.

Chapter "Ethical Practices in Digital Transformation Over the Last Half Decade: A Framework for Sustainability and Integrity in Business" research highlights the importance of robust and transparent governance, a commitment to digital inclusion, and the need for clear policies that ensure respect for human rights in the digital age.

References

Blank, H.D., Sgambati, G., Truelson, Z.: Best practices in ESG investing. J. Investing **25**(2), 103–112 (2016). https://doi.org/10.3905/joi.2016.25.2.103

De Bem Machado, A., Secinaro, S., Calandra, D., Lanzalonga, F.: Knowledge management and digital transformation for Industry 4.0: a structured literature review. Knowl. Manage. Res. Pract. **20**(2), 320–338 (2022)

Grinberga-Zalite, G., Zvirbule, A.: ESG investing issues in food industry enterprises: focusing on on-the-job training in waste management. Soc. Sci. **11**(9), 424 (2022)

Li, T.T., Wang, K., Sueyoshi, T., Wang, D.D.: ESG: Research progress and future prospects. *Sustainability*, **13**(21), 11663 (2021).

Naomi, P., Akbar, I.: Beyond sustainability: empirical evidence from OECD countries on the connection among natural resources, ESG performances, and economic development. Econ. Sociol. **14**(4), 89–106 (2021)

Pavlis-Korres, M., Barriocanal, E.G.: Promoting multiculturalism in technology based-education: a framework for the improvement of compatibility between educators of special groups and their learners and a case study on educators of Roma in Greece to prove its efficiency (2013). https://doi.org/10.4018/978-1-4666-2101-5.CH007

Perskaya, V., Starodubceva, E., Homyakova, L., Mamedov, T., Rybakov, V., Zvereva, A., Chizhova, A., Ogryzov, A.: ESG export project support tools based on international experience (2023). https://doi.org/10.12737/1906019

Sychenko, E.: Labour rights and international labour standards in the ESG Agenda. Ital. Labour Law E-J. **16**(1), 135–148 (2023)z

Tang, H.: The effect of ESG performance on corporate innovation in China: the mediating role of financial constraints and agency cost. Sustainability **14**(7), 3769 (2022). https://doi.org/10.3390/su14073769

Tiwari, S.P.: Organizational competitiveness and digital governance challenges. Arch. Bus. Res. (2022). https://doi.org/10.14738/abr.1003.12016

Ulaş, D.: Digital transformation process and SMEs. Procedia Comput. Sci. (2019). https://doi.org/10.1016/j.procs.2019.09.101

Vinodkumar, N., Alarifi, G.: Environmental social governance: a core value to responsible stakeholders and stock market sustainability in the Kingdom of Saudi Arabia. J. Sustain. Finan. Investment **12**, 1085–1101 (2020). https://doi.org/10.1080/20430795.2020.1827602

Warner, K.S.R., Wäger, M.: Building dynamic capabilities for digital transformation: an ongoing process of strategic renewal. Long Range Plann. (2019). https://doi.org/10.1016/J.LRP.2018.12.001

Williamson, B.: Governing software: networks, databases and algorithmic power in the digital governance of public education. Learning Media Technol. **40**, 83–105 (2015). https://doi.org/10.1080/17439884.2014.924527

Zaki, M.: Digital transformation: harnessing digital technologies for the next generation of services. J. Serv. Mark. (2019). https://doi.org/10.1108/JSM-01-2019-0034

Contents

About the Editors

Andreia de bem Machado (Ph.D. in Engineering and Knowledge Management) Researcher at the Innovation Intelligence, Management and Technology Research Centre—IGTI at the Federal University of Santa Catarina. Adjunct professor and research coordinator at Faculdade Anasps Brasília—Brazil. Evaluator of higher education institutions for the Brazilian Ministry of Education. Director of Education at the Municipal Department of Education in Palhoça—Santa Catarina—Brazil. She was part of the technical committee for the Ministry of Education's National Textbook Programme. She has developed teaching materials for people with low vision at UNESCO. Her research interests include: innovation, education, digital transformation, management, hybrid education, digital technologies, active methodology, knowledge media, entrepreneurship, knowledge management, digital transformation and Environmental, Social and Governance.

Maria Jose Sousa (Ph.D. in Management) is a University Professor and a research fellow at ISCTE. Her research interests currently are public policies and political science, innovation and business management issues. She is a best seller author in ICT and People Management and has co-authored over 70 articles and book chapters and published in several scientific journals (e.g. *Journal of Business Research, UAIS, Future Generations Computer Systems, Journal of Medical Systems, International Journal of Knowledge, Culture and Change Management, Knowledge Management, Wseas Transactions on Business and Economics, Information Systems Frontiers* and others), she has organized and peer-reviewed international conferences. She is also External Expert of COST Association—European Cooperation in Science and Technology.

Andrea Brambilla (Architect, Ph.D. and Researcher at Politecnico di Milano) Department of Architecture, Construction Engineering and the Built Environment (DABC), Design and Health Lab.

Assistant Professor in "Health and Sustainability for the Built Environment" at the School of Architecture Urban planning Construction engineering (AUIC), Politecnico di Milano.

Scientific Secretariat of the Joint Research Partnership Healthcare Infrastructure (JRP HI) promoted by Politecnico di Milano, Fondazione Politecnico di Milano, private companies and institutions in the healthcare sector.

Author of 36 publications in scientific journals, invited and selected speaker at over 50 national/international conferences and seminars on the relationship between the Built Environment, Technological Innovation, Sustainability and Public Health.

Antonio Pesqueira is an experienced business and academic leader in the field of healthcare and pharmaceutical operations, with a focus on technology advancements and process optimization. He has pioneered various initiatives in commercial excellence and analytical strategies, tailoring training programs for commercial teams. Pesqueira holds academic positions and is actively involved in the Ph.D. program at ISCTE/IUL in the Technology and Architecture Research Department. His research encompasses digital technologies, healthcare data management, and the intersection of innovation, capabilities, and business. Notably, he has chaired significant forums and frequently headlines major industry events. As an author, Pesqueira has co-authored over 20 articles in esteemed journals and has been a critical reviewer for prominent publications, demonstrating his vast expertise in digital transformation, healthcare digitalization, and agile project management.

Alvaro Rocha holds the title of Honorary Professor, and holds a D.Sc. in Information Science, Ph.D. in Information Systems and Technologies, M.Sc. in Information Management, and BCs in Computer Science.

He is a Professor of Information Systems at the University of Lisbon—ISEG, researcher at the ADVANCE (the ISEG Centre for Advanced Research in Management), and a collaborator researcher at both LIACC (Laboratory of Artificial Intelligence and Computer Science) and CINTESIS (Center for Research in Health Technologies and Information Systems). His main research interests are maturity models, information systems quality, online service quality, requirements engineering, intelligent information systems, e-Government, e-Health, and information technology in education.

He is also Vice-Chair of the IEEE Portugal Section Systems, Man, and Cybernetics Society Chapter, and Founder and Editor-in-Chief of two Scopus and SCIMago journals: *JISEM—Journal of Information Systems Engineering and Management*; and *RISTI—Revista Ibérica de Sistemas e Tecnologias de Informação/Iberian Journal of Information Systems and Technologies*. Additionally, he is the Scientific Manager of the *Information Systems Engineering and Management* book series (https://shorturl.at/oyBKV) at Springer-Verlag, the world's leading publisher of publications in Science, Technology and Health.

Moreover, he has served as Vice-Chair of Experts for the European Commission's Horizon 2020 Program, and as an Expert at the COST—intergovernmental framework for European Cooperation in Science and Technology, at the European

Commission's Horizon Europe Program, at the Government of Italy's Ministry of Universities and Research, at the Government of Latvia's Ministry of Finance, at the Government of Mexico's National Council of Science and Technology, at the Government of Polish's National Science Centre, and at the Government of Cyprus's Research and Innovation Foundation.

ESG and Digital Transformation in Organizations

Andreia de bem Machado, Antonio Pesqueira, João Rodrigues dos Santos, Antonio Sacavém, Maria Jose Sousa, and Fernando José Calado e Siva Nunes Teixeira

Abstract Environmental, Social and Governance (ESG) issues are increasingly permeating companies' decisions about which practices adopting and what performance and return to expect from society and its stakeholders. The concept of ESG (Environmental, Social and Governance) is a broad set of issues, from carbon footprint to labor practices and corruption, which justify the creation of criteria and practices that direct the role and responsibility of business towards environmental,

A. de bem Machado (✉)
Engineering and Knowledge Management Department, Federal University of Santa Catarina (UFSC), Florianópolis, Brazil
e-mail: andreia.bem@ufsc.br

A. Pesqueira
Technology and Architecture Research Department, ISCTE-IUL - University Institute of Lisbon, Lisboa, Portugal
e-mail: Antonio_Pesqueira@iscte-iul.pt

J. R. Santos · A. Sacavém
Universidade Europeia, Lisbon, Portugal
e-mail: Joao-Rodrigues.Santos@universidadeeuropeia.pt

A. Sacavém
e-mail: antonio.sacavem@universidadeeuropeia.pt

J. R. Santos
Centre for Transdisciplinary Development Studies (CETRAD), University of Trás-os-Montes (UTAD), Vila Real, Portugal

M. J. Sousa
Political Science and Public Policy Department, ISCTE-Instituto Universitário de Lisboa, Lisbon, Portugal
e-mail: maria.jose.sousa@ISCTE-IUL.pt

F. J. C. S. N. Teixeira
Business Sciences Department, Polytechnic Institute of Beja, Beja, Portugal

Smart Cities Research Centre (Ci2), Polytechnic Institute of Tomar, Tomar, Portugal

F. J. C. S. N. Teixeira
e-mail: fernando.teixeira@ipbeja.pt

© The Author(s), under exclusive license to Springer Nature Switzerland AG 2025
A. d. b. Machado et al. (eds.), *Environmental, Social, Governance and Digital Transformation in Organizations*, Information Systems Engineering and Management 35, https://doi.org/10.1007/978-3-031-86079-9_1

social and corporate governance factors. The aim of this chapter is therefore to map out, in the light of a systematic review, how digital transformation can contribute to the implementation of ESG. To do this, a systematic review will be carried out on the scopus and Web of Science databases, answering the following questions: (1) What is digital transformation? (2) What is ESG? (3) How can digital transformation contribute to ESG implementation? The results make it possible to organize an agenda for proposing social transformation in companies through the ESG pillars.

Keywords ESG · Systematic review · Transformation · Pillars · Companies · Digital · Governance

1 Introduction

Digital transformation has become one of the main drivers of change in contemporary organizations, impacting not only internal processes, but also the way these companies interact with their stakeholders. In this context, the integration of Environmental, Social, and Governance (ESG) practices has proved crucial for the sustainable development of organizations. The concept of ESG, which encompasses environmental, social and governance aspects, has been gaining prominence, especially when aligned with digital transformation, which provides tools and methodologies to monitor, manage and report on sustainable practices in a more efficient and transparent way.

The intersection between ESG and digital transformation reflects a paradigm shift in the way companies operate and position themselves in the market. Recent studies indicate that companies that adopt digital strategies in conjunction with ESG practices tend to achieve better financial and reputational performance [61]. Digitalization enables process automation, waste reduction and improved energy efficiency, while facilitating communication and engagement with stakeholders.

In addition, digital transformation offers new opportunities for corporate governance, enabling greater transparency and accountability through technologies such as blockchain and artificial intelligence. These technologies can be used to create more robust and transparent governance systems that ensure regulatory compliance and promote investor and public confidence.

In this context, the aim of this chapter is to map out, in the light of a systematic review, how digital transformation can contribute to the implementation of ESG. To this end, a systematic search was carried out in the Scopus and Web Of Science (WOS) databases. The research is divided into five parts. The first refers to this introduction, the second explains ESG and the Implementation Status in Different Geographical Areas, the third deals with the methodology and the results of the research, the fourth and final part presents digital transformation contributions to the implementation of ESG in organizations and the fifth part the final considerations.

2 Environmental, Social and Governance—ESG

The theoretical foundation of Environment, Social, and Governance (ESG) programs intersects with the broader discourse on sustainable development and corporate strategic initiatives. ESG principles have become essential to organizational strategy, enhancing ethical performance and long-term sustainability. The three pillars of ESG—Environmental, Social, and Governance—address distinct yet interconnected areas of corporate responsibility.

The environmental pillar focuses on the company's impact on the natural environment, including efforts to reduce carbon footprints, manage waste, and ensure sustainable resource use. This aligns with the Sustainable Development Goals (SDGs), emphasizing global efforts to combat climate change and promote ecological sustainability [5]. The social dimension encompasses the company's relationships with its stakeholders, such as employees, customers, and communities. Key areas include labor practices, diversity, inclusion, and community engagement. Socially responsible companies tend to enhance employee satisfaction and productivity while contributing positively to community development [2]. Governance involves internal systems and controls that ensure transparency, accountability, and ethical business practices, including board diversity, executive compensation, anti-corruption measures, and compliance with laws and regulations. Strong governance practices are crucial for maintaining investor trust and ensuring sustainable business operations [4].

The integration of ESG principles into corporate strategy is grounded in a rich theoretical foundation emphasizing sustainable development, corporate responsibility, and ethical governance. Current literature supports these theoretical underpinnings, providing empirical evidence and practical insights into the implementation and impact of ESG practices on organizational performance.

Empirical research demonstrates that companies engaged in environmental initiatives tend to achieve better sustainability outcomes and attract more investors. For example, Baratta et al. [5] highlight the increasing adoption of ESG practices by European companies to reduce carbon emissions and comply with new regulations, confirming that environmental sustainability is crucial for long-term corporate success.

The theoretical perspective on the social pillar emphasizes labor practices, diversity, inclusion, and community engagement. Research supports this by showing that companies with robust social policies often see enhanced employee satisfaction and productivity, contributing positively to overall organizational performance. Amaral et al. [2] found that human-centric organizations prioritizing ESG practices impact the environment by reducing health risks and workloads and improving social outcomes through labor management and diversity programs, underscoring the importance of social sustainability for corporate success.

Governance, theoretically focused on internal systems and controls ensuring transparency, accountability, and ethical business practices, is corroborated by recent studies. These studies show that strong governance practices, such as board diversity and anti-corruption measures, are critical for maintaining investor trust and sustainable business operations. Bandeira et al. [4] provide insights into how companies develop comprehensive ESG strategies that include stringent governance measures to promote ethical conduct and regulatory compliance, reinforcing the theoretical emphasis on governance for sustainable development.

Theoretical frameworks suggest that integrating ESG principles can enhance financial performance and attract investor interest. The current literature provides robust empirical support for this claim. Zamfiroiu and Pînzaru [80] highlight that companies performing well on ESG metrics often exhibit better financial performance and lower risk profiles, underscoring the financial viability of ESG integration as a strategic decision.

Research confirms that the development of robust legal frameworks has been pivotal in driving ESG adoption. Gossain [29] discusses how international guidelines, such as the TCFD, have standardized ESG reporting and practices, facilitating better implementation and compliance. This alignment highlights the critical role of regulatory frameworks in translating theoretical principles into actionable corporate practices.

Effective ESG implementation requires specific organizational competencies and maturity models, supported by current literature. Sierdovski et al. [59] discuss how organizational competencies, such as corporate social responsibility and technical skills, are essential for developing ESG criteria in the industrial sector. This empirical evidence aligns with the theoretical perspective, emphasizing the need for internal capabilities to drive ESG success.

The theoretical foundation acknowledges challenges and barriers to integrating ESG practices into corporate strategies. Current literature identifies practical challenges, including the complexity of integration and the need for standardized metrics. Clementino and Perkins [12] highlight the varied corporate responses to ESG ratings, and the difficulties firms face in aligning ESG practices with their strategies, underscoring the need for continued research and development in this area.

The connection between theoretical foundations and the current state of ESG literature is robust and mutually reinforcing. Theoretical principles provide a solid foundation validated and expanded by empirical research, highlighting the importance of ESG integration for sustainable development and long-term corporate success. This synergy emphasizes the relevance and applicability of ESG principles in contemporary corporate strategies, ensuring that organizations can effectively address environmental, social, and governance challenges while enhancing their financial performance and sustainability.

Recent literature on ESG underscores the growing importance and integration of these principles into corporate strategies. Studies highlight ESG's role in enhancing financial performance, risk management, and long-term sustainability. Investors show significant interest in companies performing well on ESG metrics, as these companies tend to exhibit better financial performance and lower risk profiles [80]. Robust legal and regulatory frameworks have promoted ESG practices, with international guidelines like the TCFD standardizing ESG reporting and practices [29]. Effective ESG implementation requires specific organizational competencies and a maturity model integrating ESG principles into everyday business operations [59]. Despite the growing adoption of ESG principles, companies face challenges integrating these practices into existing business models and standardizing metrics [12].

The conceptual framework for understanding ESG involves a multi-dimensional approach, considering environmental, social, and governance factors as interrelated components of corporate strategy. This framework can be operationalized through various methodological approaches, including qualitative case studies, quantitative analyses, and mixed-method research designs. Systematic literature reviews provide a comprehensive overview of existing research on ESG, highlighting key findings, trends, and gaps [2]. Detailed case studies of companies successfully integrating ESG principles offer practical insights and best practices [14]. Quantitative studies using regression models and econometric analyses quantify the impact of ESG practices on financial performance and sustainability outcomes [77]. Comparing ESG practices across different regions and sectors reveals variations in implementation and effectiveness, as well as contextual factors influencing these practices [53].

The evolution of ESG as a critical component of corporate strategy has been driven by regulatory pressures, investor demands, and a growing recognition of the business case for sustainability. The shift from viewing ESG as a peripheral concern to recognizing it as central to corporate strategy and risk management is significant [80]. The establishment of international guidelines and regulations has provided a structured approach to ESG reporting and compliance [29]. Numerous studies demonstrate the positive correlation between strong ESG performance and financial outcomes, reinforcing the business case for sustainability [2]. While environmental aspects have historically received more attention, there is an increasing focus on the social and governance dimensions of ESG, particularly regarding labor practices, diversity, and ethical governance [12].

Despite extensive research on ESG, gaps remain. More standardized and universally accepted metrics for measuring ESG performance are needed to facilitate better comparison and benchmarking [5]. Longitudinal studies are required to understand the long-term impacts of ESG practices on organizational performance and sustainability [80]. While some industries have been extensively studied, others, such as small and medium enterprises (SMEs), require more focused research to understand their unique challenges and opportunities in implementing ESG [53]. Research should also explore how cultural and regional differences influence ESG practices, providing a more nuanced understanding of these dynamics [77].

The integration of ESG principles into organizational strategies is essential for sustainable development and long-term corporate success. The current literature underscores the importance of ESG in enhancing financial performance, managing risks, and ensuring ethical business practices. However, further research is needed to address existing gaps, standardize metrics, and explore the long-term impacts of ESG practices.

3 ESG Implementation Status in Different Geographical Areas

The implementation of ESG principles varies significantly across different geographical regions due to regional regulatory frameworks, economic conditions, cultural factors, and the maturity of corporate governance practices. This section examines the status of ESG implementation in various regions, including emerging markets, Europe, Indonesia, Russia, China, and the United States, highlighting the unique challenges and advancements within each area. In emerging markets, the adoption of ESG practices is heavily influenced by competition and economic pressures. Martins [44] utilized a difference-in-differences model in 22 emerging markets, including Brazil, and found that firms often adjust their ESG practices negatively following competitive shocks. This contrasts with developed economies, where competition typically drives improvements in ESG performance. The study underscores the need for tailored strategies to promote ESG in these regions, considering their unique economic dynamics and competitive pressures.

European companies are generally at the forefront of ESG implementation, driven by stringent regulations and high stakeholder awareness. Iamandi et al. [33] mapped the ESG behaviors of 1,165 European companies, revealing a mature and strategic approach to ESG integration. The study found that European firms prioritize environmental and social aspects over governance, adopting customized ESG strategies to enhance competitiveness and societal well-being. This maturity is reflected in the widespread adoption of ESG reporting and compliance with international standards.

In Indonesia, ESG implementation faces both opportunities and challenges. A qualitative study by [56] on PT. Migas-North Field identified significant progress in environmental and social indicators, such as greenhouse gas emissions reduction and gender pay ratio. However, gaps remain in areas like climate oversight and gender diversity in governance. This indicates partial implementation, where some ESG aspects are well-developed while others require further attention and resources.

The adoption of ESG principles in Russia is evolving, with a focus on regional development and environmental monitoring. Zatolokin and Petrov [81] explored the use of geoportals to support ESG strategies in the Tyumen region, highlighting the integration of environmental, social, and governance data to promote sustainable development. The study emphasizes the need for advanced digital platforms to enhance transparency and effectiveness in regional ESG initiatives.

China's traditional industrial sector faces significant challenges in adopting ESG practices due to limited awareness and financial constraints. Li and [38] suggest that raising awareness, promoting education, and providing financial incentives are crucial for enhancing ESG adoption. The integration of advanced technologies like AI and blockchain could facilitate better data analysis and decision-making, helping overcome existing barriers. Standardizing ESG metrics and fostering collaboration between government and industry are essential steps forward.

The United States and Europe exhibit distinct approaches to ESG implementation, particularly in investment strategies. Research by [6] demonstrated that ESG integration through index-based portfolios yields positive returns and risk reductions in both regions. The study found significant variations in ESG profiles and performance, suggesting that while both regions benefit from ESG investments, the underlying strategies and outcomes can differ based on regional regulatory environments and market conditions.

The implementation of ESG principles varies widely across different geographical areas, reflecting diverse regulatory frameworks, economic conditions, and cultural contexts. Emerging markets often struggle with competitive pressures that hinder ESG adoption, while Europe demonstrates a mature and strategic approach driven by robust regulations. Indonesia and Russia are making strides in specific areas but face challenges in achieving comprehensive ESG integration. China's industrial sector requires significant support to overcome financial and awareness barriers. In contrast, the United States and Europe benefit from established ESG investment frameworks that enhance financial performance and risk management. These regional differences underscore the need for tailored strategies to promote ESG globally, considering the unique challenges and opportunities within each geographical context.

4 Digital Transformation

Digital transformation is the process of integrating digital technologies across all aspects of business, through which the way of doing business and how businesses render value to their clientele are changed forever. It is all about a culture shift that needs companies to continually alter the present situation, try new things, and be ok with failure. This is crucial to the organization's efforts to stay competitive in a fast-paced digital economy, and it also helps in the satisfaction of the ever-increasing consumer demand for innovative and customized products and services [32]. Such features characterize digital transformation as customer experience, operational agility, culture and leadership, workforce empowerment, and digital technologies integration. Using technologies of artificial intelligence and data analytics for customer service purposes creates a unique experience for customers, making them glad and loyal to the brand [36]. Digital tools are used to speed up and make processes more efficient which enhances operational agility. Culture and leadership play a significant role, as fostering a digital culture and educating leaders on digital skills are vital for that transformation to be successful [75]. Furthermore, workforce enablement includes furnishing employees with digital instruments that will help them get used to and deal appropriately in a digitized workplace.

Digital transformation is all about utilizing the latest technologies such as AI, IoT, or ML to create new business processes, and customer experiences and to remodel the culture of the organization [37]. The digital future lies therefore in the use of emerging technologies that will foster future efficiencies, mark real-time decision-making, and increase transparency in business operations. With the advances of digital transformation, businesses need to be flexible and responsive to innovations that will further disrupt conventional ways of doing business. Companies are becoming increasingly dependent on data analytics and automation to attain profitability and customer satisfaction. The different digital technologies that could be in the digital transformation strategy of the organization will be determined by the business requirements. There are a handful of digital instruments that make it evident that they will be around regardless of the extent of digital transformation in the future. Artificial Intelligence (AI) is a digital technology that uses computers and machines to replicate human reasoning. When AI is combined with automation, the machine becomes smart and makes intelligent, real-time decisions on any workflow. This AI revolution brings in the innovation of smart products, and the user and customer experience has become more direct. An instance is machine learning, which can allow a computer or machine to imitate the human mind. Also, artificial intelligence technology is used to simulate digital data and comprehend the physical world. The hybrid cloud ecosystem is another technology that introduces a single, good cloud solution for both public cloud, private cloud and on-premises infrastructure. It is about transforming a company's on-premises data into a private cloud infrastructure, which is then interconnected with a public cloud environment hosted by a public cloud provider. Illustrations consist of AWS®, Google Cloud Services®, IBM Cloud®, and Microsoft Azure®. Cloud computing infrastructure is a bridge for cloud resources that stretch the ability of an organization to run any workload on them. This operating model does not only streamline the operation but also simplifies the data organization. Organizations with hybrid cloud are not limited to a single platform, which creates a solid foundation for them for a successful digital transformation. Blockchain is a technology shared, immutable, public ledger or record of electronic transactions. The major advantage of blockchain is transaction transparency for these employees and security from the other employees who didn't need it. This kind of trust, for instance, illustrates how blockchain can create stronger communities both externally and internally. Finally, the internet of things (IoT) is a network of interconnected devices that exchange data with each other and the cloud. IoT devices usually have sensors as well as software and can be separate objects, machines, or devices as well as consumer electronics. More and more companies in different types of industries have started using IoT to automate their processes, provide better service to clients, make informed decisions, and increase the profitability of businesses. With IoT, data is transferable over a network without a need for human-to-human or human-to-computer interactions. An IoT thing can be a person with an implant monitoring his heart, an animal on the farm armed with a biochip transponder, a car that has built-in sensors to inform the driver when the tire pressure is low, or any other natural or man-made object that has been assigned an IP address and is able to transfer data over a network.

However, digital transformation is not just about deploying new technologies but about creating a culture where there is a restless desire for continuous change in an organization. The ongoing digital transformation of businesses entails striking a balance between the leveraging of technology as a strategic business tool and investment in long-term goals. This strategy guarantees long-term growth, innovation, and competitiveness in the digital world of the future.

In an era where digital transformation shapes the future of industries, accordingly to [22], Water Corporation, a state-owned utility based in Western Australia, has a network of pipelines that carries water, wastewater, and rainwater to an area which extends 2.6 million km approximately, the entity uses SAP architecture to run its critical resources and it realized that the server in-house which was supporting the SAP infrastructure was old. Instead of buying more hardware, the organization went to a cloud-based strategy. Water Corporation chose IBM Consulting to plan this complex migration. They opted for AWS to run SAP critical systems and IBM watsonx™ Code Assistant technology to come up with code recommendations for automation functions that include the migration and maintenance of the SAP environment. IBM Consulting and Water company estimated that automation approach has allowed the business to save about 1500 h that it used to spend on infrastructure support manually. In addition, it reduces carbon emissions by 150 tons per year approximately. The changeover to a new SAP environment was a digital transformation that required Water Corp and its partners to be on top of their game, but the result will help the latter learn how to be successful in the digital world.

Furthermore, the snack giant Frito-Lay opted for system and retailer service optimization through the salesforce. Frito-Lay's digital transformation initiatives hired user-oriented experts from IBM® Consulting and IBM Salesforce practice. Jointly, they spearheaded the creation of the Frito Lay e-commerce strategy and made a better workflow for the frontline employees. With the help of IBM Garage™ and IBM iX® team of experts that conducted extensive user research, Frito-Lay and IBM team developed 2 Salesforce solutions. These two solutions turned into new merchant tools as well as employee ones. 'Snacks to You' is an advanced e-commerce solution that helps small businesses streamline the ordering and delivery procedure. 'Sales Hub', driven by Salesforce Service Cloud, is a second solution being developed to make logistics easier, especially on the back end. These solutions allowed optimization for users and required re-engineering of the processes that were done in the previous times.

Digital transformation was shown to be a necessary strategic element for companies that want to prosper in the ever-changing business world and meet the challenges that come from issues regarding the sustainable development of society (environmental, social, and governance-related, ESG). Digital transformation profoundly affects organizations by reshaping their strategies, operations, and interactions with stakeholders, aligning with the critical imperatives of Environmental, Social, and Governance (ESG) frameworks. Nowadays, the introduction of digital technologies into business activities provides the possibility to increase the level of ESG by making operations more sustainable, societies more equitable and governance structures stronger. ESG and digital transformation represent two concepts that are mutually related.

Digital transformation with ESG initiatives can be done through digital tools and technology that can empower businesses to achieve their goals. These can help in the reduction of carbon emissions, safe storage of data, and the engagement of employees, customers, and other stakeholders. What is also important here is that the tools and technologies used will reflect the ESG considerations that such an organization has made. To make a move towards digital transformation an organization should keep in mind their ESG as an aim and should take steps related to developing policies and using tools to be able to support those. Not only do we observe various ways in which different sectors can use digital transformation to achieve their ESG goals, but we also notice that digital transformation is a multi-faced tool. For instance, in the case of the manufacturing industry digitization introduced vast improvements in ESG performance using innovations at the production level and business model improvement. One example is the study on Chinese manufacturing firms which proved that the strategies of digital transformation particularly those combined with business digitalization and platform digitalization can improve ESG results by boosting product and process innovation.

This transformative effect is especially notable in larger firms, where the scale of operations and resource consumption magnifies the benefits of digital efficiencies [38]. Moreover, the energy sector brings out an important point in connection with digital transformation's role it plays in realizing the ESG goals. The access to digital tools in this field has taken to a new level the operational efficiency combined with environmental responsibility. For instance, digital transformation strategies that encompass big data analytics and Internet of Things (IoT) technologies have been identified to run more efficient energy and lower emissions which support wider sustainability objectives like the European Green Deal and the UN's 2030 Agenda for Sustainable Development [49]. The financial sector, on the other hand, is an example that depicts how digitalization can promote ESG performance. As an example, technologies such as blockchain have taken it to another level as far as transparency and security in financial services which is now seen in improving the area of governance and compliance practice. Consequently, the digital transformation in the financial sector is found to be helpful for governance, which includes creating avenues for the inclusion and fairness in access to financial services which is a critical aspect of the social dimension of ESG [24].

Another key contributor toward digital transformation is the fact that organizations both small and large have to be effective and agile in their processes to comply with swift changes in the competitive environment and stakeholders' expectations. Organizations that incorporate digital technologies such as artificial intelligence, data analysis, cloud computing, and the Internet of Things (IoT) can improve efficiency, reduce waste, and provide timely operation reports [83]. In addition, digital transformation allows companies to incorporate the elements of ESG into their core business strategies and standardized operations. As an example, digital technologies can contribute to the sensation of environmental impact supervision and management through data collection and analysis performed in real time. This makes it easier for companies to track their carbon footprint, the amount of energy they use, and the level of waste they produce more accurately, hence supporting informed decision-making and specific interventions mainly targeted at reducing environmental damage [72].

Similarly, digital transformation would promote better governance and social responsibility with its transparency, accountability, and openness to collaboration. Through communications systems and digital platforms, organizations will receive feedback from employees, customers, suppliers, or communities. This will in turn create an atmosphere of confidence and trust [35]. Similarly, digital transformation enables organizations to harmonize ethical standards and form a supply chain that is green, each one playing the lead role within their respective business processes. Through digital tools that enable tracing product origins, monitoring the supply chain, and imposing ethical practices, firms can effectively conform to ESG regulations, avert reputational damage from unethical practices, and other debilitating consequence [79]. digital transformation represents a powerful enabler for organizations seeking to integrate ESG principles into their operations and achieve sustainable business outcomes. Through digital technology domination and the development of consciousness of innovation and responsibility, companies are able to conquer the numerous intricacies of running businesses that nevertheless contribute to all-round positive environmental, social, and governance attributes.

5 Methodology

A bibliometric analysis was conducted based on a search in two Web of Science (WoS) databases, Clarivate Analytics and Scopus, with the goal of expanding knowledge, measuring, and analyzing scientific literature on trust in Environmental, Social, and Governance (ESG) issues and digital transformation in organizations. Three steps made up the strategy used to produce the study: bibliometrics, data gathering, and execution plan. Since Bibliometrix is the most compatible software with the Web of Science and Scopus databases, it was employed to analyze the bibliometric data. Among the tools investigated for bibliometric analysis, the R Bibliometrix package Biblioshiny offers the most comprehensive and suitable set of approaches [46].

To solve this issue, a bibliometric analysis of the findings was conducted after a thorough search was conducted in two online databases, Scopus and Web of Science. A methodology from the information sciences called "bibliometrics" maps documents from bibliographic records kept in databases using statistical and mathematical techniques [41].

For the bibliometric analysis, the study was organised into three distinct stages: planning, collection and results. These stages converged to answer the study's guiding question: How can digital transformation contribute to the implementation of ESG?

Planning began in February 2024 and ended in March 2024, when the research was carried out. At this stage, some criteria were defined, such as limiting the search to electronic databases, not including physical catalogues in libraries, due to the number of documents considered sufficient in the databases chosen for this research. Considering the research problem, the search terms were defined during the planning phase, namely "esg" and "digital transformation" and "organisations".

It is considered that the variations in the expressions used for the search are presented in a larger context within the same proposal, since a concept depends on the context to which it is related. Finally, when planning the search, it was decided to use the terms defined in the "title, abstract and keyword" fields, with no time, language or any other restriction that could limit the results. In addition, the inclusion criterion

Table 1 Summary of the information found in the databases

Main information about data	Scopus	Web of science
Timespan	2023–2024	2023–2024
Documents	6	5
Authors	18	18
Author's keywords (DE)	31	29

Source Authors (2024)

was "article" type documents. As a result, 6 articles were found in the Scopus database and 5 in the WOS database.

To analyse the bibliometric data of the 11 articles (6 from the Scopus database and 5 from the WOS database), the Bibliometrix software was used, using the R Bibliometrix package, called Biblioshiny, which has the most extensive and appropriate set of techniques among the tools researched for bibliometric analysis [46]. This information is summarised in Table 1.

We began our analysis by looking at the overall growth rate in the number of scientific publications. We found that in 2023 there were the same number of publications—4 in each of the Scopus and WOS databases. In 2024, there were 2 publications in the Scopus database and 1 in the WOS database, as shown in Fig. 1.

This analysis began by verifying that there is no reference author on the subject of environmental, social and governance and digital transformation in organizations, as all of them have only one document published in the area. Thus, we can see that there is no author who stands out as a reference in the two databases.

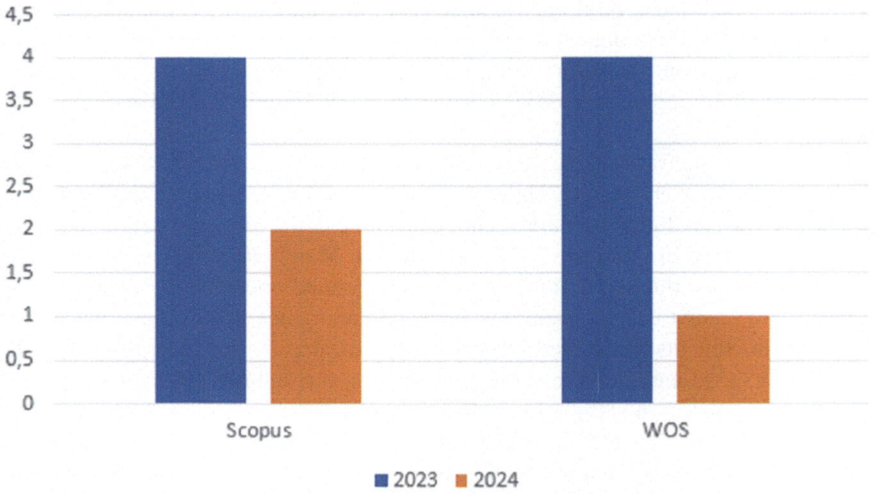

Fig. 1 Distribution of publications over the years—Scopus and WOS. *Source* Authors (2024)

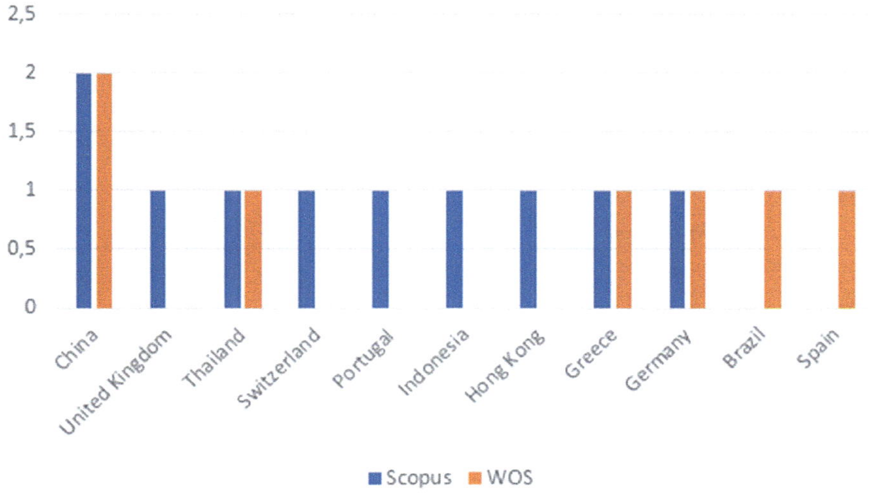

Fig. 2 Distribution of Scopus and WOS papers by country. *Source* Authors (2024)

Of the 11 papers, there is a varied list of authors, institutions and countries that stand out in research on Environmental, Social and Governance and digital transformation in organizations. When analyzing the 11 countries with the highest number of citations in the area, it can be seen that China has 20% of the total citations, a total of 2 citations, in the Scopus and WOS databases, as shown in Fig. 2.

Main scientific sources

The documents analyzed in the Scopus database were published in six different journals, all of which had one publication. In the Web Of Science database, they were published in five different journals, as shown in Table 2, with the nine most relevant scientific sources on the subject of Environmental, Social and Governance and digital transformation in organizations.

From the bibliometric analysis, based on the working group retrieved, in the two databases, 31 keywords indicated by the authors in the scopus database and 29 in the web of science database. Prominent in the first database was governance, which appeared three times. In the second WOS database, the words mentioned most often were big data, governance and performance, which appeared 2 times, as shown in Fig. 3.

The three most cited documents globally in the Scopus database and the two most cited articles in WOS are listed in Table 3.

The document that stood out with 15 citations in the Scopus database was Can digital transformation improve market and ESG performance? Evidence from Chinese by Shaofeng Wang and José Paulo Esperança, published in 2023. The article aims to present a theoretical framework for the impact of digitalization on the

Table 2 Main scientific sources in the Scopus and Web of Science Databases

Sources	Scopus	WOS
Australasian Accounting, Business And Finance Journal	1	0
Energies	1	0
Healthcare (Switzerland)	1	0
Industrial Marketing Management	1	0
Journal of Cleaner Production	1	0
Journal of Risk Management in Financial Institutions	1	0
Energies	0	1
Frontiers in Ecology and Evolution	0	1
Healthcare	0	1
Industrial Marketing Management	0	1
International Journal of Innovation	0	1

Source Authors (2024)

Fig. 3 Tag cloud. *Source* Authors (2024)

commercial and social performance of small businesses. It also examines the relationships between digital resources, digital organization, digital adoption, digital innovation culture, business competitiveness, digital management, market performance and ESG performance.

The document with the highest ranking in the Scopus (1 citation) and Web Of Science (2 citations) databases was The Implications of Triple Transformation on ESG in the Energy Sector: Fuzzy-Set Qualitative Comparative Analysis (fsQCA) and Structural Equation Modelling (SEM) Findings, by the authors Theerasak Nitlarp

Table 3 Most cited documents in the Scopus and Web of Science databases

Article title	DOI	No. of Scopus citations	No. of WOS citations
Can digital transformation improve market and ESG performance? Evidence from Chinese	https://doi.org/10.1016/j.jclepro.2023.137980	15	0
The link between ESG performance and earnings quality	https://doi.org/10.14453/aabfj.v18i1.12	2	0
The implications of triple transformation on ESG in the energy sector: fuzzy-set qualitative comparative analysis (fsQCA) and structural equation modeling (SEM) findings	https://doi.org/10.3390/en16052090	1	2
Smart digital platforms for carbon neutral management and services: business models based on ITU standards for green digital transformation	https://doi.org/10.3389/fevo.2023.1134381	0	3

Source Authors (2024)

and Theeraya Mayakul, published in 2023. In this research article, we examine the relationship between ESG and the triple transformation of business, people and technology, as well as how organizations can use digital technologies to improve their ESG performance. The next section will look at the contributions of digital transformation to the implementation of ESG in organizations.

6 Digital Transformation Contributions to the Implementation of ESG in Organizations

Digital transformation has significantly shaped the way organizations operate, directly influencing their strategies and processes.

> In the face of the strong momentum of the digital economy, businesses and organizations are integrating digital technologies to revolutionize their business models and value creation channels, aiming to enhance operational efficiency, enhance customer experience, and promote innovative activities. This transformation process is widely recognized as "digital transformation". ([52], p. 2)

In the current context, one of the areas most impacted by digital transformation is the implementation of ESG (Environmental, Social, and Governance) practices. ESG refers to a set of criteria used to measure an organization's sustainability and social impact, and is increasingly important to investors, customers and other stakeholders. Fortunately, according to the study by [42], "*theory suggests that investors seek to optimize their portfolios along the dimensions of financial return and risk while integrating environmental, social and governance (ESG) considerations*" (p. 2). This Chapter explores the conceptual and theoretical dimensions of digital transformation in the context of ESG implementation, while also providing practical examples from different countries.

Digital transformation involves the integration of digital technology into all areas of a business, fundamentally changing the way organizations operate and deliver value to their customers. Emerging technologies such as artificial intelligence (AI), the Internet of Things (IoT), big data analytics and blockchain play a crucial role in implementing ESG practices. In this regard, Wang et al. [73] state that "*every 1% increase in the degree of digital transformation will improve the ESG responsibility performance of corporation by 0.096%*" (p. 2). These technologies offer powerful tools to improve the transparency, efficiency and responsibility of organizations, allowing for better monitoring and management of their environmental and social impacts.

Transparency is one of the fundamental pillars of corporate governance and sustainability. Digital technology allows for greater transparency by facilitating reporting and disclosure of ESG data. Yin et al. [78], in the context of digital finance and ESG, highlight the central role that information transparency has in the risk of falling share prices. For example, blockchain-based platforms can be used to ensure the integrity and immutability of data, providing a reliable source of truth. This technology allows stakeholders to access detailed information about an organization's sustainability practices, increasing trust and accountability.

In the environmental domain, digital transformation enables more accurate and real-time monitoring of the environmental impacts of business operations. [34], in his study entitled "Strategic Management in the Modern World", highlights how environmental monitoring is crucial for detecting opportunities and dangers and how digital transformation strategies are relevant to the process. IoT, for example, allows the installation of sensors in industrial facilities to monitor greenhouse gas emissions, water and energy consumption, and other environmental indicators. Big data analytics can then be used to process and interpret this data, providing valuable insights that help organizations reduce their environmental footprint.

In the social field, digital technologies have the potential to significantly improve the social impact of organizations. Digital communication tools and social media platforms enable better engagement with communities and employees, promoting more inclusive and equitable work practices. Von Struensee [70], referring to [1, 15, 60, 67, 76], underline that "*AI are also emerging in the fields of environmental and social justice, including helping to increase food production and assisting people who are confronted with physical and intellectual barriers. The potential uses of AI seem endless at this time*" (p. 15). Today, there is little doubt that AI-based solutions, which can be used to analyze demographic and social data, help organizations identify areas for improvement and implement more effective social programs.

Digital transformation also has a profound impact on the governance of organizations. The use of technologies such as AI and data analysis allows for better decision-making, based on accurate and up-to-date data. In this sense, [54], regarding the suggestive case of Dubai, in his study "*Digital Transformation through Agile*

Governance – The Case of Smart Dubai", states that Dubai "is a city that is able to produce, collect and analyze data to enable 'intelligent' decisions and predictive analysis for better planning and development" (p. 8). In a broader sense, digital management tools help boards and managers monitor compliance with regulations and identify potential risks, promoting a culture of accountability and transparency.

To illustrate how digital transformation is contributing to the implementation of ESG practices, we present some examples from different countries and industries.

Siemens, one of the largest engineering companies in the world, has been using digital technologies to promote sustainability in its operations.

Siemens has an ESG Strategy and Governance. The following can be read on the ESG Strategy and Governance electronic portal:

> We help organizations develop comprehensive ESG strategies and governance frameworks tailored to their unique needs. Our approach helps businesses to identify key ESG topics, establish ambitious goals, and create a clear roadmap for implementation. With our expertise organizations can capitalize on ESG opportunities while satisfying the demands of diverse stakeholder groups.

The company implemented a digital platform called MindSphere, which uses IoT to monitor and optimize the use of resources in its factories. This platform allows Siemens to collect real-time data on energy consumption, CO_2 emissions and other environmental indicators, helping the company to reduce its environmental footprint.

The MindSphere digital platform is an open, cloud-based IoT operating system that connects products, factories, systems and machines, enabling the collection and analysis of large volumes of data in real time.

According to Sphere portal:

> Sphera is the leading provider of Environmental, Social and Governance (ESG) performance and risk management software, data and consulting services focusing on Environment, Health, Safety & Sustainability (EHS&S), Operational Risk Management (ORM), Product Stewardship and Supply Chain Risk Management (SCRM).

On the Siemens website it is possible to confirm that the program ESG Strategy and Governance Industries covers the following industries: Aerospace; Automotive & New Mobility; Energy; Food & Beverage; Industrial; Life Sciences & Healthcare; Power & Utilities; Smart Buildings, Campuses & Cities.

The program ESG Strategy and Governance Industries also includes the following capabilities: Advanced Analytics, AI & Cybersecurity; Applied Industry Analytics; Cost Transformation; Digital Implementation; Enterprise Architecture & Solution Assessment; Factory & Supply Chain Transformation; IoT Solutions; R&D Transformation; Solutions Development; Solutions Prototyping; Strategy & Digital Transformation; Sustainability.

The insights covered by the program are: Cognitive Infrastructure; Cybersecurity; Data Analytics; Digital Transformation; Digital Twin & Industrial Metaverse; Sustainability.

Through the MindSphere Documentation Area, on the Siemens portal, it is possible to consult the added value of MindSphere technology.

Using IoT sensors installed in its factories, Siemens is able to monitor various environmental parameters, such as energy and water consumption and CO_2 emissions. This data is sent to the MindSphere platform, where it is analyzed to identify patterns and inefficiencies. This continuous monitoring allows Siemens to make informed decisions to reduce resource consumption and minimize the environmental footprint.

Siemens uses the big data analysis provided by MindSphere to optimize the energy efficiency of its operations. The company has implemented energy management solutions that automatically adjust energy use based on demand and availability, reducing waste and emissions. Siemens is also integrating renewable energy sources into its operations, promoting a transition to a more sustainable energy matrix.

MindSphere not only facilitates internal resource management, but also improves transparency for stakeholders. The platform allows Siemens to report its environmental and sustainability data in detail, increasing investor and customer confidence in the company's sustainable practices.

Another case for investing in sustainability and ESG logic is Unilever (United Kingdom and the Netherlands). The following appears on the company's official portal (2024a): *"we have been driving an ambitious sustainability agenda for over two decades. Yet, in the face of ever-growing economic, environmental and social challenges, we are evolving our approach"*.

In the ESG context, Unilever (2024) attests that its Growth Action Plan in the coming years will focus on four sustainability priorities:

- to deliver net zero emissions across our value chain;
- to deliver resilient and regenerative natural and agricultural ecosystems;
- to end plastic pollution through reduction, circulation and collaboration;
- to ensure a decent livelihood for people in our global value chain, including by earning a living wage.

Unilever, a global consumer goods giant, has used big data analytics to improve the sustainability of its supply chains. In this regard, the Foodnavigator Europe portal (2024) highlights that "*Unilever is tapping into the power of AI and big data biology to optimise its current food portfolio and accelerate new product development*". The company developed an analysis system that collects data from different points in the supply chain, from agricultural production to product distribution.

Big data analysis allows Unilever to identify areas of waste and implement more sustainable practices, such as reducing water consumption and minimizing carbon emissions. On Unilever's official website (2024b) it is possible to confirm that the company's main objective is "*to reduce absolute operational GHG emissions by 100% by 2030 from a 2015 baseline*". For example, the company can monitor water and energy use at different stages of production and implement more sustainable practices to reduce consumption. In addition, Unilever can adjust logistics processes to minimize carbon emissions during the transport of products.

Unilever also uses big data to ensure the traceability of its products. The company collects and analyzes data on the origin of ingredients, ensuring that they are obtained from sustainable and ethical sources. This transparency is key to meeting ESG standards and increasing consumer confidence in Unilever's products.

On the social front, Unilever uses digital platforms to engage with local communities and promote fair and inclusive working practices. The company has launched digital training programs for farmers and small producers, helping them to improve the productivity and sustainability of their agricultural practices.

Another relevant example of good ESG practices is Tesla (United States). The brand is also widely recognized for its commitment to sustainability and the use of digital technologies to promote ESG practices. The company uses IoT and big data analysis to monitor the performance of its electric vehicles in real time. This monitoring allows Tesla to optimize vehicle performance and proactively identify problems, ensuring a more efficient and sustainable operation.

On Tesla's Impact Report 2023 (2024), it is possible to confirm that the, in 2023, Tesla "*customers avoided releasing over 20 million metric tons of* CO_2 *into the atmosphere by using our products.*"

Tesla vehicles are equipped with IoT sensors that collect detailed data on engine performance, battery efficiency, and other critical parameters. This data is transmitted to Tesla servers, where it is analyzed to optimize vehicle performance and proactively identify issues. This continuous monitoring allows Tesla to improve the energy efficiency of its vehicles and extend the life of batteries.

This highly efficient use of IoT is not surprising for the Tesla brand. Precisely in this sense, the Analyticssteps.com portal attests that "*Tesla has the most successful IoT enabled self-driving cars.*"

Big data analysis also allows Tesla to identify opportunities to reduce the carbon footprint of its operations. The company uses data on energy production and consumption in its factories to implement more sustainable practices, such as the use of renewable energy and the optimization of manufacturing processes.

Conceptually, Tesla's macrostrategic logic as a company is no different from that used in its electric vehicles. "*As a car company that is also categorized as a 'big tech' company, Tesla not only revolutionized the car industry with its EVs (Electric Vehicles) but also with its usage of Big Data*" [18].

Tesla maintains a high level of transparency by publicly disclosing its environmental and sustainability data. The company uses digital platforms to communicate its ESG initiatives and engage customers and investors in its sustainable practices. Tesla's annual Impact Reports are a good example in this regard.

Natura, a Brazilian cosmetics company, has been a pioneer in implementing sustainable practices, using digital technologies to improve transparency and traceability in its supply chain. The company uses blockchain to track the origin of its natural ingredients, ensuring they are obtained from sustainable and ethical sources. This approach not only improves environmental sustainability, but also promotes fair social practices.

Exame magazine [20] states that:

> Natura & C. updated the ambitious goals of Vision 2030: 100% of packaging will be recyclable, reusable or compostable. Incorporating Regeneration into the strategy, the CEO highlights the importance of environmental and social sustainability. In 2023, 87.8% of packaging already met the criteria.
>
> In decarbonization, six work streams for reduction, 15 million reais invested in carbon offsetting projects.

Natura uses blockchain technology to trace the origin of its natural ingredients, ensuring that they are obtained from sustainable and ethical sources. Blockchain makes it possible to record all transactions and product movements in a secure and immutable way, improving the transparency and reliability of information. Precisely to reinforce Natura's commitment to blockchain technology, the Criptofacil portal (2023) highlights the following: "*With an eye on bringing more transparency to its production process, Natura, a Brazilian cosmetics giant, has been studying the implementation of blockchain and new technologies in your business.*" This approach not only improves environmental sustainability, but also promotes fair social practices.

The company also uses IoT to monitor the environmental impact of its operations. Sensors are installed in factories to monitor energy and water consumption, as well as pollutant emissions. The data collected is analyzed to identify areas for improvement and implement more sustainable practices.

In this sense, as an example of Natura's long-standing investment in IoT, the Open Innovation Project between Natura and the company Telefônica is presented, which enhances the production of murumuru oil, by enabling the reduction of losses and increase of productivity.

> The project consisted of monitoring the process, using temperature and humidity sensors installed on drying mats in communities in Amazonas and Pará. These sensors produced data, then sent to an IoT (Internet of Things) platform.. All development of the technology also had the support of the Research Center of the University of Engineering Mauá. [48]

Natura also uses digital platforms to promote the involvement of local communities in sustainability projects. The company has developed digital training programs for small farmers, helping them to adopt more sustainable agricultural practices and improve the quality of life in rural communities.

Another important ESG company is BHP, one of the world's largest mining and natural resources companies, based in Australia. BHP has led significant efforts to implement ESG practices through digital transformation. The BHP portal [8] states that the company:

> regularly meet with investors and investors representative organizations across the world to discuss a range of strategy and governance topics including our approach to environment, social and governance matters. Some of the topics we've covered with investors in the past include Samarco, human rights, portfolio, long-term value creation, culture, diversity, purpose and executive remuneration.

The company has adopted several digital technologies to improve the sustainability of its operations, promote transparency and increase social responsibility. In terms of ESG, according to the BHP portal (2023), the main goals of the company are: Climate change; Sustainably; Planning the closure of assets; Physical climate-related risk and adaptation; Stewardship of our products; Water; Corporate Responsibility; Indigenous peoples; and Respecting human rights.

BHP uses IoT (Internet of Things) technology and drones to monitor and manage the environmental impacts of its mining operations. IoT sensors are installed in several mines to monitor air quality, greenhouse gas emissions, water usage and geotechnical stability in real time. Drones are used to carry out detailed environmental inspections, helping to quickly identify and mitigate potential negative impacts. Regarding the use of drones, according to the BHP portal (2023a), across the company's iron ore mining in Western Australia operations, "*there are close to 20 remotely operated aircraft (drone) flights a day. These flights help remove our people from potentially dangerous locations and help us innovate and re-think how we go about traditional mining tasks.*" This approach allows BHP to optimize environmental management and reduce its ecological footprint.

BHP has implemented a big data analytics platform to collect and process large volumes of environmental and operational data. Through this platform, the company can analyze trends, identify areas for improvement and make data-driven decisions to reduce resource consumption and carbon emissions. "*HP has been leveraging machine learning technologies to bring disruptive changes in its business processes and has initiated various in-house strategies to drive digital transformation. The annual ICT spending of BHP was estimated at $398.2 million in 2023.*" [28] This strategy based on advanced data analytics helps BHP meet its sustainability goals and improve operational efficiency.

To ensure transparency and traceability in its supply chain, BHP has been exploring the use of blockchain technology. This technology allows all transactions and material movements to be recorded in a secure and immutable way, ensuring that all products and materials are obtained from responsible and sustainable sources. [45] portal refers that "*BHP believes blockchain traceability is key to carbon neutrality, reducing emissions and meeting ethical considerations.*" Indeed, the use of blockchain improves compliance with ESG standards, increases investor and customer confidence, and facilitates the verification of ethical practices in the supply chain.

BHP is investing in renewable energy projects and using digital technologies to optimize energy use in its operations. The company has implemented advanced energy management systems, based on AI (artificial intelligence) and machine learning, which help to predict energy demand, adjust production according to needs and reduce waste. "*As renewables surpass the 30% global penetration mark, multiple technology options are required to provide grid flexibility of the electricity network.*" [9] These efforts contribute to greater energy efficiency and a reduction in carbon emissions, in line with the company's sustainability objectives.

BHP is thus a leading example in Australia of how digital transformation can be used to effectively implement ESG practices. Through the adoption of technologies such as IoT, drones, big data, blockchain and AI, the company has improved the monitoring and management of its environmental impacts, promoted transparency in the supply chain and encouraged social responsibility. These efforts not only help

BHP meet its sustainability goals, but also reinforce its position as a leader in the mining industry in terms of environmental and social responsibility.

Finally, the example of Portugal's Galp Energia is presented. Galp Energia is one of the country's largest energy companies. Galp Energia has demonstrated a significant commitment to implementing ESG practices through digital transformation.

> Galp formalized its adherence to the environmental, social and governance (ESG) metrics of the World Economic Forum (WEF), thus becoming the first Portuguese company to join a business community focused on creating sustainable value and responding to most pressing challenges of modern societies. [26]

The company has been adopting various digital technologies to improve the sustainability of its operations and promote transparency and social responsibility.

Galp has invested in the implementation of IoT (Internet of Things) sensors in several of its industrial facilities to monitor greenhouse gas emissions, water and energy consumption, and other environmental indicators in real time. Galp Energia has the Energy Management System for companies. This system consists of an online platform that, through sensors, allows to "*analyze, in real time, all electricity, gas and water consumption.*" (Galp [27]) This technology allows for more efficient management of resources and a faster response to potential environmental problems, contributing to the reduction of the company's ecological footprint.

The company developed a digital platform for sustainability data management, which collects and analyzes information from various internal and external sources. This platform uses big data analysis to provide detailed insights into Galp's environmental and social performance, helping to identify areas for improvement and define more sustainable strategies.

The investment in the advantages of Big Data dates back to 2020. At the time, Galp launched the company EI (Independent Energy) to boost photovoltaic self-consumption. The company "*uses the most advanced technologies on the market, such as satellite image analysis, artificial intelligence algorithms and big data*" [25].

Galp has also been exploring the use of blockchain technology to improve transparency and traceability in its supply chain. This technology allows all transactions and product movements to be recorded in a secure and immutable way, ensuring that all materials and products are obtained from responsible and sustainable sources. This not only improves compliance with ESG standards, but also increases investor and customer confidence. In 2024, Galp opened applications for the 3rd Edition of the "Startup The Future" competition. The "*company wants to support proposals that accelerate decarbonization.*" [27] One of the categories in the competition is, precisely, "AI & Blockchain for Enhanced Fuels Value Chain".

In the social sphere, Galp has used digital platforms to promote community involvement and employee participation in social responsibility initiatives.

According to [43], this are the "five strategic pillars in Galp's Social Impact area":

- Energy and sustainable mobility solutions, enhancing shared value on the path to neutrality, promoting access to energy and the adoption of energy solutions by entities and organizations that contribute to the social well-being of communities;

- Education and awareness raising of citizens, young people and children, empowering and supporting communities in the energy transition, contributing to education for the future and promoting the civic engagement that the country needs so much to overcome the challenges ahead;
- Ecosystem integration, naturalization and promotion of net positive impact on ecosystems and biodiversity;
- Support for local projects, linked to the sustainable development priorities of local communities;
- Social emergencies, from the perspective of cooperation and humanitarian response.

Through campaigns and volunteer programs advertised on social media platforms and on its corporate portal, Galp Energia encourages active participation in community and environmental projects, reinforcing its commitment to sustainable development and social inclusion.

Galp Energia is thus an outstanding example in Portugal of how digital transformation can be used to implement ESG practices effectively. Through the adoption of technologies such as IoT, big data, blockchain and AI, the company has improved the monitoring and management of its environmental impacts, promoted transparency in the supply chain and encouraged social responsibility. These efforts not only help Galp meet its sustainability goals, but also reinforce its position as a leader in the energy transition and in the implementation of ESG practices.

The examples of Siemens (Germany), Unilever (Netherlands and United Kingdom), Tesla (United States), Natura (Brazil), BHP (Australia), and Galp Energia (Portugal) illustrate how digital transformation is being used to implement ESG practices effectively in different sectors and regions. Through the adoption of emerging technologies such as IoT, big data, blockchain and AI, these companies have improved transparency, efficiency and accountability in their operations. These efforts not only help to meet sustainability objectives, but also reinforce the position of these companies as leaders in environmental and social responsibility in the global market.

Digital transformation plays a crucial role in the implementation of ESG practices in organizations, offering powerful tools to improve transparency, efficiency and accountability. Fenwick et al. [21], in their study entitled "ESG as a business model for SMEs", emphasize that digital transformation has triggered an unprecedented degree of possible alternatives. The authors even mention that *"there are infinite possibilities"*. Through the use of emerging technologies such as AI, IoT, big data and blockchain, organizations can better monitor and manage their environmental and social impacts, promoting more robust and sustainable governance. Examples of companies from different countries demonstrate how these technologies are being successfully applied to promote sustainability and social responsibility.

However, the digital transformation journey presents significant challenges, which require a strategic and integrated approach. Shapira [55], in this regard, is very clear about the difficulties:

> In the past few years, several ESG issues have become sources of major risk for companies. Companies that fail to meet evolving societal demands on issues such as cybersecurity, gender and racial diversity, and environmental sustainability may face significant reputational fallouts, and find it harder to recruit and retain top talent, charge premium prices, and access cheap capital (p. 740).

Cybersecurity, investment in infrastructure and employee training are critical areas that must be addressed to ensure effective implementation of ESG practices. With a well-designed strategy and a firm commitment to sustainability, organizations can take advantage of the opportunities offered by digital transformation to create a lasting positive impact.

7 Conclusion and Future Scope

The integration of Environmental, Social and Governance (ESG) principles in organizations has gained increasing relevance in the global corporate panorama. "*The growing emphasis on Environmental, Social, and Governance (ESG) principles, particularly in corporate sectors, shapes investment trends and operational strategies*" ([63], p. 2). These principles, which range from environmental sustainability to social responsibility and corporate governance, constitute an essential foundation for sustainable development and long-term value creation. In this context, digital transformation emerges as a crucial facilitator for the effective implementation of ESG criteria, enhancing companies' capabilities in monitoring, managing and improving their performance in relation to these dimensions. Alluding to Li et al. [39], Chen et al. [11] state that "*the rapid development of digital technologies provides effective solutions for enhancing corporate sustainability, and digital transformation has become a key technological force in promoting the enhancement of sustainable development capabilities*" (p. 3).

Digital transformation refers to the incorporation of digital technologies into all aspects of the functioning of organizations, profoundly and permanently changing the way they operate and deliver value to their customers. This process involves not only the adoption of new technologies, but also the cultural change necessary for companies to continually adapt, experiment with new approaches and accept the possibility of failure as part of the path to innovation and competitiveness. Orozco and Perilla-Maluche [51], citing Hashim et al. [31], state that "*digital transformation strategies are aimed to (a) increase the total revenue (b) enhance productivity (c) generate value through innovative practices and (e) develop a brand reputation and novelty*" (p. 9). Tools such as artificial intelligence, Big data, blockchain and the Internet of Things are fundamental in this journey, offering new ways of collecting, analyzing and using data to support more informed and effective decisions.

Digital transformation provides a series of tangible benefits for the implementation of ESG practices in organizations, distributed across its three main pillars: environmental, social and governance. In this regard, Na et al. [47], in their study entitled "Digital Transformation of Value Chain and Corporate ESG Performance in

China", explore and develop how the digital transformation of the corporate value chain influences ESG at the value chain level. More specifically, they investigate how the digitalization of various segments of the value chain impacts the environmental (E), social (S) and governance (G) dimensions of ESG.

In the environmental field, digital technologies allow for accurate, real-time monitoring of the environmental impacts of business operations. IoT sensors installed in industrial equipment can collect data on greenhouse gas emissions, water and energy consumption, among other environmental indicators. Big data analytics processes and interprets this data, providing valuable insights to reduce the environmental footprint and optimize resource utilization. Guo et al. (2024), precisely in this context, highlight "*the pivotal role of digital transformation in fostering firms' oversea business, particularly* via *the environmental responsibility and social responsibility channels*" (p. 5).

For example, Siemens uses the MindSphere platform to monitor and optimize the use of resources in its factories, contributing to reducing emissions and increasing energy efficiency. Similarly, BHP implements technologies such as IoT and drones to monitor the environmental impacts of its operations, ensuring compliance with environmental regulations and promoting sustainability.

In the social dimension, digital transformation has the potential to significantly improve the social impact of organizations. "*The digital economy, marked by innovations like artificial intelligence and blockchain, has the potential to revolutionize business operations, influencing environmental practices, social responsibility, and governance structures*" ([65], p. 4).

Digital communication tools and social media platforms enable better engagement with communities and employees, promoting inclusive and equitable working practices. AI-based solutions can analyze demographic and social data, helping to identify areas for improvement and implement more effective social programs.

A practical example is Unilever's use of Big Data to improve the sustainability of its supply chains, ensuring fair and responsible practices at all stages of the production process. In addition, companies like Natura use blockchain to trace the origin of their natural ingredients, ensuring transparency and accountability throughout the value chain.

In terms of governance, the digital transformation enables better decision-making based on accurate and up-to-date data. Zhou et al. [84], in the context of combating climate change, are very peremptory regarding the central importance of digital transformation for the quality of governance. The authors emphasize that "*regional climate governance may have heterogeneous impacts under different levels of digital transformation*" (p. 28).

Digital management tools help boards and managers monitor compliance with regulations and identify potential risks, promoting a culture of accountability and transparency. For example, Tesla uses IoT and big data to monitor the performance of its electric vehicles in real time, ensuring not only operational efficiency, but also safety and regulatory compliance.

Practical examples from companies in different regions illustrate how digital technologies are being successfully applied to promote sustainability and social responsibility:

- Siemens: The MindSphere platform is used to monitor and optimize the use of resources in factories, helping to reduce emissions and increase energy efficiency.
- Unilever: The use of big data makes it possible to improve the sustainability of supply chains, ensuring fair and responsible practices.
- Tesla: The use of IoT and big data to monitor the performance of electric vehicles in real time ensures operational efficiency, safety and regulatory compliance.
- Natura: The company uses blockchain to trace the origin of natural ingredients, ensuring transparency and accountability in the value chain.
- BHP: The implementation of technologies such as IoT and drones makes it possible to monitor the environmental impacts of operations, promoting sustainability.
- Galp Energia: The adoption of IoT sensors and big data improves the sustainability of operations, optimizing the use of resources and reducing the environmental footprint.

Despite the benefits, implementing digital transformation to support ESG practices is not without its challenges.

The European Union has the Europe's Digital Decade Program underway. With this Program, "*the EU is pursuing a human-centric, sustainable vision for digital society throughout the digital decade to empower citizens and businesses.*" (European [19]). Buzu et al. [10], in the context of Europe's Digital Decade program, highlight that "*there are several common aspects, both challenges, and opportunities, that businesses, public administrations, universities, educational organisations, and Member States, in general, will need to observe in their efforts toward reaching the goals*" (pp. 2–3). In their study, the authors outline the challenges and opportunities.

Employee resistance to change, the high costs of new technologies and the need for significant retraining of the workforce are some of the barriers faced by organizations. Security issues are also a growing concern, given the higher incidence of data breaches and cyber threats, requiring stringent cybersecurity measures to protect sensitive data and maintain customer trust. Shapira [55] develops this topic in depth in his article entitled Mission Critical ESG and the Scope of Director Oversight Duties. The article "*applies the framework to concrete ESG concerns, such as cybersecurity, climate change, and sexual misconduct*" (p. 733).

To overcome all these challenges, it is essential to adopt a holistic and strategic approach. The commitment of political and business leaders is crucial, as is clear communication and community and employee engagement, respectively. Training and continuous development programs are key to ensuring that employees are empowered to use the new digital tools and technologies. Once again, the Europe's Digital Decade Program is a good example in this regard.

Creating an environment of continuous learning and rapid adaptation to the latest technological trends and market developments is essential for maintaining competitiveness and promoting sustainable innovation.

Finally, throughout this chapter, it is concluded that the key to continuously achieving over time the potential for sustainability in the use of resources will result from future digital transformation in conjunction with ESG initiatives. This will necessarily be based on the use of emerging technologies to create new business processes, customer experiences and reshape organizational culture. Companies will be increasingly dependent on data analysis and automation to achieve profitability and customer satisfaction. AI, IoT, blockchain and big data will play an increasingly central role in promoting transparency, efficiency and accountability in public and business operations.

Digital transformation will thus increasingly be a powerful enabler for the integration of ESG principles in organizations, offering tools and technologies that significantly improve the monitoring, management and improvement of environmental, social and governance performance. This logic applies to all sectors of activity. For example, in the area of Law, DeStefano et al. [17] highlight exactly this in their study, stating that "*there are countless additional forms of potential value creation from DT of the legal department*" (p. 255). Immediately afterwards, the same authors emphasize that the dimension "*ESG is one of those new opportunities and capabilities that stems from DT*" (Digital Transformation).

Organizations that adopt digital transformation and integrate ESG principles into their business strategies will be better positioned to face the challenges and seize opportunities in the current market. The ability to monitor, manage and improve performance against ESG criteria, supported by digital technologies, will provide an increasingly significant competitive advantage, and this process will promote sustainability, innovation and long-term value creation.

References

1. Aitkenhead, M., Dalgetty, I.A., Mullins, C.E., McDonald, A.J.S., Strachan, N.: Weed and crop discrimination using image analysis and artificial intelligence method. Comput. Electron. Agric. **39**(3), 157 (2003)
2. Amaral, M., Willerding, I.A.V., Lapolli, É.M.: ESG and Sustainability: The Impact of the Pillar Social. Concilium (2023). Retrieved from https://consensus.app/papers/sustainability-impact-amaral/e2dbdeb007b6598982fd064c1d57f13a/?utm_source=chatgpt
3. Analyticssteps.com: IoT in Tesla: applications, benefits and potential risks (2024). Available at: https://analyticssteps.com/blogs/iot-tesla-applications-benefits-and-potential-risks
4. Bandeira, G.L., Trindade, D., Gardi, L.H., Sodario, M.: Developing an ESG strategy and roadmap: an integrated perspective in an O&G company. Day 2 Tue, 2 May 2023 (2023). Retrieved from https://consensus.app/papers/developing-strategy-roadmap-integrated-perspective-bandeira/d5da5147cc8359789550eaaaacf24ca3/?utm_source=chatgpt
5. Baratta, A., Cimino, A., Longo, F., Solina, V., Verteramo, S.: The impact of ESG practices in industry with a focus on carbon emissions: Insights and future perspectives. Sustainability **15**(8), 6685 (2023)
6. Ben Dor, A., Guan, J., Sun, Y.: Is Incorporating ESG Considerations Costly? J Portfolio Manage (2022). Retrieved from https://consensus.app/papers/incorporating-considerations-costly-dor/0105dffddee855bc9f19025e27b80b08/?utm_source=chatgpt

7. BHP: Drone usage across our WA iron ore sites (2023). Available at: https://www.bhp.com/news/articles/2023/09/drone-usage-across-our-wa-iron-ore-sites
8. BHP: Environmental, Social and Governance (ESG) and sustainability (2024a). Available at: https://www.bhp.com/investors/esg-sustainability
9. BHP: Renewable energy (2024b). Available at: https://www.bhp.com/about/the-future-is-clear/renewable-energy
10. Buzu, I., Ljubenkov, D., Cristea, M.F., Alic, D.: The role of automation in modernising and digitalising public services (2022). https://doi.org/10.2139/ssrn.4312092
11. Chen, N., Wu, G., Yin, Z.: Digital transformation, sustainability and MNEs' expansion: empirical evidence from Chinese listed companies (2024). https://doi.org/10.2139/ssrn.4831137
12. Clementino, E., Perkins, R.: How do companies respond to Environmental, Social and Governance (ESG) ratings? Evidence from Italy. J Bus Ethics. Retrieved from https://consensus.app/papers/companies-respond-environmental-social-governance-clementino/f732a2fae6645abfb661e21b04e660fc/?utm_source=chatgpt
13. Criptofacil: Natura estuda os impactos da tecnologia blockchain (2023). Available at: https://www.criptofacil.com/natura-estuda-os-impactos-da-tecnologia-blockchain
14. Csedő, Z., Magyari, J., Zavarkó, M.: Dynamic corporate governance, innovation, and sustainability: Post-COVID period. Sustainability **14**(6), 3189 (2022)
15. D'Alfonso, S., et al.: Artificial intelligence-assisted online social therapy for youth mental health. Front. Psychol. **8**, 796 (2017). Available at: https://www.ncbi.nlm.nih.gov/pmc/articles/PMC5454064
16. De Bem Machado, A., Dandolini, G.A., de Souza, J.A., Lytras, M.D., Sousa, M.J.: Artificial intelligence: applicability of this technology to higher education—a scoping review. Technol. Sustain. Glob. High. Educ. 211–236 (2023)
17. DeStefano, M., Tellmann, B.P., Wu, D.: Don't Let the Digital Tail Wag the Transformation Dog: A Digital Transformation Roadmap for Corporate Counsel. University of Miami Legal Studies Research Paper No. 4021593 (2022, February 6). Recuperado de https://doi.org/10.2139/ssrn.4021593
18. Digital Data Design Institute at Harvard: BIG Data in Tesla Inc. (2024). Available at: https://d3.harvard.edu/platform-digit/submission/big-data-in-tesla-inc/
19. European Commission: Europe's digital decade (2024). Available at: https://digital-strategy.ec.europa.eu/en/policies/europes-digital-decade
20. Exame: Natura & C: ESG practices seek environmental and social regeneration (2024). Available at: https://exame.com/esg/natura-co-praticas-esg-buscam-a-regeneracao-ambiental-e-social/
21. Fenwick, M., Joubert, T., van Wyk, S., Vermeulen, E.P.M.: ESG as a business model for SMEs. European Corporate Governance Institute, Law Working Paper No. 642/2022, Tilburg Law School Research Paper (2022, May 2). Recuperado de https://doi.org/10.2139/ssrn.4098644
22. Finn, T., Downie, A.: Digital Transformation Examples (2024, January 29). IBM Blog. https://www.ibm.com/blog/digital-transformation-examples/
23. Foodnavigator Europe: Unilever is using AI and big data to transform its food portfolio (2024). Available at: https://www.foodnavigator.com/Article/2023/07/25/How-Unilever-is-using-artificial-intelligence-and-big-data-biology-to-transform-food?utm_source=copyright&utm_medium=OnSite&utm_campaign=copyright
24. Fu, T., Li, J.: An empirical analysis of the impact of ESG on financial performance: the moderating role of digital transformation. Front. Environ. Sci. **11**, 1256052 (2023). https://doi.org/10.3389/fenvs.2023.1256052
25. Galp Energia: Galp launches company EI—Energia Independente to Boost Photovoltaic Self-consumption (2020). Available at: https://www.galp.com/corp/pt/media/comunicados-de-imprensa/comunicado/id/1156/galp-lanca-empresa-ei-energia-independente-para-impulsionar-o-autoconsumo-fotovoltaico

26. Galp Energia: Galp is the first Portuguese company to adhere to the world economic forum's ESG metrics (2021). Available at: https://www.galp.com/corp/pt/investidores/publicacoes-e-comunicados/comunicados-investidor/comunicado/id/1236/galp-e-a-primeira-empresa-portuguesa-a-aderir-as-metricas-de-esg-do-world-economic-forum
27. Galp Energia: Do you already know the energy management system for galp companies? (2023). Available at: https://galp.com/pt/pt/empresas/blog/blog-post/sistema-de-gestao-de-energia-galp
28. Globaldata: BHP—Digital transformation strategies (2023). Available at: https://www.globaldata.com/store/report/bhp-enterprise-tech-analysis/
29. Gossain, R.: Legal Underpinnings Drive ESG Amidst Global Flux (2023). Paradigm. Retrieved from https://consensus.app/papers/underpinnings-drive-amidst-global-flux-gossain/d6abb73d0a8a52baa2253478b410b87c/?utm_source=chatgpt
30. Guo, L., Zhong, Q., Wang, H.: Digital transformation ESG responsibility and corporate's export performance. Finan. Res. Lett. (2024). https://doi.org/10.2139/ssrn.4806801
31. Hashim, M.A., Tlemsani, I., Matthews, R.: Higher education strategy in digital transformation. Educ. Inf. Technol. **27**, 3171–3195 (2022). https://doi.org/10.1007/s10639-021-10739-1
32. Hess, T., Matt, C., Benlian, A., Wiesböck, F.: Options for formulating a digital transformation strategy. MIS Q. Execut. **15**(2), 123–139 (2016). https://aisel.aisnet.org/misqe/vol15/iss2/6
33. Iamandi, I., Constantin, L.-G., Munteanu, S., Cernat-Gruici, B.: Mapping the ESG Behavior of European Companies. Sustainability (2019). Retrieved from https://consensus.app/papers/mapping-behavior-companies-holistic-kohonen-approach-iamandi/c26a7dd2322d51a4be360c634161bd82/?utm_source=chatgpt
34. Jamilzade, I.: Strategic management in the modern world (2024). Available at: https://ssrn.com/abstract=4741352
35. Jiang, J.: Is digital transformation beneficial for ESG performance? Evidence from Chinese Listed Companies. BCP Bus. Manage. **50**, 306–314 (2023). https://doi.org/10.54691/bcpbm.v50i.5618
36. Kane, G.C., Palmer, D., Nguyen-Phillips, A., Kiron, D., Buckley, N.: Strategy, not technology, drives digital transformation. MIT Sloan Manage. Rev. (2015). https://sloanreview.mit.edu/projects/strategy-drives-digital-transformation
37. Kwiliński, A., Lyulyov, O., Pimonenko, T.: Unlocking sustainable value through digital transformation: an examination of ESG performance. Information **14**(8), 444 (2023). https://doi.org/10.3390/info14080444
38. Li, J., Li, Y.: Literature study of ESG in corporate investment and ESG rating status. Adv. Econ. Manage. Polit. Sci. (2023). Retrieved from https://consensus.app/papers/literature-study-corporate-investment-rating-status-li/3bbdcc4a9b1f5d219fb61091d7f876b4/?utm_source=chatgpt
39. Li, Y., Dai, J., Cui, L.: The impact of digital technologies on economic and environmental performance in the context of Industry 4.0: a moderated mediation model. Int. J. Prod. Econ. **229**, 107777 (2020)
40. Liao, H.T., Pan, C.L., Zhang, Y.: Smart digital platforms for carbon neutral management and services: business models based on ITU standards for green digital transformation. Front. Ecol. Evol. **11**, 1134381 (2023)
41. Linnenluecke, M.K.: Resilience in business and management research: a review of influential publications and a research agenda: Resilience in business and management research. Int. J. Manag. Rev. **19**(1), 4–30 (2017). https://doi.org/10.1111/ijmr.12076
42. Lobe, S., & Halbritter, G. (2023). Finding value in sustainable and responsible investments. Corporate Ownership and Control **20**(4), 84–96. Available at: https://doi.org/10.2139/ssrn.4875140
43. Marketeer: Galp: Social impact as a measure of success (2024). Available at: https://marketeer.sapo.pt/galp-o-impacto-social-como-uma-medida-de-sucesso/
44. Martins, H.: Competition and ESG practices in emerging markets: Evidence from a difference-in-differences model. Finan. Res. Lett. (2021). Retrieved from https://consensus.app/papers/competition-practices-emerging-markets-evidence-martins/afd7b54b1360502abc597bbfbf749508/?utm_source=chatgpt

45. Miningdigital: BHP pushes blockchain traceability to drive sustainability (2022). Available at: https://miningdigital.com/technology/bhp-pushes-blockchain-traceability-to-drive-sustainability
46. Moral-Muñoz, J.A., Herrera-Viedma, E., Santisteban-Espejo, A., Cobo, M.J.: Software tools for conducting bibliometric analysis in science: an up-to-date review. El profesional de la información **29**(1), e290103 (2020)
47. Na, C., Ma, A., Zhu, J., He, H., Ni, Z.: Digital transformation of value chain and corporate ESG performance in China (2024). https://doi.org/10.2139/ssrn.4819919
48. Naturacampus: Open innovation project between Natura and Telefônica benefits production of murumuru oil (2024). Available at: http://www.naturacampus.com.br/cs/naturacampus/post/2015-09/inovacao-aberta-murumuru
49. Nitlarp, T., Mayakul, T.: The implications of triple transformation on ESG in the energy sector: fuzzy-set qualitative comparative analysis (fsQCA) and structural equation modeling (SEM) findings. Energies **16**(5), 2090 (2023)
50. Nitlarp, T., Mayakul, T.: The implications of triple transformation on ESG in the energy sector: fuzzy-set qualitative comparative analysis (fsQCA) and structural equation modeling (SEM) findings. Energies **16**(5), 2090 (2023). https://doi.org/10.3390/en16052090
51. Orozco, L.A., Perilla-Maluche, R.: Be similar or be different—organizational and business model innovations in the universities' digital transformation (2024). https://doi.org/10.2139/ssrn.4875719
52. Peng, Y., Chen, G.: New intelligent empowerment for digital transformation (2024). https://doi.org/10.2139/ssrn.4877072
53. Petre, A., Pleşea, D.: Human capital as a key factor for organizations' ESG goals. In: European Conference on Knowledge Management (2023). Retrieved from https://consensus.app/papers/human-capital-factor-organizations-goals-petre/ec6726c44dab5eb4b6b6d764e6f62b64/?utm_source=chatgpt
54. Salem, F.: A smart city for public value: Digital Transformation Through Agile Governance—The Case of 'Smart Dubai'. World Government Summit Publications (2016). Available at: https://ssrn.com/abstract=2733632
55. Shapira, R.: Mission Critical ESG and the scope of director oversight duties. Columbia Bus. Law Rev. **732** (2022, May 12). Recuperado de https://ssrn.com/abstract=4107748
56. Sibarani, S.: ESG (Environmental, Social, and Governance) implementation to strengthen business sustainability Pt. Migas–north field. Eur. J. Bus. Manage. Res. (2023). Retrieved from https://consensus.app/papers/social-governance-implementation-strengthen-business-sibarani/f4d9eab30ee35bdd8aea6802b07bebf7/?utm_source=chatgpt
57. Siemens: ESG Strategy and Governance (2024). Available at: https://www.siemens-advanta.com/capabilities/offers/esg-strategy-governance?gclid=CjwKCAjw4f6zBhBVEiwATEHFVhYvHr2tvLj9gDLciMxLEjJ6xk3xvcZdrYaZ49-xxtpRQk9YBv-zIxoCRxwQAvD_BwE&acz=1&gad_source=1
58. Siemens: MindSphere documentation area: documentation for insights hub and industrial IoT capabilities (2024). Available at: https://documentation.mindsphere.io/MindSphere/index.html
59. Sierdovski, M., Pilatti, L., Rubbo, P.: Organizational competencies in the development of Environmental, Social, and Governance (ESG) criteria in the industrial sector. Sustainability (2022). Retrieved from https://consensus.app/papers/competencies-development-environmental-social-sierdovski/1c116356114b5801b57f2b60dace89ff/?utm_source=chatgpt
60. Simonite, T.: Machine learning opens up new ways to help people with disabilities. Technol. Rev. (2017). Available at: https://www.technologyreview.com/s/603899/machine-learning-opens-up-newways-to-help-disabled-people
61. Smith, J.: Digital transformation and ESG: enhancing sustainability and governance. J. Bus. Ethics **169**(1), 1–20 (2024)
62. Sphera: Who we are (2024). Available at: https://sphera.com/company/who-we-are/
63. Tan, W., Yan, H.E., Chen, X.H., Hoang, H.Y.: Positive press, greener progress: the role of ESG media reputation in corporate energy innovation (2024). https://doi.org/10.2139/ssrn.4880611

64. Tesla: Impact Report 2023 (2024). Available at: https://www.tesla.com/ns_videos/2023-tesla-impact-report-highlights.pdf
65. Tian, L., Sun, K., Yang, J., Zhao, Y.: Does digital economy affect corporate ESG performance? New insights from China (2024). https://doi.org/10.2139/ssrn.4700677
66. Tohang, V., Hutagaol-Martowidjojo, Y., Pirzada, K.: The link between ESG performance and earnings quality. Austr. Account. Bus. Finan. J. **18**(1), 187–204 (2024)
67. Trice, R.: Can artificial intelligence help feed the world? Forbes (2017). Available at: https://www.forbes.com/sites/themixingbowl/2017/09/05/can-artificial-intelligence-helpfeed-the-world/#3f99736846db
68. Unilever: Leading in the next era of corporate sustainability (2024a). Available at: https://www.unilever.com/sustainability/
69. Unilever: Our ambition is to deliver net zero emissions across our value chain (2024b). Available at: https://www.unilever.com/sustainability/climate/
70. Von Struensee, S.: Artificial intelligence and corporate social responsibility (2021). https://doi.org/10.2139/ssrn.3873097
71. Wang, H., Wang, S., Lu, J., Luo, H.: Does digital transformation improve corporate ESG responsibility performance? Evidence from China (2024). https://doi.org/10.1016/j.frl.2023.104310
72. Wang, J., Hong, Z., Long, H.: Digital transformation empowers ESG performance in the manufacturing industry: from ESG to DESG. SAGE Open **13**(4), 21582440231204160 (2023). https://doi.org/10.1177/21582440231204158
73. Wang, Q., Zhang, C., Li, R.: Towards carbon neutrality by improving carbon efficiency-a system-GMM dynamic panel analysis for 131 countries' carbon efficiency. Energy **258**, 124880 (2022)
74. Wang, S., Esperança, J.P.: Can digital transformation improve market and ESG performance? Evidence from Chinese SMEs. J. Clean. Prod. **419**, 137980 (2023)
75. Westerman, G., Bonnet, D., McAfee, A.: Leading Digital: Turning Technology into Business Transformation. Harvard Business Press (2014). https://www.hbs.edu/faculty/Pages/item.aspx?num=48035
76. World Economic Forum: Shaping the future of global food systems: a scenarios analysis (2017). Available at: http://www3.weforum.org/docs/IP/2016/NVA/WEF_FSA_FutureofGlobalFoodSystems.pdf
77. Ye, J., Moslehpour, M., Tu, Y., Vinh, N.T., Ngo, T.Q., Nguyen, S.: Investment on environmental social and governance activities and its impact on achieving sustainable development goals: evidence from Chinese manufacturing firms. Econ. Res.-Ekonomska Istraživanja (2022). Retrieved from https://consensus.app/papers/investment-governance-activities-impact-achieving-ye/13625db4a49751f2b494d58293dee3ab/?utm_source=chatgpt
78. Yin, L., Li, R., Xuan, S.: Digital finance's impact on corporate stock price crash risk: the mediating roles of digital transformation and ESG performance (2023). https://doi.org/10.2139/ssrn.4573715
79. Yuan, Y.: Corporate digital transformation and ESG performance. Evidence from Chinese-Listed Companies. BCP Bus. Manage. **45**, 407–415 (2023). https://doi.org/10.54691/bcpbm.v45i.4962
80. Zamfiroiu, T., Pînzaru, F.: Advancing strategic management through sustainable finance. Manage. Dyn. Knowl. Econ. (2021). Retrieved from https://consensus.app/papers/advancing-strategic-management-sustainable-finance-zamfiroiu/d75ec0ccdb88545483d9d1ce1041ed33/?utm_source=chatgpt
81. Zatolokin, A., Petrov, Y.: Principles of implementation of the ESG agenda as part of geoportals. InterCarto. InterGIS (2023). Retrieved from https://consensus.app/papers/principles-implementation-agenda-part-geoportals-zatolokin/30cb945d6a2452c3ab309cb41d76cae8/?utm_source=chatgpt
82. Zhao, Q., Li, X., Li, S.: Analyzing the relationship between digital transformation strategy and ESG performance in large manufacturing enterprises: the mediating role of green innovation. Sustainability **15**(13), 9998 (2023). https://doi.org/10.3390/su15139998

83. Zhong, Y., Zhao, H., Yin, T.: Resource bundling: How does enterprise digital transformation affect enterprise ESG development? Sustainability **15**(2), 1319 (2023). https://doi.org/10.3390/su15021319
84. Zhou, Z., Yang, Z., Li, H., Liu, J.: Climate governance action and corporate ESG performance—evidence from China (2024). https://doi.org/10.2139/ssrn.4693239

Digital Transformation in ESG Programs: Understanding Environmental, Social, and Governance Factors

João Farinha and Maria de Fátima Pina

Abstract In the contemporary business landscape, the integration of Environmental, Social, and Governance (ESG) considerations has become paramount for organizations striving to navigate a complex terrain of sustainability challenges and stakeholder expectations. Concurrently, the advent of digital transformation is revolutionizing the way businesses operate and innovate. This chapter explores the dynamic convergence of these two critical domains: Digital Transformation in ESG Programs—Understanding Environmental, Social, and Governance Factors. It delves into the foundational understanding of ESG factors, examining their significance in shaping corporate strategy and reporting frameworks. Furthermore, it explores the transformative potential of digital technologies, from artificial intelligence to blockchain, in augmenting ESG initiatives. Drawing upon recent research and industry insights, the chapter navigates through the challenges and opportunities inherent in this intersection, while offering best practices and recommendations for organizations seeking to harness digital innovation for sustainable impact. Looking towards the future, it anticipates emerging trends and directions in the ongoing evolution of digital transformation and ESG integration within organizations, positioning readers at the forefront of responsible governance and sustainable innovation.

J. Farinha (✉) · M. de Fátima Pina
ISLA Santarém, Santarém, Portugal
e-mail: joao.farinha@islasantarem.pt

M. de Fátima Pina
e-mail: maria.pina@islasantarem.pt

J. Farinha
CINTURS - University of Algarve, Faro, Portugal

GOVCOPP - University of Aveiro, Aveiro, Portugal

M. de Fátima Pina
ISCTE - University Institute of Lisbon, Lisbon, Portugal

ISR - UC Institute of Systems and Robotics, University of Coimbra, Coimbra, Portugal

© The Author(s), under exclusive license to Springer Nature Switzerland AG 2025
A. d. b. Machado et al. (eds.), *Environmental, Social, Governance and Digital Transformation in Organizations*, Information Systems Engineering and Management 35, https://doi.org/10.1007/978-3-031-86079-9_2

Keywords ESG integration · Digital transformation · Sustainability · Innovation · Governance

1 Introduction

In today's rapidly evolving business landscape, organisations are increasingly recognising the imperative of integrating environmental, social, and governance (ESG) considerations into their strategic agendas [14]. Simultaneously, the pervasive influence of digital technologies is reshaping the way businesses operate and innovate [39]. This chapter explores the dynamic convergence of these two critical domains: Digital Transformation in ESG Programs—Understanding Environmental, Social, and Governance Factors.

The importance of integrating digital technologies into ESG initiatives is crucial. Indeed, the complex interplay between environmental sustainability, social responsibility, and effective governance has become paramount, not only for meeting regulatory requirements but also for addressing stakeholder expectations and ensuring long-term viability in a competitive marketplace [14]. Concurrently, the pervasive influence of digital technologies is reshaping the way businesses operate and innovate, presenting both challenges and opportunities in the pursuit of sustainable development [2]. As businesses grapple with navigating a complex terrain of sustainability challenges and stakeholder expectations, the integration of ESG considerations has emerged as a strategic imperative. From mitigating environmental risks and reducing carbon footprints to fostering inclusive workplaces and upholding ethical standards, organisations are increasingly recognising the interconnectedness of ESG factors with corporate performance and reputation [4]. Simultaneously, the advent of digital transformation is revolutionising traditional business models and processes, offering unprecedented opportunities for innovation and efficiency gains [39]. By leveraging technologies such as artificial intelligence, blockchain, and data analytics, organisations can augment their ESG initiatives, driving transparency, accountability, and responsible governance practices.

Drawing upon recent research and industry insights, this chapter explores the foundational understanding of ESG factors and their significance in shaping corporate strategy and reporting frameworks [12]. It further examines the transformative potential of digital technologies in enhancing ESG performance, from streamlining data collection and analysis to enabling real-time monitoring of sustainability metrics [17]. By navigating through the challenges and opportunities inherent in this intersection, organisations can unlock new pathways for sustainable impact and competitive advantage. Looking towards the future, this chapter anticipates emerging trends and directions in the ongoing evolution of digital transformation and ESG integration within organisations, positioning readers at the forefront of responsible governance and sustainable innovation in the digital age.

As previously underlined, the present study investigates the critical intersection of digital transformation and ESG considerations in today's business landscape.

It highlights the imperative for organisations to integrate ESG factors into their strategic agendas amidst the evolving business environment. The study explores how digital technologies, including AI, data analytics, blockchain, and IoT, are reshaping business operations and innovation processes, offering transformative potential for sustainable growth. Examining the role of digital technologies in augmenting ESG programs, the study shows their application in data collection, analysis, and decision-making to drive positive changes across environmental, social, and governance dimensions. Additionally, it addresses challenges such as data privacy concerns and regulatory complexities, while outlining best practices for organisations to navigate digital transformation effectively.

In the end, the study also anticipates future trends, including AI-driven ESG analytics and sustainable fintech solutions, positioning organisations to lead in sustainable innovation and responsible governance in the digital era.

2 Understanding ESG Factors

We start this section by presenting a definition and analyzing the significance of Environmental, Social, and Governance factors. In fact, ESG factors comprise a diverse range of concerns, spanning from environmental sustainability and social responsibility to governance practices [4]. Organizations are faced with the challenge of navigating intricate frameworks and reporting standards while endeavouring to showcase their dedication to sustainability and ethical behaviour. Originating from the concept of sustainability, ESG has evolved its focus from the external impacts of business activities on society and the environment towards the financial implications for investors when ESG factors are overlooked [5].

Further, ESG factors constitute a comprehensive framework that encompasses various aspects of corporate responsibility and sustainable business practices. Environmental factors entail the assessment of an organization's impact on natural resources, such as energy consumption, greenhouse gas emissions, water usage, and biodiversity preservation [4, 8]. Social factors extend beyond internal operations to consider the broader societal impact of an organization, including its treatment of employees, human rights practices, community engagement, and supply chain management [16, 32]. Governance factors focus on the structures and processes that govern corporate decision-making, encompassing issues such as board composition, executive compensation, risk management practices, and transparency in financial reporting [4, 8].

Environmental factors encompass the ecological impact of an organization's activities, focusing on resource consumption, pollution, and climate change mitigation. This includes assessing carbon emissions, energy efficiency, water usage, waste management, and efforts towards biodiversity conservation. Organizations are increasingly expected to adopt sustainable practices to minimize their environmental footprint and contribute to global efforts for environmental preservation [13, 28].

Social factors encompass the impact of an organization's operations on society, including its employees, communities, and broader stakeholder groups. This involves considerations such as labour rights, employee well-being, diversity and inclusion, community engagement, and human rights practices. Socially responsible organizations prioritize fair labour practices, promote diversity and equality, support community development initiatives, and uphold human rights standards throughout their supply chains [4, 9].

Governance factors relate to the structures, processes, and practices governing an organization's decision-making and accountability mechanisms. This includes board composition, executive compensation, risk management practices, shareholder rights, and adherence to legal and regulatory requirements. Effective corporate governance ensures transparency, integrity, and ethical conduct, fostering trust among stakeholders and enhancing long-term sustainability [8, 38].

It is important to note that to effectively assess ESG performance, organizations utilize a range of key metrics and indicators tailored to each dimension. Environmental metrics may include carbon emissions intensity, renewable energy adoption rates, waste diversion ratios, and water efficiency measures [8, 16]). Social metrics encompass employee turnover rates, diversity and inclusion indices, community investment levels, and stakeholder engagement metrics [4, 9]. Governance metrics often involve board diversity statistics, executive remuneration transparency, regulatory compliance records, and the presence of anti-corruption policies [8, 16]. These metrics provide stakeholders with quantifiable insights into an organization's ESG performance, facilitating informed decision-making and accountability.

To finish, we would like to mention that the landscape of ESG reporting standards and frameworks has undergone significant development in recent years, driven by increasing investor demand for transparent and comparable ESG information. Frameworks such as the Global Reporting Initiative (GRI), Sustainability Accounting Standards Board (SASB), and Task Force on Climate-related Financial Disclosures (TCFD) have gained prominence for their efforts to standardize ESG reporting practices [4, 8]. These frameworks provide organizations with structured guidelines for disclosing material ESG-related information, enhancing transparency and comparability across industries and regions. Moreover, regulatory initiatives and stock exchange listing requirements worldwide are increasingly mandating ESG disclosures, further accelerating the adoption of standardized reporting practices and driving corporate accountability [16, 32].

3 Digital Transformation Trends

The digital revolution is indeed unfolding at an unprecedented pace, marked by significant advancements in artificial intelligence, data analytics, blockchain, and the Internet of Things (IoT) [2]. Building upon the advancements of previous industrial revolutions, the Digital revolution is characterized by the pervasive integration of technologies such as artificial intelligence, robotics, the Internet of Things (IoT), cloud computing, and advanced analytics. These technologies enable interconnected systems, smart machines, and digital platforms to revolutionize production processes, business models, and societal interactions [11]. These technological innovations hold transformative potential, offering organizations the means to elevate operational efficiency, drive innovation, and explore novel avenues for sustainable growth. For example, artificial intelligence empowers organizations to automate routine tasks, analyse vast datasets, and derive actionable insights, thereby enhancing decision-making processes and driving operational excellence [19]. Similarly, data analytics enables organizations to extract valuable insights from complex datasets, enabling them to make informed decisions, optimize processes, and uncover new opportunities for value creation [22]. Furthermore, recent research suggests that corporations may benefit from initiating their AI transformation sooner rather than later, with a bias towards exploiting AI capabilities for powerful new innovations [1].

Digital transformation is fundamentally reshaping organizations at an unprecedented pace, driven by these advancements in technology [2]. It represents a multifaceted process that goes beyond mere technological adoption; it encompasses the integration of digital technologies into all facets of business operations and interactions [39]. This integration extends from internal processes to customer experiences, with the overarching goal of enhancing efficiency, agility, and competitiveness. For example, blockchain technology enables organizations to establish transparent and secure transaction records, fostering trust and integrity in various business processes, such as supply chain management and financial transactions [36]. Moreover, the Internet of Things (IoT) facilitates real-time monitoring and control of connected devices, enabling organizations to optimize resource utilization, improve asset management, and enhance operational efficiency [18].

In essence, digital transformation represents a paradigm shift in how organizations operate, innovate, and engage with stakeholders [27]. It enables organizations to leverage digital technologies to streamline workflows, enhance decision-making, and drive continuous improvement. By fostering a culture of innovation and collaboration, digital transformation empowers organizations to adapt to changing market dynamics, seize new opportunities, and achieve sustainable growth in today's rapidly evolving business landscape.

Furthermore, the ongoing digital transformation across industries is driven by various emerging technologies, each playing a pivotal role in reshaping organizational processes and strategies. Artificial intelligence (AI) and machine learning (ML) stand out as frontrunners in this transformation, revolutionizing business operations by automating repetitive tasks, analyzing vast datasets, and delivering actionable insights [19]. Additionally, data science employs complex machine learning algorithms to construct predictive models, making it an essential component of AI and machine learning [10]. These technologies empower organizations to optimize processes, enhance decision-making, and foster innovation, thereby enabling them to remain competitive in the digital age.

Moreover, cloud computing emerges as a fundamental enabler of digital transformation, providing organizations with scalable infrastructure and resources to support digital initiatives [2]. Cloud solutions offer flexibility, cost-efficiency, and rapid deployment of applications and services, empowering organizations to adapt quickly to changing market demands and customer expectations. Organizations can streamline their operations, decrease infrastructure expenses, and expedite the introduction of new products and services by harnessing cloud computing capabilities [1].

In addition to AI, ML, and cloud computing, the Internet of Things (IoT) plays a crucial role in driving digital transformation efforts. IoT technologies connect devices and sensors, enabling real-time monitoring, data collection, and predictive maintenance [18]. This connectivity enables the seamless integration of physical and digital systems, thereby optimizing operations, enhancing customer experiences, and unlocking new opportunities for innovation. Insights derived from the Internet of Things (IoT) empower organizations to make data-driven decisions, optimize resource utilization, and deliver personalized services, driving both operational efficiency and customer satisfaction. In this digital age, the significance of data cannot be overstated, underscoring the importance of proper storage and error-free processing [10].

Furthermore, blockchain technology emerges as a disruptive force in supply chain management, offering transparent and secure transactions while enhancing traceability and accountability across the value chain [36]. By leveraging blockchain, organizations can establish immutable records of transactions, ensuring transparency and integrity in complex supply chain networks. This technology enables organizations to mitigate risks, reduce fraud, and build trust among stakeholders, ultimately enhancing the efficiency and resilience of supply chain operations.

Consequently, emerging technologies such as AI, ML, cloud computing, IoT, and blockchain are driving digital transformation efforts across industries, revolutionizing business operations, and enabling organizations to innovate, adapt, and thrive in the digital age. In this sense, we finish this section with some examples of digital tools and platforms used for sustainability and responsible governance.

Indeed, in the pursuit of sustainability and responsible governance, digital tools and platforms have become indispensable assets for organizations aiming to address environmental, social, and governance (ESG) challenges. One example is sustainability reporting software, which facilitates the systematic collection, analysis, and reporting of ESG data, aiding organizations in monitoring their performance against sustainability objectives and complying with reporting standards [33]. By leveraging this software, organizations can gain valuable insights into their ESG performance, identify areas for improvement, and communicate their sustainability efforts transparently to stakeholders.

Another essential digital tool in the realm of sustainability is environmental management systems, which harness data analytics and Internet of Things (IoT) sensors to monitor various environmental parameters such as energy consumption, waste generation, and emissions [7]. These systems provide organizations with real-time visibility into their environmental footprint, enabling them to pinpoint inefficiencies, track progress towards sustainability goals, and implement targeted initiatives to reduce their environmental impact. By integrating data-driven insights into their sustainability strategies, organizations can optimize resource utilization, minimize waste, and enhance their overall environmental performance.

Furthermore, blockchain-based platforms have emerged as a powerful tool for enhancing supply chain transparency and traceability, particularly in industries where ethical sourcing and fair labor practices are paramount [30]. By leveraging blockchain technology, organizations can establish immutable records of transactions, ensuring transparency and accountability throughout the supply chain. This enables stakeholders to trace the origins of products, verify their authenticity, and ensure compliance with sustainability standards and ethical guidelines. Additionally, blockchain enhances trust and collaboration among supply chain partners, fostering a more sustainable and responsible approach to procurement and production processes.

In addition to sustainability reporting software, environmental management systems, and blockchain-based platforms, data analytics platforms play a crucial role in driving sustainability initiatives within organizations [33]. These platforms enable organizations to derive actionable insights from ESG data, identify trends, and make informed decisions to drive sustainability performance. By harnessing the power of data analytics, organizations can optimize resource allocation, identify opportunities for efficiency improvements, and demonstrate tangible progress towards their sustainability goals.

Therefore, digital tools and platforms are essential enablers of sustainability and responsible governance practices within organizations. By leveraging these technologies, organizations can enhance transparency, accountability, and performance across environmental, social, and governance dimensions, ultimately driving long-term value creation and stakeholder trust.

4 The Role of Digital Technologies in ESG Programs

Digital technologies play a pivotal role in augmenting ESG programs, offering powerful tools for data collection, analysis, and decision-making (Loebbecke et al., 2020). From leveraging big data analytics to monitor environmental impact to implementing blockchain for transparent supply chains, organizations are harnessing digital innovations to drive positive change across ESG dimensions.

The integration of digital technologies into Environmental, Social, and Governance (ESG) programs has emerged as a pivotal area of research and practice, reflecting a growing recognition of the complexities involved in corporate sustainability and accountability. The evolution of ESG from traditional corporate social responsibility (CSR) frameworks highlights the need for more nuanced approaches to understanding corporate roles in environmental and social contexts.

Park et al. [23] emphasizes that ESG encompasses a broader and more detailed evaluation than CSR, which has historically focused on accountability. This foundational understanding sets the stage for exploring how digital technologies can enhance ESG measurement and reporting, thereby addressing stakeholder concerns regarding transparency and efficacy in corporate practices. As organizations grapple with the pressing challenges of climate change and resource management, the role of digital transformation becomes increasingly significant. Qing and Jin [25] argue that effective ESG practices are essential for achieving corporate sustainability, particularly in the face of abnormal climate conditions and the imperative for carbon neutrality. Their research underscores the importance of leveraging artificial intelligence in digital transformation efforts to bolster ESG initiatives. By providing a strategic direction for integrating digital technologies, this study illuminates how companies can enhance their ESG activities to mitigate negative environmental impacts while fostering sustainable growth.

Further advancing the discourse on digital technologies in ESG, Xu [40] explores the potential of AI and machine learning in the financial sector's approach to ESG. The establishment of centralized data platforms that integrate ESG metrics with existing financial frameworks is essential for achieving a unified source of truth. Xu [40] also highlights the importance of advanced analytics capabilities for predictive modelling and trend analysis, which can significantly enhance the accuracy and accessibility of ESG data. The emphasis on automated reporting tools also addresses the need for compliance in an increasingly regulated environment, while robust cybersecurity measures ensure the protection of sensitive information. This comprehensive approach to digital-ESG integration not only facilitates better decision-making but also aligns technological advancements with the overarching goals of sustainability and corporate governance.

Therefore, it is important to notice that literature seems to reveal a clear trajectory toward the utilization of digital technologies as instrumental in enhancing ESG frameworks. The insights provided by [23, 25, 40], collectively underscore the necessity of embracing digital transformation to effectively measure, report, and manage ESG initiatives, thereby contributing to the sustainable evolution of corporate practices in a rapidly changing global landscape.

5 Challenges and Opportunities

Despite the promise of digital transformation in ESG programs, organizations face a myriad of challenges, including data privacy concerns, digital skills gaps, and regulatory complexities [12]. However, these challenges also present opportunities for collaboration, innovation, and continuous improvement in sustainability practices.

The integration of digital technologies into Environmental, Social, and Governance (ESG) programs presents both challenges and opportunities (see, Fig. 1). These technologies enhance efficiency, transparency, and sustainability but also raise critical concerns around data privacy, inclusivity, and adoption barriers, while offering potential for innovation and collaboration.

As organizations increasingly use digital tools to collect ESG-related data, concerns over data privacy and security become crucial. Technologies like big data analytics, artificial intelligence (AI), and the Internet of Things (IoT) generate sensitive data on aspects such as energy use and employee welfare [31]. Without robust data protection, organizations risk breaches, which can result in penalties and loss of

Fig. 1 Challenges and opportunities tracks

trust [24]. Blockchain technology offers a solution by enhancing data transparency and security through immutable and traceable records [37].

Digital ESG programs must prioritize inclusivity to avoid reinforcing social inequalities. While these tools can democratize access, marginalized communities' risk being excluded due to limited digital access or literacy [19]. Failure to address this can deepen inequities, undermining ESG objectives, especially regarding social outcomes [22]. Ensuring inclusivity requires accessible platform design, digital skills training, and collaboration with communities to co-create solutions [29]. A diverse workforce in ESG roles also fosters equitable outcomes [34].

Many organizations face barriers to digital adoption, such as high costs and a lack of technical expertise. Smaller organizations or those in developing regions may struggle to invest in advanced technologies like AI and IoT [35]. To address this, companies can start with scalable solutions and collaborate with technology providers to share resources. Employee training programs can also close the skills gap, enabling organizations to sustain digital transformation [24].

Despite challenges, digital transformation presents opportunities for ESG innovation. AI and machine learning can optimize sustainability processes, while blockchain enhances supply chain transparency [31]. Collaboration between businesses, governments, and NGOs can further accelerate digital ESG efforts through shared resources and knowledge [5]. Public–private partnerships in areas like renewable energy and smart cities can also drive sustainable progress [37]. Therefore, while digital integration into ESG programs raises challenges around data privacy, inclusivity, and adoption, it also offers opportunities for innovation and collaboration, ultimately driving sustainable growth.

6 Best Practices and Recommendations

Drawing upon industry insights and case studies, this section outlines best practices for integrating digital technologies into ESG strategies. From setting clear objectives and KPIs to fostering stakeholder engagement and transparency, organizations can navigate the complexities of digital transformation while advancing their sustainability agendas. In fact, digital transformation within Environmental, Social, and Governance (ESG) programs presents organizations with significant opportunities to enhance sustainability, transparency, and operational efficiency. To maximize the benefits of digital ESG efforts, organizations must adopt best practices that address strategic planning, stakeholder engagement, and continuous evaluation. Below are listed several key recommendations for effectively integrating digital tools into ESG initiatives (see, for instance, Fig. 2).

A. Establishing Clear Goals and KPIs for Digital ESG Initiatives: Setting clear, measurable goals and Key Performance Indicators (KPIs) is essential for ensuring the success of digital ESG initiatives. Well-defined goals aligned

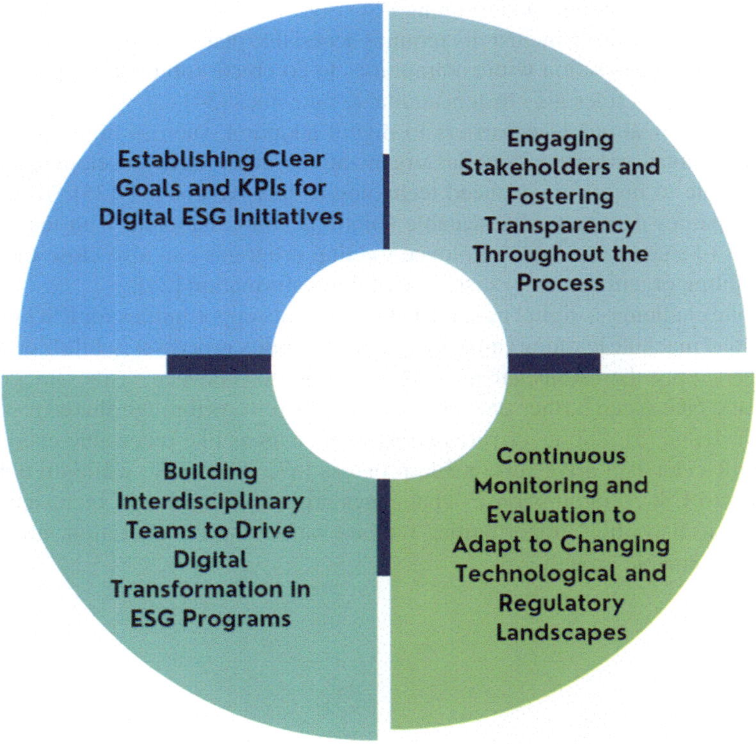

Fig. 2 Key recommendations for effectively integrating digital tools into ESG initiatives

with broader corporate sustainability objectives can provide direction and accountability. KPIs allow organizations to track performance, identify areas for improvement, and demonstrate progress to stakeholders. For example, KPIs could include metrics for carbon reduction, resource efficiency, or supply chain transparency [34]. These indicators should be integrated into digital platforms that can automate data collection and real-time reporting, facilitating ongoing evaluation and adjustment [5].

B. Building Interdisciplinary Teams to Drive Digital Transformation in ESG Programs: The successful implementation of digital ESG initiatives requires collaboration across various disciplines. Building interdisciplinary teams that include experts in technology, sustainability, data analytics, and governance ensures a holistic approach to ESG transformation. These teams can bring diverse perspectives and expertise to solve complex problems and design effective solutions [22]. Such collaboration fosters innovation, particularly when integrating emerging technologies such as artificial intelligence (AI), blockchain, and the Internet of Things (IoT) into ESG programs [19]. Engaging diverse skill sets allows for the alignment of technological solutions with sustainability goals, creating more effective and inclusive ESG programs [26].

C. Engaging Stakeholders and Fostering Transparency Throughout the Process: Transparency and stakeholder engagement are crucial components of any successful ESG initiative. By engaging stakeholders (including employees, investors, customers, and community members) organizations can build trust and ensure that ESG programs reflect the interests and concerns of those affected [29]. Digital platforms, such as blockchain, can enhance transparency by providing immutable records of ESG-related activities, from sourcing practices to carbon emissions tracking [37]. Furthermore, maintaining open communication channels and involving stakeholders in the decision-making process strengthens the credibility of ESG efforts [24]. This approach not only improves organizational accountability but also enhances the social impact of ESG initiatives by ensuring they meet the needs of all stakeholders.

D. Continuous Monitoring and Evaluation to Adapt to Changing Technological and Regulatory Landscapes: In the rapidly evolving digital and regulatory environment, continuous monitoring and evaluation are essential for keeping ESG initiatives relevant and effective. Technologies such as AI and IoT enable real-time monitoring of ESG metrics, providing organizations with the ability to respond quickly to changes in performance or compliance requirements [35]. Additionally, organizations must stay informed of regulatory changes related to ESG reporting, data privacy, and sustainability standards, such as the European Union's Corporate Sustainability Reporting Directive (CSRD) or the California Consumer Privacy Act (CCPA) [5]. Regular audits, reviews of performance data, and adapting strategies based on emerging best practices will ensure that digital ESG programs remain resilient and agile in the face of technological advancements and regulatory shifts [24].

To conclude, it is important to notice that by establishing clear goals and KPIs, building interdisciplinary teams, fostering transparency, and continuously adapting to changing landscapes, organizations can enhance the success of their digital ESG initiatives. These best practices enable organizations to leverage digital tools effectively, driving sustainability efforts and ensuring long-term value creation.

7 Future Directions

Looking ahead, the intersection of digital transformation and ESG programs holds immense potential for driving positive societal and environmental impact. Anticipated trends include the widespread adoption of AI-driven ESG analytics, the proliferation of sustainable fintech solutions, and the emergence of digital ecosystems for collaborative sustainability initiatives.

Further, as digital transformation continues to evolve, its integration with Environmental, Social, and Governance (ESG) programs is anticipated to be shaped by several emerging trends. These trends will play a critical role in enabling organizations to enhance sustainability efforts, promote responsible governance, and respond to growing stakeholder expectations for transparency and accountability. In what follows we present some of these anticipated trends.

1. AI and Data Analytics for Predictive ESG Management: One of the most significant anticipated trends is the increased use of Artificial Intelligence (AI) and data analytics to manage ESG programs proactively. AI-driven predictive analytics will enable organizations to assess risks and opportunities related to sustainability in real time. For example, AI can be used to forecast environmental impacts, identify inefficiencies in resource use, and optimize supply chain operations for greater sustainability [19]. By leveraging AI, organizations can shift from reactive to proactive ESG management, enhancing their ability to meet regulatory requirements and achieve sustainability goals more effectively [22].
2. Blockchain for Enhanced Supply Chain Transparency: Blockchain technology is expected to continue playing a pivotal role in ensuring supply chain transparency, especially in industries such as agriculture, fashion, and electronics, where ethical sourcing is a growing concern. Blockchain allows for immutable and transparent records of transactions, which can verify the origin of raw materials and ensure that suppliers adhere to sustainable practices [37]. As consumer demand for ethical products grows, the adoption of blockchain for ESG purposes is likely to expand, providing companies with tools to improve accountability across complex supply chains [31].
3. Internet of Things (IoT) for Real-Time Environmental Monitoring: The integration of the Internet of Things (IoT) into ESG initiatives is expected to increase significantly. IoT devices can provide real-time data on energy consumption, emissions, and waste, enabling organizations to make more informed decisions to reduce their environmental footprint. According to Delgado-Ceballos et al.

Fig. 3 Challenges that organizations must address

[5], IoT-driven insights will allow businesses to monitor and optimize resource utilization continuously, driving operational efficiency while minimizing environmental harm. As IoT technology advances, its applications in ESG programs will expand, creating new opportunities for data-driven sustainability efforts.

In this scope, it is also important to weigh up some potential challenges and opportunities on the horizon. Indeed, while the integration of digital technologies into ESG programs presents numerous opportunities, it also introduces a range of challenges (see, for instance, Fig. 3) that organizations must address, such as:

1. Data Privacy and Security: As digital ESG initiatives rely heavily on data collection and sharing, ensuring data privacy and security will become an increasingly pressing challenge. With large volumes of sensitive information, such as employee welfare and environmental impact data, being processed through digital platforms, the risk of data breaches and cyberattacks is heightened [15]. To address this, organizations must invest in robust cybersecurity measures and comply with evolving data protection regulations [24]. While these challenges may pose initial hurdles, they also present opportunities to develop innovative solutions for secure data management.
2. Bridging the Digital Divide: One of the primary challenges of digital transformation in ESG is ensuring inclusiveness. The digital divide the gap between those with access to digital technologies and those without and could exacerbate social inequalities if not adequately addressed [35]. Marginalized communities may

lack access to the digital tools required to benefit from ESG initiatives. However, this challenge also offers opportunities for organizations to develop programs that increase digital literacy and expand access to technology, fostering more equitable ESG outcomes [29].

3. Regulatory Complexity: As the digital landscape evolves, regulatory frameworks governing ESG reporting and digital technology use are becoming more complex. Organizations will need to navigate an increasing number of regulations, from the European Union's Corporate Sustainability Reporting Directive (CSRD) to the California Consumer Privacy Act (CCPA). While keeping up with these regulations may be challenging, it also creates an opportunity for companies to lead in compliance and set new standards for responsible digital transformation [5].

Before finishing this section, an approach about the role of Research and Development (R&D) in advancing digital solutions for sustainability and responsible governance are presented. In this sense we notice that research and development will play a crucial role in advancing digital technologies that support sustainability and responsible governance. Further, innovations in AI, blockchain, IoT, and other emerging technologies will be critical to solving complex ESG challenges and driving progress toward long-term sustainability goals. Below two considerations are given, namely, AI for sustainability innovation, and open innovation and collaboration:

1. AI for Sustainability Innovation: Continued investment in AI research will lead to the development of more sophisticated tools for sustainability management. For example, machine learning algorithms can be designed to optimize energy consumption, predict environmental risks, and model the impacts of climate change. Manyika et al. [19] highlight the potential for AI to revolutionize industries by offering new capabilities for resource management and decision-making.
2. Open Innovation and Collaboration: Collaboration between academia, industry, and government institutions will be essential to advancing digital ESG solutions. Open innovation, where organizations share knowledge and resources, can accelerate the development of technologies that address sustainability challenges [26]. Public–private partnerships, in particular, offer opportunities for pooling resources to tackle issues like climate change and supply chain transparency on a global scale [37].

The integration of digital technologies into ESG programs is expected to grow, with trends such as AI, blockchain, and IoT playing pivotal roles. While these advancements present challenges related to data security, inclusivity, and regulatory complexity, they also offer opportunities for innovation and collaboration. Research and development will be critical in driving the next wave of digital solutions for sustainability and responsible governance.

8 Conclusion

In this chapter, we have explored the profound impact of digital technologies on Environmental, Social, and Governance (ESG) programs, as well as the challenges and opportunities they present. Digital transformation is driving significant changes across industries, enabling organizations to enhance their ESG performance by leveraging technologies such as Artificial Intelligence (AI), blockchain, data analytics, and the Internet of Things (IoT). These tools are reshaping how organizations manage sustainability, improving transparency, and addressing stakeholder demands for greater accountability.

A key point discussed is the potential of digital technologies to streamline ESG data collection and analysis, providing organizations with real-time insights into their environmental and social impacts. AI and data analytics allow for predictive capabilities, while blockchain technology offers enhanced transparency in supply chains. However, these advancements also bring challenges, particularly regarding data privacy and security. Organizations must adopt robust cybersecurity measures to protect sensitive information while adhering to regulatory frameworks such as the General Data Protection Regulation (GDPR) and the California Consumer Privacy Act (CCPA).

Another critical theme is the importance of inclusivity and equity in digital ESG initiatives. While digital tools can democratize access to information, there is a risk of exacerbating social inequalities if marginalized communities are left behind. Therefore, it is vital for organizations to design accessible and equitable digital platforms, invest in digital literacy, and foster diverse teams that bring a variety of perspectives to their ESG efforts.

The discussion also highlighted the barriers to digital adoption, such as the financial cost of implementing advanced technologies and the skills gap in many organizations. Overcoming these barriers requires a phased approach to digital transformation, partnerships with technology providers, and a strong focus on employee training and development.

As a final reflection, it is clear that digital transformation plays an indispensable role in enhancing ESG performance. By adopting innovative digital tools, organizations can not only improve operational efficiency and sustainability but also build stronger relationships with stakeholders by demonstrating transparency and accountability. As ESG considerations become increasingly central to corporate strategy, the integration of digital technologies will be a key differentiator for organizations striving to lead in sustainability.

The call to action for organizations is clear: embrace digital innovation in sustainability efforts. Investing in the right digital tools, fostering collaboration across disciplines, and prioritizing inclusivity will enable businesses to achieve long-term ESG goals while contributing to global efforts to combat climate change, promote social equity, and ensure responsible governance. By harnessing the potential of digital transformation, organizations can drive meaningful change and position themselves for success in a future where sustainability is paramount.

References

1. Bughin, J.: Does artificial intelligence kill employment growth: the missing link of corporate AI posture. Front. Artif. Intell. **6** (2023). https://doi.org/10.3389/frai.2023.1239466
2. Bughin, J., Hazan, E., Sree Ramaswamy, P., Dc, W., Chu, M. (2017). Artificial intelligence—the next digital frontier? In: Network Security (vol. 2017, Issue 4). McKinsey Global Institute. https://doi.org/10.1016/S1353-4858(17)30039-9
3. Bughin, J., Hazan, E., Ramaswamy, S., Chui, M., Allas, T., Dahlström, P., Trench, M., et al.: Skill Shift: Automation and the Future of the Workforce. McKinsey Global Institute (2018). https://www.mckinsey.com
4. Clark, G.L., Feiner, A., Viehs, M.: From the Stockholder to the Stakeholder—How Sustainability Can Drive Financial Outperformance (Arabesque Partners (ed.), Issue March). University of Oxford (2015)
5. Delgado-Ceballos, J., Ortiz-De-Mandojana, N., Antolín-López, R., Montiel, I.: Connecting the sustainable development goals to firm-level sustainability and ESG factors: the need for double materiality. BRQ Bus. Res. Q. **26**(1), 2–10 (2023). https://doi.org/10.1177/23409444221140919
6. Delgado-Ceballos, J., Aragon-Correa, J.A., Sharma, S.: Corporate sustainability in the digital age: toward a smart and inclusive approach. Bus. Soc. (2023). https://doi.org/10.1177/00076503221150718
7. Delmas, M.A., Pekovic, S.: Corporate sustainable innovation and employee behavior. J. Bus. Ethics **150**(4), 1071–1088 (2018). https://doi.org/10.1007/s10551-016-3163-1
8. Eccles, R.G., Serafeim, G., Krzus, M.P.: The Sustainable Accounting Standards Board: The Case for Sustainability Standards. Harvard Business School (2019)
9. Elkington, J.: Green Swans: The Coming Boom in Regenerative Capitalism (vol. 17, pp. 296). Greenleaf Book Group (2020)
10. Farinha, J., & Pina, M.: Data science in the employee recruitment process. In: Figueiredo, P. (Ed.), Complex AI Dynamics and Interactions in Management. IGI Global (2023)
11. Farinha, J., Thomaz, J., Ramos P., Coelho, B.: Applications of augmented reality in industrial unit processes. In: Rouco, J., Figueiredo, P. (Eds.), Business Continuity Management and Resilience: Theories, Models, and Processes. IGI Global (2023)
12. Fosfuri, A., Giarratana, M.S., Luzzi, A.: Big data analytics, digital transformation and sustainable development: the promise and perils. J. Manage. Stud. **55**(1), 52–87 (2018)
13. Jackson, T.: Prosperity Without Growth: Foundations for the Economy of Tomorrow. Taylor & Francis Group (2016)
14. Kolk, A.: The social responsibility of international business: From ethics and the environment to CSR and sustainable development. J. World Bus. **51**(1), 23–34 (2016). https://doi.org/10.1016/j.jwb.2015.08.010
15. Kshetri, N.: Blockchain and sustainable supply chain management. Sustainability **13**(9), 5248 (2021). https://doi.org/10.3390/su13095248
16. Laine, M., Tregidga, H., Unerman, J.: Sustainability Accounting and Accountability. Routledge (2021). https://doi.org/10.4324/9781003185611
17. Loebbecke, C., Picot, A.: Reflections on societal and business model transformation arising from digitization and big data analytics: a research agenda. J. Strat. Inf. Syst. **24**(3), 149–157 (2015). https://doi.org/10.1016/j.jsis.2015.08.002
18. Manyika, J., Chui, M., Bisson, P., Woetzel, J., Dobbs, R., Marrs, A.T.: The Internet of Things: Mapping the Value Beyond the Hype. McKinsey Global Institute (2015)
19. Manyika, J., Chui, M., Bisson, P., Woetzel, J., Dobbs, R., Bughin, J., Aharon, D.: A Future that Works: Automation, Employment, and Productivity. McKinsey Global Institute (2017).
20. Manyika, J., Chui, M., Bughin, J., Dobbs, R., Bisson, P., Marrs, A.: Harnessing Automation for a Future that Works. McKinsey Global Institute (2017). https://doi.org/10.2139/ssrn.3583017
21. McAfee, A., Brynjolfsson, E., Davenport, T.H., Patil, D.J., Barton, D.: How AI boosts industry profits and innovation. Harv. Bus. Rev. **96**(1), 30–39 (2018)

22. McAfee, A., Brynjolfsson, E., Davenport, T.H., Patil, D., Barton, D.: Data science and the future of innovation. MIT Sloan Manag. Rev. **60**(1), 22–26 (2018). https://doi.org/10.1162/inov_a_00253
23. Park, J., Choi, W., Jung, S.U.: Exploring Trends in Environmental, Social, and Governance Themes and Their Sentimental Value Over Time (2022). ncbi.nlm.nih.gov
24. PWC: ESG data and privacy: balancing risks and opportunities. PWC Report (2020). https://doi.org/10.1787/4d9f0e74-en
25. Qing, C., Jin, S.: Does ESG and Digital Transformation affects Corporate Sustainability? The Moderating Role of Green Innovation (2023)
26. Ross, J.W., Beath, C.M., Mocker, M.: Designed for Digital: How to Architect Your Business for Sustained Success. MIT Press (2019)
27. Ross, J.W., Sebastian, I. M., Beath, C.M.: Digital Transformation: A Roadmap for Billion-Dollar Organizations. MIT CISR Working Paper (2019). https://doi.org/10.2139/ssrn.3429146
28. Sachs, J.D., Kroll, C., Lafortune, G., Fuller, G., Woelm, F.: Sustainable Development Report 2022. Cambridge University Press (2022)
29. Santos, F.M., Eisenhardt, K.M.: Constructing markets and shaping boundaries: entrepreneurial action in nascent fields. Acad. Manag. J. **63**(2), 407–436 (2020). https://doi.org/10.5465/amj.2017.0035
30. Sarkis, J., Cohen, M.J., Dewick, P., Schröder, P.: A brave new world: lessons from the COVID-19 pandemic for transitioning to sustainable supply and production. Resour. Conserv. Recycl. **159**, 104894 (2020). https://doi.org/10.1016/j.resconrec.2020.104894
31. Sarkis, J., Zhu, Q., Lai, K.H.: An organizational theoretic review of green supply chain management literature. Int. J. Prod. Econ. **228**, 107668 (2019). https://doi.org/10.1016/j.ijpe.2020.107668
32. Savitz, A.: The Triple Bottom Line: How Today's Best-Run Companies are Achieving Economic, Social and Environmental Success-and How You Can Too. Wiley (2013)
33. Schaltegger, S., Lüdeke-Freund, F., Hansen, E.G.: Business models for sustainability: a co-evolutionary analysis of sustainable entrepreneurship, innovation, and transformation. Organ. Environ. **29**(3), 264–289 (2016)
34. Schaltegger, S., Hansen, E.G.: Sustainability balanced scorecards and their architectures: irrelevant or misunderstood? J. Bus. Ethics **150**(4), 937–952 (2019). https://doi.org/10.1007/s10551-018-3863-4
35. Schneider, A., Schmidpeter, R.: Digitalisation and sustainability in small and medium-sized enterprises (SMEs): an overview. In: Digitalization and Sustainability (pp. 21–38). Springer, Cham (2021). https://doi.org/10.1007/978-3-030-63296-9_2
36. Tapscott, D., Tapscott, A.: Blockchain Revolution. Senai-SP Editora (2018)
37. Tapscott, D., Tapscott, A.: How blockchain is changing finance. Harv. Bus. Rev. **1**(9), 2–5 (2017). https://doi.org/10.2139/ssrn.2999287
38. Tricker, R.I.: Corporate governance: Principles, Policies, and Practices. Oxford University Press (2015)
39. Westerman, G., Bonnet, D., McAfee, A.: Leading Digital: Turning Technology into Business Transformation. Harvard Business Press (2011)
40. Xu, J.: AI in ESG for Financial Institutions: An Industrial Survey (2024)

Does Corporate Governance Impact ESG Performance Reporting in Healthcare?

Amélia Ferreira da Silva, Francisco Carreira, and Catarina Cepêda

Abstract This paper aims to analyze factors that influence Environmental, Social and Governance (ESG) performance reporting by healthcare sector companies in European Union (EU) countries. To achieve the goal, an empirical analysis was conducted with healthcare companies, listed in the EU, for 2017–2021. Our research model considered nine research hypotheses regarding the relationship between corporate governance variables and ESG performance reporting. The moderate effect of stakeholder engagement on these relations was tested. Statistical analysis was performed using Partial Least Squares based Structural Equation. We confirmed eight hypotheses regarding the effect of corporate governance variables on ESG performance reporting. The moderate effect of stakeholder engagement was confirmed for only three of the corporate governance variables. Our Results are useful for all ESG disclosure users, governments, and organizations working on developing sustainability reporting standards.

Keywords ESG reporting · Corporate governance · Gender · Stakeholder engagement

1 Introduction

Corporate governance, Environmental, Social, and Governance (ESG) performance, and healthcare are interconnected areas that have gained increasing attention from researchers, policymakers, and practitioners (e.g. [19, 28, 58]) The integration of

A. F. da Silva (✉) · C. Cepêda
CEOS.PP, ISCAP, Polytechnic of Porto, Porto, Portugal
e-mail: acfs@iscap.ipp.pt

C. Cepêda
e-mail: clmc@iscap.ipp.pt

F. Carreira
CICF-IPCA, Instituto Politécnico de Setúbal, Setúbal, Portugal
e-mail: francisco.carreira@esce.ips.pt

© The Author(s), under exclusive license to Springer Nature Switzerland AG 2025
A. d. b. Machado et al. (eds.), *Environmental, Social, Governance and Digital Transformation in Organizations*, Information Systems Engineering and Management 35, https://doi.org/10.1007/978-3-031-86079-9_3

ESG factors into corporate governance practices has become crucial for organizations, including those in the healthcare sector, aiming for sustainable and responsible business practices.

Research suggests that the composition of a company's board, including the presence of independent directors with expertise in sustainability and healthcare, can positively influence ESG performance reporting [28]. Shareholder and stakeholder activism and engagement are key components of corporate governance. Organizations that actively engage with shareholders and stakeholders on ESG matters may demonstrate better ESG performance reporting [58].

The regulatory environment plays a crucial role in shaping corporate governance practices. Stringent ESG reporting requirements, influenced by governance mechanisms, can lead healthcare organizations to enhance their ESG disclosures [19].

Healthcare companies focusing on ESG performance may emphasize patient and community engagement. Corporate governance structures that prioritize stakeholder inclusion can positively impact the reporting of ESG initiatives related to community health and well-being [7]. In the same line, ethical decision-making is inherent in healthcare. Strong governance frameworks that prioritize ethical considerations can contribute to the transparent reporting of social and governance aspects of ESG performance [69].

Despite the growing emphasis on ethical behavior and social responsibility within the healthcare sector [56], there is a noticeable lack of research exploring the relationship between on ESG performance reporting and the structure of governing boards. This gap in the literature represents significant research. Understanding how corporate structures influence ESG reporting could provide valuable insights into improving corporate governance, enhancing stakeholder trust, and ultimately, driving positive social impact within the industry. Hence, there is a relevance regarding stakeholder engagement role in shaping ESG reporting practices within healthcare. Understanding how different stakeholders, including patients, employees, and communities, influence ESG disclosures [47].

This study represents a comprehensive effort to delve deeply into the intricate web of factors influencing Environmental, ESG performance reporting within the healthcare sector across EU countries. Recognizing the growing importance of ESG considerations in today's corporate landscape, particularly within the healthcare domain, our research aimed to shed light on the underlying mechanisms driving ESG reporting practices.

Spanning the years 2017–2021, our empirical analysis focused on scrutinizing healthcare companies listed in the EU, leveraging a robust research model comprising nine meticulously crafted hypotheses. These hypotheses were designed to explore the multifaceted relationships between various corporate governance variables (Audit Committee Expertise; Audit Committee Independence; Board Cultural Diversity (%); Board Gender Diversity (%); Board Size; CEO Board Membership; Executive Members' Gender Diversity; Females on the board; Independent Board Members Score) and ESG performance reporting metrics, aiming to uncover the factors that shape and influence ESG disclosures within this sector.

Employing sophisticated statistical techniques, specifically Partial Least Squares based Structural Equation Modelling, we subjected the gathered data to rigorous examination.

The implications of our research extend far beyond academia, offering valuable insights to a diverse array of stakeholders. ESG disclosure users stand to benefit from a more nuanced understanding of the factors shaping the reliability and comprehensiveness of ESG reports. Governmental bodies can leverage these insights to inform policy decisions and regulatory frameworks aimed at promoting transparency and accountability within the healthcare sector. Additionally, organizations involved in the formulation of sustainability reporting standards can utilize our findings to refine existing frameworks and foster more meaningful ESG disclosures.

The subsequent sections of the paper are organized as follows: Sect. 2 provides the theoretical framework, reviews pertinent literature, and outlines the study's hypotheses. Section 3 delineates the research methodology employed in this study. The empirical findings and their discussions, along with their implications, are presented in Sect. 4. Finally, the study is concluded in the last section, which also includes a summary of our findings and recommendations for future research endeavors.

2 Conceptual Framework and Research Hypotheses

Corporate governance theories, ranging from agency theory to stewardship theory, underscore the significance of effective oversight mechanisms within organizations. According, organizations engage in activities and responsible business practices to maintain societal congruence [59]. In this section, the association between various governance variables and ESG performance reporting is explored through nine hypotheses grounded in corporate governance literature and agency theory.

Corporate governance theories, such as agency theory and stewardship theory, emphasize the importance of effective oversight mechanisms within organizations [1].

Audit committees play a critical role in corporate governance, overseeing financial reporting and ensuring compliance [66]. There is growing interest in understanding how audit committee characteristics, including independence, influence the quality and transparency of ESG reporting [27].

An expert audit committee is expected to enhance corporate governance by providing better scrutiny of financial and non-financial disclosures, including ESG reporting [61]. Legitimacy theory proposes that organizations engage in activities to maintain congruence with societal expectations. Audit committee expertise may contribute to the legitimacy of ESG reporting, as it demonstrates a commitment to transparency, accountability, and responsible business practices [57, 64, 72]. Institutional theory focuses on the impact of institutional pressures on organizational behavior. The presence of audit committee expertise in ESG reporting can be seen as a response to institutional demands for greater sustainability disclosure [20]. By drawing on these theoretical frameworks, we formulate hypothesis H1: *H1: The audit committee expertise score is positively associated with ESG performance reporting in healthcare companies.*

Regarding audit committee independence, some research suggests that a more independent audit committee is associated with improved corporate governance practices, including enhanced oversight of ESG-related risks and reporting [37]. The relationship with audit committee independence and ESG performance reporting can be theoretically grounded by drawing on insights from corporate governance literature and agency theory. Agency theory emphasizes the importance of governance mechanisms in mitigating agency conflicts between managers and shareholders. Audit Committee Independence is a key governance mechanism, and its association with ESG reporting can be explained by its role in enhancing transparency, accountability, and oversight [41]. Corporate governance literature consistently argues that independent boards contribute to better decision-making and oversight. Audit Committee Independence, as part of the broader governance structure, is expected to positively influence the quality of ESG reporting [6, 10, 57, 68, 70, 16]. Also, an independent audit committee enhances the legitimacy of ESG reporting, as it signals a commitment to objective and unbiased oversight [64]. Moreover, an independent audit committee is expected to act as a steward for shareholders, ensuring that the organization's non-financial performance is accurately disclosed [21].

Given these fundaments, we formulate hypothesis H2.

H2: The audit committee independence is positively associated with ESG performance reporting in healthcare companies.

A diverse board is expected to bring different perspectives and ideas, fostering innovation and better decision-making, which may extend to ESG reporting [17]. A culturally diverse board may be more attuned to the diverse interests of stakeholders, leading to a more comprehensive and inclusive approach to ESG reporting [26]. Moreover, a culturally diverse board may drive innovation in corporate reporting practices, including ESG reporting [46, 53].

Research on diversity, particularly in the context of boards, highlights the potential benefits of cultural diversity regarding ESG reporting [10, 39, 71]. The theoretical foundation lies in the understanding that a culturally diverse board may contribute to more comprehensive, innovative, and socially responsible decision-making, which could manifest in the quality of ESG performance reporting.

By integrating these theoretical perspectives, we formulate the hypothesis H3:

H3: The board cultural diversity is positively associated with ESG performance reporting in healthcare companies.

Gender Diversity on the board is expected to bring diverse perspectives and enhance the effectiveness of governance structures, which may extend to ESG reporting [2]. Literature on corporate social responsibility (CSR) suggests that diverse boards, including gender diversity, may be more attuned to social issues [36]. A gender-diverse board may be more sensitive to the interests of diverse stakeholders, and this may be reflected in the organization's approach to ESG reporting (Coles et al. 2008) [10, 26, 71]. Moreover, the growing emphasis on gender diversity in corporate boards reflects institutional demands and may be associated with a company's commitment to societal expectations [20]. Considering these fundaments, we formulate the hypothesis H4:

H4: The board gender diversity is positively associated with ESG performance reporting in healthcare companies.

The size of the board is considered a fundamental governance attribute. A larger or smaller board size may influence decision-making processes, which can extend to the quality of ESG reporting [70]. Agency theory suggests that governance mechanisms, such as board composition, are designed to mitigate agency conflicts. The size of the board may affect the efficiency of monitoring and oversight. Although the existence of different perspectives on the ideal board size, and how this affects ESG reporting [22], research findings suggest that smaller businesses may encounter comparatively less public scrutiny and shareholder pressure concerning their Environmental, Social, and Governance (ESG) practices, as indicated by DeNicola (2022). Conversely, larger corporations tend to allocate more resources to ESG initiatives, driven by the economies of scale they possess, aiming to align with the expectations of stakeholders, as highlighted in the literature (Coles et al. 2008) [11, 15].

Considering these fundaments, we formulate the hypothesis H5:

H5: The board size is positively associated with ESG performance reporting in healthcare companies.

CEO Board Membership refers to the situation where the CEO is also a member of the board. Corporate governance literature often explores the impact of CEO duality, where the CEO also serves as the board chair. The presence of the CEO on the board may influence decision-making and governance processes, including those related to ESG reporting [18]. Agency theory suggests that the interests of managers (CEOs) may differ from those of shareholders [40]. CEO Board Membership, or duality, can create a potential conflict of interest as the CEO is involved in both the management and oversight functions. This may impact the quality and transparency of ESG reporting. Research on leadership structures in organizations indicates that the separation of CEO and board chair roles is associated with better governance. CEO Board Membership may affect the independence and effectiveness of the board in overseeing ESG reporting practices [24]. CEO Board Membership may indicate a higher level of CEO influence on the governance and reporting processes, impacting the organization's approach to ESG reporting [21, 50, 60]. According to agency theory, this phenomenon occurs due to the concentration of power in the CEO's hands, which tends to undermine the oversight function of other board members and shareholders. Consequently, organizational decisions may not always prioritize the long-term growth of business value and the protection of stakeholders' wealth. Moreover, CEO duality also has negative implications for ESG (Environmental, Social, and Governance) performance reporting. Considering these fundaments, we formulate the hypothesis H6:

H6: The CEO board membership is positively associated with ESG performance reporting in healthcare companies.

Previously literature differentiated between outside directors (independent or proprietary directors [9, 52]. However, as far we are aware, none of the analyses carried out up to this point have concluded that these two variables have different impacts on ESG performance reporting. Research on gender diversity in executive positions emphasizes its potential impact on governance and decision-making. Gender Diversity among executive members is expected to bring diverse perspectives, which may extend to the quality of ESG reporting [2], and be more attuned to social issues [36]. Executive Members' Gender Diversity may contribute to more comprehensive, innovative, and socially responsible decision processes, including those related to ESG reporting [53]. Considering the research background we formulate the hypothesis H7:

H7: The Executive member's gender diversity is positively associated with ESG performance reporting in healthcare companies.

Having females on the board is expected to contribute to a broader range of perspectives and improved governance practices, which can extend to ESG reporting [2]. The presence of women on boards is often linked to improved decision-making processes. Women on boards may prioritize social and environmental concerns, influencing the organization's commitment to ESG practices and reporting [14]. Moreover, female representation on boards may lead to a more comprehensive understanding of stakeholder concerns, influencing the organization's approach to ESG reporting [26]. Also, research on CSR suggests that women on boards are associated with higher CSR performance. Literature extended this to ESG reporting as a part of broader CSR initiatives [5, 14, 15, 38, 67] (Torchia et al. 2011; Fernandez-Feijoo et al. 2014). Hence, we formulate the hypothesis H8:

H8: The presence of females on the board is positively associated with ESG performance reporting in healthcare companies.

Finally, corporate governance literature also emphasizes the importance of independent directors in enhancing board effectiveness. Independent Board Members are expected to contribute to better governance practices, which may extend to the oversight of ESG reporting [61]. Independent Board Members serve as a check on managerial behavior, potentially influencing the organization's commitment to ESG practices [40] Independent directors are often associated with increased accountability and transparency. This transparency may extend to ESG reporting, as Independent Board Members can play a crucial role in ensuring accurate and comprehensive disclosure [12, 15, 68, 65, 45] and more likely to engage in responsible corporate practices [23]. Considering the theoretical fundaments we formulate the hypothesis H9:

H9: The independent board members score is positively associated with ESG performance reporting in healthcare companies.

Following, we argue that Stakeholder Engagement plays a crucial role in moderating the relationship between various governance variables and ESG reporting. Stakeholder Engagement is proposed as a moderating factor that reflects the level of interaction and responsiveness to stakeholder concerns, aligning organizational interests with stakeholders, considering the diverse interests of stakeholders [41], and enhancing transparency and accountability.

Considering these fundaments, we formulate the following hypothesis:

H1a: Stakeholder Engagement can moderate the effect of Audit Committee Expertise on ESG performance reporting.

H2a: Stakeholder Engagement can moderate the effect of Audit Committee Independence on ESG performance reporting.

H3a: Stakeholder Engagement can moderate the effect of Board Cultural Diversity (%)on ESG performance reporting.

H4a: Stakeholder Engagement can moderate the effect of Board Gender Diversity (%) on ESG performance reporting.

H5a: Stakeholder Engagement can moderate the effect of Board Size on ESG performance reporting.

H6a: Stakeholder Engagement can moderate the effect of CEO Board Members on ESG performance reporting.

H7a: Stakeholder Engagement can moderate the effect of Executive Members' Gender Diversity on ESG performance reporting.

H8a: Stakeholder Engagement can moderate the effect of Females on Board on ESG performance reporting.

H9a: Stakeholder Engagement can moderate the effect of Independent Board Members on ESG performance reporting.

We assume that corporate variables' impact on ESG reporting may be contingent on the degree of engagement with stakeholders [17, 26, 53]. These hypotheses reflect the intricate interplay between corporate governance mechanisms and the diverse interests of stakeholders. Stakeholder Engagement is crucial in the context of ESG reporting as it reflects the extent to which an organization involves and responds to the needs and expectations of its stakeholders [25]. Stakeholder Engagement, in this context, can serve as a mechanism to align the interests of the organization with those of its stakeholders, moderating the relationship between corporate variables and ESG reporting. Research in corporate governance and social responsibility highlights the importance of stakeholder engagement in promoting responsible business practices. A moderating role for Stakeholder Engagement in the relationship between the corporate governance model and ESG reporting aligns with the broader literature on governance mechanisms and sustainability [3]. Moreover, stakeholder engagement can be seen as a response to societal and institutional demands for greater transparency and responsiveness, moderating these relationships.

We are aware that our dependent and independent variables and the relationship between them can be affected by other variables. To reduce Confounding Bias we use five control variables: Leverage; Total Current Assets (size); Return On Equity (ROE); Return on Average Total Assets (RATA); Dloss.

3 Methodology

Our sample is made up of healthcare companies, listed in the EU. Data were collected from the Refinitiv database, now named LSEG Data & Analytics. For the years 2017 and 2021, 207 companies were obtained. Our data was panelled, but we chose to treat each record as a different firm because a large part of the repeated records was missing, ending up with 1000 observations.

Statistical analysis was performed using Partial Least Squares based Structural Equation Modelling (PLS-SEM), according to [29] recommendations. PLS choice, instead of covariance-based Structural Equation Modelling (CB-SEM) was supported by our sample size, low to the standards of CB-SEM, and the absence of the need to satisfy the normality assumption of variables (e.g. [63]). The statistical software was SmartPLS, version 4.0.

In this study, we included moderation analysis to investigate the potential damping or strengthening of moderators into the associations of explanatory variables and outcomes.

Introducing moderator variables in a statistical model is a valuable approach to understanding the complexity of relationships between variables and improving the accuracy and generalization of results [35]. Indeed, moderator variables allow for investigating whether the relationship between independent and dependent variables varies depending on the value of the moderator variable. Also, including moderator variables allows for identifying whether the effect of independent variables on the dependent variable is consistent across different subgroups of the sample. This allows for a more sophisticated analysis and a more comprehensive understanding of the phenomenon under study [8, 33, 49].

Data was analysed according to [31] recommendations. We did not assess reliability, convergent and discriminant validity because our analysis fell in the category of path analysis, with a single variable per construct.

Regarding the explanatory power of the model, we analysed endogenous latent variables' coefficient of determination (R^2). Model fit was assessed with Standardized Root Mean Square Residual (SRMR), considering values below 0.10 as threshold to accept model specification and prevent the bias of estimates [34]. Tuker-Lewis index (TLI) was calculated and the threshold for good adjustment was set at > 0.90.

Collinearity diagnostics were verified by calculating the Variance Inflation Factor (VIF), accepting results below 5. The contribution of each construct to the explanatory power of the latent variables was assessed with f squared (f^2), an effect size measure. Threshold recommendations for f^2 followed [29] recommendations, namely 0.02, 0.15 and 0.35 for low, medium and high effect size, respectively.

Predictive relevance was assessed according to [62] recommendations. So, we calculated root mean square error (RMSE) and mean absolute error (MAE) with PLS Predict algorithm with a tenfold validation.

A bootstrap method for 1000 samples was used to calculate p-values and confidence intervals for the calculated estimates. The strength of association was assessed with non-standardized coefficients (β); T-student's statistics were calculated by dividing β by the standard deviation (e.g. [30]).

To reduce Confounding Bias we use five control variables: Leverage; Total Current Assets (size); Return On Equity (ROE); Return on Average Total Assets (RATA); Dloss. These control variables help reduce confounding bias by controlling factors that may influence both the independent and dependent variables [42]. Additionally, controlling for confounding variables enhances the internal validity of the study, ensuring that conclusions about the relationship between the independent and dependent variables are valid and reliable within the study context [48, 54].

4 Results Presentation and Discussion

A total of 210 complete records were included in the study considering all variables studied, namely the independent variables of corporate governance (Audit Committee Independence, Board Size, Board Gender Diversity, Board Cultural Diversity, CEO Board Member, Female on Board, Members Gender Diversity, Independent Board Members, Audit Committee Expertise) the dependent variable of ESG performance reporting, the moderator Stakeholder Engagement and the control variables (Return on Equity (mean), Leverage, Return on Average Total Assets (%) TTM, Dloss and Total Current Assets (€). Figure 1 presents the research model. Results for composite variables of corporate governance were not viable. Hence, a total of nine different models were run, one per each independent variable of corporate governance.

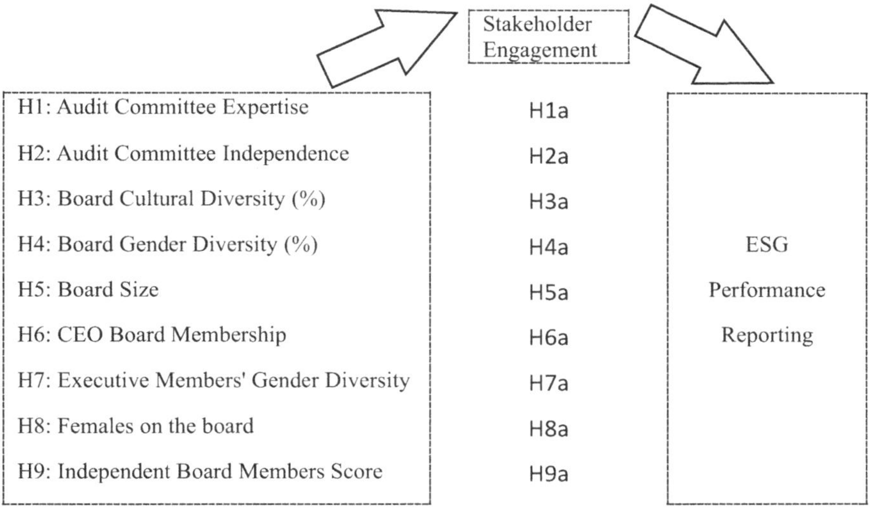

Fig. 1 The moderating effect of stakeholders' engagement

Table 1 shows results for all nine models and their respective hypotheses, regarding the association of each corporate governance variable with the dependent variable of ESG performance reporting and the moderator effect of Stakeholder Engagement in the association of each corporate governance variable with the dependent variable of ESG performance reporting. R-squared results were all close to or higher than 0.50. SRMR was lower than 0.10, except for model 6. TLI was always higher than 0.80, in some models higher than 0.85 for models 4, 8 and 9 and higher than 0.90 for model 5. These results suggest moderate to adequate fit. Considering all models, between 48 and 64% of the indicators in each model had lower RMSE and lower MAE when compared to the linear regression model. So, we concluded that medium does good prediction power.

Direct positive associations of corporate governance variables with ESG performance reporting were confirmed for Audit Committee Expertise ($\beta = 0.150$, $p < 0.001$, $f^2 = 0.047$), Audit Committee Independence ($\beta = 0.192$, $p < 0.001$, $f^2 = 0.077$), Board Cultural Diversity (%) ($\beta = 0.061$, $p = 0.008$, $f^2 = 0.008$), Board Gender Diversity (%) ($\beta = 0.230$, $p < 0.001$, $f^2 = 0.105$), Board Size ($\beta = 0.300$, $p < 0.001$, $f^2 = 0.147$), Female on Board ($\beta = 0.230$, $p < 0.001$, $f^2 = 0.105$) and Independent Board Members ($\beta = 0.251$, $p < 0.001$, $f^2 = 0.131$). A negative association was found for CEO Board Member ($\beta = -0.152$, $p = 0.026$, $f^2 = 0.007$) association with ESG performance reporting. These results confirm hypothesis H1, H2, H3, H4, H5, H6, H8 and H9.

Moderator effects were found for Stakeholder Engagement on Audit Committee Expertise association with ESG performance reporting ($\beta = -0.067$, $p = 0.003$, $f^2 = 0.014$), for Stakeholder Engagement on Board Cultural Diversity (%) with ESG performance reporting ($\beta = 0.039$, $p = 0.035$, $f^2 = 0.005$) and Stakeholder Engagement on Board Size with ESG performance reporting ($\beta = -0.101$, $p < 0.001$, $f^2 = 0.126$). These results confirm hypotheses H1a, H3a and H5a. Details on moderation effects are found in Figs. 2, 3 and 4.

Concerning the moderation effect of Stakeholder Engagement on Audit Committee Expertise association with ESG performance reporting (Fig. 1) a higher slope was found for Stakeholder Engagement at – 1 SD, suggesting that the association of Audit Committee Expertise association with ESG performance reporting is strengthened for lower scores of Stakeholder Engagement.

Regarding the moderation effect of Stakeholder Engagement on Board Cultural Diversity (%) association with ESG performance reporting (Fig. 2) a higher slope was found for Stakeholder Engagement at + 1 SD, suggesting that the association of Board Cultural Diversity (%) with ESG performance reporting is strengthened for higher scores of Stakeholder Engagement.

The moderation effect of Stakeholder Engagement on Board Size association with ESG performance reporting (Fig. 3) showed a higher slope for Stakeholder Engagement at – 1 SD, suggesting that the association of Board Cultural Diversity (%) with ESG performance reporting is strengthened for lower scores of Stakeholder Engagement.

Table 1 summarizes the results for each of the hypotheses. Regarding the main hypotheses, eight within nine hypotheses were confirmed H1; H2; H3; H4; H5; H6;

Table 1 PLS-SEM for effects on the dependent variable of ESG performance reporting

	β	SD	*p*-value	f^2	R^2 for DV	SRMR/NFI
(H1/H1a)						
IV effect: Audit Committee Expertise	0.150	0.027	< 0.001*	0.047	0.531	0.097/0.842
M effect: Stakeholder Engagement	– 0.067	0.023	0.003*	0.014		
(H2/H2a)						
IV effect: Audit Committee Independence	0.192	0.024	< 0.001*	0.077	0.532	0.098/0.822
M effect: Stakeholder Engagement	0.006	0.019	0.761	0.000		
(H3/H3a)						
IV effect: Board Cultural Diversity (%)	0.061	0.023	0.008*	0.008	0.502	0.098/0.835
M effect: Stakeholder Engagement	0.039	0.018	0.035*	0.005		
(H4/H4a)						
IV effect: Board Gender Diversity (%)	0.230	0.029	< 0.001*	0.105	0.548	0.095/0.859
M effect: Stakeholder Engagement	– 0.010	0.025	0.698	0.000		
(H5/H5a)						
IV effect: Board Size	0.300	0.034	< 0.001*	0.147	0.566	0.067/0.923
M effect: Stakeholder Engagement	– 0.101	0.029	< 0.001*	0.126		
(H6/H6a)						
IV effect: CEO Board Member	– 0.152	0.068	0.026*	0.007	0.500	0.103/0813
M effect: Stakeholder Engagement	– 0.025	0.054	0.668	0.000		
(H7/H7a)						
IV effect: Executive members' Gender Diversity	0.020	0.027	0.465	0.001	0.497	0.096/0.841
M effect: Stakeholder Engagement	0.006	0.022	0.796	0.000		
(H8/H8a)						
IV effect: Female on Board	0.230	0.029	< 0.001*	0.105	0.548	0.095/0.859
M effect: Stakeholder Engagement	– 0.010	0.025	0.698	0.000		
(H9/H9a)						

(continued)

Table 1 (continued)

	β	SD	*p*-value	f^2	R^2 for DV	SRMR/NFI
IV effect: Independent Board Members	0.251	0.026	< 0.001*	0.131	0.555	0.098/0.852
M effect: Stakeholder Engagement	0.019	0.021	0.381	0.001		

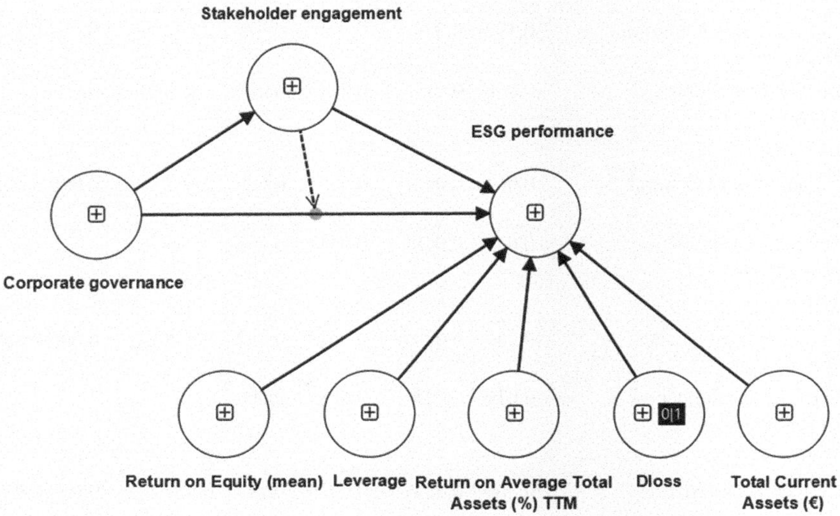

Fig. 2 Research model

H8; H9. Only the H7 was not confirmed, which means that executive members' gender diversity is not associated with ESG performance reporting.

RATA, Return on Average Total Assets (%) TTM; ROE, Return on Equity (mean); IV, independent variable; M, moderator variable.

Figures 2, 3, and 4 exhibit the moderate effect of stakeholders' engagement for three of the corporate variables. Thus, the effect of Audit Committee Expertise, Board Cultural Diversity and Board Size on ESG performance reporting vary depending on the value of the moderator variable (Stakeholder Engagement). Agency theory plays a pivotal role in deciphering our findings. Integrated reporting disclosure serves as a valuable approach to diminish information asymmetry and thereby mitigate agency expenses. Therefore, following agency theory, managers operate in the best interest of shareholders. The level of ESG reporting acts as a mechanism capable of alleviating the agency dilemma, as high-quality information disclosure tends to synchronize the objectives of shareholders and managers [15].

Table 2 shows Control variables' associations with ESG performance reporting for all models. Return on Average Total Assets (%) TTM, Stakeholder engagement

Fig. 3 Moderation effect of stakeholder engagement on the effect of stakeholder engagement on ESG performance reporting

score and Total Current Assets (€) were consistently and positively associated with ESG performance reporting, meaning that when results in these variables are high, ESG performance reporting is also expected to be high. In the same way, when results in these variables are low, ESG performance reporting is also expected to be high. Dloss results association with ESG performance reporting were mixed, with a negative coefficient for model 2 (− 0.258, $p < 0.05$) and a positive coefficient for model 8 (0.225). These results mean that this variable is not consistently associated with ESG performance reporting. Leverage and return on equity (mean) found no significant associations.

As we can see in Table 3, our study confirmed eight hypotheses regarding the effect of corporate governance variables on ESG performance reporting. Notably, our findings revealed that Audit Committee Expertise (ACE); Audit Committee Independence (ACI); Board Cultural Diversity (BCD %); Board Gender Diversity (BGD %); Board Size (BS); CEO Board Membership (CEO BM); Females on the Board (FB); and Independent Board Members (IBMs), play a pivotal role in shaping ESG reporting practices, promoting more comprehensive and transparent disclosures. Interestingly, our study did not find a significant association between Executive Member Gender Diversity (EMGD) and ESG reporting, contrary to some prior research. Moreover, we confirmed that the impact of ACE, BCD, and BS on ESG performance reporting varies based on the level of stakeholder engagement.

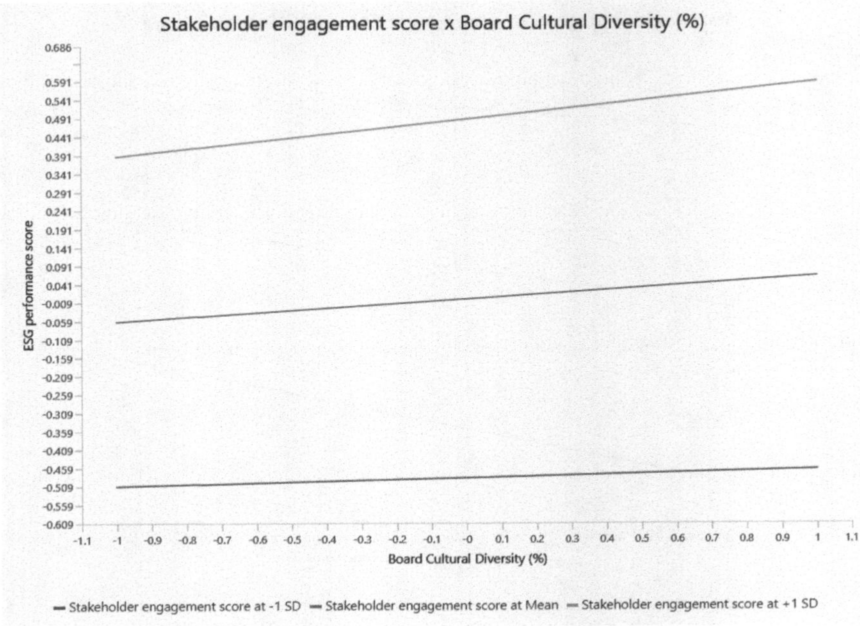

Fig. 4 Moderation effect of stakeholder engagement on the effect of board cultural diversity (%) on ESG performance reporting

In line with those achieved by [57, 72], our findings confirmed that Audit Committee Independence are associated with ESG performance reporting. Regarding H2, our study confirmed the results of [6, 10, 57, 68], which means that Audit Committee Independence is associated with ESG performance reporting. However, it contrasts with the results obtained by [13], because these authors report a negative association between financial expertise and sustainability reporting. We also confirmed that the board's Cultural Diversity is associated with ESG performance reporting. A similar association was found by [10, 39, 71]. Different findings were described by [4], according to the board cultural diversity negatively moderates the impact of ESG performance on investment efficiency. The association between the Board Gender Diversity (%) and ESG performance reporting was also confirmed. This result is comparable to the findings of [10, 71] but they contradict the findings of [47].

The Board Size appears to be higher in ESG performance reporting, consistent with Coles et al. (2008) and [15]. In line with the literature, our study indicates that the dynamics and decision-making processes of larger boards may positively affect the effectiveness of ESG reporting [11].

According to Table 2, H6 was confirmed, meaning that CEO Board Member is negatively associated with ESG performance reporting. Our results are in line with literature that suggests that CEOs can shape organizational practices, including

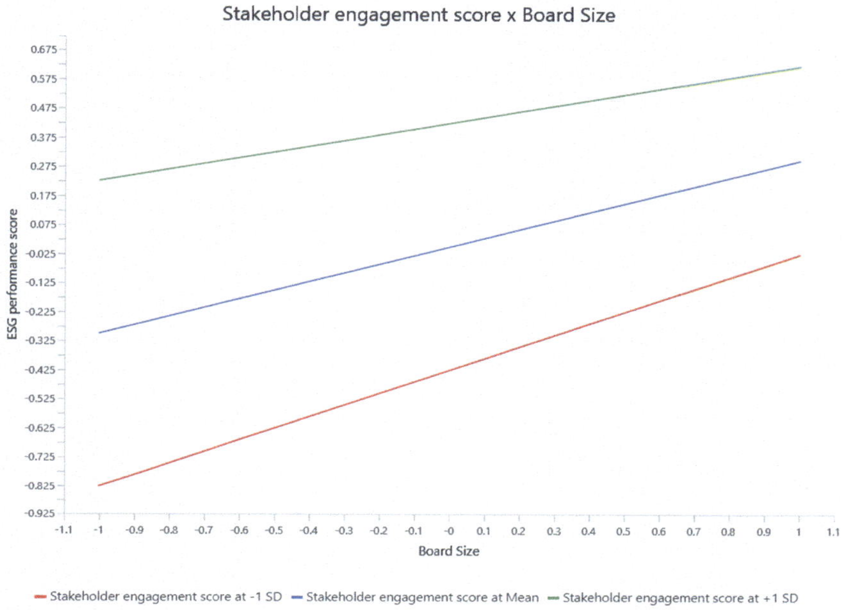

Fig. 5 Moderation effect of stakeholder engagement on the effect of board size on ESG performance reporting

reporting. Scholars such as [50, 60] have pointed out that when the roles of CEO and board chairperson are combined, it can lead to less effective monitoring and accountability mechanisms, thereby affecting the quality and transparency of ESG reporting. This underscores the importance of separating these roles to ensure robust corporate governance practices and enhance both financial and non-financial performance outcomes.

Surprisingly, the executive member's gender diversity didn't reveal a significative association with ESG performance reporting. This finding contradicts the evidence obtained by [9].

In line with Fernandez-Feijoo et al. (2014), we confirm that the presence of females on Board is positively and significantly with ESG performance reporting. Thus, our results are in line with those found by previous works (Torchia et al. 2011; Fernandez-Feijoo et al. 2014) [5, 14, 67]. It shows that Females on Board put great significance t CRS reporting According to [15], this finding can be explained by the gender difference in leadership styles and communication skills.

At last, the findings confirm hypothesis 9, indicating a positive correlation between Independent Board Members and ESG performance reporting. As a result, a higher level of board independence means greater accountability and transparency regarding ESG. This outcome aligns with previous research [15, 68, 65, 45]. This outcome is consistent with insights from agency theory, supporting that a higher proportion of

Table 2 Control variables associations with ESG performance reporting

	H1	H2	H3	H 4	H5	H6	H7	H 8	H9
IV→	Audit committee expertise score	Audit committee independence	Board cultural diversity	Board gender diversity	Board size	CEO board member	Executive members gender diversity (%)	Female on board	Independent board members score
Dloss	− 0.173	**− 0.258***	− 0.175	− 0.129	− 0.132	− 0.147	− 0.090	**0.225*****	0.006
Leverage	− 0.001	− 0.008	− 0.001	0.000	− 0.004	− 0.001	− 0.003	0.000	− 0.004
Return on average total assets (%) TTM	**0.077****	**0.070****	**0.073****	**0.084*****	**0.046***	**0.074****	**0.074****	**0.084*****	**0.100*****
Return on equity (mean)	0.019	0.005	0.006	0.035	0.007	0.014	0.009	0.035	0.023
Stakeholder engagement score	**0.476*****	**0.496*****	**0.485*****	**0.431*****	**0.424*****	**0.492*****	**0.488*****	**0.431*****	**0.494*****
Total current assets (€)	**0.328*****	**0.309*****	**0.329*****	**0.328*****	**0.266*****	**0.335*****	**0.333*****	**0.328*****	**0.268*****

* $p < 0.05$; ** $p < 0.01$; *** $p < 0.001$; IV, independent variable

Table 3 Results for the research hypotheses

Research hypothesis	Results
H1: Audit Committee Expertise is positively associated with ESG performance reporting	Confirmed
H1a: Stakeholder Engagement can moderate the effect of Audit Committee Expertise on ESG performance reporting	Confirmed
H2: Audit Committee Independence is positively associated with ESG performance reporting	Confirmed
H2a: Stakeholder Engagement can moderate the effect of Audit Committee Independence on ESG performance reporting	–
H3: Board Cultural Diversity (%) is positively associated with ESG performance reporting	Confirmed
H3a: Stakeholder Engagement can moderate the effect of Board Cultural Diversity (%) on ESG performance reporting	Confirmed
H4: Board Gender Diversity (%) is positively associated with ESG performance reporting	Confirmed
H4a: Stakeholder Engagement can moderate the effect of Board Gender Diversity (%) on ESG performance reporting	–
H5: Board Size is positively associated with ESG performance reporting	Confirmed
H5a: Stakeholder Engagement can moderate the effect of Board Size on ESG performance reporting	Confirmed
H6: CEO Board Member is negatively associated with ESG performance reporting	Confirmed
H6a: Stakeholder Engagement can moderate the effect of CEO Board Member on ESG performance reporting	–
H7: Executive member's gender Diversity is positively associated with ESG performance reporting	–
H7a: Stakeholder Engagement can moderate the effect of Executive Members Gender Diversity on ESG performance reporting	–
H8: Female on Board is positively associated with ESG performance reporting	Confirmed
H8a: Stakeholder Engagement can moderate the effect of Female on Board on ESG performance reporting	–
H9: Independent Board Members is positively associated with ESG performance reporting	Confirmed
H9a: Stakeholder Engagement can moderate the effect of Independent Board Members on ESG performance reporting	–

independent board members, coupled with increased monitoring, better quality of reporting and enhanced oversight levels.

5 Conclusion and Future Scope

In this study, we conducted an extensive examination of the factors influencing Environmental, Social, and Governance (ESG) performance reporting within the healthcare sector across EU countries. Our research aimed to elucidate the intricate relationships between various corporate governance variables and the effectiveness of ESG reporting practices in healthcare companies. To achieve the goal, an empirical analysis was conducted with healthcare companies, listed in the EU, for 2017–2021. The database, currently known as LSEG Data & Analytics, consists of information from 207 companies for the years 2017 and 2021. While our data was organized in panels, we opted to consider each entry as a distinct entity due to significant gaps in a considerable portion of the duplicated entries. Our research model considered nine research hypotheses regarding the relationship between corporate governance variables and ESG performance reporting. The moderate effect of stakeholder engagement on these relations was tested.

Drawing on a review of existing literature and employing robust empirical analysis techniques, we explored the impact of several key factors on ESG reporting outcomes. Statistical analysis was performed using Partial Least Squares based Structural Equation.

We confirmed eight hypotheses regarding the effect of corporate governance variables on ESG performance reporting. Notably, our findings revealed that Audit Committee Expertise; Audit Committee Independence; Board Cultural Diversity (%); Board Gender Diversity (%); Board Size; CEO Board Membership; Females on the Board; and Independent Board Members, play a pivotal role in shaping ESG reporting practices, promoting more comprehensive and transparent disclosures. Interestingly, our study did not find a significant association between Executive Member Gender Diversity and ESG reporting, contrary to some prior research.

Moreover, we confirmed that the impact of Audit Committee Expertise, Board Cultural Diversity, and Board Size on ESG performance reporting varies based on the level of stakeholder engagement. Regarding the control variables, Return on Average Total Assets (%), Stakeholder engagement score, and Total Current Assets (€) consistently exhibited positive associations with ESG performance reporting, while Dloss showed an inconsistent relationship with ESG reporting, and Leverage and return on equity (mean) did not show significant associations.

One limitation of this study lies in its empirical nature, which may have been influenced by uncontrolled or unconsidered factors. Additionally, the analysis focused solely on healthcare companies listed in the EU, which may limit the generalizability of the findings to other sectors or geographical regions. The lack of complete data in certain variables may have also impacted the accuracy of the conclusions, despite efforts to mitigate this issue. Future research may consider including a broader set of companies and exploring other relevant variables that may influence ESG performance disclosure and consider a panel data analysis.

Additionally, exploring the role of stakeholders and their involvement in corporate governance and ESG performance disclosure could provide a deeper understanding of the mechanisms underlying these relationships. Longitudinal studies could also be conducted to examine changes over time in ESG performance disclosure practices and the factors driving these changes, providing a more dynamic view of the current landscape.

In conclusion, our study offers valuable insights into the complex dynamics shaping ESG performance reporting within the healthcare sector. By highlighting the importance of corporate governance structures, diversity, and independence in driving sustainable business practices, our findings provide actionable recommendations for healthcare companies, policymakers, and stakeholders seeking to enhance their ESG reporting practices and promote responsible corporate behavior.

References

1. Abdullah, H., Valentine, B.: Fundamental and ethics theories of corporate governance. Middle Eastern Finan. Econ. **4**(4), 88–96 (2009)
2. Adams, R.B., Ferreira, D.: Women in the boardroom and their impact on governance and performance. J. Financ. Econ. **94**(2), 291–309 (2009)
3. Aguilera, R.V., Jackson, G.: The cross-national diversity of corporate governance: dimensions and determinants. Acad. Manag. Rev. **28**(3), 447–465 (2003)
4. Al-Hiyari, A., Ismail, A.I., Kolsi, M.C., Kehinde, O.H.: Environmental, social and governance performance (ESG) and firm investment efficiency in emerging markets: the interaction effect of board cultural diversity. Corporate Governance Int. J. Bus. Soc. **23**(3), 650–673 (2023)
5. Arayssi, M., Jizi, M., Tabaja, H.H.: The impact of board composition on the level of ESG disclosures in GCC countries. Sustain. Account. Manage. Policy J. **11**(1), 137–161 (2020)
6. Arif, M., Sajjad, A., Farooq, S., Abrar, M., Joyo, A.S.: The impact of audit committee attributes on the quality and quantity of environmental, social and governance (ESG) disclosures. Corporate Governance: Int. J. Bus. Soc. **21**(3), 497 (2021)
7. Bansal, P., Song, H.C.: Similar but not the same: differentiating corporate sustainability from corporate responsibility. Acad. Manag. Ann. **11**(1), 105–149 (2017)
8. Baron, R.M., Kenny, D.A.: The moderator-mediator variable distinction in social psychological research: conceptual, strategic, and statistical considerations. J. Pers. Soc. Psychol. **51**(6), 1173 (1986)
9. Bhatia, S., Marwaha, D.: The influence of board factors and gender diversity on the ESG disclosure score: a study on Indian companies. Glob. Bus. Rev. **23**(6), 1544–1557 (2022)
10. Bigelli, M., Mengoli, S., Sandri, S.: ESG score, board structure and the impact of the non-financial reporting directive on European firms. J. Econ. Bus. **127**, 106133 (2023)
11. Bissoondoyal-Bheenick, E., Brooks, R., Do, H.X.: ESG and firm performance: the role of size and media channels. Econ. Model. **121**(106203), 1–19 (2023). https://doi.org/10.1016/j.econmod.2023.106203
12. Brown, W.O., Helland, E., Smith, J.A.: Corporate Governance and the monitoring role of debt: evidence from corporate tax avoidance. J. Corporate (2006)
13. Buallay, A., Al-Ajmi, J.: The role of audit committee attributes in corporate sustainability reporting: evidence from banks in the Gulf Cooperation Council. J. Appl. Acc. Res. **21**(2), 249–264 (2020)
14. Carter, D.A., Simkins, B.J., Simpson, W.G.: Corporate Governance, board diversity, and firm value. Financ. Rev. **38**(1), 33–53 (2003)
15. Chouaibi, S., Chouaibi, Y., Zouari, G.: Board characteristics and integrated reporting quality: evidence from ESG European companies. EuroMed J. Bus. **17**(4), 425–447 (2022)

16. Coles, J.L., Daniel, N.D., Naveen, L.: Boards: does one size fit all? J. Financ. Econ. **87**(2), 329–356 (2008)
17. Cox, T.: Cultural Diversity in Organizations: Theory, Research, and Practice. Berrett-Koehler, San Francisco (1994)
18. Dalton, D.R., Daily, C.M., Ellstrand, A.E., Johnson, J.L.: Meta-analytic reviews of board composition, leadership structure, and financial performance. Strateg. Manag. J. **19**(3), 269–290 (1998)
19. Das, N., Chatterjee, S., Sunder, A., Ruf, B.: ESG ratings and the performance of socially responsible mutual funds: a panel study. J. Finan. Iss. **17**(1), 49–57 (2018)
20. DiMaggio, P.J., Powell, W.W.: The iron cage revisited: institutional isomorphism and collective rationality in organizational fields. Am. Sociol. Rev. **48**(2), 147–160 (1983)
21. Donaldson, L., Davis, J.H.: Stewardship theory or agency theory: CEO governance and shareholder returns. Aust. J. Manag. **16**(1), 49–64 (1991)
22. Fama, E.F., Jensen, M.C.: Separation of ownership and control. J. Law Econ. **26**(2), 301–325 (1983)
23. Filatotchev, I., Jack, S.L., Wright, M.: The impact of executive and non-executive directors on firm value: evidence from the UK. J. Corp. Finan. **13**(4), 474–506 (2007)
24. Finkelstein, S., D'Aveni, R.A.: CEO duality as a double-edged sword: how boards of directors balance entrenchment avoidance and unity of command. Acad. Manag. J. **37**(5), 1079–1108 (1994)
25. Freeman, R.E.: Strategic Management: A Stakeholder Approach. Pitman Publishing (1984)
26. Freeman, R.E., Harrison, J.S., Wicks, A.C., Parmar, B.L., De Colle, S.: Stakeholder Theory: The State of the Art. Cambridge University Press (2010).
27. Ghafran, C., O'Sullivan, N.: The governance role of audit committees: reviewing a decade of evidence. Int. J. Manag. Rev. **15**(4), 381–407 (2013)
28. Hahn, R., Reimsbach, D., Schiemann, F.: Organizations, climate change, and transparency: reviewing the literature on carbon disclosure. Organ. Environ. **28**(1), 80–102 (2015)
29. Hair, J., Hult, G., Ringle, C., Sarstedt, M.: A Primer on Partial Least Squares Structural Equation Modeling (PLS-SEM), 2nd edn. Sage, Thousand Oaks (2017)
30. Hair, J., Risher, J., Sarstedt, M., Ringle, C.: When to use and how to report the results of PLS-SEM. Eur. Bus. Rev. **31**(1), 2–24 (2019). https://doi.org/10.1108/EBR-11-2018-020
31. Hair, J.F., Sarstedt, M., Ringle, C.M., Gudergan, S.P.: Advanced Issues in Partial Least Squares Structural Equation Modeling (PLS-SEM). Sage, Thousand Oaks (2018). https://doi.org/10.3926/oss.37
32. Hambrick, D.C.: Presidential address: what if the academy actually mattered? Acad. Manag. Rev. **21**(1), 16–26 (1996)
33. Hayes, A.F.: Introduction to Mediation, Moderation, and Conditional Process Analysis: A Regression-Based Approach. Guilford Publications (2018)
34. Henseler, J., Dijkstra, T.K., Sarstedt, M., Ringle, C.M., Diamantopoulos, A., Straub, D.W., Ketchen, D.J., Hair, J.F., Hult, G.T.M., Calantone, R.J.: Common beliefs and reality about PLS: comments on Ronkko and Evermann (2013). Organ. Res. Methods **17**(2), 182–209 (2014). https://doi.org/10.1177/1094428114526928
35. Henseler, J., Fassott, G.: Testing moderating effects in PLS path models: an illustration of available procedures. In: Handbook of Partial Least Squares: Concepts, Methods and Applications, pp. 713–735 (2010)
36. Hoogendoorn, S., Oosterbeek, H., van Praag, M.: The impact of gender diversity on the performance of business teams: evidence from a field experiment. Manage. Sci. **59**(7), 1514–1528 (2013)
37. Hossain, S., Yazawa, K., Monroe, G.S.: The relationship between audit team composition, audit fees, and quality. Audit. J. Pract. Theory **36**(3), 115–135 (2017)
38. Ionescu-Somers, A., Mathieu, J.E.: Female representation on corporate boards of directors: a cross-national analysis of the correlates of female presence on corporate boards. J. Bus. Ethics **74**(2), 141–155 (2007)

39. Ismail, A.M., Latiff, I.H.M.: Board diversity and corporate sustainability practices: evidence on environmental, social and governance (ESG) reporting. Int. J. Finan. Res. **10**(3), 31–50 (2019)
40. Jensen, M.C.: The modern industrial revolution, exit, and the failure of internal control systems. J. Finan. **48**(3), 831–880 (1993)
41. Jensen, M.C., Meckling, W.H.: Theory of the firm: managerial behavior, agency costs and ownership structure. J. Financ. Econ. **3**(4), 305–360 (1976)
42. Kraemer, H.C., Stice, E., Kazdin, A., Offord, D., Kupfer, D.: How do risk factors work together? Mediators, moderators, and independent, overlapping, and proxy risk factors. Am. J. Psychiatry **158**(6), 848–856 (2001)
43. Li, Y., Zhang, X.Y.: Impact of board gender composition on corporate debt maturity structures. Eur. Financ. Manag. **25**(5), 1286–1320 (2019)
44. Li, Y., Zhang, X.-Y.: Impact of Board Gender Composition on Corporate Debt Maturity Structures (2019)
45. Liao, L., Lin, T., Zhang, Y.: Corporate board and corporate social responsibility assurance: evidence from China. J. Bus. Ethics **150**, 211-225 (2018). https://doi.org/10.1007/s10551-016-3176-9
46. Maak, T., Pless, N.M.: Responsible leadership in a stakeholder society—a relational perspective. J. Bus. Ethics **66**(1), 99–115 (2006)
47. Manita, R., Bruna, M.G., Dang, R., Houanti, L.H.: Board gender diversity and ESG disclosure: evidence from the USA. J. Appl. Acc. Res. **19**(2), 206–224 (2018)
48. Morton, R.B., Williams, K.C.: Experimental Political Science and the Study of Causality: From Nature to the Lab. Cambridge University Press (2010)
49. Muller, D., Judd, C.M., Yzerbyt, V.Y.: When moderation is mediated and mediation is moderated. J. Pers. Soc. Psychol. **89**(6), 852 (2005)
50. Naciti, V.: Corporate governance and board of directors: The effect of a board composition on firm sustainability performance. J. Clean. Prod. **237**, 117727 (2019)
51. Nicolo, G., Zampone, G., Sannino, G., Tiron-Tudor, A.: Worldwide evidence of corporate governance influence on ESG disclosure in the utilities sector. Utilities Policy **82**(101549), 1–14 (2023). https://doi.org/10.1016/j.jup.2023.101549
52. Ouni, Z., Ben Mansour, J., Arfaoui, S.: Board/Executive gender diversity and firm financial performance in Canada: the mediating role of environmental, social, and governance (ESG) orientation. Sustainability **12**(20), 8386 (2020)
53. Page, S.E.: The Difference: How the Power of Diversity Creates Better Groups, Firms, Schools, and Societies. Princeton University Press (2007)
54. Pedhazur, E.J., Schmelkin, L.P.: Measurement, Design, and Analysis: An Integrated Approach. Psychology Press (1991)
55. Pfeffer, J., Salancik, G.R.: The External Control of Organizations: A Resource Dependence Perspective. Stanford Business Books (2003)
56. Piechocka-Kałużna, A., Tłuczak, A., Łopatka, P.: The impact of CSR/ESG reporting on the cost of capital: an example of US healthcare entities. Eur. Res. Stud. J. **24**(Special 3) (2021)
57. Pozzoli, M., Pagani, A., Paolone, F.: The impact of audit committee characteristics on ESG performance in the European Union member states: Empirical evidence before and during the COVID-19 pandemic. J. Clean. Prod. **371**, 133411 (2022)
58. Qin, B., Yang, L.: CSR contracting and performance-induced CEO turnover. J. Corp. Finan. **73**, 102173 (2022)
59. Randel, A.E.: The maintenance of an organization's socially responsible practice: a cross-level framework. Bus. Soc. **41**(1), 61–83 (2002)
60. Shahbaz, M., Karaman, A.S., Kilic, M., Uyar, A.: Board attributes, CSR engagement, and corporate performance: What is the nexus in the energy sector? Energy Policy **143**, 111582 (2020)
61. Shleifer, A., Vishny, R.W.: A Survey of Corporate Governance. J. Finan. **52**(2), 737–783 (1997)
62. Shmueli, G., Ray, S., Estrada, J.M.V., Chatla, S.B.: The elephant in the room: predictive performance of PLS models. J. Bus. Res. **69**(10), 4552–4564 (2016)

63. Silva, G.M., Styles, C., Lages, L.F.: Breakthrough innovation in international business: the impact of tech-innovation and market-innovation on performance. Int. Bus. Rev. **26**(2), 391–404 (2017). https://doi.org/10.1016/j.ibusrev.2016.10.001
64. Suchman, M.C.: Managing legitimacy: strategic and institutional approaches. Acad. Manag. Rev. **20**(3), 571–610 (1995)
65. Sun, N., Salama, A., Hussainey, K., Habbash, M.: Corporate environmental disclosure, corporate governance and earnings management. Manag. Audit. J. **25**(7), 679-700 (2010). https://doi.org/10.1108/02686901011061351
66. Turley, S., Zaman, M.: The corporate governance effects of audit committees. J. Manage. Governance **8**, 305–332 (2004)
67. Velte, P.: Women on management board and ESG performance. J. Glob. Respons. **7**(1), 98–109 (2016)
68. Vitolla, F., Raimo, N., Rubino, M.: Board characteristics and integrated reporting quality: an agency theory perspective. Corp. Soc. Responsib. Environ. Manag. **27**(2), 1152–1163 (2020)
69. Wright, C., Nyberg, D., Grant, D.: "Hippies on the third floor": climate change, narrative identity and the micro-politics of corporate environmentalism. Organ. Stud. **33**(11), 1451–1475 (2012)
70. Yermack, D.: Higher market valuation of companies with a small board of directors. J. Financ. Econ. **40**(2), 185–211 (1996)
71. Yilmaz, M.K., Hacioglu, U., Nantembelele, F.A., Sowe, S.: Corporate board diversity and its impact on the social performance of companies from emerging economies. Glob. Bus. Organ. Excell. **41**(1), 6–20 (2021)
72. Yorke, S.M., Donkor, A., Appiagyei, K.: Experts on Boards' Audit Committee and sustainability performance. J. Clean. Prod. 137553 (2023)

Environment, Social, and Governance Business: Challenges and Opportunities

Albérico Travassos Rosário and Rui Nunes Cruz

Abstract This article provides an in-depth examination of the Environment, Social, and Governance (ESG) criteria within the business context, delineating the intricate balance between emerging challenges and opportunities. As the global business landscape evolves, ESG factors have become pivotal in shaping sustainable business practices, influencing stakeholder perceptions, and determining financial performance. We identify key challenges in ESG implementation, including regulatory compliance, data transparency, and integration of ESG into strategic decision-making. Additionally, the research highlights significant opportunities such as innovation in sustainable technologies, enhancement of brand reputation, and access to green finance. The findings suggest that while businesses face substantial hurdles in aligning operations with ESG criteria, there exists a vast potential for leveraging ESG for competitive advantage, stakeholder engagement, and long-term sustainability. The study contributes to the burgeoning literature on ESG in business by providing an evidence-based analysis of the current state of ESG integration, shedding light on the dynamic interplay between its challenges and opportunities. Implications for policymakers, business leaders, and stakeholders are discussed, emphasizing the need for a collaborative approach to foster a conducive ESG ecosystem. This research not only expands our understanding of ESG in the business context but also charts a path forward for sustainable and responsible business practices. Based on the above, it is intended to systematically review the bibliometric literature on Environment, Social, and Governance business in the Scopus database with the analysis of 113 academic and/or scientific documents.

Keywords Environment · Social · Governance · Business

A. T. Rosário
GOVCOPP, Universidade Europeia—Lisbon, Lisbon, Portugal
e-mail: alberico@ua.pt

R. N. Cruz (✉)
UNIDCOM/IADE, Universidade Europeia—Lisbon, Lisbon, Portugal
e-mail: rui.cruz@universidadeeuropeia.pt

© The Author(s), under exclusive license to Springer Nature Switzerland AG 2025
A. d. b. Machado et al. (eds.), *Environmental, Social, Governance and Digital Transformation in Organizations*, Information Systems Engineering and Management 35, https://doi.org/10.1007/978-3-031-86079-9_4

1 Introduction

In recent years, there has been an increase in interest in sustainable investments and developments due to increased concerns relating to climate change. Regulators, leaders, and the general public increasingly encourage businesses to embrace sustainable practices and promote green consumerism [41]. These issues have led to the emergence of the environmental, social, and governance (ESG) concept. Qu and Zhang [90] define ESG as the "inclusion of environment (environmental), society (social) and corporate governance (governance) in the decision-making process, pushing firms to go beyond the short-term financial benefits and develop strategies for improving social and environmental performance." Businesses have embraced ESG as a strategy for improving their firm value by ensuring their practices comply with appropriate policies and benefit society and the environment. They increasingly recognize that sustainable and ethical practices are a moral imperative and a source of competitive advantage [59]. For instance, ESG factors are now integral to risk management. They help identify potential vulnerabilities related to environmental degradation, social unrest, and governance deficiencies.

There are multiple benefits associated with the adoption of ESG in business. For instance, robust ESG practices can enhance a company's reputation, foster customer loyalty, and attract investment. Modern-day stakeholders, including consumers, investors, and regulators, emphasize sustainability and ethical conduct. Consequently, [41] explains that businesses that proactively embrace ESG principles are better positioned to achieve sustainable growth and resilience in the global business environment. In addition, ESG addresses some of the most pressing global challenges, such as climate change, social inequalities, and the need for robust corporate governance. For example, [115] found that the ESG frameworks encourage businesses to adopt sustainable practices that reduce carbon footprints, promote energy efficiency, and support renewable energy initiatives [97]. Its social aspect fosters diversity and inclusion, ensures fair labour practices, and contributes positively to community growth. Modern consumers and employees are socially conscious and demand ethical behaviours from corporations [59]. Therefore, adopting the ESG framework and principles helps build trust and legitimacy. The governance practices advocated through ESG principles promote transparency, accountability, and ethical decision-making within organizations. The holistic approach in ESG frameworks enhances organizational resilience and protects the interests of all stakeholders.

This systematic bibliometric literature review (LRSB) synthesizes data from 131 sources to explore the ESG concept in business. It also identifies opportunities, challenges, and practical implications of ESG.

2 Methodological Approach

The researchers conducted a systematic literature review using the PRISMA (Preferred Reporting Items for Systematic Reviews and Meta-Analyses) model. Page et al. [84] indicates that the PRISMA 2020 model is an effective tool for systematic reviews since it allows reviewers to transparently report why the research was done, what was done, and the findings. In addition, the framework provides a structured methodology that guides researchers through the stages of identification, screening, eligibility, and inclusion of studies. It encourages researchers to document the executed research protocol, allowing readers to replicate and validate the research methods.

This insightful analysis contributes to the refinement of future research directions and methodologies within the field, fostering a more robust and informed academic discourse [99, 100]. Additionally, the findings of this comprehensive literature review can offer practical insights for industry professionals and policymakers. In contrast to conventional literature reviews, the LRSB adopts a thorough approach in examining the published literature about the research topic [99, 100]. This method aims to condense the number of articles, concentrating on the most relevant ones. Additionally, the researcher furnishes an audit trail, enabling readers to evaluate the research's methodologies, conclusions, and the quality of the amalgamated studies.

Consequently, the LRSB encompasses a meticulous screening and selection of information sources to ensure the precision and validity of the presented data. This process unfolds through three phases and six steps, as delineated by [99, 100], as illustrated in Table 1.

The investigators utilized the Scopus database to pinpoint and choose reputable sources esteemed in academic and scientific communities. It is crucial to underscore that this study has limitations since it concentrated solely on the Scopus database, overlooking other academic and scientific databases.

The literature search began with identifying the database, in this case, the Scopus database. This was then followed by defining the inclusion and exclusion criteria, which included: (i) the sources had to primarily focus on ESG and its related

Table 1 Process of systematic LRSB

Fase	Step	Description
Exploration	Step 1	Formulating the research problem
	Step 2	Searching for appropriate literature
	Step 3	Critical appraisal of the selected studies
	Step 4	Data synthesis from individual sources
Interpretation	Step 5	Reporting findings and recommendations
Communication	Step 6	Presentation of the LRSB report

Source adapted [99, 100]

components and applications, (ii) peer-reviewed sources, including journal articles, academic books, and authoritative industry reports, were prioritized to ensure the credibility and rigor of the information (iii) only sources available in English were included to maintain consistency and accessibility of the research material, (iv) sources selected must have a detailed and clear methodology. After defining the inclusion/exclusion criteria, the researchers searched for relevant studies on the Scopus database. During the initial stage, the keywords "environment, social, and governance" resulting in 284 document results. The search was then limited to the subject area "business" alongside other filters like "peer-reviewed," "full text," and "English" for the language. These reduced the search results to 131, N = 131, as shown in Fig. 1.

We used content and thematic analysis techniques to identify, examine, and present various documents, as detailed by [99] and [100]. Subsequently, the 131 scientific and academic documents cataloged in Scopus were subjected to both narrative and bibliometric analysis. This approach aimed to explore the content and uncover recurring themes that directly addressed the research questions, as described by [99] and

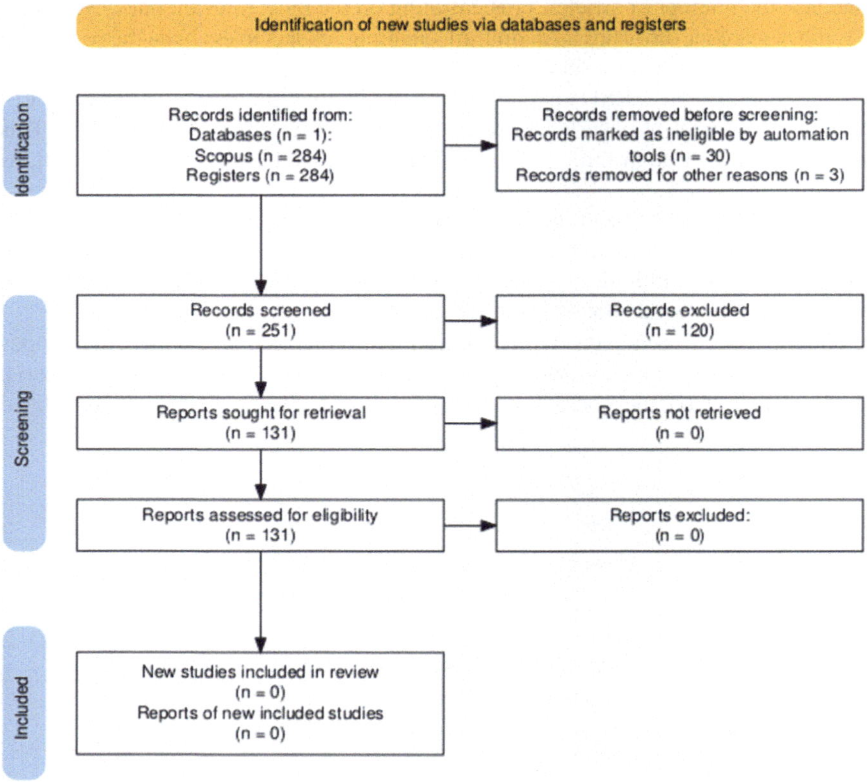

Fig. 1 PRISMA 2020 flow diagram for the literature search process. *Source* own elaboration

[100]. This whole process resulted in 131 documents, classified as 113 articles, 13 Book chapter, and 5 conference papers.

3 Publication Distribution

Peer-reviewed articles on Environment, Social, and Governance business until June 2024. The year 2023 had the highest number of peer-reviewed publications on the subject, reaching 37 (Fig. 2). Investigations into this subject are still scarce.

Figure 3 shows the countries with the most active scientific output in research areas, focusing on India, the USA, and China, which have the most publications.

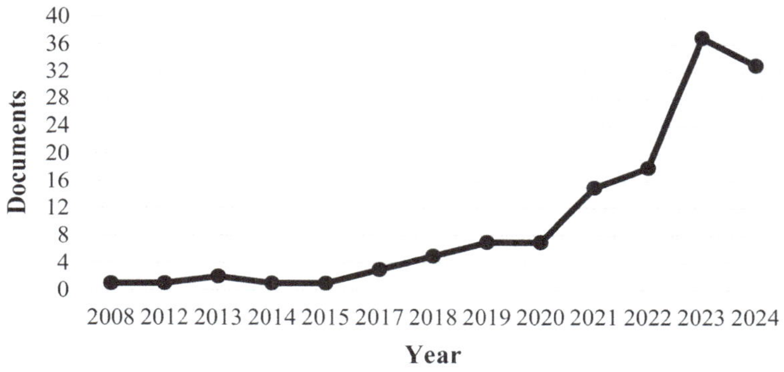

Fig. 2 Documents by year. *Source* own elaboration

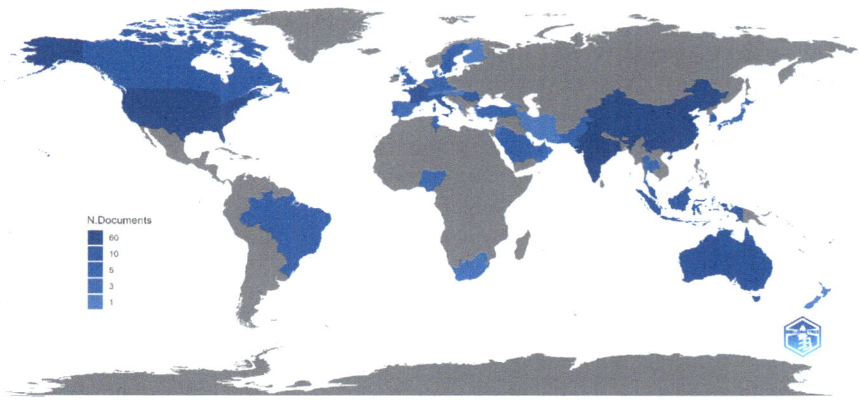

Fig. 3 Scientific production by country

Table 2 Top 10 countries by number of publications

Country	Number of publications
India	60
USA	42
China	41
France	23
Australia	17
Malaysia	16
Indonesia	14
Italy	13
Romania	10
South Korea	9

Source own elaboration

Table 2 and Fig. 3 show the top 10 countries that have made the most noteworthy scientific contributions in the fields under study. This analysis aims to understand whether these countries prioritize utilizing artificial intelligence in the consumer behavior process in a business or if they emphasize boosting their productivity.

The publications were sorted out as follows: Journal Of Cleaner Production (8); Journal Of Portfolio Management (7); Journal Of Sustainable Finance And Investment (5); Business Strategy And The Environment (5); Sustainability Accounting Management And Policy Journal (4); Global Business Review (4); Corporate Social Responsibility And Environmental Management (4); Journal Of Business Ethics (3); and with 2 (Journal Of Risk And Financial Management; Journal Of Property Investment And Finance; Journal Of Asset Management; Corporate Governance Bingley; Cogent Business And Management; Business Ethics The Environment And Responsibility; Asian Review Of Accounting; Accounting Finance Sustainability Governance And Fraud; and the remaining publications with 1 document.

The subject areas covered by the 131 scientific or academic documents were: Limited to Business, Management and Accounting (131); Economics, Econometrics and Finance (66); Social Sciences (25); Environmental Science (23); Decision Sciences (17); Energy (15); Engineering (14); Computer Science (9); Arts and Humanities (6); Psychology (2); Agricultural and Biological Sciences (2); Mathematics (1); and Earth and Planetary Sciences (1).

The most quoted article was *"How Do Companies Respond to Environmental, Social and Governance (ESG) ratings? Evidence from Italy"* by [25], with 159 quotes published Journal of the Journal of Business Ethics 2620 (SJR), the best quartile (Q1) and with H index (253), this paper is to explore the in current understanding by examining how firms react to ESG ratings, and the factors influencing their response.

Figure 4 presents citation changes for documents published until June 2024. The period 2014–2024 shows a positive net growth in citations with an R2 of 69%, reaching 1547 citations in June 2024.

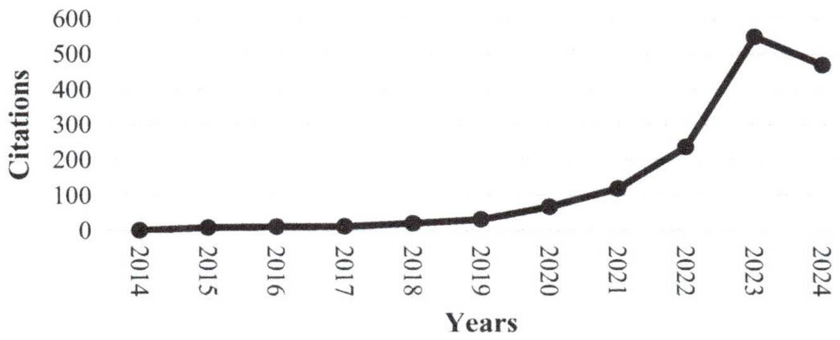

Fig. 4 Evolution of citations between 2014 and 2024. *Source* own elaboration

Citations of all scientific and/or academic documents from the period ≤ 2013 to until June 2024, with a total of 1288 citations, of the 131 documents 10 were not cited. The self-citation of documents in the period ≤ 2014 to June 2024 was self-cited 1430 times.

Using the primary keywords, the bibliometric study delved into and uncovered markers of the dynamics and evolution of scientific and/or academic information within documents, as depicted in Fig. 4.

These findings were extracted utilizing the scientific software VOS viewer, with a focus on identifying the primary search keywords Environment, Social, Governance, and business (Fig. 5).

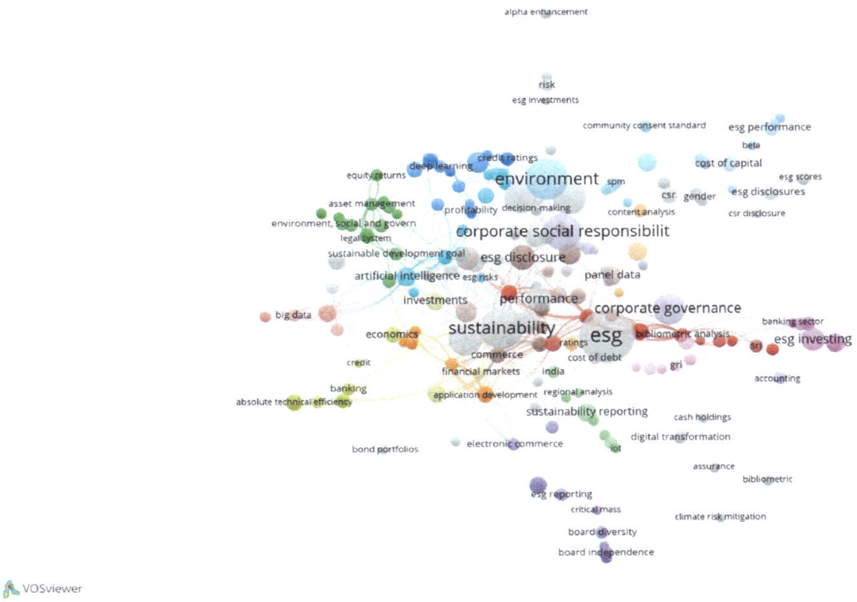

Fig. 5 Network of all keywords

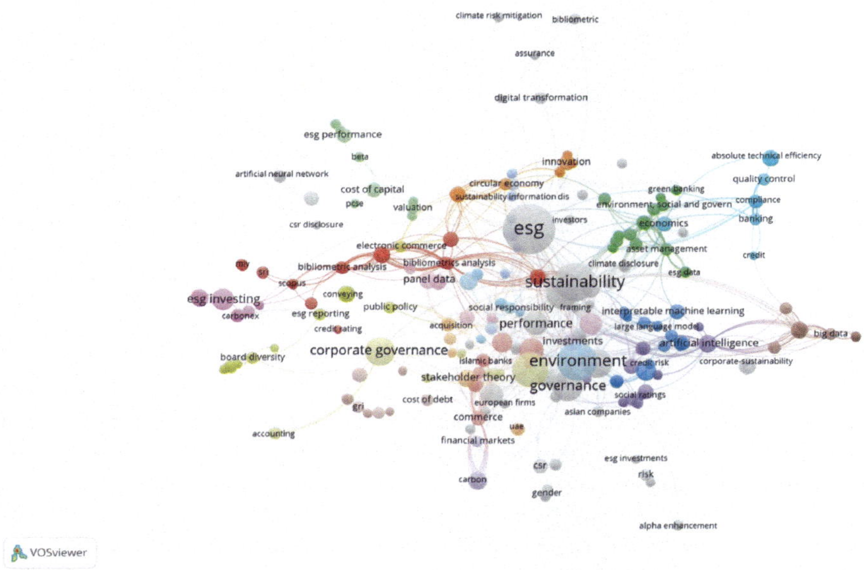

Fig. 6 Network of linked keywords

This study is rooted in scientific and academic documents that explore the potential of Environment, Social, and Governance business. In Fig. 6, we can scrutinize the interconnected keywords.

This enables us to emphasize the interrelationships among keywords within each scientific article, aiding in the identification of subjects explored in the studies and providing insights into future research trends (Fig. 7).

4 Theoretical Perspectives

The ESG framework helps stakeholders understand how an organization manages risks and opportunities related to environmental, social, and governance criteria. It has become a significant consideration in investment decisions. Investors are seeking to align their portfolios with their values and mitigate risks associated with unsustainable business practices [112]. Abdul Rahman et al. [2] explains that the rise of ESG can be attributed to a growing recognition of the significant impact of environmental and social issues on financial performance and corporate reputation. Besides, the mandatory disclosures related to ESG factors adopted by regulatory bodies worldwide reinforce the framework's significance and its associated practices and principles.

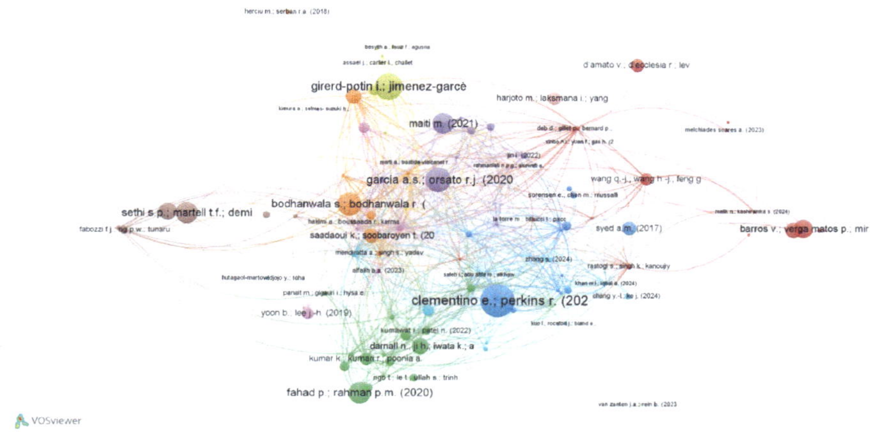

Fig. 7 Network of co-citation

4.1 ESG Concept

ESG is a comprehensive framework used to evaluate an organization's sustainability and ethical impact. This framework encompasses three core dimensions: environmental, social, and governance [1]. The ecological component assesses a company's initiatives to protect nature. For instance, it focuses on efforts to minimize ecological footprints through initiatives like reducing carbon emissions, managing waste, conserving water, and ensuring sustainable resource use [14, 56, 98]. The social dimension examines how a company manages relationships with its stakeholders, including employees, suppliers, customers, and its communities. This includes labour practices, diversity and inclusion, human rights, community engagement, and health and safety standards [47]. The governance aspect pertains to the internal policies and procedures that govern an organization's decision-making processes. This involves the structure and composition of the board of directors, executive compensation, audit practices, transparency, anti-corruption measures, and shareholder rights.

The three dimensions provide a holistic view of an organization's long-term sustainability and ethical impact. Unlike the traditional financial metrics, the ESG framework includes considerations of risk management, corporate responsibility, and social license to operate [50]. As a result, ESG's role in guiding investment decisions has become increasingly important in recent years. For example, [4] found that 70% of investors use sustainable investment criteria to make decisions. They also use ESG policies to assess risk management and enhance risk-adjusted returns. Girerd-Potin et al. [39] explain that this shift towards responsible investments has accelerated stakeholders' demand for greater accountability and sustainability from businesses. In addition, investors recognize that strong ESG performance can correlate with superior financial performance, resilience, and long-term value creation.

4.2 Key Drivers of ESG

There are various factors driving the adoption of ESG in business. Ahmad et al. [4] found that 52% of investors and asset managers push for ESG adoption to achieve long-term benefits, while 47% aim to use it to improve firm reputation. Other drivers include investment risks, disclosure, stakeholders, new opportunities, altruistic values, investor pressure, and diversification, as shown in Fig. 2. Governments and regulatory bodies have introduced strict policies to increase accountability and encourage businesses to adopt sustainable business practices [119]. This has led to the introduction of disclosure policies, which require companies to transparently report on their ESG activities and performance [18, 103]. The reporting provides investors and stakeholders consistent, comparable, and reliable ESG information [75]. This transparency pressures companies to enhance their ESG strategies, as failure to comply can result in reputational damage, financial penalties, or diminished investor confidence [81, 95]. Moreover, the requirement to disclose ESG metrics fosters a culture of continuous improvement and encourages companies to adopt best practices in sustainability, social responsibility, and governance to meet regulatory standards and stakeholder expectations (Fig. 8).

Governments worldwide are offering incentives to encourage the adoption of ESG in businesses. These incentives often take various forms, such as tax breaks, grants, subsidies, and preferential treatment in public procurement processes for companies demonstrating strong ESG performance [74, 134, 135]. Such financial benefits make it economically advantageous for businesses to invest in sustainable technologies, improve their social impact, and establish strong governance structures [87]. Therefore, the low costs of implementing ESG initiatives and financial rewards for compliance from governments drive adoption.

Mergers and acquisitions (M&A) activities are also crucial drives of ESG adoption. ESG factors are increasingly becoming a key criterion for evaluating potential

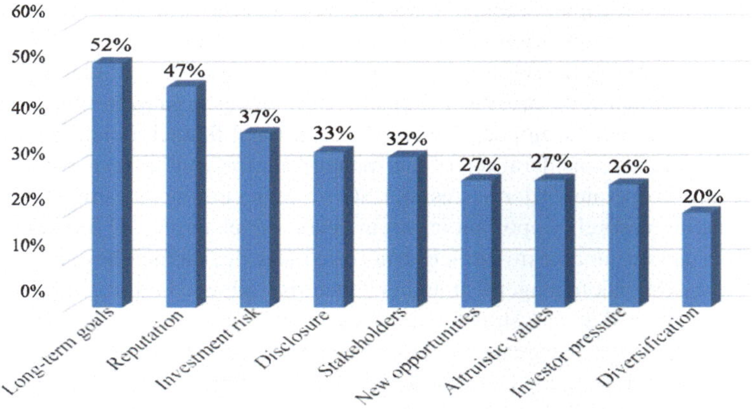

Fig. 8 Factors driving ESG incorporation in business decision-making [16]

targets in M&A [11]. For instance, acquirers are more likely to seek out companies with strong ESG portfolios. They recognize that these companies have higher chances of long-term success, regulatory compliance, and risk mitigation [71, 92]. In addition, companies involved in M&A transactions often use the process as an opportunity to harmonize and strengthen their ESG policies [136]. The integration of ESG considerations in due diligence and valuation processes reinforces their importance in achieving sustainable growth and creating value [94]. As a result, the focus on ESG in M&A activities promotes adopting these practices in acquired companies and improves the overall standards within the industry.

4.3 Opportunities of ESG in Business

Adopting ESG in business creates numerous opportunities for growth and improved performance. For instance, accurate ESG reporting can improve an organization's reputation and build investors' and other stakeholders' confidence and trust, paving the way for long-term benefits [72]. This section synthesizes findings on other opportunities associated with implementing ESG in business.

Innovation in Sustainable Technologies

ESG principles encourage companies to develop and adopt environmentally friendly solutions that reduce their ecological footprint and enhance operational efficiency. This focus on sustainability leads to the creation of new products, processes, and business models that address pressing environmental challenges, such as climate change, resource depletion, and pollution [57, 131]. For instance, companies are investing in renewable energy technologies, such as solar and wind power, to reduce reliance on fossil fuels. In addition, advancements in energy storage, electric vehicles, and green building materials are examples of how ESG priorities are fostering technological innovation, [106 109, 110]. These innovations help companies comply with environmental regulations and meet stakeholder expectations. They also provide a competitive advantage by opening new markets and revenue streams [125]. As consumers and businesses increasingly demand sustainable solutions, companies that lead in this space can position themselves as industry pioneers, attract eco-conscious customers, and establish long-term growth trajectories.

4.4 Enhancement of Brand Reputation

Implementing ESG practices significantly enhances a company's brand reputation by demonstrating a commitment to ethical and sustainable operations. According to [24], this commitment resonates with a growing segment of consumers, investors, and employees who prioritize corporate responsibility. Companies with strong ESG portfolios are often perceived as more trustworthy, transparent, and forward-thinking

[113, 131]. These perceptions increase customer loyalty, higher employee satisfaction, and greater investor confidence. In addition, a positive ESG reputation also helps mitigate risks associated with negative publicity, such as social controversies, by showcasing proactive measures and responsible governance [61, 120]. According to [33], accurate ESG ratings can be used as a differentiation element. Companies that effectively communicate their ESG achievements can differentiate themselves in the marketplace, attract premium customers, and build long-lasting relationships based on shared values. This reputational capital can be a significant asset, fostering resilience and support from stakeholders during challenging times.

4.5 Access to Green Finance

Businesses with strong ESG profiles can access green finance more efficiently and benefit from financial instruments designed to support sustainable projects. Green bonds, sustainability-linked loans, and other forms of green financing offer favourable terms and conditions for businesses committed to environmental and social goals [12, 19, 65]. These financial products often come with lower interest rates, longer repayment periods, and increased investor interest, reflecting sustainable practices' reduced risk and enhanced growth potential. Pereira et al. [86] found that the allocation of bond investing under socially responsible investment assets in Europe increased from 40% in December 2013 to 64% by December 2015. With initiatives such as the United Nations Principles for Responsible Investment (PRI), most investors are incorporating ESG in their investment decisions, thereby increasing access to green financing [30, 48]. Access to these financial resources enables companies to fund large-scale projects, such as renewable energy installations, energy efficiency upgrades, and sustainable infrastructure developments, without straining their capital reserves [20, 54]. Green finance enables businesses to accelerate their ESG initiatives, achieve long-term cost savings, and contribute positively to the global sustainability agenda [108]. This flexibility supports immediate project implementation and enhances overall financial health and operational resilience.

4.6 Gender Diversity Boards

Promoting gender diversity on corporate boards is a crucial aspect of ESG that can drive significant business benefits. Diverse boards bring various perspectives, experiences, and skills to decision-making, leading to more innovative and effective strategies [13]. Research has shown that companies with diverse boards are better at problem-solving, risk management, and navigating complex business environments [123, 128]. Gender diversity also enhances corporate governance by reducing groupthink and fostering a more inclusive corporate culture. Moreover, diverse boards reflect the broader society and marketplace, improving stakeholder relations and enhancing corporate image [3]. Companies prioritising gender diversity can attract top talent, increase employee engagement, and strengthen customer and investor relations [5, 121]. This commitment to inclusivity and equality is not only a moral imperative but also a strategic advantage that contributes to improved business performance and resilience.

4.7 Improved Organizational Financial Performance

ESG principles and practices can enhance operational efficiencies, reduce costs, and drive revenue growth, improving a firm's financial performance. Sustainable business practices, such as energy efficiency, waste reduction, and resource optimization, can lower operational expenses and increase profitability [6, 29]. According to [37], social initiatives, such as fair labor practices and community engagement, can boost employee morale and productivity, reducing turnover and recruitment costs. Strong governance practices, including transparent reporting and robust risk management, can prevent financial scandals and regulatory fines, safeguarding company assets [45]. Jasuja et al. [53] found that companies with strong ESG performance are often more attractive to investors, leading to increased access to capital and lower costs of equity. In addition, studies have shown that companies with high ESG ratings tend to outperform their peers regarding stock market performance and long-term financial stability [25, 114]. Ultimately, integrating ESG considerations into business strategies can enhance firms' financial health, achieve sustainable growth, and deliver superior value to shareholders.

4.8 Employee Relations and Innovation

Implementing ESG frameworks can result in positive employee relations, driving innovation and business success. Companies that prioritize social responsibility, such as fair wages, safe working conditions, and diversity and inclusion, create a supportive and engaging workplace culture [17, 58]. This positive environment attracts top talent, reduces turnover, and enhances employee satisfaction and productivity. Engaged employees are more likely to contribute innovative ideas and solutions, driving continuous improvement and competitive advantage [46]. In addition, businesses that actively involve employees in their ESG initiatives, such as sustainability projects and community programs, can leverage their collective creativity and commitment to achieve organizational goals [42, 66]. Promoting a culture of social responsibility and innovation can empower brands to develop quality products and services, improve operational efficiencies, and respond more effectively to market changes and customer demands. This approach enhances business performance and strengthens the company's reputation, positioning the company as an employer of choice.

4.9 Promotes Sustainability

ESG promotes sustainability by aligning business operations with long-term environmental and social goals. Companies committed to ESG principles are more likely to implement sustainable practices that minimize environmental impact, such as reducing greenhouse gas emissions, conserving water, and promoting circular economy models [7, 122]. These practices contribute to the preservation of natural resources and the reduction of ecological footprints. As a result, they ensure that business activities do not compromise the ability of future generations to meet their needs [8, 21]. In addition, socially responsible practices, such as supporting local communities, ensuring fair labor conditions, and fostering diversity, enhance social sustainability by promoting equity, inclusion, and human well-being [15]. Governance practices that emphasize transparency, accountability, and ethical behaviour ensure that businesses operate in a manner that is consistent with societal values and expectations [68]. Consequently, firms that integrate sustainability into their core strategies build resilient and adaptable business models that thrive in a competitive economic environment, securing long-term success and contributing to global sustainable development goals.

4.10 Improved CSR Reporting

Improved corporate social responsibility (CSR) reporting is a critical outcome of ESG. Detailed and transparent CSR reports provide stakeholders comprehensive information on a company's environmental, social, and governance activities and impacts [34, 85]. Jamali et al. [52] explain that enhanced reporting practices ensure that companies effectively communicate their ESG initiatives, achievements, and challenges, fostering transparency and accountability. In addition, high-quality CSR reports strengthen stakeholder trust as they demonstrate a company's commitment to ethical and sustainable business practices [73, 101]. These reports also provide valuable insights for investors, helping them assess the company's ESG performance and risks. Improved CSR reporting can highlight areas for improvement, guide strategic planning, and drive continuous improvement in ESG practices [28, 31]. Resilient reporting frameworks, such as the Global Reporting Initiative (GRI) and the Sustainability Accounting Standards Board (SASB), provide standardized metrics that enable benchmarking and comparability across industries. Excelling in CSR reporting can improve organizational reputation and enable companies to meet regulatory requirements and attract socially responsible investors.

4.11 Long-Term Value

Focusing on ESG principles creates long-term value for businesses by ensuring sustainable growth and resilience. Companies that integrate ESG considerations into their strategies are better equipped to manage risks and seize opportunities [51]. For example, environmental stewardship reduces exposure to regulatory changes, resource scarcity, and climate-related risks [5, 118]. This safeguards operational continuity and cost stability. Social responsibility helps establish a loyal and supportive stakeholder base by creating strong relationships with employees, customers, and communities [64]. Finally, good governance practices reduce the likelihood of scandals and enhance corporate stability by ensuring ethical conduct, transparency, and accountability [63]. Collectively, these ESG practices contribute to a positive corporate reputation, attract long-term investors, and create a sustainable competitive advantage [9, 127]. Prioritizing ESG can help companies navigate uncertainties, adapt to changing societal expectations, and ensure longevity and prosperity. This focus on long-term value creation aligns business success with broader societal goals, contributing to a more sustainable and equitable world.

4.12 Challenges of ESG

Implementing ESG initiatives can be subject to numerous challenges that can hinder progress. For instance, some ESG programs require additional investments in technological solutions. This hinders implementation among small and medium businesses with limited resources [26]. In addition, some business leaders struggle with implementing strategies that promote sustainability without affecting the company's profitability. These issues slow the adoption and implementation of ESG frameworks. Other challenges identified in the research include:

- **Regulatory Compliance**

Regulatory compliance is one of the most common challenges hindering ESG implementation. Governments and regulatory bodies worldwide are recognizing the importance of sustainability [93]. As a result, they are enacting various ESG-related regulations and reporting requirements. These regulations can vary significantly across jurisdictions, making it challenging for multinational corporations to comply with all relevant laws [82, 88]. For instance, the European Union's Sustainable Finance Disclosure Regulation (SFDR) and the Non-Financial Reporting Directive (NFRD) impose strict reporting requirements that may differ from those in the United States or Asia [36]. These regulatory issues require companies to invest substantial resources to understand and comply with these diverse regulations, often necessitating specialized legal and compliance expertise. This additional investment further increases the cost of implementing ESG initiatives, thereby causing slow adoption

[62]. In addition, staying up to date with changes in ESG regulations requires continuous monitoring and adaptation, which can be resource intensive. Non-compliance can result in hefty fines, legal repercussions, and damage to a company's reputation. These issues show the critical importance of rigorous compliance strategies with ESG policies.

- **Data Transparency**

Data transparency is another significant challenge in ESG. Accurate and comprehensive data collection is crucial for effectively measuring and reporting ESG performance. However, [55] found that ESG data lacks "materiality, accuracy and reliability." These challenges result in inaccurate reporting and hinder the optimal use of ESG data in decision-making. In addition, many companies struggle with the complexity of gathering reliable ESG data due to the lack of standardized metrics and methodologies [111]. This challenge is exacerbated by the need to collect data from a wide range of sources, including internal operations, supply chains, and external stakeholders, [96, 126]. Moreover, companies must navigate issues related to data privacy and security, particularly when dealing with sensitive information [23]. The lack of a universally accepted framework for ESG reporting means that companies often face difficulties benchmarking their performance against peers [22, 107]. This leads to challenges in setting and achieving meaningful ESG targets. Addressing these data transparency issues requires significant investment in technology, training, and process improvements to ensure that ESG data is reliable and actionable.

- **Integration of ESG in Decision-Making**

Integrating ESG principles into core business decision-making processes presents a substantial challenge for many organizations. Traditionally, business decisions have been primarily driven by financial performance metrics, such as profitability, return on investment, and shareholder value [67, 91]. Shifting to a more holistic approach that includes environmental, social, and governance considerations requires a fundamental change in corporate culture and strategic priorities. This integration involves aligning ESG goals with overall business objectives, which can be complex and time-consuming [130, 132]. Additionally, it requires buy-in from all levels of the organization, including top management, middle managers, and frontline employees. There can be resistance to change, particularly if ESG initiatives are perceived as conflicting with short-term financial goals. Khan and Iqbal [60] explained that integrating ESG into decision-making necessitates the development of new tools and frameworks to assess the long-term impacts of business activities on the environment, society, and governance structures. This approach demands extensive training programs to equip employees with the necessary skills and knowledge to incorporate ESG considerations into their daily operations and strategic planning [133]. This can pose a challenge to businesses with limited resources to allocate to employee training, thereby hindering their ability to leverage the opportunities that ESG presents.

- **ESG-Washing**

ESG-washing, or greenwashing, is a significant challenge that undermines the credibility and effectiveness of ESG initiatives. It refers to the practice of companies exaggerating or misrepresenting their ESG performance to appear more environmentally and socially responsible than they are [129]. This can occur through misleading marketing campaigns, selective disclosure of favourable data, or omitting negative ESG impacts. Consequently, ESG-washing deceives stakeholders and erodes trust in genuine sustainability efforts [117]. It creates a misleading perception of a company's commitment to ESG principles, potentially attracting investments and customers under false pretences. Detecting and preventing ESG-washing requires strict regulatory oversight, third-party audits, and transparent reporting standards [38, 49]. Companies should ensure that their ESG communications are accurate, comprehensive, and verifiable to reflect a true picture of their performance. In addition, stakeholders, including investors, customers, and employees, must remain vigilant and demand accountability and transparency [104]. Addressing ESG-washing is crucial for maintaining the integrity of ESG practices and ensuring that efforts to promote sustainability and ethical governance achieve their intended outcomes.

- **Technological and Operational Constraints**

Businesses implementing ESG initiatives often encounter significant technological and operational constraints that can negatively affect progress and effectiveness. Many companies, particularly those in traditional or resource-intensive industries, operate with legacy systems and infrastructure that are not optimized for sustainability [26, 27]. Upgrading these systems to more environmentally friendly technologies require substantial capital investment and can be logistically challenging [106, 134, 135].

For example, transitioning to renewable energy sources or adopting advanced waste management systems requires financial resources, time and expertise [124]. Besides, the pace of technological advancement in areas such as energy efficiency, carbon capture, and sustainable materials can outrun a company's ability to implement these innovations quickly.

This lag can leave companies struggling to keep up with best practices and regulatory standards. Operational constraints such as supply chain dependencies and production processes can limit the ability to integrate ESG initiatives effectively [32]. Complex supply chains often span multiple countries with varying regulatory environments. This makes it difficult for businesses to ensure consistent ESG standards across all tiers [10]. In some cases, business leaders worry that operational disruptions when implementing new technologies or processes can impact productivity and profitability in the short term. This fear creates resistance within the organization.

4.13 Implications

- **For Policymakers**

Policymakers play a crucial role in shaping the regulatory environment that governs ESG practices, and their actions have significant implications for businesses and society at large.

For instance, they can drive corporate behaviour towards greater sustainability and ethical governance by implementing and reinforcing stringent ESG regulations [83].

Environmental regulations can compel companies to reduce carbon emissions, manage waste more effectively, and adopt renewable energy sources [77].

These regulations not only help mitigate climate change but also promote innovation in green technologies and sustainable practices. Policymakers can further influence corporate social responsibility by mandating fair labor practices, diversity and inclusion initiatives, and community engagement programs.

Such policies ensure that businesses contribute positively to social welfare, enhancing the quality of life for employees and communities.

Policymakers can create standardized reporting frameworks that enhance transparency and comparability in ESG performance. Standardization helps investors and other stakeholders make informed decisions based on consistent and reliable data. It also reduces the risk of greenwashing.

They can set clear guidelines for ESG disclosures, thereby fostering a culture of accountability and continuous improvement within the corporate sector [43].

Furthermore, aligning ESG regulations with international standards can facilitate global trade and investment. This practice helps businesses operating in multiple jurisdictions to adhere to a unified set of expectations [70].

Consequently, this harmonization can drive global sustainability efforts, ensuring that companies contribute to broader environmental and social goals beyond their immediate operational boundaries.

- **For Business Leaders**

The integration of ESG principles into corporate strategy represents both a significant opportunity and a complex challenge for business leaders. Embracing ESG can lead to enhanced risk management by empowering companies to proactively address environmental and social risks that could otherwise result in financial liabilities, reputational damage, or regulatory penalties [40, 69].

In addition, business leaders can embed ESG considerations into decision-making processes. This practice can help them anticipate and mitigate potential disruptions, ensuring long-term operational resilience. In addition, this proactive approach can safeguard the company's assets and strengthen stakeholder trust and confidence, which are critical for sustained success.

Moreover, integrating ESG into corporate strategy can drive competitive advantage. Companies that lead in ESG performance often benefit from increased customer loyalty. Today, most consumers are increasingly favouring brands that demonstrate a commitment to sustainability and social responsibility [35] and business leaders who prioritize ESG can attract and retain top talent among employees seeking to work for organizations that align with their values. This can lead to improved firm performance, reduced employee turnover, and costs associated with hiring and training [44].

As more investors are increasingly incorporating ESG criteria into their investment decisions, a strong ESG performance can enhance access to capital. Companies should demonstrate leadership in ESG to position themselves as industry pioneers, capitalize on emerging market opportunities and build a sustainable and successful brand.

Successful adoption of ESG principles requires a paradigm shift in leadership and corporate governance. Business leaders must create a culture of sustainability and social responsibility at all levels of the organization. This involves setting clear ESG goals, integrating them into the company's mission and values, and ensuring they are reflected in everyday business operations [115].

Leaders must establish governance structures to oversee ESG initiatives, including dedicated committees and reporting mechanisms. This governance framework should ensure accountability, monitor progress, and facilitate continuous improvement. Effective leadership in ESG requires transparency, communication, and collaboration with all stakeholders, including employees, customers, suppliers, and investors.

- **For Other Stakeholders**

ESG considerations are becoming an integral part of the investment decision-making process. ESG criteria provide insights into a company's long-term sustainability and risk profile. Investors use these metrics to identify companies likely to deliver stable and sustainable returns [76]. In this regard, companies with strong ESG performance are perceived as lower risk with a higher probability of future growth, attracting more investment [116].

This shift in investor behaviour also pressures companies to improve their ESG practices and reporting, creating a virtuous cycle of enhanced corporate responsibility and investor confidence [80, 102]. Moreover, investors can use their influence to advocate for better ESG practices within the companies they invest in, driving broader industry changes.

For employees, ESG initiatives create a more engaging and inclusive workplace environment. Companies that prioritize social responsibility and ethical governance are likely to foster a positive organizational culture, enhancing employee morale, productivity, and retention [105]. For example, ESG initiatives that focus on diversity and inclusion can lead to a more dynamic and innovative workforce by bringing together diverse perspectives and skills. Furthermore, environmental sustainability efforts can instil a sense of pride and purpose among employees, motivating them to contribute to the company's goals (Singh et al., 2024). Engaged and motivated employees are critical to the successful implementation of ESG strategies. They are more likely to support and drive these initiatives.

Customers are increasingly prioritizing sustainability and ethical considerations in their purchasing decisions. Newell [79] found that companies that are transparent about their ESG efforts and demonstrate a genuine commitment to these principles can build stronger relationships with their customers.

This loyalty can translate into increased sales, brand advocacy, and a competitive edge in the marketplace. In addition, businesses that engage customers in their ESG initiatives, such as through sustainable product offerings or community projects, can enhance customer engagement and satisfaction. These interactions strengthen the brand and contribute to the company's social impact.

Promoting sustainable consumption within communities is another critical aspect. Companies can engage with local populations to raise awareness about the benefits of sustainable practices and encourage more responsible consumption habits. This can involve community outreach programs, partnerships with local organizations, and educational campaigns to promote recycling, energy conservation, and the use of sustainable products [89].

This proactive approach of fostering a culture of sustainability within the communities they serve can help companies create a supportive environment for their ESG initiatives. This can drive collective action towards broader environmental and social goals [78]. The collaborative strategy enhances the impact of ESG efforts, strengthens community relations and contributes to the overall sustainability of the business ecosystem.

5 Conclusions

This research findings reflect the complicated nature of ESG principles and their significant importance in the contemporary business environment. ESG is a comprehensive framework designed to support sustainability, ethical governance, and social responsibility within corporate strategies. The evolution of ESG has been driven by a convergence of factors, including growing environmental concerns, heightened social awareness, and the demand for robust corporate governance. In addition, regulatory mandates, government incentives, and market dynamics such as mergers and acquisitions encourage companies to adopt ESG initiatives. This creates numerous opportunities for businesses, employees, investors, customers, and communities.

For instance, companies can leverage innovation in sustainable technologies to achieve competitive advantage, enhance their brand reputation, and access green finance. ESG also supports gender diversity on boards, which contributes to gender equality movements and improves overall decision-making processes. Companies implementing ESG frameworks are more likely to experience improved overall financial performance through enhanced risk management and operational efficiency. Finally, ESG contributes to enhanced employee relations, promotes sustainability, and improves corporate CSR reporting. These practices result in long-term value creation.

Despite its advantages, implementing ESG principles presents several significant challenges. For instance, regulatory compliance is a primary concern associated with the diverse and evolving nature of ESG-related regulations across different jurisdictions. Most businesses struggle with data transparency due to the complexity of collecting reliable and accurate ESG data. In addition, integrating ESG into core business decision-making requires a cultural and strategic shift that many organizations find daunting. The prevalence of ESG-washing undermines the credibility of genuine ESG efforts. This issue reflects the need for strict regulatory oversight and accountability. Finally, technological and operational constraints, such as outdated infrastructure and complex supply chains, undermine the effective adoption of ESG practices. Businesses looking to leverage ESG to improve performance and competitiveness must find ways to address these challenges.

The implications of ESG extend to various stakeholders. For instance, policymakers must implement comprehensive and harmonized regulations to guide corporate behavior. Business leaders are tasked with embedding ESG into corporate culture and balancing short-term financial performance and long-term sustainability goals.

For other stakeholders, including investors, employees, and customers, ESG represents a framework for informed decision-making, enhanced engagement, and promoting sustainable consumption. Collectively, these stakeholders can collaborate to leverage ESG opportunities and address its associated challenges, thereby paving the way for a more sustainable and ethically responsible business environment.

Contributions to theories and practices in Environment, Social, and Governance business are significant and diverse, reflecting the multifaceted nature of ESG itself.

Environment: (i) The Triple Bottom Line concept, which measures a company's social, environmental, and financial performance, has been pivotal in integrating sustainability into business practices; (ii) Natural Capitalism introduced the idea, of advocating for the efficient use of resources and the restoration of natural ecosystems; (iii) A Circular Economy has been promoted, which aims to eliminate waste and continuously use resources.

Practical Contributions: (i) Green Supply Chain Management (Integration of environmental thinking into supply chain management, including product design, material sourcing, and end-of-life management); (ii) Renewable Energy Adoption (Companies like Tesla and Ørsted have significantly contributed to the adoption and advancement of renewable energy technologies); (iii) Corporate Environmental Reporting (Development of frameworks like the Global Reporting Initiative for companies to disclose their environmental impacts.

Social: (i) Expanded the concept of stakeholders beyond shareholders to include employees, customers, suppliers, and communities; (ii) Carroll's Corporate Social Responsibility pyramid categorizes a corporation's responsibilities as economic, legal, ethical, and philanthropic.

Practical Contributions: (i) Programs aimed at increasing workforce diversity and ensuring inclusive work environments, led by companies; (ii) Implementation of fair labor practices and protection of human rights in supply chains by companies such as Unilever and Patagonia; (iii) Initiatives like foundation's community development programs.

Governance: (i) Examines the relationship between principals (shareholders) and agents (company executives), emphasizing the need for alignment of interests; (ii) Proposes that managers, as stewards, are motivated to act in the best interests of the company and its shareholders.

Practical Contributions: (i) Promoting diverse boards to improve decision-making and reflect a broader range of perspectives, supported by policies from companies like Nasdaq; (ii) Adoption of practices ensuring transparent reporting and accountability; (iii) Companies emphasized ethical leadership and decision-making frameworks.

ESG Reporting Standards: (i) Provides industry-specific standards for ESG reporting; (ii) Recommends disclosures for climate-related financial risks; (iii) Encourages businesses worldwide to adopt sustainable and socially responsible policies.

ESG Investment Practices: (i) Investments made to generate positive, measurable social and environmental impact alongside a financial return; (ii) Incorporation of ESG factors into investment analysis and portfolio construction by asset managers like BlackRock and Vanguard.

Theoretical and Practical Integration: (i) companies can create economic value by addressing societal challenges; (ii) Certification for companies that meet high standards of social and environmental performance, accountability, and transparency.

In summary, contributions to ESG theories and practices span academic research, corporate strategies, regulatory frameworks, and grassroots initiatives, reflecting a comprehensive and evolving approach to integrating environmental, social, and governance considerations into business operations and strategy.

Exploring future lines of investigation in Environment, Social, and Governance business is essential for understanding how these practices can evolve and adapt to meet emerging challenges and opportunities: (i) Research into new renewable energy technologies, such as advanced solar panels, wind turbines, and bioenergy.

Carbon Capture and Storage (CCS): (ii) Development of more efficient and cost-effective CCS technologies to reduce greenhouse gas emissions; (iii) Investigating strategies for enhancing the resilience of businesses and communities to climate change impacts; (iv) Innovations in water conservation, purification, and management techniques, especially in arid regions; (v) Expanding the principles of the circular economy to new industries and sectors, with a focus on reducing waste and enhancing resource efficiency; (vi) Developing smart infrastructure that minimizes environmental impact while enhancing urban living conditions; (vii) Creating and scaling business models that promote social inclusion and economic equity; (vii) Investigating the relationship between board diversity and company performance.

By focusing on these areas, researchers and practitioners can contribute to the ongoing evolution of ESG practices, ensuring they remain relevant and effective in addressing the critical environmental, social, and governance challenges of the future.

References

1. Abdul Razak, L., Ibrahim, M.H., Ng, A.: Environment, social and governance (ESG) performance and CDS spreads: the role of country sustainability. J. Risk Financ. **24**(5), 585–613 (2023). https://doi.org/10.1108/JRF-10-2022-0278

2. Abdul Rahman, R., Alsayegh, M.F.: Determinants of corporate environment, social and governance (ESG) reporting among Asian firms. J. Risk Financ. Manag. **14**(4), C7–167 (2021). https://doi.org/10.3390/jrfm14040167
3. Agustina, R.P., Barokah, Z.: Women on boards and ESG performance: empirical evidence before and during the COVID-19 pandemic in Indonesia and Malaysia. Corp. Gov. (Bingley) **24**(3), 489–508 (2024). https://doi.org/10.1108/CG-10-2023-0415
4. Ahmad, H., Yaqub, M., Lee, S.H.: Environmental-, social-, and governance-related factors for business investment and sustainability: a scientometric review of global trends. Environ. Dev. Sustain. **26**(2), 2965–2987 (2024). https://doi.org/10.1007/s10668-023-02921-x
5. Ahn, K.H.: Main contents and implications of the 2023 revision of the OECD Guidelines for multinational enterprises. J. Int. Logist. Trade **22**(2), 80–92 (2024). https://doi.org/10.1108/JILT-12-2023-0081
6. Alfalih, A.A.: ESG disclosure practices and financial performance: a general and sector analysis of SP-500 non-financial companies and the moderating effect of economic conditions. J. Sustain. Financ. Invest. **13**(4), 1506–1533 (2023). https://doi.org/10.1080/20430795.2022.2150511
7. Ali, S.A.M., Hassan, R., Othman, A.A.: Al-Musharakah-led Islamic sustainable finance. In: Islamic Finance and Sustainable Development: A Global Framework for Achieving Sustainable Impact Finance, pp. 148–161. Taylor and Francis (2024). https://doi.org/10.4324/9781003468653-17
8. An, E.: Accelerating sustainability through better reporting. Sustain. Account. Manag. Policy J. **14**(4), 904–914 (2023). https://doi.org/10.1108/SAMPJ-08-2022-0453
9. Asl, M.G., Adekoya, O.B., Oliyide, J.A.: Carbon market and the conventional and Islamic equity markets: where lays the environmental cleanliness of their utilities, energy, and ESG sectoral stocks? J. Clean. Prod. **351**, C7-131523 (2022). https://doi.org/10.1016/j.jclepro.2022.131523
10. Assael, J., Carlier, L., Challet, D.: Dissecting the explanatory power of ESG features on equity returns by sector, capitalization, and year with interpretable machine learning. J. Risk Financ. Manag. **16**(3), C7–159 (2023). https://doi.org/10.3390/jrfm16030159
11. Barros, V., Verga Matos, P., Miranda Sarmento, J., Rino Vieira, P.: M&A activity as a driver for better ESG performance. Technol. Forecast. Soc. Chang. **175**, C7-121338 (2022). https://doi.org/10.1016/j.techfore.2021.121338
12. Basyith, A., Fauzi, F., Agusria, L.: Green finance and governance: the effect of climate change. Corp. Bus. Strat. Rev. 5(1), 16–29 (2024). https://doi.org/10.22495/cbsrv5i1art2
13. Bhatia, S., Marwaha, D.: The influence of board factors and gender diversity on the ESG disclosure score: a study on Indian companies. Glob. Bus. Rev. **23**(6), 1544–1557 (2022). https://doi.org/10.1177/09721509221132067
14. Bian, Y., Gao, H., Wang, R., Xiong, X.: Sustainable development for private equity: integrating environment, social, and governance factors into partnership valuation. Bus. Strateg. Environ.. Strateg. Environ. **32**(6), 3359–3370 (2023). https://doi.org/10.1002/bse.3304
15. Bodhanwala, S., Bodhanwala, R.: Does corporate sustainability impact firm profitability? Evidence from India. Manag. Decis.. Decis. **56**(8), 1734–1747 (2018). https://doi.org/10.1108/MD-04-2017-0381
16. Boffo, R., Patalano, R.: ESG Investing Practices, Progress and Challenges. Éditions OCDE, Paris (2020). https://doi.org/10.1787/5504598c-en
17. Brander, J.A., Zhang, W.: Employee relations and innovation: an empirical analysis using patent data. Econ. Innov. New Technol.Innov. New Technol. **26**(4), 368–384 (2017). https://doi.org/10.1080/10438599.2016.1202523
18. Buallay, A.M., Abu Wadi, R.M., Kukreja, G., Hassan, A.A.: Evaluating ESG disclosures of Islamic banks: evidence from the organization of Islamic cooperation members. Int. J. Innov. Sustain. Dev. **14**(3), 266–287 (2020). https://doi.org/10.1504/IJISD.2020.108045
19. Bukhari, S.A.A., Hashim, F., Amran, A.B., Hyder, K.: Green banking and Islam: two sides of the same coin. J. Islam. Mark. **11**(4), 977–1000 (2020). https://doi.org/10.1108/JIMA-09-2018-0154

20. Cai, L., Cooper, R., He, D.: Socially responsible investing and factor investing, is there an opportunity cost? J. Portf. Manag.Portf. Manag. **48**(2), 181–197 (2022). https://doi.org/10.3905/JPM.2021.1.307
21. Chang, Y.L., Ke, J.: Socially responsible artificial intelligence empowered people analytics: a novel framework towards sustainability. Hum. Resour. Dev. Rev.Resour. Dev. Rev. **23**(1), 88–120 (2024). https://doi.org/10.1177/15344843231200930
22. Chen, M., von Behren, R., Mussalli, G.: The unreasonable attractiveness of more ESG data. J. Portf. Manag.Portf. Manag. **48**(1), 147–162 (2021). https://doi.org/10.3905/JPM.2021.1.281
23. Chen, M.H., Wu, Y.H.: The association among ESG external assurance、negative media coverage of social responsibility and cost of debt capital. J. Account. Rev. **78**(1), 1–44 (2024). https://doi.org/10.6552/JOAR.202401_(78).0001
24. Cheng, L.T., Sharma, P., Broadstock, D.C.: Interactive effects of brand reputation and ESG on green bond issues: a sustainable development perspective. Bus. Strateg. Environ.Strateg. Environ. **32**(1), 570–586 (2023)
25. Clementino, E., Perkins, R.: How do companies respond to environmental, social and governance (ESG) ratings? Evidence from Italy. J. Bus. Ethics **171**(2), 379–397 (2021). https://doi.org/10.1007/s10551-020-04441-4
26. Cox, A.: Fit-for-purpose and effective environment, social and governance (ESG) management: ESG implementation challenges, concepts, methods and tips for improvement. In: CSR, Sustainability, Ethics and Governance, pp. 43–58. Springer Nature (2015). https://doi.org/10.1007/978-3-319-10311-2_2
27. D'Amato, V., D'Ecclesia, R., Levantesi, S.: ESG score prediction through random forest algorithm. CMS **19**(2), 347–373 (2022). https://doi.org/10.1007/s10287-021-00419-3
28. Darnall, N., Ji, H., Iwata, K., Arimura, T.H.: Do ESG reporting guidelines and verifications enhance firms' information disclosure? Corp. Soc. Responsib. Environ. Manag.Responsib. Environ. Manag. **29**(5), 1214–1230 (2022). https://doi.org/10.1002/csr.2265
29. Deb, D., Gillet, P., Bernard, P., De, A.: Examining the impact of corporate social responsibility on the financial performance of Indian companies. FIIB Bus. Rev. (2022). https://doi.org/10.1177/23197145221099682
30. Dorfleitner, G., Utz, S., Wimmer, M.: Patience pays off–corporate social responsibility and long-term stock returns. J. Sustain. Financ. Invest. **8**(2), 132–157 (2018). https://doi.org/10.1080/20430795.2017.1403272
31. Dunne, T.C., McBrayer, G.A.: In the interest of small business' cost of debt: a matter of CSR disclosure. J. Small Bus. Strateg.Strateg. **29**(2), 58–71 (2019)
32. Eklund, M.A., Svensson, O.: Executive compensation and corporate governance post COVID-19: the Nordic perspective. In: Accounting, Finance, Sustainability, Governance and Fraud, pp. 53–64. Springer Nature (2023). https://doi.org/10.1007/978-981-19-9079-3_4
33. Fabozzi, F.J., Ng, P.W., Tunaru, D.E. The impact of corporate social responsibility on corporate financial performance and credit ratings in Japan. In: Risks related to environmental, social and governmental issues (ESG), pp. 3–20. Springer Nature (2022). https://doi.org/10.1007/978-3-031-18227-3_2
34. Fahad, P., Rahman, P.M.: Impact of corporate governance on CSR disclosure. Int. J. Discl. Gov.Discl. Gov. **17**(2–3), 155–167 (2020). https://doi.org/10.1057/s41310-020-00082-1
35. Fatimah, Y.A., Kannan, D., Govindan, K., Hasibuan, Z.A.: Circular economy e-business model portfolio development for e-business applications: impacts on ESG and sustainability performance. J. Clean. Prod. **415**, C7-137528 (2023). https://doi.org/10.1016/j.jclepro.2023.137528
36. Frecautan, I., NITA, A. Who is going to win: the eu esg regulation or the rest of the world?—a critical review. Ann. Univ. Oradea, Econ. Sci. Ser. **31**(2), 109–120 (2022)
37. Garcia, A.S., Orsato, R.J.: Testing the institutional difference hypothesis: a study about environmental, social, governance, and financial performance. Bus. Strateg. Environ.Strateg. Environ. **29**(8), 3261–3272 (2020). https://doi.org/10.1002/bse.2570
38. Gidwani, B.: Some issues with using ESG ratings in an investment process. J. Invest. **29**(6), 76–84 (2020). https://doi.org/10.3905/JOI.2020.1.147

39. Girerd-Potin, I., Jimenez-Garcés, S., Louvet, P.: Which Dimensions of social responsibility concern financial investors? J. Bus. Ethics **121**(4), 559–576 (2014). https://doi.org/10.1007/s10551-013-1731-1
40. Gupta, S., Aggarwal, D.: Shrinking the capital costs and beta risk impediments through ESG: study of an emerging market. Asian Rev. Account. **32**(2), 249–277 (2024). https://doi.org/10.1108/ARA-05-2023-0130
41. Habib, A.M., Mourad, N.: The influence of environmental, social, and governance (ESG) practices on US firms' performance: evidence from the coronavirus crisis. J. Knowl. Econ. 1–22 (2023). https://doi.org/10.1007/s13132-023-01278-w
42. Hakimi, A., Boussaada, R., Karmani, M.: Corporate social responsibility and firm performance: a threshold analysis of European firms. Eur. J. Manag. Bus. Econ.Manag. Bus. Econ. (2023). https://doi.org/10.1108/EJMBE-07-2022-0224
43. Halder, A., Batra, S.: Navigating the ethical discourse in construction: a state-of-the-art review of relevant literature. J. Constr. Eng. Manag. **150**(3), C7–03124001 (2024). https://doi.org/10.1061/JCEMD4.COENG-13880
44. Harjoto, M., Laksmana, I., Yang, Y.W.: Why do companies obtain the B corporation certification? Soc. Responsib. J. **15**(5), 621–639 (2019). https://doi.org/10.1108/SRJ-07-2018-0170
45. Hassan, Y., Roychowdhury, S.: Nexus between sustainability management and financial performance—study on manufacturing firms from global emerging market. Int. J. Environ. Work. Employ. **5**(3), 206–219 (2019). https://doi.org/10.1504/IJEWE.2019.103387
46. Herciu, M., Serban, R.A.: Measuring firm performance: testing a proposed model. Stud. Bus. Econ. **13**(2), 103–114 (2018). https://doi.org/10.2478/sbe-2018-0023
47. Hes, T., Srnec, K., Drasarová, M., Neradová, A.: Proposal for establishing an environment, social, and governance (ESG) groundwork: creating a closed system within the microfinance sector. J. Internet Bank. Commer. **17**(1) (2012)
48. Ho, M.: The social construction perspective on ESG issues in SRI indices. J. Sustain. Financ. Invest. **3**(4), 360–373 (2013). https://doi.org/10.1080/20430795.2013.772889
49. Huang, K.J., Bui, D.G., Hsu, Y.T., Lin, C.Y.: The ESG washing in banks: evidence from the syndicated loan market. J. Int. Money Financ. 103043 (2024)
50. Hutagaol-Martowidjojo, Y., Tohang, V., Payung, E.P.T.: The moderating role of policy intervention on the relationship of environment, social, and governance (ESG) and cost of equity capital: a study in basic materials companies in Asia. Australasian Account. Bus. Financ. J. **17**(5), 4–14 (2023). https://doi.org/10.14453/aabfj.v17i5.02
51. Jagyasi, D., Raut, A.R.: Implementation of ESG index on long-term value and performance of organizations using AI and ML. In: 2022 OPJU International Technology Conference on Emerging Technologies for Sustainable Development, OTCON 2022 (2023)
52. Jamali, D.R., Ahmad, I., Aboelmaged, M., Usman, M.: Corporate social responsibility in the United Arab Emirates and globally: a cross-national comparison. J. Clean. Prod. **434**, C7-140105 (2024). https://doi.org/10.1016/j.jclepro.2023.140105
53. Jasuja, D., Prosad, J.M., Nautiyal, N.: Understanding socially responsible investing—a scientific mapping and bibliometric analysis. Int. J. Sustain. Econ. **14**(4), 349–379 (2022). https://doi.org/10.1504/ijse.2022.125981
54. Jin, I.: Systematic ESG risk and passive ESG investing. J. Portf. Manag.Portf. Manag. **48**(5), 71–86 (2022). https://doi.org/10.3905/JPM.2022.1.344
55. Jonsdottir, B., Sigurjonsson, T.O., Johannsdottir, L., Wendt, S.: Barriers to using ESG data for investment decisions. Sustainability **14**(9), 5157 (2022)
56. Kacanski, S.: How environment, social and governance scores impact company financial performance indicators: evidence from Denmark. In: Handbook of Sustainability-Driven Business Strategies in Practice, pp. 338–351. Edward Elgar Publishing Ltd (2021)
57. Kang, A.S., Arikrishnan, S.: Sustainability reporting and total quality management post-pandemic: the role of environmental, social, governance (ESG), and smart technology adoption. J. Asia Bus. Stud. (2024). https://doi.org/10.1108/JABS-03-2022-0080

58. Kartal, M.T., Kılıç Depren, S., Pata, U.K., Taşkın, D., Şavlı, T.: Modeling the link between environmental, social, and governance disclosures and scores: the case of publicly traded companies in the Borsa Istanbul Sustainability Index. Financ. Innov. **10**(1) C7–80 (2024). https://doi.org/10.1186/s40854-024-00619-1
59. Khamisu, M.S., Paluri, R.A., Sonwaney, V.: Environmental social and governance (ESG) disclosure motives for environmentally sensitive industry: an emerging economy perspective. Cogent Bus. Manag. **11**(1), 2322027 (2024). https://doi.org/10.1080/23311975.2024.2322027
60. Khan, M.I., Iqbal, A.: Integrating ESG with corporate investment decision-making. In: The Emerald Handbook of Ethical Finance and Corporate Social Responsibility: A Framework for Sustainable Development, pp. 329–350. Emerald Group Publishing Ltd. https://doi.org/10.1108/978-1-80455-406-720241014
61. Koh, H.K., Burnasheva, R., Suh, Y.G.: Perceived ESG (environmental, social, governance) and consumers' responses: the mediating role of brand credibility, Brand Image, and perceived quality. Sustainability **14**(8), 4515 (2022)
62. Krueger, P., Sautner, Z., Tang, D.Y., Zhong, R.: The effects of mandatory ESG disclosure around the world. J. Account. Res. **62**(5), 1795–1847 (2021)
63. Kumar, K., Kumari, R., Kumar, R.: The state of corporate sustainability reporting in India: evidence from environmentally sensitive industries. Bus. Soc. Rev. **126**(4), 513–538 (2021). https://doi.org/10.1111/basr.12247
64. Kumawat, R., Patel, N.: Are ESG disclosures value relevant? A panel-corrected standard error (PCSE) approach. Glob. Bus. Rev. **23**(6), 1558–1573 (2022). https://doi.org/10.1177/09721509221128637
65. Kuo, I., Rocafort, J., Bland, E., Fish, S.: Bond ladders—unlocking direct indexing opportunities in fixed income. J. Beta Invest. Strat. **14**(3), 150–163 (2023). https://doi.org/10.3905/jbis.2023.1.044
66. Kwon, H.G., Shin, H.: A study on the relationship between ESG performance information and value-relevance: focusing on travel & tourism industry in Korea. J. Glob. Bus. Trade **18**(6), 79–96 (2022). https://doi.org/10.20294/jgbt.2022.18.6.79
67. Lee, D.D., Fan, J.H., Wong, V.S.H.: No more excuses! Performance of ESG-integrated portfolios in Australia. Account. Financ. **61**(S1), 2407–2450 (2021). https://doi.org/10.1111/acfi.12670
68. Lichtenthaler, U.: Sustainability skills and sustainable natives: key competencies and maturity model for sustainability management. J. Innov. Manag. **11**(3), 95–113 (2023). https://doi.org/10.24840/2183-0606_011.003_0005
69. Maiti, M.: Is ESG the succeeding risk factor? J. Sustain. Financ. Invest. **11**(3), 199–213 (2021). https://doi.org/10.1080/20430795.2020.1723380
70. Malik, N., Kashiramka, S.: Impact of ESG disclosure on firm performance and cost of debt: empirical evidence from India. J. Clean. Prod. **448**, C7-141582 (2024). https://doi.org/10.1016/j.jclepro.2024.141582
71. Martellini, L., Vallée, L.S.: Measuring and managing ESG risks in sovereign bond portfolios and implications for sovereign debt investing. J. Portf. Manag.Portf. Manag. **47**(9), 198–223 (2021). https://doi.org/10.3905/JPM.2021.1.290
72. Marti, A., Bastida-Vialcanet, R., Marimon, F.: A systematic literature review: ESG criteria implementation in the insurance industry. Intang. Cap. **20**(1), 193–214 (2024). https://doi.org/10.3926/IC.2426
73. Melchiades Soares, A.: Climate change framing in the communication of CSR policies: the Secil Group example. J. Commun. Manag. **27**(2), 226–240 (2023). https://doi.org/10.1108/JCOM-04-2022-0048
74. Mendiratta, A., Singh, S., Yadav, S.S., Mahajan, A.: ESG controversies and firm performance in India: the moderating impact of government effectiveness. Glob. Bus. Rev. (2023). https://doi.org/10.1177/09721509231151490
75. Mohamed Buallay, A., Al Marri, M., Nasrallah, N., Hamdan, A., Barone, E., Zureigat, Q.: Sustainability reporting in banking and financial services sector: a regional analysis. J. Sustain. Financ. Invest. **13**(1), 776–801 (2023). https://doi.org/10.1080/20430795.2021.1978919

76. Mohanty, S.S., Mohanty, O., Ivanof, M.: Alpha enhancement in global equity markets with ESG overlay on factor-based investment strategies. Risk Manage. **23**(3), 213–242 (2021). https://doi.org/10.1057/s41283-021-00075-6
77. Mulchandani, K., Iyer, G., Lonare, A.: Do equity investors care about environment, social and governance (ESG) disclosure performance? Evidence from India. Glob. Bus. Rev. **23**(6), 1336–1352 (2022). https://doi.org/10.1177/09721509221129910
78. Narang, P.: Fragmentation of community consent standard in the era of friendshoring and nearshoring. Glob. Trade Cust. J. **19**(3), 198–204 (2024). https://doi.org/10.54648/gtcj2024033
79. Newell, G.: Real estate insights the increasing importance of the "S" dimension in ESG. J. Prop. Invest. Financ. **41**(4), 453–459 (2023). https://doi.org/10.1108/JPIF-01-2023-0003
80. Newell, G., Marzuki, M.J.: The increasing importance of environmental sustainability in global real estate investment markets. J. Prop. Invest. Financ. **40**(4), 411–429 (2022). https://doi.org/10.1108/JPIF-01-2022-0005
81. Ngo, T., Le, T., Ullah, S., Trinh, H.H.: Climate risk disclosures and global sustainability initiatives: a conceptual analysis and agenda for future research. Bus. Strateg. Environ.Strateg. Environ. **32**(6), 3705–3720 (2023). https://doi.org/10.1002/bse.3323
82. Ni, J.: Research on compliance risk identification and prevention mechanism of enterprises in the ESG field. Highlights Sci. Eng. Technol. **85**, 1294–1299 (2024)
83. Noja, G.G., Baditoiu, B.R., Buglea, A., Munteanu, V.P., Gligor Cimpoieru, D.C.: The impact of environmental, social and governance policies on companies' financial and economic performance: a comprehensive approach and new empirical evidence. E M Econ. Manag. **27**(1), 121–144 (2024). https://doi.org/10.15240/tul/001/2024-1-008
84. Page, M.J., McKenzie, J.E., Bossuyt, P.M., Boutron, I., Hoffmann, T.C., Mulrow, C.D., Shamseer, L., Tetzlaff, J.M., Akl, E.A., Brennan, S.E., Chou, R., Moher, D.: The PRISMA 2020 statement: an updated guideline for reporting systematic reviews. BMJ **372** (2021). https://doi.org/10.1136/bmj.n71
85. Panait, M., Gigauri, I., Hysa, E., Raimi, L.: Corporate social responsibility and environmental performance: reporting initiatives of oil and gas companies in Central and Eastern Europe. In: Corporate Governance for Climate Transition, pp. 167–186. Springer International Publishing (2023). https://doi.org/10.1007/978-3-031-26277-7_6
86. Pereira, P., Cortez, M.C., Silva, F.: Socially responsible investing and the performance of Eurozone corporate bond portfolios. Corp. Soc. Responsib. Environ. Manag.Responsib. Environ. Manag. **26**(6), 1407–1422 (2019). https://doi.org/10.1002/csr.1756
87. Pisoni, G., Molnár, B.: AI-based solution for sustainability tracing for companies. Int. J. Knowl. Manag. **20**(1) (2024). https://doi.org/10.4018/IJKM.340723
88. Pollman, E.: Corporate social responsibility, ESG, and compliance. In: Daniel Sokol, D., van Rooij, B. (eds.) Forthcoming, Cambridge Handbook of Compliance, (pp. 2019–35). Loyola Law School, Los Angeles Legal Studies Research Paper (2019)
89. Qi, X., Wang, B., Gao, Q.: Environment, social and governance research of infrastructure investment: a literature review. J. Clean. Prod. **425**, C7-139030 (2023). https://doi.org/10.1016/j.jclepro.2023.139030
90. Qu, W., Zhang, J.: Environmental, social, and corporate governance (esg), life cycle, and firm performance: evidence from China. Sustainability **15**(18), 14011 (2023). https://doi.org/10.3390/su151814011
91. Rabhi, F., Yu, M., Ng, A., Lim, E., Tan, F., Hsiao, A.: Building an ESG decision-making system challenges and research directions. In: Advances in Complex Decision Making: Using Machine Learning and Tools for Service-Oriented Computing, pp. 38–51. CRC Press (2023). https://doi.org/10.1201/9781003340621-3
92. Rahman, F.A., Nainggolan, Y.A.: The effect of ESG rating on credit rating. In: Proceedings of the International Conference on Industrial Engineering and Operations Management (2021)
93. Rahmaniati, N.P.G., Ekawati, E.: The role of Indonesian regulators on the effectiveness of ESG implementation in improving firms' non-financial performance. Cogent Bus. Manag. **11**(1), C7–2293302 (2024). https://doi.org/10.1080/23311975.2023.2293302

94. Rastogi, S., Singh, K.: ESG and dividend distribution decisions: evidence of moderation by shareholder activism. J. Glob. Responsib. (2023). https://doi.org/10.1108/JGR-11-2022-0129
95. Rastogi, S., Singh, K., Kanoujiya, J.: Firm's value and ESG: the moderating role of ownership concentration and corporate disclosures. Asian Rev. Account. **32**(1), 70–90 (2024). https://doi.org/10.1108/ARA-10-2022-0266
96. Rehman, A., Ganesan, Y., Haron, H.: What is priority for organization: environmental, social and governance or sustainable corporate governance: literature review. In: Accounting, Finance, Sustainability, Governance and Fraud, vol. Part F1255, pp. 167–191. Springer Nature (2023). https://doi.org/10.1007/978-981-99-3346-4_8
97. Renwick, N.: The belt and road initiative: Infrastructure and sustainable development. In: Sustainable Economy and Emerging Markets, pp. 65–78. Taylor and Francis (2019). https://doi.org/10.4324/9780429325144-5
98. Roncalli, T., Guenedal, T.L., Lepetit, F., Sekine, T.: The market measure of carbon risk and its impact on the minimum variance portfolio. J. Portf. Manag.Portf. Manag. **47**(9), 54–68 (2021). https://doi.org/10.3905/JPM.2021.1.285
99. Rosário, A.T., Dias, J.C.: How industry 4.0 and sensors can leverage product design: opportunities and challenges. Sensors **23**(3), 1165 (2023). https://doi.org/10.3390/s23031165
100. Rosário, A.T., Lopes, P.R., Rosário, F.S.: Metaverse in marketing: challenges and opportunities. In: Khang, A., Shah, V., Rani, S. (eds.) Handbook of Research on AI-Based Technologies and Applications in the Era of the Metaverse, pp. 204–227. IGI Global (2023). https://doi.org/10.4018/978-1-6684-8851-5.ch010
101. Saadaoui, K., Soobaroyen, T.: An analysis of the methodologies adopted by CSR rating agencies. Sustain. Account. Manag. Policy J. **9**(1), 43–62 (2018). https://doi.org/10.1108/SAMPJ-06-2016-0031
102. Sakuma-Keck, K., Hensmans, M.: A motivation puzzle: can investors change corporate behavior by conforming to ESG pressures? Crit. Stud. Corp. Responsib. Gov. Sustain. **5**, 367–393 (2013). https://doi.org/10.1108/S2043-9059(2013)0000005023
103. Saleh, I., Abu Afifa, M., Alkhawaja, A.: Environmental, social, and governance (ESG) disclosure, earnings management and cash holdings: evidence from a European context. Bus. Ethics Environ. Responsib. (2023). https://doi.org/10.1111/beer.12650
104. Scatigna, M., Xia, F.D., Zabai, A., Zulaica, O.: Achievements and challenges in ESG markets. BIS Q. Rev. Dec. (2021)
105. Sethi, S.P., Martell, T.F., Demir, M.: An evaluation of the quality of corporate social responsibility reports by some of the world's largest financial institutions. J. Bus. Ethics **140**(4), 787–805 (2017). https://doi.org/10.1007/s10551-015-2878-8
106. Shamsuddoha, M., Woodside, A.G.: Achieving radical process innovations by applying technology-mindset transformations via second-order system-dynamics engineering. J. Bus. Res. **147**, 37–48 (2022). https://doi.org/10.1016/j.jbusres.2022.04.006
107. Sharma, R.K., Abidi, N., Misra, R.K.: Assessment of agricultural sustainability—a study of farmers growing basmati rice under conventional and fair-trade systems in India. Int. J. Sustain. Agric. Manag. Inform. **6**(1), 94–105 (2020). https://doi.org/10.1504/IJSAMI.2020.106549
108. Shen, Q., Zhu, J., Qi, M., Tao, X.: Can environment, social, and governance performance improve the high-quality development of equipment manufacturing companies? Transform. Bus. Econ. **22**(3), 625–652 (2023)
109. Singh, S., Irfan, M., Saxena, A., Muhammad, A.D.: Framework for circular economy industrial cluster ecosystems to evaluate the sustainability of ESG performance. In: Issues of Sustainability in AI and New-Age Thematic Investing, pp. 240–255. IGI Global (2024). https://doi.org/10.4018/979-8-3693-3282-5.ch015
110. Singh, S., Tejasmayee, P., Singh Gautam, R., Rastogi, S.: Impact of ESG on financial distress in the healthcare sector using debt-equity ratio as a moderator. In: 2024 ASU International Conference in Emerging Technologies for Sustainability and Intelligent Systems, ICETSIS 2024 (2024)

111. Sokolov, A., Caverly, K., Mostovoy, J., Fahoum, T., Seco, L.: Weak supervision and Black-Litterman for automated ESG portfolio construction. J. Financ. Data Sci. **3**(3), 129–138 (2021). https://doi.org/10.3905/jfds.2021.1.070
112. Sood, K., Arijit, K., Pathak, P., Purohit, H.C.: Did ESG portfolio augment investors' wealth during Covid19? Evidence from Indian stock market. Sustain. Account., Manag. Policy J. **14**(5), 922–944 (2023). https://doi.org/10.1108/SAMPJ-02-2022-0087
113. Sorensen, E., Chen, M., Mussalli, G.: The quantitative approach for sustainable investing. J. Portf. Manag.Portf. Manag. **47**(8), 1–12 (2021). https://doi.org/10.3905/JPM.2021.1.267
114. Svanberg, J., Ardeshiri, T., Samsten, I., Öhman, P., Neidermeyer, P.E., Rana, T., Danielson, M.: Must social performance ratings be idiosyncratic? An exploration of social performance ratings with predictive validity. Sustain. Account. Manag. Policy J. **14**(7), 313–348 (2023). https://doi.org/10.1108/SAMPJ-03-2022-0127
115. Syed, A.M.: Environment, social, and governance (ESG) criteria and preference of managers. Cogent Bus. Manag. **4**(1), 1340820 (2017). https://doi.org/10.1080/23311975.2017.1340820
116. Tiwari, R., Sharma, N., Sharma, N.K.: Categorizing and understanding the evolution of literature on ESG investments: a bibliometric analysis. Vision (2023). https://doi.org/10.1177/09722629231197574
117. Todaro, D.L., Torelli, R.: From greenwashing to ESG-washing: a focus on the circular economy field. Corp. Soc. Responsib. Environ. Manag. **31**(5), 4034–4046 (2024)
118. Tokat-Acikel, Y., Aiolfi, M., Johnson, L., Hall, J., Jin, J.Y.: Top-down portfolio implications of climate change. J. Portf. Manag.Portf. Manag. **47**(9), 69–91 (2021). https://doi.org/10.3905/JPM.2021.1.279
119. La Torre, M., Bittucci, L., Paccione, C., Palma, A.: Evaluating the sustainability profile of banks: a comprehensive benchmarking analysis in the Italian context. Bus. Strateg. Environ.Strateg. Environ. **33**(4), 3654–3668 (2024). https://doi.org/10.1002/bse.3673
120. Tripopsakul, S., Puriwat, W.: Understanding the impact of ESG on brand trust and customer engagement. J. Hum. Earth Futur. **3**(4), 430–440 (2022)
121. Tseng, M.L., Chang, C.H., Lin, C.W., Nguyen, T.T.H., Lim, M.K.: Environmental responsibility drives board structure and financial and governance performance: a cause-and-effect model with qualitative information. J. Clean. Prod. **258**, C7-120668 (2020). https://doi.org/10.1016/j.jclepro.2020.120668
122. van Zanten, J.A., Rein, B.: Who owns (un)sustainable companies? Examining institutional determinants of sustainable investing. J. Clean. Prod. **422**, C7-138542 (2023). https://doi.org/10.1016/j.jclepro.2023.138542
123. Vyas, D.P., Oza, S.: Gender representation in ESG communication of Indian companies: observations and insights. In: Infrastructure Development—Theory, Practice and Policy: Sustainability and Resilience: 2021 Conference Compendium, pp. 16–21. Taylor and Francis (2022). https://doi.org/10.4324/9781003311157-4
124. Wan, G., Dawod, A.Y., Chakpitak, N.: A bibliometric and visual analysis in the field of environment, social and governance (ESG) between 2004 and 2021. Int. J. Inf. Sci. Manag. **21**(2), 103–125 (2023). https://doi.org/10.22034/ijism.2023.1977765.0
125. Wang, R., Duan, Y.: Dynamic comparison on the technical efficiency between China's EPEs and PEs: two-dimensional measurement based on SFA. J. Clean. Prod. **406**, C7-136986 (2023). https://doi.org/10.1016/j.jclepro.2023.136986
126. Wang, Q.J., Wang, H.J., Feng, G.F., Chang, C.P.: Impact of digital transformation on performance of environment, social, and governance: empirical evidence from China. Bus. Ethics Environ. Responsib. **32**(4), 1373–1388 (2023). https://doi.org/10.1111/beer.12573
127. Xinbo, N.I., Yuan, F., Gao, H.: Research on the impact of environment, social and governance performance on corporate valuation: evidence from China. Corp. Soc. Responsib. Environ. Manag.Responsib. Environ. Manag. **31**(2), 948–961 (2024). https://doi.org/10.1002/csr.2612
128. Yadav, P., Prashar, A.: Board gender diversity: implications for environment, social, and governance (ESG) performance of Indian firms. Int. J. Product. Perform. Manag.Manag. **72**(9), 2654–2673 (2023). https://doi.org/10.1108/IJPPM-12-2021-0689

129. Yoo, J.S., Song, W.J., Ku, J.E.: CEO turnover, ESG-washing, and firm value. Manag. Decis. Econ.. Decis. Econ. **45**(5), 2801–2819 (2024). https://doi.org/10.1002/mde.4123
130. Yoon, B., Lee, J.H.: Corporate social responsibility and information asymmetry in the Korean market: implications of chaebol affiliates. J. Asian Financ., Econ. Bus. **6**(1), 21–31 (2019). https://doi.org/10.13106/jafeb.2019.vol6.no1.21
131. Yu, W., Gu, Y., Dai, J.: Industry 4.0-enabled environment, social, and governance reporting: a case from a Chinese energy company. J. Emerg. Technol. Account. **20**(1), 245–258 (2023). https://doi.org/10.2308/JETA-2022-014
132. Zaro, E., Flores, E., Fasan, M., Murcia, F.D.R., Zaro, C.S.: Voluntary adoption of integrated reporting, effective legal system and the cost of equity. Corp. Gov. (Bingley) **22**(6), 1197–1221 (2022). https://doi.org/10.1108/CG-03-2021-0096
133. Zhang, S.: The impact of digital transformation on ESG performance and the moderation of mixed-ownership reform: the evidence from Chinese state-owned enterprises. Corp. Soc. Responsib. Environ. Manag.Responsib. Environ. Manag. **31**(3), 2195–2210 (2024). https://doi.org/10.1002/csr.2656
134. Zhang, C., Li, X., Tamoni, A., van Beek, M., Ang, A.: ESG risk and returns implied by demand-based asset pricing models. J. Asset Manag.Manag. (2024). https://doi.org/10.1057/s41260-024-00354-3
135. Zhang, M., Shen, Q., Zhao, Z., Wang, S., Huang, G.Q.: Commitment or rent seeking? Government incentive policies for ESG reporting in sustainable e-commerce logistics. Int. J. Prod. Econ. **268**, C7-109134 (2024). https://doi.org/10.1016/j.ijpe.2023.109134
136. Zrigui, M., Khanchel, I., Lassoued, N.: Does environmental, social and governance performance affect acquisition premium? Rev. Int. Bus. Strat. (2024). https://doi.org/10.1108/RIBS-07-2023-0076

AI and the Next Generation: Protecting Childhood in the Digital Age

Gabriel Osório de Barros and Odete Severino Soares

Abstract Artificial Intelligence (AI) has impacted many aspects of modern life, including in the field of childhood. This chapter aims to explore the complex impact of AI on children, examining both the benefits and the risks. Initially, we summarize ways in which AI enhances educational instruments and learning experiences, by adapting them to individual learners' needs. Subsequently, we explore AI's role in promoting safety and accessibility in children's environments, including the ability to support vulnerable groups of children. We address the impact of AI on children's cognitive, social, and emotional development, and mental health, underlining opportunities to improve these areas while also warning for potential negative effects such as reduced human interaction. We also analyse the main risks, as well as ethical and privacy concerns that are due to the integration of AI into children's everyday lives, play and learning. Particular attention is paid to the balance between technological benefits and excessive dependence. Finally, we reflect on the importance of ensuring that children actively contribute to the development of AI technologies, emphasizing ethical and inclusive use to benefit all children. Children's meaningful participation in shaping AI would be important to close the theory, policy, and practice gaps, challenging the way to increase children's agency and what would that mean for AI policy and practice. Considering how AI can be applied in a child-centric approach, this chapter seeks to contribute to deepen the understanding of the role of AI in the future of childhood.

Keywords Artificial Intelligence · Childhood · Children's rights · Child safety · Education · Ethical concerns · Digital privacy

G. Osório de Barros (✉)
ISCTE-IUL, Lisbon, Portugal
e-mail: m3255@iscte-iul.pt

O. Severino Soares
NOVA School of Law, Universidade NOVA de Lisboa, Lisbon, Portugal
e-mail: odete.soares@novalaw.unl.pt

© The Author(s), under exclusive license to Springer Nature Switzerland AG 2025
A. d. b. Machado et al. (eds.), *Environmental, Social, Governance and Digital Transformation in Organizations*, Information Systems Engineering and Management 35, https://doi.org/10.1007/978-3-031-86079-9_5

1 Introduction

Manuela Eanes, former First Lady of Portugal and founder of the Portuguese Child Support Institute (Instituto de Apoio à Criança—IAC), has long been an advocate for children's rights. Her profound statement, "*A criança é o projecto mais belo da Humanidade*" ("The child[1] is the most beautiful project of Humanity"), summarizes her vision of the importance of children in defining the future. Eanes' dedication to protecting children, particularly through her work with the IAC, highlights the need for safeguarding their rights, namely in today's digital age, where new challenges regarding privacy, security, and digital well-being emerge [30].

Artificial Intelligence (AI) has progressively and increasingly spread through many aspects across society [36], playing varied and significant roles, including in the field of childhood. This chapter provides a thorough review of the opportunities and challenges that AI presents for children, drawing from the literature and the existing international policy initiatives on AI and children's rights, which are more aligned to a certain extent regarding the risks and the opportunities of AI for children, but they differ in terms of goals and priorities [7].

The analysis begins by exploring how AI can enhance educational tools and learning experiences, adapting them to individual needs and promoting inclusive educational environments. It also explores the role of AI in promoting child safety, with advanced monitoring systems and accessibility enhancements for vulnerable groups of children [15].

Conversely, the chapter explores the ethical concerns and risks associated with integrating AI into children's daily lives, particularly through increased interaction with AI-driven applications, devices, and spaces [14]. As AI is integrated into more areas of children's lives, several questions emerged related to such as privacy issues, overreliance on technology, and potential impacts on cognitive, social, and emotional development.

The chapter highlights the importance of balancing the benefits of AI and the growing awareness of the risks presented by AI with the necessity of safeguarding children's rights and wellbeing, following the recommendations as outlined in the existing international policy frameworks and initiatives on AI and children's rights and according to the rights and principles of the United Nations Convention on the Rights of the Child (UNCRC or Convention), adopted in November 1989. In 2021, the United Nations confirmed that the UNCRC applies to the digital environment through its adoption of General Comment No. 25 [47] that provides concrete guidance on this, governments, businesses, and communities take specific actions to ensure that every child has every right in the technologically changing environment children face [7].

In general terms, the existing international policy frameworks and initiatives are aligned to certain extent regarding the risks and opportunities of AI for children but differ in terms of goals, opportunities and priorities. For example, the OECD

[1] CHILD—Any person under the age of 18 is considered a "child" as defined in Article 1 of the United Nations Convention on the Rights of the Child [45]

initiatives concentrate mainly on education [27], while UNICEF has a more generic approach to children‘s rights [43].

In summary, through the review of existing international policy frameworks and initiatives on AI and children’s rights, as well as relevant research and reports, this chapter aims to contribute to increasing the understanding of AI’s dual-edged impact on the future of childhood. It includes key recommendations for policy development, future dialogue, design, and research in this crucial area, seeking to ensure that AI serves as a tool for empowerment rather than harm, promoting and protecting children‘s rights. The need to address children’s rights and best interests *vis-à-vis* AI is thus critical [14].

2 AI in the Context of the United Nations Convention on the Rights of the Child

In the context of “ESG and Digital Transformation in Organizations”, it is essential to understand how the integration of AI into childhood interconnects with the environmental, social and governance (ESG) principles. With the organizations and the society increasingly adopting AI technologies, they have a fundamental role in ensuring that these tools protect the vulnerable groups of children and support the opportunities AI provides children while mitigating its risks.

The current scientific evidence about the impact of AI-based applications on a child’s development and the kind of opportunities and risks they bring is still limited. Even less explored is how the rights of the child, as defined in the UNCRC and the General Comment No. 25, can be realised in the AI context [20].

Technological advances have radically changed the world in which children develop, play and learn. The rapid development of digital technologies, particularly AI, presents both exceptional opportunities and risks for the promotion and protection of children’s rights. Thus, one of the fundamental concerns of this chapter is to analyse the extent to which children‘s rights, established in the UNCRC, can be employed as an effective benchmark in the various AI technology-related frameworks, as a starting point to establish a timely child-centric and rights-based framework for a responsible AI and children, that is applicable across technological devices and innovations and adaptable to diverse contexts of childhood, in areas as education, healthcare and personal development [47].

The UNCRC is a legally binding international agreement setting out the civil, political, economic, social and cultural rights of every child, regardless of their race, religion or abilities. The Convention has become the most widely ratified human rights treaty in history, helping transform children’s lives around the world.

While the rights and principles established in the UNCRC are applicable to the digital world, the United Nations, in 2021, reinforced this by adopting General Comment No. 25, following extensive consultations with children and other stakeholders.

The General Comment No. 25 provides concrete guidance on this, governments, businesses and communities systematically take specific actions to ensure that every child has every right in the technologically changing environment children face. It considers children's possible interactions within digital networks, content, services and applications, connected devices and environments, virtual reality (VR) and augmented reality (AR), AI, robotics, automated systems, algorithms and data analytics, biometrics, and implant technology and invites governments to take the appropriate actions in order to mitigate the possible risks and ensure that the new opportunities for the realization of children's rights are accessible to all children in the digital world [7, 31].

This Comment elaborates on the four basic principles of the Rights of the child namely, non-discrimination; best interest of the child; right to life survival and development; and respect for the views of the child. It proposes a set of general measures of implementation by States parties, such as the review and possible adoptions, amendments, and updates in the national legislation in line with the international human rights standards and the implementation of national policies relating to children's rights to specifically address the evolving digital environments [32, 45].

The General Comment No. 25 invites governments to implement regulation, industry codes, standards and plans to provide children with beneficial opportunities from engaging with the digital environment in a safe manner. The document indicates that the cross-border and transnational nature of the digital environment requires strong international and regional cooperation to ensure that all stakeholders, including businesses and other actors, effectively respect, protect, and fulfil children's rights in relation to the digital environment [15, 47].

Additionally, the General Comment No. 25 recognizes that in developed countries, the digital world is part of children's everyday life, and the global community has started to understand the challenges and the opportunities digital environments bring to children. However, it indicates that special attention should be paid to AI-based technologies that have been shown to change fundamentally children's interaction with the digital and physical world [45].

For example, the AI used in educational backgrounds should be intended to serve different children's needs, providing fair opportunities for learning without any form of discrimination. Additionally, the use of AI in the collection, monitoring or management of data should prioritize the child's right to privacy and dignity, guaranteeing that children's data is used ethically, wisely and in line with their best interest—Article 3 of the UNCRC [45].

Additionally, AI increasingly plays a role in how children's rights are understood, protected, and implemented. As AI becomes a tool through which children's rights are mediated (e.g., AI mediates the right to education by providing personalized learning experiences or by identifying children with special educational needs), the principles of participation and inclusion become essential. Article 12 of the UNCRC, which gives children the right to express their views freely in matters affecting them, does not refer this subject specifically, considering the time it was written, but underscores the need for child-centric policies that not only listen to but also incorporate the opinions and perspectives of children.

In integrating the UNCRC with AI, society must take into consideration how digital platforms and AI technologies can be leveraged to foster the wellbeing of all children, guaranteeing that no child is left behind. Through this chapter, we will explore these intersections, highlighting the opportunities, the risks and the necessity for policies that harness the benefits of AI in upholding children's rights and mitigate the threats that these technologies may create.

2.1 The Prevalence of AI in Contemporary Society

The importance of AI in contemporary society, in recent years, is unquestionable. AI technologies are quickly entering in every aspect of our lives, in areas such as health, education, entertainment or social interactions. These technologies increasingly influence how people—particularly children—learn, communicate, and engage with the world around them [1, 9, 49]. The widespread use of AI means that, more than just a tool, it is shaping the next generations, often in ways that are both profound and wide-ranging.

As AI becomes more integrated in our daily routines and activities, the impact on childhood is becoming increasingly relevant. Nowadays, children interact with AI through social media algorithms, virtual assistants, and educational platforms that select and organize content based on user data [16, 17, 49]. This increasing integration raises serious questions about how AI technologies influence children's development, behaviour, and well-being. The presence of AI also increases the need for greater attention to the ethical implications and the potential risks associated with the use of those technologies in environments where children are exposed to them.

2.2 Rationale for Investigating AI's Impact on Childhood

Considering the growing adoption of AI, it is essential to analyse its impact on childhood, recognising that children are one of the most vulnerable groups of the population and are particularly exposed to both the benefits and the risks associated with digital technologies, including AI technologies [38].

If it is true that AI has the potential to improve learning experiences, expand access to information, and offer tailored support, it also true that it poses important risks, including the exposure to inappropriate content, threats to privacy, and the erosion of critical thinking.

Understanding the impact of AI on childhood is essential because children are in a critical developmental stage where their social, emotional, and cognitive development is highly influenced by their environment, including the technologies they interact with [16, 17, 49]. The integration of AI into learning and social environments can influence their developmental in ways that, today, are not fully understood. Hence, it is essential to study both the positive and the negative consequences of the use of such technologies.

Secondly, researching the impact of AI on childhood is in line with the principles set forth in the General Comment No 25. In this regard, it highlights the need to protect children from potential harm and to ensure that their rights are upheld in all contexts, including digital environments and, nowadays, particularly AI [9, 18, 47]. Only by analysing the impact of AI in children's rights, namely the right to privacy, to education, and to participation, the society can design strategies to guarantee that those technologies are used to benefit (rather than harm) children's rights. This is particularly important considering that most national AI strategies do not sufficiently address the impact of AI on children, especially regarding rights such as protection from discrimination, abuse, and exploitation [42].

Finally, as AI is not immutable, on the contrary it is evolving every day, it will probably become even more relevant in children's lives in the future. Only exploring and addressing today's implications of AI on childhood we can create frameworks and policies that promote the best use of AI technologies, guaranteeing that they contribute in a positive way to children's development [26, 49].

In conclusion of Sect. 2, this chapter analyses the crossroads between AI and childhood, which are aligned with ESG principles, underlining the importance of balancing the opportunities that are presented by AI technologies with the need to protect children's rights. To work towards a future where AI technologies are tools for safety, protection, and empowerment, and to guarantee that they don't become sources of risks for the protection of children's rights, it is crucial to align AI practices according to the rights and principles outlined in the UNCRC.

3 AI in Education

Although few countries are already integrating AI in their education systems [41], the integration of AI into the educational sector, including both formal (school-based programming) and informal (community-based programming) education, is deeply transforming traditional teaching and pedagogical methods, characterised by teacher-centred instruction, offering advanced tools that improve learning in different contexts.

The influence of AI is particularly obvious when we consider its capacity to provide personalized and adaptive educational contents and resources that accommodate the needs of different students, ensuring that each student can learn and reach their full potential.

3.1 Leveraging AI for Personalized Learning

By enabling the development of systems that can analyse individual learning patterns, preferences, and development in real-time, AI is transforming personalized learning [25, 32, 41]. These systems adjust in a dynamic way the educational contents, providing students with adapted learning processes that are adapted to their specific requirements and capabilities [37]. AI-driven platforms, for example, utilize machine learning algorithms to accompany and monitor student's performance, providing tailored feedback and recommendations that can be useful to improve understanding and retention of information and knowledge, including for children with disabilities [5, 49, 51]. This is particularly relevant considering that children with disabilities less likely to complete primary school than the average child [41].

One of the most important applications is the use of technology in educational tools/equipment, allowing the assessment of student's engagement and comprehension. Those technologies can have great importance and be particularly effective regarding the creation of learning methods in which the students are immersed into a virtual dialogue, with the AI adapting the content in response to the learner's interaction, making education more attractive, more engaging, and more effective [16, 17, 51].

AI has the potential to revolutionize elementary education by providing personalized learning experiences tailored to individual student needs, promoting creative engagement through interactive and gamified content, and fostering digital literacy skills, although nowadays more than 50% of the countries have no digital skills standards in their education systems [41]. AI-powered tools support educators by offering supplementary tutoring resources, allowing them to focus on individualized instruction, while also helping students develop foundational skills in subjects like math and coding [10].

3.2 Adaptive Educational Tools and Platforms

The role of AI goes beyond the referred personalization. It also includes the creation of adaptive educational environments that can adjust to the varying needs of students. These adaptive systems are designed to provide feedback, which is essential for maintaining and increasing student's engagement and to ensure that the learning objectives defined are achieved. For instance, platforms powered by AI can help in the identification of students struggling with a particular subject, problem or concept and can offer additional resources or alternative explanations, preventing learning gaps [3, 5, 17, 49].

In the context of special education, meaning the method of educating students in a way that meets their individual differences, AI has the potential to support children with learning disabilities. Resorting to the use of adaptive learning technologies, educators/teachers can provide tailored educational experiences that adjust to the specific challenges that are faced by those students [25]. Additionally, AI systems are able to support the monitoring of progress, making the required adjustments to the learning plan (includes what, how and when will be learned, as well the criteria and validation used to evaluate the learning), aiming to guarantee that all students can succeed according to their potential [5, 16, 17, 51].

3.3 Interactive and Immersive Learning Environments

Another disruptive influence of AI in education is the development of interactive and immersive learning environments, with AI creating simulations and gamified learning experiences that aims to make complex subjects more accessible. Examples of this are AI-driven VR and AR platforms that provide students with hands-on learning experiences. These new approaches would be difficult to replicate in a traditional classroom setting. Those technologies are not only able to improve engagement but can also enhance students' ability to retain and apply the subjects they have learned [3, 17, 49, 51].

In addition to conventional educational backgrounds, the referred immersive technologies are nowadays being used in specialized contexts, namely the education of children with autism spectrum disorder. Immersive environments that are driven by AI are being developed to provide safe and controlled spaces where children can engage in social interactions and develop skills that are challenging for them in real-world scenarios. These scenarios are aimed to be adjustable to the individual needs of each child, providing a level of personalization that is essential to guarantee that they can effectively evolve in the learning process [5, 16, 51].

3.4 Supporting Vulnerable Groups of the Population

Another significant impact of AI is the capacity to support vulnerable groups of the population, namely children, such as migrant/refugee children. Barriers in terms of language and culture can create considerable difficulties to the learning process of these children. Nevertheless, educational tools driven by AI can help in reducing those gaps by providing adapted language education resources, tailored to the specific needs of each children/student. These tools can evaluate each child's current language ability and offer targeted lessons that are based on their already existing knowledge but aiming to facilitate effective and smoother integration in the educational system [8, 17].

3.5 Ethical Considerations

Although there is a perspective that AI can offer benefits in educational contexts, it also raises significant ethical concerns, in particular regarding data privacy and the psychological impact of the use of AI by younger learners. The collection and analysis of student data, considering that data is a key factor of any AI system, must be operated and controlled with the maximum caution to protect the privacy and the rights of children, considering that only 16% of the countries guarantee data protection in education in their legislation [41, 44]. Furthermore, robust ethical principles are needed to ensure that AI-driven educational tools are beneficial rather than harmful to students. These considerations are crucial for maintaining the trust of society (students, parents, and educators) in the use of AI in education [9, 16, 26, 38].

To address these concerns, considering different levels of risk posed to education, the European Union, in the AI Act [13], considers:

- Among the prohibited AI systems, those that are intended "to infer emotions of a natural person in the areas of workplace and education institutions, except where the use of the AI system is intended to be put in place or into the market for medical or safety reasons";
- Among the high-risk AI systems, the following in the context of education and vocational training that are intended to be used:
  - "to determine access or admission or to assign natural persons to educational and vocational training institutions at all levels";
  - "to evaluate learning outcomes, including when those outcomes are used to steer the learning process of natural persons in educational and vocational training institutions at all levels";
  - "for the purpose of assessing the appropriate level of education that an individual will receive or will be able to access, in the context of or within educational and vocational training institutions at all levels"; and
  - "for monitoring and detecting prohibited behaviour of students during tests in the context of or within educational and vocational training institutions at all levels".

3.6 Preventive Health Education

AI is also playing a valuable role in promoting health education among children, especially in areas where early intervention can have a substantial impact. For instance, some AI tools are being used to spot early signs of health problems, such as dental caries [35], and provide interactive educational content that helps children and their families in taking preventive measures. By using AI in health education, we are going beyond the traditional boundaries of academic learning, exploring the potential to contribute to the overall well-being of students and helping students in staying healthy and thriving in all areas of their lives [33, 50].

Concluding Sect. 3, the incorporation of AI in education represents an important improvement in how educational content is provided and in how students engage with that content. By providing personalized, adaptive, and inclusive educational tools, AI has the potential to make learning more accessible and effective, regardless of the students' individual needs or contexts.

4 Child Safety and Well-Being

The application of AI in child safety is transforming how we monitor, protect, and increase the accessibility of resources for children. AI technologies provide advanced mechanisms not only for the safeguard of children in digital and physical environments, but also to contribute to the creation of safer and more supportive environments for their development.

4.1 Monitoring and Protection Mechanisms

AI systems are increasingly being used to monitor children's activities, not only online but also offline, providing evaluation and warning to potential threats. These systems can detect inappropriate or harmful content, unsafe interactions, cyberbullying, predatory behaviour, and other online threats to ensure that children are protected. This is particularly relevant when considering that over 10% of primary students in most middle-income and high-income countries have experienced cyberbullying [41]. Nonetheless, only 16% of countries have adopted legislation to prevent and act on cyberbullying focused on education, a growing issue as children increasingly engage with digital platforms for education [41].

Technologies driven by AI are improving the ability to monitor and protect children from various threats. These systems can also monitor patterns of behaviour, identifying signs of distress or dangerous behaviour that may indicate a need for intervention [1, 26, 38].

Again, it is important to recall that AI technologies may be effective but raise concerns about privacy and autonomy. It is essential to balance the need for protection with respect for children's rights, ensuring that AI systems are designed and implemented to guarantee that they do not infringe on children's privacy or freedom. Hence, ethical frameworks are essential to guide the deployment of those technologies, to ensure that they serve the best interests of the child [1, 9, 38, 47].

One of the most important applications regards the development of surveillance systems, which can detect and react to signs of abuse, neglect, or other forms of harm. For instance, those technologies can analyse large volumes of data from social media platforms and online games to recognise patterns of harmful behaviour, to allow timely intervention. Through real-time monitoring of online interactions and content, AI systems can flag harmful behaviour and content, alerting administrators, moderators, or parents to act accordingly.

AI is increasingly used to moderate online content on platforms like social media to address cyberbullying and abuse, with platforms integrating Natural Language Processing (NLP), machine learning, and deep learning to proactively detect harmful content [23]. In this sense, AI allows for proactive moderation, identifying harmful content before it is reported (contrasting with reactive moderation, where content is flagged by users), although detecting sarcasm, irony, and context-based abuse is difficult for AI.

This approach is essential to prevent harm before it rises, thus providing a safer online experience for children regarding digital contact and engagement [1, 9, 26, 38].

Additionally, monitoring tools that use AI can be employed in physical spaces such as schools and playgrounds, where they can trace unusual activities or unauthorized access. These systems can be equipped with the abilities to recognize faces and to analyse behaviour, enhancing security measures and protecting children from potential physical harm [1, 26, 38, 47].

To address this problem, in April 2024, several leading AI companies, including OpenAI, Microsoft, Google, and Meta, joined an initiative led by Thorn and All Tech Is Human to protect children online. This initiative aims to prevent AI tools from being used to generate or distribute CSAM. The initiative's goal is to enforce the integration of safety-by-design principles in AI technologies, ensuring that children are safeguarded from online exploitation. Thorn emphasizes the importance of selecting AI training datasets carefully to avoid mixing CSAM with adult content, as AI systems have the potential to combine harmful elements [34].

Additionally, global cooperation for child protection is also crucial. The WePROTECT Global Alliance unites over 150 governments, tech companies, and civil society organizations to combat online child exploitation through coordinated actions, shared resources, and policy harmonization [2]. This initiative highlights the importance of collaborative efforts at the international level, ensuring AI tools for child protection are standardized and effective globally.

4.2 Accessibility Improvements

AI can also play an important role in improving the accessibility for children, considering in particular those with disabilities or special needs (e.g., physical, cognitive, or sensory impairment). By using AI-driven technologies, educational institutions can provide adapted learning that tailor to the specific needs of each child. For example, AI can be used to create personalized learning plans that adapt to children with autism spectrum disorder, providing them tools and resources that are specifically designed to support their learning process in a way that traditional methods are not able to [5, 16, 51].

AI-based applications are also being developed to improve communication and interaction for children with speech and language difficulties. These tools use NLP to interpret and reply to children's speech in real-time, facilitating communication and reducing the frustration frequently experienced by children with such impairments [16, 17, 33].

Moreover, AI is also playing a fundamental role in creating more comprehensive environments, as already referred, by personalizing learning and communication tools to respond to the specific needs of each child, including the use of adaptive learning platforms that adjust content delivery based on the child's progress and interaction, ensuring that all children, regardless of their abilities, can engage, fully, safely, and properly in their educational process [5, 17, 49].

AI enhances accessibility for children, particularly those with disabilities or special needs. For example, educational tools using AI can be customized to be adapted for children with visual, hearing, communication, learning or motor impairments [41], ensuring that they can access and benefit from educational content on an equal footing with their peers, so that all children have an equal chance to succeed. The AI's ability to adapt interfaces and content to individual needs can be very important in making digital environments more inclusive [5, 17].

Additionally, AI tools that assist children in using the digital space safely are becoming increasingly important, not only in helping children to access appropriate content, but also educating them about digital literacy and empowering them to make informed decisions about their online activities [26, 38].

4.3 Child Safety

AI also contributes to child safety through preventive health measures. AI technologies are increasingly being used in healthcare to monitor and predict potential health issues among children, such as through early detection systems that alert parents and caregivers to the inception of dental problems or other health concerns. These systems not only ensure timely medical intervention but also promote a complete approach to child, providing personalized interventions or alert caregivers to search for professional help, thus addressing mental health concerns before they become critical [5, 16, 50].

Additionally, AI's role in predictive analytics is being leveraged to identify children at risk of school dropping out or falling into patterns of absenteeism. Analysing factors such as attendance records, school performance, and social interactions, AI can help educators and policymakers in developing targeted interventions that maintain children engaged [8, 49].

4.4 Ethical Considerations

The ethical implications of AI on development justifies attention, as children increasingly interact with AI, there are concerns about data privacy, autonomy, and the potential for AI to shape children's worldviews in ways that are not always transparent or ethical. Ensuring that AI is used trustworthy, with safeguards to protect children's rights, is crucial for promoting healthy development [16], Child Rights International Network, 2022; [38].

The use of AI in protecting children has introduced innovative monitoring and protection mechanisms, enhanced accessibility, and addressed safety concerns. Although AI is offering important benefits in educating and in enhancing child safety, it also raises important ethical implications and potential risks. The use of AI in monitoring children, both online and in physical spaces, must be balanced with the need to protect their privacy and autonomy. Nowadays, there is an increasing acknowledgement that AI systems should be designed with ethical safeguards to ensure that data is used responsibly, and that children's rights are protected in all applications of the AI technologies [1, 26].

Ethical AI development for children must consider a child-centric design approach that aligns with established frameworks like the OECD AI Principles [4, 26]:

- These principles emphasize the importance of safety, security, and privacy by integrating robust data protection mechanisms, ensuring age-appropriate content filtering, and avoiding risks to children's privacy;
- In conjunction with the child rights by design framework, transparency and explainability play crucial roles, ensuring children and their families receive clear, understandable information about how AI systems function;
- Additionally, inclusivity and non-discrimination must be foundational in AI design to ensure fair treatment and representation of all children, regardless of their background, with models trained on diverse datasets;
- Furthermore, ethical AI systems for children should be built around human-centred values, respecting children's autonomy and dignity, while ensuring accountability and oversight through strict governance, regular audits, and avenues for redress;
- Finally, fostering digital literacy is essential, empowering children to interact safely with AI technologies.

The main goal is to create AI systems that not only ensure safety by design but also actively contribute to children's well-being and development, mitigating potential biases or exploitation, aligning with the growing consensus that ethical safeguards must be integrated in AI systems.

Concluding Sect. 4, the integration of AI in child safety measures is changing the way society monitors, protects, and ensures the well-being of children in various environments. Through innovative and responsive technologies, from monitoring and protection mechanisms, AI offers tools for safeguarding children, enhancing accessibility, and addressing critical safety concerns. However, this requires that the society keeps in mind the developmental perspective and the importance of privacy and ethical concerns, ensuring that such monitoring does not become intrusive or violate the rights of the child.

5 Developmental Perspectives

The integration of AI technologies in children's lives brings opportunities and challenges regarding their cognitive, social, and emotional development and mental health. While AI has the potential to enhance learning and developmental, it also raises concerns about possible effects on critical aspects of childhood development, particularly in relation to the importance of human interaction.

The role of AI is progressively influencing the cognitive, social, and emotional development of children. As AI-driven technologies become more integrated into educational models and in the daily life, it is essential to consider how these technologies shape children's development and the importance of maintaining human interaction along with AI.

5.1 Cognitive Development

AI's prevalent presence in children's lives influences their cognitive development in both positive and negative ways. While AI can enhance learning experiences and social skills through interactive and personalized content, as already referred, it can also hinder development if not used appropriately. Overexposure to AI-driven tools may lead to a reduction in face-to-face interactions, which are essential in the process of development of social skills such as empathy, communication, and emotional intelligence [1, 16, 17, 49].

As already referred, AI technologies have been increasingly used to support cognitive development of children by providing tailored and interactive learning experiences. For example, AI-powered educational tools can adapt to a child's learning pace, providing tailored challenges and exercises that promote cognitive development [49].

These tools are particularly useful to reinforce knowledge and skills in areas such as language acquisition, mathematics, and problem-solving [32], offering feedback in real-time and providing appealing content that stimulates intellectual curiosity and cognitive development.

For example, AI-powered educational tools can help children develop critical thinking skills by presenting challenges that adjust in difficulty based on their progress [25]. This continuous feedback encourages deeper cognitive engagement and promotes critical thinking [16, 17, 49].

Moreover, AI's ability to provide information in different formats—e.g., visual or audio formats—ensures that children with different learning skills can understand

and learn contents more effectively. This capacity to adapt supports cognitive development by allowing children to engage with material in ways that are aligned with their strengths, thereby enhancing understanding and retention [5, 16, 49].

However, the reliance on AI for cognitive development also raises concerns. There is a risk that overexposure to AI learning tools could lead to a decrease in critical thinking if children become excessively dependent on technology to provide answers, rather than developing those skills autonomously [1, 16]. Moreover, the quality of AI interactions may lack the depth and richness of human interaction, potentially inhibiting cognitive development if it is not balanced with human-guided learning experiences [16, 29].

5.2 *Social Development*

The role of AI in social development is complex. On the one hand, AI technologies, such as social robots and interactive apps, can provide children with opportunities to practice social skills, especially for those with developmental challenges like autism [5, 17, 51]. These AI systems can be programmed to simulate social scenarios, helping children to face social interactions in a controlled environment. For instance, AI-driven simulations and role-playing games can help children practice social interactions, which is particularly positive for children with social difficulties in integration, such as those on the autism spectrum, who may struggle with real-world social scenarios. By providing a safe space to explore social behaviours, AI can complement traditional social learning methods [16, 51].

Conversely, there are relevant concerns that AI may damage social development by reducing the human interaction that children experience. Human interaction is very important for developing empathy, emotional intelligence, and communication skills—qualities that are best developed through real-world social experiences [1, 9, 16, 19]. Excessive use of AI-driven technologies may isolate children from their colleagues and caregivers, leading to deficits in these essential areas of social development.

Hence, in environments where AI is broadly used, there is a risk that children may miss essential social learning opportunities. This lack of interaction can hamper the development of crucial social skills, such as cooperation, teamwork, conflict resolution, and the ability to exert control over their own emotional state [9, 16, 19].

Human interaction remains essential for healthy cognitive, social, and emotional development. Human teachers, parents, and colleagues provide emotional support, cultural context, and ethical guidance that AI cannot replicate. These interactions help children develop a sense of belonging, understand social norms, and build emotional intelligence.

Research shows that children learn best in environments where they can interact with adults and colleagues, engaging in conversations, collaborative play, and other activities that promote social connections and emotional development [9, 16, 18]. AI should, therefore, be seen as a complement to, rather than a replacement for, those human interactions.

Ensuring that AI is integrated in a way that supports rather than replaces human interaction is vital. For example, AI can be used to augment the educational process by providing teachers with tools that help them better understand their students' needs, allowing for more tailored and focused interactions between teachers and students [16, 49].

5.3 Emotional Development and Mental Health

In addition to cognitive and social development, AI's impact on emotional development is another critical area of concern. AI systems are not yet capable of recognizing, understanding, or responding to emotions in the detailed and complex ways that humans can. As such, children who engage significantly with AI may miss developing the full range of emotional skills, such as empathy and emotional regulation (control over their own emotional state), which are typically cultivated through human interaction [1].

Furthermore, there is an ongoing debate about the long-term impacts of AI on children's mental health. While AI can provide benefits such as early detection of mental health issues through behavioural monitoring, it can also contribute to stress and anxiety if not controlled adequately. The ethical use of AI in monitoring and supporting children's mental health requires careful consideration to avoid unintended negative consequences [1, 5, 16].

Concluding Sect. 5, while AI offers promising tools for cognitive, social, and emotional development, and mental health, it is essential to incorporate those technologies with much care to ensure that children receive a balanced education that includes both technological and human elements. This complete approach will better support children's development and prepare them for the complexities of the modern world.

6 Risks of AI in Childhood: Protection, Privacy and Ethical Considerations

As AI becomes increasingly integrated in children's daily lives, it is essential to analyse the potential risks associated with its use. While AI offers significant benefits in education and development, as already stated, its influence also poses various dangers, including exposure to manipulative content, risks of cyberstalking, loss of critical thinking and decision-making skills, and negative impacts on well-being and social skills.

While some national strategies focus on preparing children for the future workforce by promoting science, technology, engineering, and mathematics (STEM) education and digital literacy, there is a lack of emphasis on teaching the ethical and privacy aspects of AI and developing critical thinking skills [42]. Additionally, just about 35% of tertiary STEM graduates (post-secondary education) are women [41]. With the integration of AI in many aspects of children's lives, the ethical and privacy implications associated with its use are receiving increased attention.

This section deepens the analysis of these risks and highlights the ongoing debate surrounding AI's role regarding data protection, privacy, the potential for manipulative content and advertising, the dangers of cyberstalking, the risk of overreliance on AI, and the impact of AI on children's mental well-being and social skills.

6.1 Child Abuse

The proliferation of digital technologies has intensified the distribution of child sexual abuse material (CSAM), posing severe challenges to child protection efforts. AI-generated CSAM has become a growing issue, with AI tools increasingly being used to create realistic images and videos of abuse, including deepfakes [15]. This highlights a significant risk for online child protection. Although the misuse of AI for generating CSAM represents a small (but growing) proportion of content, indicating that the broader potential of AI is positive, specific harmful uses must be addressed with urgency [15].

Despite the implementation of several legal instruments, as such the Optional Protocol of the UNCRC on the sale of children, child prostitution, and child pornography, the Council of Europe Convention on Protection of Children against Sexual Exploitation and Sexual Abuse, also known as "the Lanzarote Convention" [11], and the European Directive and initiatives [12], there is a growing concern that the existing legal frameworks focus too heavily on criminalization without adequately addressing prevention.

This gap has allowed CSAM to become more accessible and widespread than ever before [15], highlighting the urgent need for strategies that prioritize prevention and safeguard children's rights in the digital age. An effective response requires not only severe law enforcement but also child-centred initiatives that prevent exploitation and protect children from harm, including in digital environments [22, 47].

6.2 Data Protection and Privacy for Children.

One of the most important concerns in the use of AI technologies for children is the protection of their personal data. Children are particularly vulnerable to privacy violations because they may not fully understand the long-term implications of data sharing. AI systems require large amounts of data to work effectively. When used in educational or entertainment settings, this can include sensitive information about a child's location, behaviour, and preferences [1, 26, 47].

Ensuring that AI systems comply with stringent data protection regulations, such as the European Union General Data Protection Regulation (GDPR), is essential to safeguard children's privacy. The GDPR recognizes that children need special protection regarding their personal data, and it contains provisions aimed at securing the processing of their data and to ensure that children understand and can exercise their data protection rights. Additionally, there is a growing need for transparency in how children's data is collected, used, and stored, as well as for strong mechanisms that allow parents to control and monitor data usage [1, 9, 26, 47].

6.3 Inadvertent Sharing of Personal Information

AI's capacity to track and predict user behaviour introduces substantial risks related to cyberstalking and digital predation. Children who interact with AI-driven platforms may accidentally or inadvertently share personal information that could be exploited by predators. AI tools that monitor online activity can also be misused to track children's location and behaviours, increasing the risk of exploitation [9, 20, 26, 38, 47].

In this subject, the ethical imperative is to design AI systems that prioritize the safety and security of children, incorporating features that prevent unauthorized tracking and alert authorities to suspect activities.

6.4 The Debate on Surveillance and Autonomy

The use of AI in monitoring and surveillance has triggered an important debate regarding children's autonomy and privacy. While surveillance technologies can protect children by preventing harmful behaviours and ensuring their safety, they also raise concerns about the loss of privacy and the impact on children's autonomy. Continuous monitoring can create an environment of control and surveillance that may hinder children's ability to explore, learn, and develop independently. The ethical implications of using AI for surveillance purposes must be carefully balanced against the benefits, ensuring that children's rights to privacy and autonomy are respected [1, 38, 47].

6.5 Manipulative Content

One of the most pressing concerns regarding AI's interaction with children is its potential to expose them to manipulative content. AI algorithms, especially those used in social media and online gaming, are designed to capture, and retain users' attention, often by promoting content that produces strong emotional responses. This can lead to children being exposed to inappropriate or manipulative content that shapes their beliefs, behaviours, and values in harmful ways [1, 9, 38, 44]. The lack of adequate supervision and regulation in these AI-driven environments exacerbates the risk, making children vulnerable to content that may not align with their developmental requirements.

6.6 Dangers of Cyberstalking

More than 6.8% of children in OECD countries reported having problematic social media use in 2017–2018 [24]. The rise of AI-powered tools has also introduced new threats to children's online safety, particularly in the form of cyberstalking. AI-driven social media platforms can accidentally enable predators to track, monitor, and interact with children in the absence of adequate safeguards. The ability of AI to analyse and predict user behaviour increases the risk of targeted cyberstalking, where perpetrators can exploit the detailed data profiles generated by these platforms to manipulate or harm children [1, 9, 26, 38]. Hence, to mitigate these risks, it is essential to ensure that robust protective measures are integrated into AI systems.

6.7 Loss of Critical Thinking and Decision-Making Skills

Another important risk associated with the use of AI in childhood is the potential loss of critical thinking and decision-making skills. As children become more dependent on AI for information, there is a danger that they may lose the ability to think independently and critically evaluate the information they receive. AI systems often present information in a way that appears convincing, which can lead to passive acceptance rather than active questioning.

In summary, this over-reliance on AI can diminish children's ability to develop essential cognitive skills, such as problem-solving and critical analysis, which are crucial for their overall intellectual growth [1, 16, 49].

6.8 Social Skills and Human Interaction

The emotional impact of AI on children is another area of concern. AI can provide personalized support that promotes mental well-being, such as through educational apps designed to reduce stress or anxiety. Conversely, constant interaction with AI-driven systems can lead to social isolation and a reduction in face-to-face interactions, which are vital for the development of social skills [1, 16, 49]. Moreover, the impersonal nature of AI interactions may fail to provide the emotional support that children need, potentially exacerbating feelings of loneliness or depression. Ethical considerations must therefore include the development of AI systems that promote positive social interactions and that do not replace the human connections that are essential for healthy emotional development [1, 16].

The constant interaction with AI-driven platforms can lead to problems such as anxiety, depression, and social isolation, particularly if children substitute real-world interactions with digital ones [16]. AI's role in promoting addictive behaviours, such as excessive screen time, also contributes to these issues, negatively affecting children's mental health and social development [1, 9, 16].

Despite the advances in AI, human interaction remains irreplaceable in child development. Children learn and grow through social interactions that involve empathy, exchange of emotions, and real-world experiences. AI, notwithstanding all its complex capabilities, cannot replicate the depth and quality of human relationships. It is crucial that AI is used to supplement, rather than replace, human interaction, ensuring that children continue to develop the social and emotional skills necessary for their well-being [16, 19, 49].

The lack of human empathy and emotional understanding in AI interactions further aggravates these risks, as children may not receive the emotional support they need during critical developmental stages. Encouraging balanced use of AI, combined with active participation in social activities, is essential for developing a healthy environment for the development of children.

6.9 Overreliance

While AI can enhance learning and provide support in decision-making, children may become overly reliant on AI for answers, leading to a decline in critical thinking and decision-making skills. AI systems often present information in a way that seems authoritative, which can discourage children from questioning or analysing the information critically. This overreliance can weaken the development of cognitive skills that are necessary for independent thinking and problem-solving [16, 49]. It is crucial to encourage the use of AI as a tool that supports learning, rather than as a solution that replaces the need for independent thought.

6.10 The Potential for Manipulative Content and Advertising

Platforms that use AI are often used to deliver personalized content and advertisements to users [35]. While this can improve user's experience, it also raises ethical concerns, particularly when targeting children. AI algorithms are designed to maximize engagement, which can lead to the promotion of manipulative content that influences children's purchasing decisions or exposes them to harmful material [49]. The sensitivity of such manipulations is particularly worrying as children may not be able to distinguish advertisings from genuine contents, making them susceptible to undue influence. This demands stricter regulations and ethical guidelines that limit targeted advertising to children and ensure that content delivered by AI systems aligns with the best interests of young users [1, 38].

Concluding Sect. 6, the risks associated with AI in childhood are complex and require a thorough approach to address the ethical considerations, particularly data privacy, manipulative content, cyberstalking, and overreliance on AI. This requires a combined approach involving developers, policymakers, educators, and parents. By understanding and mitigating these risks, we can ensure that AI serves as a tool that enhances rather than hinders children's development. Balancing AI's benefits with the need for human interaction, critical thinking, and autonomy must be key to safeguard children's rights, well-being and future.

7 Balancing Advantages and Reliance

The integration of AI into children's daily lives, striking a balance between leveraging its advantages and preventing overreliance is crucial for fostering healthy development of AI technologies. This section discusses strategies for encouraging healthy technology usage among children and addressing the digital divide to ensure equitable access to AI's benefits.

7.1 Encouraging Healthy Use of Technology Among Children

With the growing use of AI-driven technologies in education, entertainment, and social interactions, it is essential to promote healthy technology usage among children. Overreliance on AI can lead to negative consequences, such as diminished critical thinking skills, social isolation, and increased exposure to manipulative content [1, 9, 16, 38]. To mitigate these risks, parents, educators, and policymakers must work together to establish guidelines that encourage a balanced use of technology.

One approach is to integrate AI in a way that complements rather than replaces traditional learning methods. For example, AI tools can be used to supplement classroom instruction by providing tailored feedback and adaptive learning experiences, while still emphasizing the importance of human interaction and critical thinking [1, 49].

Additionally, promoting digital literacy and teaching children how to critically evaluate the information they receive from AI systems is essential for helping them in developing the skills needed to face the digital world [32].

Moreover, setting limits to on screen time and encouraging children to engage in offline activities—such as sports, reading, and face-to-face social interactions—can help prevent overreliance on AI-driven technologies. By fostering a balanced approach to technology use, children can benefit from the use of AI while maintaining a healthy lifestyle [9, 19].

7.2 Addressing the Digital Divide and Equitable Access

The digital divide remains a significant barrier to ensuring that all children have equitable access to the benefits of AI. Internet connectivity among countries is very unequal, with the disadvantaged groups having fewer devices and being less connected to the internet [29, 41].

Disparities in access to technology, driven by socioeconomic factors, can exacerbate existing inequalities and prevent certain groups of children from fully benefiting from AI's potential [9, 25, 47]. For instance, young women in poor and rural households are more likely to be educationally more disadvantaged [41]. Additionally, AI algorithms applied in education can reproduce or deepen inequality [41]. Addressing this divide requires targeted efforts to provide equal access to digital resources and AI technologies, particularly in underserved communities (considering groups that have limited or no access to these resources).

AI holds significant potential to bridge educational gaps and address the digital divide, particularly by supporting marginalized children. AI can provide personalized learning experiences and enhance access to education for children in emergencies or those outside formal systems, including the 244 million children globally out of school [6]. However, despite these advancements, a recent UNESCO report warns that by 2030, 84 million children and young people will still be out of school if additional measures are not taken, highlighting the need for increased efforts to achieve universal access to education [40].

Additionally, challenges such as inadequate digital infrastructure and unequal access to devices and the internet remain barriers, particularly in low-income regions. UNICEF emphasizes the importance of ensuring inclusivity by integrating AI solutions for children with disabilities and addressing gender digital divides [6].

Governments and organizations must prioritize initiatives that provide affordable and reliable internet access, as well as devices and educational tools powered by AI, to all children, regardless of their background [2, 47, 49].

Furthermore, it is important to consider the ethical implications of AI use in education and ensure that AI-driven tools do not inadvertently reinforce existing biases or discrimination [1, 16, 26]. Algorithmic bias occurs when algorithms unintentionally encode societal biases, leading to discriminatory outcomes, including in education, where certain groups (e.g., non-native speakers or students from specific demographic backgrounds) may be disadvantaged and while technical solutions to mitigate bias are advancing, challenges remain [25, 44].

AI can complement teacher roles, enhancing learning outcomes without replacing human interaction. Successful integration of AI in education requires global collaboration among governments, the private sector, and civil society to ensure equitable access and avoid exacerbating existing inequalities [6, 29, 41].

Developing AI-driven educational platforms that are culturally and linguistically inclusive can help ensure that children from diverse backgrounds have access to content that is relevant and accessible to them. Additionally, it is important to design AI systems that are transparent, fair, and accountable, and that actively work to reduce disparities.

Concluding Sect. 7, balancing the advantages of AI with the need to prevent overreliance and ensure equitable access is critical for supporting the healthy development of all children. By promoting healthy tech usage, fostering digital literacy, and addressing the digital divide, society can support children benefit from AI while safeguarding the well-being of all children.

8 Inclusive AI: Upholding Children's Rights to Participation

Children's right to participation is a fundamental right globally (Article 12 of the UNCRC). As mentioned previously in 1.2., since 2021, the General Comment No. 25 of the UNCRC clarifies that children's fundamental rights apply equally to online interactions as well as offline [47]. As AI evolves, it is essential to ensure that children's rights, particularly their right to participate, are protected but also empowered in their interactions with AI. This section explores the existing international policy frameworks and initiatives advocating for children's participation rights in AI, examines the benefits and challenges of involving children in the AI development process, children are actively using AI-driven emerging technology products for their education, health, and entertainment across the globe [43], and highlights the importance of AI education in empowering children to participate effectively. Children scarcely participate in shaping AI policy and practice beyond being users [43].

8.1 Exploration of International Policy Frameworks and Initiatives

The existing international policy frameworks and initiatives emphasize the importance of supporting children's rights, including their right to participate in decisions that affect them. The UNCRC is one of the most significant documents in this regard, establishing the foundation for children's rights to express their views freely in all matters affecting them, including in the digital space [9, 18, 45, 47]. As AI increasingly shapes children's lives, it is important that these rights are extended to include participation in the development and implementation of AI technologies. However, it is challenged by gaps in theory, policy and practice, leading to the question to how to increase children's agency and what would that mean for AI policy and practice [21].

In this regard, the meaningful participation of children needs to be reviewed in line with the guiding principles of the General Comment No. 12 of the UNCRC for child participation [46]. These principles hold that child participation needs to be: "(1) transparent and informative, (2) voluntary; (3) respectful; (4) relevant; (5) child-friendly; (6) inclusive; (7) supported by training; (8) safe and sensitive to risks, and (9) accountable."

Additionally, international policy frameworks such as UNICEF's Policy Guidance on AI for Children [28,43] offer specific recommendations on how to integrate children's rights, in particular child participation into AI system design and deployment. The policy guidance advocates for the inclusion of children's perspectives in AI development processes to ensure that technologies serve their best interests and promote their well-being [1, 26, 32, 38, 43], but neglects to offer models for what the increased agency of involving children could mean for AI systems and policies [43].

Other international policy frameworks have the same approach. For example, the World Economic Forum's AI for Children Toolkit—FIRST (Fair, Inclusive, Responsible, Safe, and Transparent) provides the principles for ethical AI product development promoting children's healthy development [48], the Children's Online Safety Toolkit of the 5Rights Foundation explains several potential online harms that children can face [1]. However, neither UNICEF's Policy guidance [43] nor the World Economic Forum's or the 5Rights Foundation's toolkit requires for meaningful child participation in AI policy and practices.

Currently, there is a significant practice gap approaches and a noticeable absence of a comprehensive framework that promotes children's participation, which is essential to assure an inclusive participation of children in the development and implementation of AI policies and practices [21].

Therefore, it is relevant to assure that designing AI systems are not only protective of children but also inclusive, ensuring children's meaningful participation in shaping the technologies they use [1, 47] and to close the policy and practice gaps.

8.2 *Benefits and Challenges of Engaging Children in the AI Development Process*

Engaging children in the AI development process can bring significant benefits. When children participate in the design and development of AI tools, the resulting technologies are more likely to meet their needs, respect their rights, and enhance their well-being. Their unique perspectives contribute to the creation of AI systems that are more intuitive, user-friendly, and relevant to their everyday experiences [1, 16, 19].

However, there are also challenges associated with involving children in AI development. Children may lack the technical knowledge or experience to fully engage in complex AI-related discussions. Additionally, there is a risk that their contributions might be undervalued, rather than being meaningfully integrated into the development process [1, 9, 18, 38].

Overcoming these challenges, it is needed to identify best practices for involving children in the design and the development of AI technologies as well as the need to shift regulatory efforts into children‘s rights and wellbeing approach [14] rather than focusing only on privacy issue as a form of consumer data protection. It also requires a commitment to providing children not simply be protected but also empowered in their interactions with AI with the necessary education and support to participate effectively, ensuring children‘s meaningful participation in AI decision-making processes.

8.3 Importance of AI Education in Empowering Children

To enable children to participate meaningfully and effectively in AI ecosystems, it is crucial to provide them with a robust education in AI. This education should not only cover the technical aspects of AI but also address ethical considerations, data privacy, and the societal impacts of AI technologies [38, 49]. By equipping children with a comprehensive understanding of AI, we empower them to engage critically with the technologies they face, advocate for their rights, and step out of their role as only users and become contributors and innovators to the responsible development of AI systems, achieving their fullest potential, foster their healthy development, reduce harm, and promote safety in AI products and services [21].

AI education initiatives should also be inclusive, ensuring that all children, regardless of their background, have access to the knowledge and skills needed to participate in the digital world [37]. Addressing the digital divide and promoting equitable access to AI education are essential for fostering an inclusive environment where every child can participate in the future of AI [18, 29, 47, 49]. There is also an urgent need to build critical AI literacy among children from kindergarten to grade and beyond [39].

Concluding Sect. 8, upholding children's rights to participation in the context of AI at all levels is complex but critical for the present and future generations for ethical AI innovations. By exploring existing international policy frameworks and initiatives, promoting children‘s meaningful participation in the development process, and providing comprehensive AI education, we can have a comprehensive framework that could assure that AI technologies are more inclusive, ethical, and beneficial for all children. As AI continues to evolve, it is imperative that children's voices are not only heard but also empowered and act as contributors in shaping and in the governance of AI systems.

9 Conclusion and Future Scope

The intersection of AI and children's rights is a domain filled with both huge potential and considerable challenges. However, research is currently scarce regarding the impacts of AI on children's rights. The need for research on these impacts is clear, as is the need for proactive regulatory and ethical industry standards that support the opportunities AI provides children while mitigating its risks [14].

AI technologies have the capacity to transform education, health, entertainment, and social contexts for children, enabling more personalized and effective interventions. However, these benefits are accompanied by substantial risks and online harm, such as the exacerbation of existing inequalities, threats to privacy and safety, profiling, cyberbullying harassment, abuse, the increased distribution of harmful content like CSAM, and negative effects on physical and mental health (e.g., technology addiction).

The dual nature of AI's impact on children requires a balanced approach in policymaking, research, and implementation. On one hand, there is a pressing need to leverage AI's skills to enhance learning and development opportunities for all children. On the other, there is an equally urgent need to mitigate the risks associated with AI, ensuring that the rights and well-being of children are safeguarded in this rapidly evolving digital landscape.

9.1 Key Recommendations

As AI is integrated into more areas of children's lives (e.g., through increasing interaction with AI-driven applications, devices, and spaces), excluding child users becomes increasingly unethical. The need to address children's rights and best interests vis-à-vis AI is thus critical [14]. It is essential to establish a clear set of recommendations that can guide future policy, research, and educational initiatives.

The following key recommendations have been identified to ensure that AI design, development and deployment are aligned with the best interest of the child:

- Child-centred AI policies: Governments and international institutions must prioritize the development of AI policies that place children's rights at the forefront. This includes integrating safety-by-design principles into AI systems used in environments used by children;
- Global cooperation: Given the transnational nature of digital technologies, there is a need for international collaboration to harmonize regulations and standards related to AI and child protection, also ensuring a coordinated global effort to combat crimes against children, particularly in preventing online exploitation and abuse;
- Enhanced research and development: Further research is required to develop AI tools that can effectively detect and prevent the spread of harmful content, as well as to explore the ethical implications of AI in children's lives;

- Education and awareness: Integrating AI literacy into education systems is crucial to empower children with the skills needed to participate in digital spaces safely and responsibly;
- Guidance through General Comment No. 25: The General Comment No. 25 [47] provides a guide for setting priority areas and flagging concerns that a multi-stakeholder, such as children, experts, caregivers, educators, companies, and child advocates worldwide have already identified as fundamental to supporting children's rights and wellbeing in the digital environment. The General Comment applies to all digital technologies, including AI;
- Child-centred design methodologies: Strengthening the efforts for a child-centred design methodology and emphasize age-appropriate hands-on activities that enable children to learn, design, and evaluate AI technologies and policies to better understand their implications and promote responsible innovation.

These recommendations aim to create a safe, equitable, and empowering digital environment for children, supported by strong international cooperation, continuous research, and enhanced educational efforts.

9.2 Future Directions for Research and Policy

The future of AI in relation to children's rights must focus on creating a more inclusive, safe, and equitable digital environment.

Key areas for future research include:

- AI in education: Further investigation into how AI can be used to support personalized learning while ensuring equitable access for all children, regardless of their socio-economic background;
- Ethical AI development: Research into the ethical implications of AI, particularly in relation to bias, privacy, and consent, is essential to inform the development of AI systems that respect children's rights;
- AI for protection: The development of advanced AI-driven tools that can proactively identify and mitigate risks to children (not only online but also offline), such as the detection of CSAM and other harmful content;
- Children's participation in AI Policy and practices: An engagement of children at all levels of the AI system process, would help to achieve a more children-friendly AI policy and practice, assuring a more meaningful and practical participation of children in the end-to-end process of AI design, development, and deployment;
- Interdisciplinary research on children and AI: Promoting and funding a new and interdisciplinary research on children and AI is critical. There is a growing need for research that considers the wider socio-cultural and political economic impacts of the interaction of AI technologies across children's lives—at home and at school, in public and private, and online and off-line.

9.3 Final Remarks

As we conclude this analysis of AI's intersection with children's rights, it is clear that AI holds both transformative potential and substantial risks. The rapid integration of AI into various aspects of children's lives—from education to social interactions—demands a balanced approach that maximizes benefits while mitigating harms. The dual-edged nature of AI calls for cautious oversight, ethical considerations, and a commitment to upholding the fundamental rights of children as established in international legal frameworks, particularly in the UNCRC.

The future of AI in the context of childhood centres on our capacity to create systems that are safe, equitable, and inclusive. Policymakers, educators, and scientists must cooperate to ensure that AI technologies are developed and implemented regarding the best interest of the child. This involves not only refining the technical skills of AI but also fostering a global dialogue on the ethical implications and societal impacts of these technologies.

Despite the growing interest in AI's impact on children, there remain few studies on its long-term cognitive, social, and emotional developmental effects, and mental health. This is an area that requires further exploration and should be prioritized in future research to better understand and mitigate potential risks.

The path forward must be informed by continuous research, robust international cooperation, and an unwavering commitment to protect children's rights and to assure their best interest as established in the UNCRC.

References

1. 5Rights Foundation: Artificial Intelligence and Children's Rights. Retrieved from https://5rightsfoundation.com/our-work/childrens-rights/artificial-intelligence-and-childrens-rights.html (2024)
2. Abecassis, A.: Global Governance of Digital Economy: Artificial Intelligence (Centre on Regulation in Europe). Retrieved from https://cerre.eu/wp-content/uploads/2024/09/CERRE_GGDE_AI-Global-Governance_FINAL.pdf (2024)
3. Aguilera, C.A., Castro, A., Aguilera, C., Raducanu, B.: Voice-Controlled Robotics in Early Education: Implementing and Validating Child-Directed Interactions Using a Collaborative Robot and Artificial Intelligence. Retrieved from https://www.mdpi.com/2076-3417/14/6/2408 (2024)
4. Baird, D.E.: Aligning OECD AI Principles with Child Rights by Design for Ethical AI in Children's Products—Crafting a Framework for Responsible AI in Children's Tech. Retrieved from https://derekebaird.medium.com/aligning-oecd-ai-principles-with-child-rights-by-design-for-ethical-ai-in-childrens-products-9d86d20bc270 (2024)
5. Barua, P.D., Vicnesh, J., Gururajan, R., Oh, S.L., Palmer, E., Azizan, M.M., Kadri, N.A., Acharya, R.: Artificial Intelligence Enabled Personalised Assistive Tools to Enhance Education of Children with Neurodevelopmental Disorders—A Review. Retrieved from https://www.mdpi.com/1660-4601/19/3/1192 (2021)
6. Cappelle, F.V.: Can AI transform learning for the world's most marginalized children? Emerging Technologies (World Economic Forum). Retrieved from https://www.weforum.org/agenda/2023/10/ai-education-learning-marginalized-unicef/ (2023)

7. Charisi, V., Chaudron, S., Di Gioia, R., Vuorikari, R., Escobar-Planas, M., Sanchez, I., Gomez, E.: Artificial Intelligence and the Rights of the Child: Towards an Integrated Agenda for Research and Policy. Retrieved from https://op.europa.eu/en/publication-detail/-/publication/b7d0196a-eb8c-11ec-a534-01aa75ed71a1/language-en (2022)
8. Chen, Z., Song, Z., Yuan, S., Chen, W.: Influence Analysis of Education Policy on Migrant Children's Education Integration Using Artificial Intelligence and Deep Learning. Retrieved from https://www.ncbi.nlm.nih.gov/pmc/articles/PMC9247459/ (2022)
9. Child Rights International Network: "Briefing: Children's Rights in the Digital Age. Retrieved from https://home.crin.org/issues/digital-rights/childrens-right-digital-age (2022)
10. Constantino, T.: 5 Ways AI Can Help Students Master Elementary School Subjects. Retrieved from https://www.forbes.com/sites/torconstantino/2024/10/03/5-ways-ai-can-help-enrich-your-child-in-primary-school/?ss=ai (2024)
11. Council of Europe: Convention on the Protection of Children against Sexual Exploitation and Sexual Abuse (CETS No. 201). https://www.coe.int/en/web/conventions/full-list?module=treaty-detail&treatynum=201 (2010)
12. European Commission: Child Sexual Abuse. https://home-affairs.ec.europa.eu/policies/internal-security/child-sexual-abuse_en (2024)
13. European Parliament and Council: Regulation (EU) 2024/1689 of the European Parliament and of the Council of 13 June 2024 Laying Down Harmonised Rules on Artificial Intelligence and Amending Regulations (EC) No 300/2008, (EU) No 167/2013, (EU) No 168/2013, (EU) 2018/858, (EU) 2018/1139 and (EU) 2019/2144 and Directives 2014/90/EU, (EU) 2016/797 and (EU) 2020/1828 (Artificial Intelligence Act). Retrieved from https://eur-lex.europa.eu/eli/reg/2024/1689/oj (2024)
14. Grimes, M.S., Antle, N.A., Steeves, V., Natalie, C.: Responsible AI and Children: Insights, Implications, and Best Practices. Retrieved from https://cifar.ca/wp-content/uploads/2024/04/CIFAR-Responsible-AI-and-Children-EN_Final.pdf (2024)
15. Internet Watch Foundation: AI CSAM Report Update: What has Changed in the AI CSAM Landscape? Retrieved from https://www.iwf.org.uk/media/opkpmx5q/iwf-ai-csam-report_update-public-jul24v11.pdf (2024)
16. Kurian, N.: AI's Empathy Gap: The Risks of Conversational Artificial Intelligence for Young Children's Well-Being and Key Ethical Considerations for Early Childhood Education and Care. Retrieved from https://journals.sagepub.com/doi/full/10.1177/14639491231206004 (2023)
17. Li, Q.: A Study on Mobile Resources for Language Education of Preschool Children Based on Wireless Network Technology in Artificial Intelligence Context. Retrieved from https://pubmed.ncbi.nlm.nih.gov/35712005/ (2022)
18. Lievens, E., Livingstone, S., McLaughlin, S., O'Neill, B., Verdoodt, V.: Children's Rights and Digital Technologies. Retrieved from https://eprints.lse.ac.uk/84871/1/Children%27srightsanddigitaltech_revised_clean%20final.pdf. (2018)
19. Livingstone, S.: Can We Realise Children's Rights in A Digital World? A Provocation Paper. Retrieved from https://medium.com/reframing-childhood-past-and-present/can-we-realise-childrens-rights-in-a-digital-world-d4f5f19f298f (2020)
20. Mahomed, S., Aitken, M., Atabey, A., Wong, J., Briggs, M.: AI, Children's Rights, & Well-being: Transnational Frameworks: Mapping 13 Frameworks at the Intersections of Data-Intensive Technologies, Children's Rights, and Wellbeing, The Alan Turing Institute. Retrieved from https://www.turing.ac.uk/news/publications/ai-childrens-rights-wellbeing-transnational-frameworks (2023)
21. Mathiyazhagan, S., Fors, K.L.: Children's Right to Participation in AI: Exploring Transnational Co-Creative Approaches to Foster Child-Inclusive AI Policy and Practice. Retrieved from https://dl.acm.org/doi/abs/10.3233/IP-229014 (2023)
22. Maxwell, F.: Children's Rights, The Optional Protocol and Child Sexual Abuse Material in the Digital Age—Moving from Criminalisation to Prevention. Retrieved from https://www.academia.edu/95871940/Children_s_Rights_The_Optional_Protocol_and_Child_Sexual_Abuse_Material_in_the_Digital_Age (2023)

23. Milosevic, T., Van Royen, K., Davis, B.: Artificial Intelligence to Address Cyberbullying, Harassment and Abuse: New Directions in the Midst of Complexity. Retrieved from https://link.springer.com/article/10.1007/s42380-022-00117-x (2022)
24. OECD—Organisation for Economic Cooperation and Development: OECD Child Well-Being Dashboard. Retrieved from https://www.oecd.org/en/data/dashboards/oecd-child-well-being-dashboard.html (2022)
25. OECD—Organisation for Economic Cooperation and Development: OECD Digital Education Outlook 2023—Towards an Effective Digital Education Ecosystem. Retrieved from https://www.oecd-ilibrary.org/education/oecd-digital-education-outlook-2023_c74f03de-en (2023)
26. OECD—Organisation for Economic Cooperation and Development: Towards Digital Safety by Design for Children. Retrieved from https://www.oecd.org/en/publications/towards-digital-safety-by-design-for-children_c167b650-en.html (2024a)
27. OECD—Organisation for Economic Cooperation and Development: Recommendation of the Council on OECD Legal Instruments Children in the Digital Environment. Retrieved from https://legalinstruments.oecd.org/en/instruments/OECD-LEGAL-0389 (2024b)
28. OECD—Organisation for Economic Cooperation and Development: Recommendation of the Council on OECD Legal Instruments Artificial Intelligence. Retrieved from https://legalinstruments.oecd.org/en/instruments/oecd-legal-0449. (2024c)
29. OECD—Organisation for Economic Cooperation and Development: The Potential Impact of Artificial Intelligence on Equity and Inclusion in Education. Retrieved from https://www.oecd.org/en/publications/the-potential-impact-of-artificial-intelligence-on-equity-and-inclusion-in-education_15df715b-en.html (2024d)
30. Osório de Barros, G.: A Economia da Cibersegurança (*Tema Económico* 54, GEE). Retrieved from https://www.gee.gov.pt/pt/en/studies-and-seminars/estudos-de-temas-economicos-category/17739-a-economia-da-ciberseguranca (2018)
31. Osório de Barros, G.: Digitalisation, Skills and Cybersecurity in Portugal—Critical Factors in a Digital Economy driven by Covid-19 (*Tema Económico* 89, GEE). Retrieved from https://www.gee.gov.pt/pt/estudos-e-seminarios/estudos-de-temas-economicos-category/31552-te-89-digitalisation-skills-and-cybersecurity-in-portugal-critical-factors-in-a-digital-economy-driven-by-covid-19 (2021)
32. Osório de Barros, G.: Forging AI Pathways: Portugal's Journey within the EU Digital Landscape (GEE Paper 177). Retrieved from https://www.gee.gov.pt/pt/estudos-e-seminarios/gee-papers-category/32990-forging-ai-pathways-portugal-s-journey-within-the-eu-digital-landscape (2023)
33. Rose-Davis, B., Van Woensel, W., Stringer, E., Abidi, S., Abidi, S.S.R.: Using an Artificial Intelligence-Based Argument Theory to Generate Automated Patient Education Dialogues for Families of Children with Juvenile Idiopathic Arthritis. Retrieved from https://pubmed.ncbi.nlm.nih.gov/31438143/ (2019)
34. Sapo Tek: Empresas de inteligência artificial juntam-se à iniciativa Thorn para proteger crianças online. Retrieved from https://tek.sapo.pt/noticias/internet/artigos/empresas-de-inteligencia-artificial-juntam-se-a-iniciativa-thorn-para-proteger-criancas-online (2024)
35. Sousa, M.J., Osório de Barros, G., Tavares, N.: Artificial Intelligence a Driver for Digital Transformation. Retrieved from https://www.igi-global.com/chapter/artificial-intelligence-a-driver-for-digital-transformation/271781 (2021)
36. Sousa, M.J., Osório de Barros, G., Tavares, N.: Artificial Intelligence Trends: Insights for Digital Economy Policymakers. Retrieved from https://link.springer.com/chapter/10.1007/978-3-030-75123-4_8 (2022a)
37. Sousa, M.J., Dal Mas, F., Osório de Barros, G., Tavares, N.: Artificial Intelligence: Technologies, Applications, and Policy Perspectives. Insights from Portugal. Retrieved from https://link.springer.com/chapter/10.1007/978-3-030-82763-2_6 (2022b)
38. UC Berkeley Human Rights Center: Memorandum on Artificial Intelligence and Child Rights. Retrieved from https://www.unicef.org/innovation/reports/memoAIchildrights (2019)
39. UNESCO—United Nations Educational, Scientific and Cultural Organization: Recommendation on the Ethics of Artificial Intelligence. Retrieved from https://www.unesco.org/en/articles/recommendation-ethics-artificial-intelligence (2022a)

40. UNESCO—United Nations Educational, Scientific and Cultural Organization: 84 Million Children at Risk of Still Being Out of School by 2030. Retrieved from https://www.unesco.org/gem-report/en/articles/84-million-children-risk-still-being-out-school-2030 (2022b)
41. UNESCO—United Nations Educational, Scientific and Cultural Organization: Global Education Monitoring Report—Technology in Education: A Tool on Whose Terms? Retrieved from https://www.unesco.org/gem-report/en (2023)
42. UNICEF—United Nations Children's Fund: National AI strategies and Children—Reviewing the Landscape and Identifying Windows of Opportunity. Retrieved from https://www.unicef.org/innocenti/documents/national-ai-strategies-and-children (2020)
43. UNICEF—United Nations Children's Fund: Policy Guidance on AI for Children 2.0. Retrieved from https://www.unicef.org/innocenti/reports/policy-guidance-ai-children (2021a)
44. UNICEF—United Nations Children's Fund: The Case for Better Governance of Children's Data: A Manifesto. Retrieved from https://www.unicef.org/innocenti/reports/case-better-governance-childrens-data-manifesto (2021b)
45. United Nations: Convention on the Rights of the Child. Retrieved from https://www.ohchr.org/en/instruments-mechanisms/instruments/convention-rights-child (1989)
46. United Nations: General Comment No. 12 on the Right of the Child to be Heard. Retrieved from https://digitallibrary.un.org/record/671444?v=pdf (2009)
47. United Nations: General Comment No. 25 on Children's Rights in Relation to the Digital Environment, Office of the United Nations High Commissioner for Human Rights (OHCHR). Retrieved from https://digitallibrary.un.org/record/3906061?v=pdf (2021)
48. World Economic Forum: Artificial Intelligence for Children—Toolkit. Retrieved from https://www3.weforum.org/docs/WEF_Artificial_Intelligence_for_Children_2022.pdf (2022)
49. World Economic Forum: Shaping the Future of Learning: The Role of AI in Education 4.0. Retrieved from https://www.weforum.org/publications/shaping-the-future-of-learning-the-role-of-ai-in-education-4-0/ (2024)
50. Xiao, J., Luo, J., Ly-Mapes, O., Wu, T.T., Dye, T., Al Jallad, N., Hao, P., Ruan, J., Bullock, S., Fiscella, K.: Assessing a Smartphone App (AICaries) That Uses Artificial Intelligence to Detect Dental Caries in Children and Provides Interactive Oral Health Education: Protocol for a Design and Usability Testing Study. Retrieved from https://pubmed.ncbi.nlm.nih.gov/34529582/ (2021)
51. Yuchen, X.: Application of Immersive Artificial Intelligence Based on Machine Vision in Education Management of Children with Autism. Retrieved from https://link.springer.com/article/10.1007/s13198-023-02141-6 (2023)

Political Governance and State Modernization: Bridging AI Research with e-Democracy in Portugal

João Reis, Manuel do Carmo, David Pascoal Rosado, Pedro Alexandre Marques, Pedro Carmona Marques, and Nuno Melão

Abstract Artificial Intelligence (AI) has undergone significant advancements in recent years, showing its potential impact on political decisions and state modernization efforts. The integration of AI into democratic governance systems appears inevitable, given the collection of vast amounts of citizen data and its utility. The capacity of AI to analyze extensive datasets unveils hitherto inaccessible insights, revolutionizing knowledge acquisition. Ultimately, AI facilitates informed political decision-making, resource management and enhances citizens' quality of life. Clarifying the mentioned points, this article presents two cases to explain the strategies adopted by the Portuguese government to pioneer innovative policies and services. Through empirical investigation, we shown the AI technologies developed in Portugal, which enables a comprehensive understanding of how the modernization of state institutions facilitates political decision-making while addressing citizens' needs. Prospective research endeavors can explore citizen feedback concerning the

J. Reis (✉) · P. A. Marques · P. C. Marques
Industrial Engineering and Management, Faculty of Engineering, Lusófona University and RCM2+, Lisbon, Portugal
e-mail: joao.reis@ulusofona.pt

J. Reis
Research Unit on Governance, Competitiveness and Public Policies (GOVCOPP), Aveiro University, Aveiro, Portugal

J. Reis · M. do Carmo · D. P. Rosado
Center for Research, Development, and Innovation (CINAMIL), Academia Militar, Lisbon, Portugal

M. do Carmo · D. P. Rosado
Faculdade de Ciências Sociais E Tecnologia (FCST) and IADE (FDTC), Europeia University, Lisbon, Portugal

N. Melão
CISeD—Research Center in Digital Services, Polytechnic Institute of Viseu, Lisbon, Portugal

M. do Carmo
CIMA/IIFA—Center for Research in Mathematics and Applications, University of Évora, Évora, Portugal

© The Author(s), under exclusive license to Springer Nature Switzerland AG 2025

A. d. b. Machado et al. (eds.), *Environmental, Social, Governance and Digital Transformation in Organizations*, Information Systems Engineering and Management 35, https://doi.org/10.1007/978-3-031-86079-9_6

identified technologies, discerning the most pertinent tools from a political and social standpoint.

Keywords Artificial intelligence · e-Democracy · Political governance · State modernization

1 Introduction

The influence of Artificial Intelligence (AI) on political governance and the modernization of democratic states has become a central focus of academic discourse and public policy development. This growing interest is identified in several prominent publications. For instance, the Journal of Democracy devoted a special issue to "Artificial Intelligence and Democracy" [7], while Business and Politics explored "The Future of AI Politics, Policy, and Business" [4], and the Journal of European Public Policy addressed the topic in "Governing AI—Attempting to Herd Cats" [5]. These examples highlight the increasing attention AI is receiving in the context of democratic governance.

Drawing from historical precedents in technological revolutions, AI presents significant challenges to e-democracy and political governance. These challenges include intensifying social inequalities and widening the political and economic power gap. For example, while AI-driven innovations are creating new forms of work, they also have the potential to undermine traditional social security systems, as automation and robotics may not contribute to tax revenues to the same extent as human labor.

Considering these challenges, it is unsurprising that concepts like "technology anxiety" are gaining prominence, highlighting the urgent need to anticipate and address the implications of AI implementation in public administration. This calls for rigorous scientific inquiry into the political ramifications of AI, particularly within the discipline of political science. Such research is vital because political decisions influenced by AI can have profound social consequences at multiple levels—local, regional, national, and supranational. In Portugal, for example, discussions on the political impact of AI remain limited and are predominantly confined to academic circles.

Although limited in number, there are a few examples of research on digitalization in Portugal. One relevant study is by Jalali [14], who investigated the extent to which digitalization has become a subject of inquiry within Portuguese political science. Jalali's findings are significant, revealing that PhDs focusing on digitalization, completed between 2009—when this topic first emerged in Portuguese academia—and 2018, constituted only 4% of the 299 political science PhDs awarded or registered in the country during that period [14]. Furthermore, Jalali [14] noted that the proportion of publications by Portugal-based researchers on digitalization topics, as indexed in the Web of Science (WoS), shows the low engagement seen among Ph.D. researchers. Up until 2017, there were only 26 such publications, representing only 6% of the total political science articles by Portuguese scholars indexed in WoS from 2008 to 2017. Interestingly, this small percentage, despite its modest size, appears to exceed the global average [14]. Another significant contribution to this field is the study by Cardoso et al. [6] titled "Bridging the e-Democracy Gap in Portugal". The authors argued that there was weak vertical communication between elected representatives and voters and that neither political-institutional integration methods nor media efforts alone were sufficient to enhance public participation. Cardoso et al. [6] also suggested that the Internet could bolster citizens' political participation by altering both the representation and attitudes of politicians toward the public. Nearly 15 years after the publication of Cardoso et al.'s article, it is now pertinent to shift the focus from the Internet to AI research, which we believe holds the potential to further bridge the e-democracy gap in Portugal. Despite the importance of this research domain, few academics have pursued research on AI in the context of e-democracy. However, recent efforts, such as the doctoral thesis by Reis et al. [22] titled "From Artificial Intelligence Research to e-Democracy: Contributions to Political Governance and State Modernization", provides a modest contribution to the theoretical and practical implications of AI research for Portuguese e-democracy. Nevertheless, more extensive research is urgently needed, especially considering the rapid advancements in AI and its inevitable impact on public institutions and European Union politics.

This time, a scientific contribution to this field is the one by Wirtz et al. [33], proposes an integrated AI governance framework rooted in regulatory theory. This framework identifies key elements of AI governance and regulation from a public administration perspective. The authors argue that their framework provides a structure for the diverse and interdisciplinary field of AI, laying the groundwork for its practical application in public administration, societal acceptance, and beneficial use. However, they acknowledge that their study has limitations, particularly in focusing on the risks and challenges of AI without fully addressing its potential benefits in the public sector.

This leaves several critical questions unanswered: How effectively can these tools support political decision-making? Why are AI technologies currently being deployed in public administration? To address this gap, it is essential to (1) classify the AI technologies that mediate the relationship between national politics and citizens to enhance the political decision-making process, and (2) evaluate the benefits and challenges that these new tools bring to e-democracy.

As preliminary results of this study, we have seen that AI technologies are increasingly being leveraged in public administration to modernize state functions. These AI technologies suggest improved citizen participation, greater transparency, and the safeguarding of rights and freedoms. Despite these advancements, AI's impact on national politics has not yet met expectations. Citizen engagement remains predominantly utilitarian, with participation largely motivated by direct personal benefits rather than a proactive interest in political involvement or strategic decision-making. In contrast, AI has yielded more promising outcomes at the local and regional levels, where municipalities and frontline public services have played a pivotal role in fostering civil mobilization and participation. The responsiveness of local authorities has particularly exceeded that of the national government, likely due to their closer proximity to the population and greater influence over local issues. However, preliminary findings also indicate that AI applications supporting political decision-making are mainly concentrated at the local/regional levels in Portugal. However, this information has essentially facilitated the formulation of policies and measures at a limited national level. The integration of AI into these processes highlights its potential to improve governance by bridging the gaps between different levels of government and improving responsiveness to citizens' needs.

This paper follows the IMRaD format (Introduction, Methods, Results, and Discussion), a widely recognized and standard framework for organizing scientific manuscripts [23]. Consequently, this manuscript is divided into four distinct sections, each fulfilling a specific purpose. The Introduction sets the stage by offering an overview and contextual background, identifying gaps in the current literature, and establishing the research questions that this study seeks to address–why and how questions, as recommended by Yin [34]. The Methods section provides a comprehensive explanation of the analytical tools employed, detailing the methodological process in a step-by-step manner. The Results section presents the findings of the study, directly addressing the research questions and fulfilling the objectives outlined. Lastly, the Discussion section explores the theoretical and policy implications of the findings, acknowledges the study's limitations, and offers suggestions for future research to further advance the field.

2 Method

This study follows a qualitative multiple case study research design [34], chosen for its descriptive and exploratory nature. A qualitative approach was selected given the complexity of the research topic, which needs a deep analysis of underlying patterns and interrelationships among concepts. The lack of existing research on this topic further requires a qualitative approach, as it allows for a more flexible and in-depth analysis of phenomena that are not yet well understood [34]. The exploratory nature of this research also is central, as it seeks to investigate a phenomenon that has been minimally studied in the academic literature, thereby contributing to the expansion of knowledge in this area. We found the case study research appropriate since it enables an investigation of a phenomenon where the boundaries between the context and the subject are blurred, and where the complexities involved are not easily defined [16]. This approach assists the researcher in exploring the relationships of the subject, capturing the rich, contextual details that might be overlooked by other research methods. To ensure a comprehensive understanding, the study employs multiple data collection sources, including semi-structured interviews, documentary analysis, and direct observations. These sources were chosen for their ability to capture a wide range of data, providing a holistic view of the phenomenon under study. The integration of these diverse sources not only enriches the data but also enhances the validity of the findings through triangulation, which was the process of cross-verifying data from multiple sources [19]. This methodological rigor ensures that the findings are not only reliable but also offer a well-rounded perspective on the research question. Moreover, by employing a multiple case study design, the research allows for the comparison of different instances of the phenomenon across varied contexts, which can lead to the identification of common patterns or unique variations. This comparative approach is particularly useful in exploratory research, as it helps to build a more comprehensive understanding of the phenomenon and generates insights that might apply to other settings or contexts. The context-rich data gathered through these methods improves the credibility of the findings and by drawing on multiple sources of evidence, this study aims to produce findings that are both novel and relevant.

3 Results

3.1 Current State of the Art

Recent literature highlights the growing application of AI across several issues of individual and organizational life. Despite its vast potential, many of AI's implications and challenges remain underexplored, indicating a significant research gap [21, 28]. Sousa et al. [28] further notes the increasing number of AI research initiatives and start-ups in Portugal (e.g., Feedzai, James, Heptasense), yet the link between AI, political governance, and state modernization, particularly concerning e-democracy, remains understudied.

Starke and Lünich [29] emphasize that the EU's political legitimacy is undermined by its inability to effectively address crises, conflicts, and systemic threats. They argue that integrating AI into political processes could enhance the EU's decision-making by grounding it in solid evidence, thereby increasing political legitimacy, identifying urgent social issues, predicting policy outcomes, and evaluating policy effectiveness. Reis et al. [26] also highlights the importance of AI-driven decision-making for political leaders at local, regional, and national levels, especially in managing complex scenarios involving multiple lobby groups. Such scenarios require the analysis of structured and logical processes to facilitate assertive and equitable decisions. Given that AI influences societies and political systems in diverse ways, the EU is investigating common challenges among member states and exploring opportunities for collaboration, cost savings, and synergy [25]. However, challenges persist, particularly due to the limited understanding of AI research potential within EU countries, necessitating further investigation. Furthermore, EU-funded research has often struggled to interpret findings into successful outcomes, highlighting the need for future studies to examine the relationship between AI-driven decision-making and perceptions of legitimacy within the EU [29].

Starke and Lünich [29] propose a comparative study between the Netherlands and Portugal to address this gap. Given that public support for Algorithmic Decision-Making (ADM) in the Netherlands (43%) far exceeds that in Portugal (19%) [27], there is the argument that AI can enhance policymaking, public service delivery, and internal management [31], and that Portugal should not be left behind in this AI-driven transformation. Van Noordt and Misuraca [31] identified three primary applications of AI within public administration: (1) detecting social issues to inform new policy development or monitor existing policies and their impacts; (2) enhancing information delivery to citizens and businesses within public services; and (3) supporting internal management functions, including public recruitment and cyber-security. However, the concept of e-democracy, the focus of this research, extends beyond mere technological implementation. As Mahrer and Krimmer [15] points out, e-democracy also encompasses societal and stakeholder engagement. Sundberg [30] presents a simplified model that illustrates the interconnectedness of e-democracy, e-administration, and e-government, emphasizing that these concepts form a technological ecosystem that supports societal and political interests.

Figure 1 highlights various e-government concepts, including electronic services developed by public administrations and made available online, enabling citizens to complete tasks without needing to visit an agency in person. E-democracy, a key area of interest for governments aiming to foster a developed society, encourages citizen engagement in formal politics using innovative technologies. Strategic documents, such as the EU's Digital Agenda, reflect these political objectives for (e-)public administration and have been adopted by national, regional, and local governments across EU member states [30]. E-democracy seeks to motivate younger citizens to actively participate in their communities, whether by voting on local decisions through web applications or by engaging in discussions about policy decisions in formal politics [32]. This focus on community-oriented decision-making is vital, emphasizing that both citizens and government should prioritize the common good over individual interests. Although AI in e-democracy is still in its infancy in Portugal, several European initiatives are paving the way forward. One example, CitizenLab, is discussed in the next Sect. 3.2, alongside numerous other cases in Portugal (Sect. 3.3)

Arana-Catania et al. [2] explored the application of Natural Language Processing (NLP) and machine learning to enhance citizen experiences on digital platforms

Fig. 1 E-Democracy, e-Administration and e-Government [30]

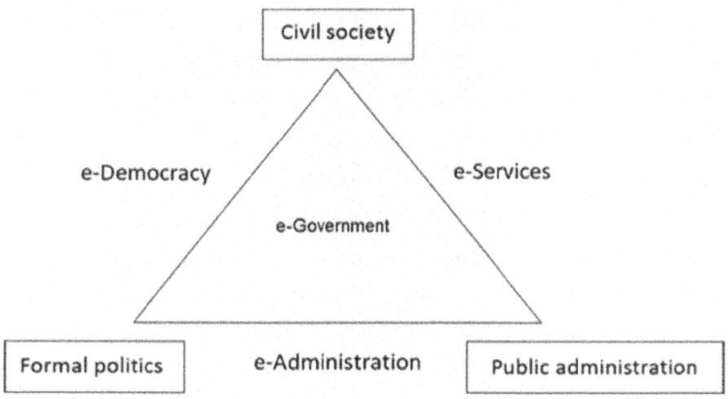

designed for public participation. Their study focused on the "Decide Madrid" platform by Consul, where citizens could propose policies for adoption by the city council. NLP and machine learning were utilized to: (a) suggest relevant proposals to users, (b) group citizens by shared interests to facilitate interaction; (c) summarize responses to proposals; and (d) assist citizens in developing and consolidating their proposals. According to the study, these technologies play a crucial role in addressing challenges currently faced by users of such platforms.

Moreover, Nielsen et al. [18] discusses a significant challenge for the European Union in their book European E-democracy in Practice: Achieving Full Inclusion within the Context of E-participation. The EU faces specific challenges due to its unique institutional structures, including language barriers, multi-level governance complexities, and the digital divide [18]. However, AI has made notable contributions in overcoming language-related obstacles. For many years, the promise of digital translation seemed elusive—always on the horizon but never quite meeting expectations. Today, there is renewed optimism surrounding AI-driven translation software, particularly those based on deep learning technologies (www.ec.europa.eu/citizens-initiative) [18]. Although the EU is far from being a fully multilingual region, significant investments in digital translation for the world's major languages are underway. This progress is expected to greatly enhance the accessibility of e-participation mechanisms across the EU, allowing citizens to have a greater say in the policies that impact their lives within the European community. The European Citizens' Initiative is a prime example of this, enabling EU citizens to influence policy by petitioning the European Commission to propose new laws [9]. When an initiative garners one million signatures, the Commission determines the necessary actions. Such initiatives are increasingly feasible due to advancements in AI-powered translation technologies.

While there are many international examples discussed in current literature, this thesis primarily focuses on the Portuguese case and its implications for political governance and state modernization.

As democracies continue to digitize core civic functions like voting and expand their "e-democratic" initiatives, unresolved fundamental questions about these policies persist. One critical issue is the concept of equality in e-democratic contexts. In a democracy, equality is foundational—every citizen is considered significant, possessing basic rights and the ability to vote, thereby expressing their political will [20]. However, disparities in access to education and information remain prevalent, even in "developed" democracies like the United States. In Estonia, where democratic functions have begun to digitize, the principle of equality has explicitly shaped national policies [8]. These policies, such as the Tiger Leap campaign, aim to educate citizens about IT and ensure access to necessary technologies [20]. Such efforts continue today as governments integrate increasingly sophisticated technologies into their digital democratic infrastructures. These policies are justified as they ensure equal access and opportunity for all citizens regarding the technologies used in democratic processes.

3.2 What is New? Why AI Technologies Are Relevant?

AI technologies have increasingly become influential globally and significantly impact various sectors–including civic engagement [12]. In Belgium, AI technologies have played a crucial role in bridging the gap between citizens and national politics. An example is CitizenLab [17] (currently called "Go Vocal" [13], a civic technology company highlighted in a 2019 report by the Organization for Economic Co-operation and Development (OECD) [3].

CitizenLab's mission was to empower public servants by providing advanced machine-learning tools that enhance their ability to analyze citizen feedback and make informed decisions. Belgian citizens were encouraged to submit ideas, comments, and votes on various initiatives through the CitizenLab platform. The platform employed AI to categorize these inputs, identify emerging topics, summarize trends, and cluster similar contributions based on themes and demographic characteristics. For instance, an analysis might reveal that one neighborhood prioritizes road improvements, while a neighboring area demands additional traffic stops. These citizen-generated insights are then utilized as a foundation for political decision-making.

Another example is the Youth4Climate case, where young Belgians utilized CitizenLab's Natural Language Processing (NLP) technology to distill thousands of citizen contributions into actionable insights. In early 2019, a dedicated CitizenLab platform (Youth4Climate) was launched as part of the climate protest movement, allowing citizens to submit ideas on combating climate change. Over three months, the platform received 1700 ideas, 2600 comments, and more than 32,000 votes on various initiatives [11]. After collecting and organizing the data, the Youth4Climate organizers presented these proposals to elected officials, demonstrating the platform's effectiveness in facilitating civic participation.

Furthermore, AI4 Belgium [1] is a grassroots initiative that enables Belgian citizens and organizations to harness the potential of AI while guiding the country's transition toward responsible AI adoption. Experts involved in AI4Belgium contribute to developing AI guidelines for Belgian policymakers, ensuring that AI's integration into society is both beneficial and ethical.

Given these examples, we believe it is crucial to explore how such technological advancements can be leveraged by the Portuguese government and its citizens. The research presented in this paper aims to provide a comprehensive understanding of this phenomenon, offering insights into how AI technologies can enhance civic engagement and inform policymaking in Portugal.

3.3 Supporting Political Decision-Making

In this section, we focus on two technologies utilized within Portuguese public administration that partially support political decision-making processes.

The first technology our respondents identified is SIGMA, an AI-powered virtual assistant (chatbot) deployed on the eportugal.gov.pt portal to enhance citizens' interactions with public services. SIGMA uses Natural Language Processing (NLP) to interpret and respond to user queries, facilitating efficient communication between the public and government services. The system currently serves thousands of users each month, reflecting a consistent increase in both the volume of interactions and the system's sophistication. Despite its limitations in addressing complex queries, SIGMA is continually evolving, aiming to improve the efficiency and accuracy of public service delivery. SIGMA continuously evolves by monitoring conversations and recording user feedback. This information is used to train the machine learning models, enhancing the chatbot's ability to interpret and respond to user queries. Over time, and as programmers update the system, SIGMA becomes more effective at providing services. Regular training is crucial to improving its accuracy and efficiency. By collecting specific data and crafting relevant queries, SIGMA can better understand user interactions and incorporate new types of questions into its training data. SIGMA's services are organized into two primary phases: the informational phase, where it assists users in gathering useful information, and the performative phase, where it executes services after obtaining the required information (such as updating the address on a citizen's ID card). If SIGMA is unable to answer a question, it will register a contact request to forward the user to the call center. While this technology represents a significant step forward in modernizing public administration through AI-driven service optimization—such as enabling services like address changes—it does not directly contribute to political decision-making, unlike some tools used in Belgium that have been previously discussed.

The second technology identified by the respondents were BUPi, a digital platform in Portugal designed to identify and manage the geographic boundaries of citizens' properties. BUPi was developed through collaboration across multiple public administration sectors, including finance, environment, agriculture, and justice [24]. One of BUPi's primary objectives is to enable citizens to register their properties by submitting geographic data via the platform. This data is then validated by a qualified technician and cross-referenced with other public records. By providing accurate information on property boundaries and ownership, BUPi aids the state in resource management, deforestation prevention, and regulating natural resource exploitation.

Although both SIGMA (national) and BUPi (local/regional) have been successful in their respective domains, neither tool is explicitly designed to foster citizen participation in national-level public policy formulation. SIGMA, despite its AI foundation, primarily enhances public service efficiency rather than directly influencing political decision-making. However, BUPi directly impacts policy formulation at the local, regional, and even national levels. For instance, BUPi allows authorities to identify and hold property owners accountable for issues such as land misuse, abandonment, and inadequate land maintenance. From a judicial perspective, it helps resolve disputes involving overlapping property boundaries. In the context of municipal management, BUPi facilitates the identification of unowned properties that may fall under state responsibility. This capability supports political decision-making by enabling local policymakers to collaborate with national authorities, such as civil protection agencies, in developing policies and measures to prevent forest fires.

3.4 Addressing the Research Questions

AI technologies are gradually being implemented in the Portuguese public administration to address a range of challenges. One of the primary motivations for this integration is to enhance the quality and efficiency of public services. By automating routine tasks and improving data processing capabilities (vide SIGMA), AI enables public administrators to focus on more strategic issues, such as collaboration with national authorities as we saw previously. The next step for AI is to further help administrations respond more effectively to societal needs and enhance their responsiveness to citizens' concerns, thereby strengthening democratic governance. However, what is currently lacking is a comprehensive model of e-democracy, like the one in Belgium, where citizens have a direct influence on the development of public policies. While AI holds significant potential to improve political decision-making and public administration, its current impact at the national level remains limited, and the integration of AI into national politics has not fully met expectations. Citizen engagement often remains utilitarian, motivated by direct personal benefits rather than proactive political involvement. Despite this, AI has demonstrated promise in local governance, where closer proximity to citizens allows for more direct influence and improved responsiveness to local issues.

4 Discussion

4.1 Theoretical Contributions

This paper contributes theoretically by exploring the connection of AI with political governance and state modernization, particularly in the context of e-democracy. It introduces new knowledge into how AI technologies can be integrated into democratic governance systems to enhance decision-making processes. This study's originality lies in its empirical investigation of AI technologies developed in Portugal and their application within public administration. Thus, filling a research gap in the literature on the role of AI in political science and e-democracy.

Building on existing literature, the article draws from previous research on digitalization in political science and expands it by focusing specifically on AI's potential. It critiques and extends the discussions in studies like those by Wirtz et al., which proposed an integrated AI governance framework, and Starke and Lunich, who emphasized AI's role in improving political legitimacy through evidence-based decision-making. By applying these theoretical frameworks to the Portuguese context, the article offers a case-specific examination that enriches the broader understanding of AI's implications for political governance and state modernization. The study's methodological approach, utilizing two case studies, further supports its contribution by providing an analysis of AI's impact at various governmental levels, thereby proposing new avenues for future research in this emergent field.

Despite the article's originality in exploring such topic, the reader might feel somewhat disappointed due to the lack of strong results. This is primarily for two reasons: first, Portugal remains a country with a liberal democracy, which, while interesting because it partially supports policies that promote the use of AI in a political context, limits the scope of the study's implications. Second, the application of AI in public administration is relatively underexplored compared to other areas, such as the business sector. However, we believe this issue reflects a broader, global challenge.

4.2 Political Contributions

To explain political contributions well, it is necessary to understand that there are two main types of democratic systems, each with its characteristics and forms of governance. The first is liberal democracy, a system of government characterized by regular elections in which citizens elect representatives to govern on their behalf. In this model, citizen participation is indirect, as elected officials make decisions on their behalf. Liberal democracy is widely practiced in countries such as the USA, Germany, and Portugal.

The second type is direct democracy, a system in which citizens play a more active role in political decision-making. Instead of electing representatives to make decisions, citizens vote directly on laws and public policies through mechanisms like referendums and popular initiatives. Switzerland is a prime example of a country where direct democracy is a significant part of the political process, with frequent referendums.

This distinction is crucial for our discussion on Political Governance and State Modernization through e-democracy and AI. AI technologies tend to be more compatible with direct democratic systems, where citizens actively express their opinions and will. In contrast, liberal democracies often show greater resistance to adopting new technologies and incorporating popular voting into public policy development.

However, this article is particularly significant in the context of Portugal, as there appears to be a growing political willingness to embrace technology to amplify the voice of its citizens, especially in policy-making and public administration decisions.

Hence, although Belgium is considered a liberal democracy, it has adopted policies that lean toward direct democracy, driven by certain movements (e.g., Youth4Climate). In practical terms, Belgian citizens elect representatives who make political decisions on their behalf, rather than directly participating in drafting laws and policies. The Belgian political system is also characterized by federalism, with communities and regions having legislative powers. While Belgium includes some elements of direct democracy, such as regional or community-level referendums, it primarily operates as a liberal representative democracy.

Similarly, Portugal has a representative parliamentary system in which voters elect deputies to the Assembly of the Republic, the country's main legislative body. These deputies are responsible for legislating and overseeing the government. Although Portugal does have some mechanisms for citizen participation, such as petitions to parliament and referendums, these direct democracy elements are infrequently used and are not central to the political system. Therefore, Portugal predominantly follows a liberal representative democracy model. However, there are signs that Portugal is beginning to adopt some practices more aligned with the Belgian framework. While the main system in both countries remains a liberal representative democracy, Portugal appears to be gradually incorporating limited elements of direct democracy, like Belgium.

4.3 Relevance and Originality

This paper adopts a multidisciplinary approach, meeting several scientific domains, particularly within the fields of social sciences and computer science. It explores contributions to political governance and state modernization through the expanding research on AI and e-Democracy within the context of the Portuguese political context. The relevance of this study is shown by the growing importance of e-Democracy in contemporary politics, exemplified by the use of digital technologies in Barack Obama's 2008 U.S. presidential campaign, which significantly transformed the political scene. A similar impact was observed in the 2016 U.S. election, indicating that the integration of technology in political processes is not a one-time occurrence but a developing trend. While these cases were not universally perceived as beneficial for political campaigns, genuine e-Democracy can enhance the political process and foster citizen engagement.

In Portugal, a study conducted by the University of Aveiro [10] examined e-participation practices within the government, revealing citizen involvement in e-collaboration initiatives such as the "My Movement" program. This study highlighted that, although there was a willingness among Portuguese citizens to engage in national-level initiatives, the development of local and regional electronic platforms for e-collaboration remained limited. The present work is particularly timely and relevant as it demonstrates the evolution of Portuguese citizen participation at local and regional levels since the findings of [10]. Additionally, it illustrates the Portuguese State's investment in the research and application of AI technologies for advancing e-Democracy, an area previously underexplored in the literature.

4.4 Limitations and Future Research

One significant limitation is the narrow geographical focus on Portugal, which may limit the generalizability of the results to other countries with different political, cultural, and administrative contexts. Additionally, the study primarily relies on qualitative methods, such as case studies and interviews, which, while providing in-depth contributions, may not capture the full scope of AI's impact across a more extensive and varied set of cases. This reliance on qualitative data also introduces the potential for bias, as the findings are contingent on the perspectives of a limited number of stakeholders. Another limitation is the relatively early stage of AI integration in public administration in Portugal, which means that the study's findings are more abstract than definitive regarding the long-term impacts of AI on governance and democratic processes. The article also does not fully explore the ethical implications of AI deployment in the public sector, such as issues related to privacy, bias, and accountability, which are crucial for a comprehensive understanding of AI's role in political governance.

For future research, it would be helpful to expand the study's scope to include comparative analyses across different EU countries and political systems to better understand how AI integration in governance varies in a certain region in the globe.

Employing a mixed-methods approach that combines qualitative and quantitative data could also provide a more robust analysis of AI's impact, allowing for more generalizable and statistically significant findings. Additionally, future studies should explore the ethical and societal implications of AI in governance, addressing concerns about transparency, fairness, and the potential for unintended consequences. Research could also explore the role of different stakeholders, including civil society, the private sector, and citizens, in determining the development and implementation of AI-driven governance policies.

Acknowledgements This article originates from the doctoral thesis entitled "From Artificial Intelligence Research to e-Democracy: Contributions to Political Governance and State Modernization", discussed at the University of Aveiro–Portugal in November 2023. Although the thesis was discussed through a compilation of articles, it lacked a comprehensive overview that synthesized the entire body of work. This article aims to fill this need by providing a cohesive summary of the main subjects of the thesis. This research was partially funded by the Research Centre in Digital Services (CISeD) with national funds provided by FCT—Foundation for Science and Technology, I.P., under project reference UIDB/05583/2020. We would also like to extend our gratitude to the Polytechnic Institute of Viseu for their support.

Credit Authorship Contribution Statement João Reis: Writing—review and editing, Writing—original draft, Visualization, Validation, Supervision, Software, Resources, Project administration, Methodology, Investigation, Funding acquisition, Formal analysis, Data curation, Conceptualization. Manuel do Carmo: Validation. David Pascoal Rosado: Validation. Pedro Alexandre Marques: Validation. Pedro Carmona Marques: Validation. Nuno Melão: Writing—original draft, Validation, Supervision, Conceptualization.

Data Availability Data will be made available on request.

Declaration of Competing Interest The authors declare that they have no known competing financial interests or personal relationships that could have appeared to influence the work reported in this paper.

References

1. AI4 Belgium: AI4Belgium. BOSA (2024). http://bosa.belgium.be/nl/themas/digitale-overheid/samenwerking-en-kennisdeling/nationale-samenwerking/ai4belgium
2. Arana-Catania, M., Lier, F.-A.V., Procter, R., Tkachenko, N., He, Y., Zubiaga, A., Liakata, M.: Citizen participation and machine learning for a better democracy. Digit. Gov. Res. Pract. **2**(3), 27:1–27:22 (2021). https://doi.org/10.1145/3452118
3. Berryhill, J., Heang, K.K., Clogher, R., McBride, K.: Hello, World: artificial intelligence and its use in the public sector. OECD (2019). https://doi.org/10.1787/726fd39d-en
4. Best, E., Robles, P., Mallinson, D.J.: The future of AI politics, policy, and business. Bus. Polit. **26**(2), 171–179 (2024). https://doi.org/10.1017/bap.2024.6
5. Büthe, T., Djeffal, C., Lütge, C., Maasen, S., von Ingersleben-Seip, N.: Governing AI—attempting to herd cats? Introduction to the special issue on the governance of artificial intelligence. J. Eur. Publ. Policy **29**(11), 1721–1752 (2022). https://doi.org/10.1080/13501763.2022.2126515
6. Cardoso, G., Cunha, C., Nascimento, S.: Bridging the e-democracy gap in Portugal: MPs, ICTs and political mediation. Inf. Commun. Soc. **9**(4), 452–472 (2006). https://doi.org/10.1080/13691180600858630
7. Editors: Artificial intelligence and democracy. J. Democracy 109–110 (2023)
8. Ernsdorff, M., Berbec, A.: Estonia: the short road to e-government and e-democracy. In: E-government in Europe. Routledge (2007)
9. European Commission: European Citizens' Initiative—European Commission (2024). https://commission.europa.eu/get-involved/engage-eu-policymaking/european-citizens-initiative_en
10. Fedotova, O., Teixeira, L., Alvelos, H.: E-participation in Portugal: evaluation of government electronic platforms. Procedia Technol. **5**, 152–161 (2012). https://doi.org/10.1016/j.protcy.2012.09.017
11. Fillet, S.: Artificial intelligence and citizen participation, a case-study—CitizenLab's Blog (2023). https://www.govocal.com/en-uk/case-studies/youth-for-climate-case-study
12. Firmstone, J., Coleman, S.: Public engagement in local government: the voice and influence of citizens in online communicative spaces. Inf. Commun. Soc. **18**(6), 680–695 (2015). https://doi.org/10.1080/1369118X.2014.986180
13. Go Vocal: Go Vocal: your next digital community engagement platform (2024). https://www.govocal.com/
14. Jalali, C.: Digitalization and political science in Portugal. In: Kneuer, M., Milner, H.V. (eds.) Political Science and Digitalization—Global Perspectives, 1st edn, pp. 175–196. Verlag Barbara Budrich (2019). https://doi.org/10.2307/j.ctvm7bc05.16
15. Mahrer, H., Krimmer, R.: Towards the enhancement of e-democracy: identifying the notion of the 'middleman paradox.' Inf. Syst. J. **15**(1), 27–42 (2005). https://doi.org/10.1111/j.1365-2575.2005.00184.x
16. Mills, A., Durepos, G., Wiebe, E.: Encyclopedia of Case Study Research. SAGE Publications, Inc. (2010) https://methods.sagepub.com/reference/encyc-of-case-study-research
17. Müller, A.P.R.: Citizens engagement in policy making: Insights from an e-participation platform in Leuven, Belgium. In: Engaging Citizens in Policy Making, pp. 180–195. Edward Elgar Publishing (2022). https://www.elgaronline.com/edcollchap-oa/edcoll/9781800374355/9781800374355.00020.xml
18. Nielsen, R.Ø., Hennen, L., Korthagen, I., Aichholzer, G., Lindner, R.: Options for improving e-participation at the EU level. In: Hennen, L., van Keulen, I., Korthagen, I., Aichholzer, G.,

Lindner, R., Nielsen, R.Ø. (eds.) European E-Democracy in Practice, pp. 329–359. Springer International Publishing, Berlin (2020). https://doi.org/10.1007/978-3-030-27184-8_13
19. Olsen, W.K., Haralambos, M., Holborn, M.: Triangulation in social research: qualitative and quantitative methods can really be mixed. In: Developments in Sociology. Causeway Press Ltd. (2004). https://research.manchester.ac.uk/en/publications/triangulation-in-social-research-qualitative-and-quantitative-met
20. Owens, C.: E-Democracy: Fundamentals and Applications, University of Delaware (2020). https://udspace.udel.edu/handle/19716/27727
21. Palomares, I., Martínez-Cámara, E., Montes, R., García-Moral, P., Chiachio, M., Chiachio, J., Alonso, S., Melero, F.J., Molina, D., Fernández, B., Moral, C., Marchena, R., de Vargas, J.P., Herrera, F.: A panoramic view and swot analysis of artificial intelligence for achieving the sustainable development goals by 2030: progress and prospects. Appl. Intell. **51**(9), 6497–6527 (2021). https://doi.org/10.1007/s10489-021-02264-y
22. Reis, J.: From artificial intelligence research to e-Democracy: contributions to political governance and state modernization (doctoral Thesis) (2023). https://ria.ua.pt/handle/10773/39753
23. Reis, J.: EU space security–an 8-Step online discourse analysis to decode hybrid threats. PLoS ONE **19**(7), e0303524 (2024). https://doi.org/10.1371/journal.pone.0303524
24. Reis, J., Melão, N.: E-democracy: artificial intelligence politics and state modernization. Proelium **8**(11), 399–432 (2023)
25. Reis, J., Santo, P., Melão, N.L: Artificial intelligence research and its contributions to the European Union's political governance: comparative study between member states. Soc. Sci. **9**(11), Article 11. https://doi.org/10.3390/socsci9110207
26. Reis, J., Santo, P., Melão, N.: Impact of artificial intelligence research on politics of the European Union member states: the case study of Portugal. Sustainability **12**(17), 6708 (2020). https://doi.org/10.3390/su12176708
27. Rubio, D., Lastra, C.: European Tech Insights 2019: Mapping European Attitudes to Technological Change and its Governance. Center for the Governance of Change. CGC (2019). https://www.ie.edu/cgc/research/european-tech-insights/
28. Sousa, M.J., Dal Mas, F., Osório De Barros, G., Tavares, N.: Artificial intelligence: technologies, applications, and policy perspectives. Insights from Portugal. In: Jeyanthi, P.M., Choudhury, T., Hack-Polay, D., Singh, T.P., Abujar, S. (eds.) Decision Intelligence Analytics and the Implementation of Strategic Business Management, pp. 69–84. Springer International Publishing (2022). https://doi.org/10.1007/978-3-030-82763-2_6
29. Starke, C., Lünich, M.: Artificial intelligence for political decision-making in the European Union: effects on citizens' perceptions of input, throughput, and output legitimacy. Data Policy **2**, e16 (2020). https://doi.org/10.1017/dap.2020.19
30. Sundberg, L.: Electronic government: towards e-democracy or democracy at risk? Saf. Sci. **118**, 22–32 (2019). https://doi.org/10.1016/j.ssci.2019.04.030
31. van Noordt, C., Misuraca, G.: Artificial intelligence for the public sector: results of landscaping the use of AI in government across the European Union. Gov. Inf. Q. **39**(3), 101714 (2022). https://doi.org/10.1016/j.giq.2022.101714
32. Vassiliades, A., Bassiliades, N., Patkos, T.: Argumentation and explainable artificial intelligence: A survey. Knowl. Eng. Rev. **36**, e5 (2021). https://doi.org/10.1017/S0269888921000011
33. Wirtz, B.W., Weyerer, J.C., Sturm, B.J.: The dark sides of artificial intelligence: an integrated AI governance framework for public administration. Int. J. Public Adm. **43**(9), 818–829 (2020). https://doi.org/10.1080/01900692.2020.1749851
34. Yin, R.K.: Case study research: Design and methods, 4th edn. Sage Publications (2009)

ESG and Stakeholder Engagement: Guidelines for Sustainable Organizational Development

Marta Silva Neves and Patrícia de Sá Freire

Abstract Organizations must consider socio-environmental indicators for survival, aligning their objectives with stakeholders' expectations through good governance. This chapter explores the integration of multiple actors via stakeholder equity and sustainable organizational development (SOD). SOD is an organizational macro-process involving strategies and practices aligned with corporate governance principles, addressing internal and external factors, and embracing transformation and future opportunities. Using the Knowledge Acquisition Design (KAD) framework, the authors propose twelve guidelines and a method for SOD based on stakeholder equity, validated by experts.

Keywords Organizational development · Sustainable organizational development · Stakeholders · Equity · Stakeholder equity

1 Introduction

The complex context of society, characterized by globalization, sustainability risks, geopolitical changes and immediacy [51], requires a reorientation of organizational development (OD) towards sustainable organizational development (SOD). This need is driven by the dynamics of the organization's internal and external environment, new ways of producing, consuming and innovating, new perceptions of work, changes in attitudes and behaviors, an understanding of being integrated and inseparable from the external context, and an awareness of the impact generated [10, 20, 21, 28, 41, 42, 52, 57–59, 66, 73].

In 2004, the *Who Care Wins* report presented the ESG guidelines, encouraging environmental, social, and governance criteria to be present in organizations'

M. S. Neves (✉) · P. de Sá Freire
Graduate Program in Knowledge Engineering and Management, Federal University of Santa Catarina (UFSC), Florianópolis, Brazil
e-mail: martaneves242@gmail.com

© The Author(s), under exclusive license to Springer Nature Switzerland AG 2025
A. d. b. Machado et al. (eds.), *Environmental, Social, Governance and Digital Transformation in Organizations*, Information Systems Engineering and Management 35, https://doi.org/10.1007/978-3-031-86079-9_7

strategies, productivity, and results, and encouraging stakeholders to monitor their soundness, costs, impact, and reputation [7, 36–40, 58, 67, 88].

However, Fernandes et al. [23] it points out that implementing ESG criteria is a long and arduous process that depends on the commitment of senior management and the awareness and commitment of the organizational community (p. 19). This complexity is exacerbated by the need for organizations to look beyond their own interests in order to exercise good governance, taking into account an organizational culture that is aligned with society's expectations.

This chapter aims to provide researchers with twelve guidelines and methodologies for sustainable organizational development based on stakeholder equity, validated by academic experts and practitioners in the field.

The main contributions of this chapter are as follows:

- Identify barriers and initiatives that move toward an integrative perspective on the relationship between stakeholder equity.
- Offer 12 guidelines and 38 steps to deepen the relationship between stakeholder equity and sustainable organizational development.
- Reflect on the socio-environmental impact, based on the interconnections between individuals, groups, and organizations in an integrative way for the sake of innovation, efficiency, responsiveness, engagement of managers in the creation of value, generating a sense of belonging in the search for alternatives and solutions of common interest for something new and transformative.

2 Methodology

This is an applied study with a qualitative approach developed using exploratory, descriptive [16, 32, 93] and propositional [4, 69] strategies. In the exploratory phase, the bibliographic research will be carried out using narrative literature review techniques [25], integrative literature review [91] and documentary research, in the descriptive and propositional phases, it will be carried out through the analysis of organizational and regulatory documents and case studies. To this end, a detailed and intensive evaluation of various cases in which one of the authors participated or observed [16, 33] as carried out.

In order to comply with the vision of applied research, the methodology of transdisciplinary co-production research [71, 72] was respected, using the Knowledge Acquisition Design (KAD) Framework [3], so that it would be possible to integrate academic knowledge and experts in the field during the course of the research, making it possible to consider the different instances of validation [64]. The verifications are related to the phases of formulation of the research (prior), development of the research (internal) and the results obtained (external), through interviews with experts in the field, making it possible to propose guidelines applicable to the Brazilian organizational context.

3 Sustainable Organizational Development

The literature review in search of an understanding of the trajectory of Organizational Development has revealed two main trends: one that focuses primarily on the relationship processes between people and groups within the organization [80], based on humanistic, democratic, and self-realization values; and another trend that, without abandoning the human system, focuses on the organization and its criteria, such as the need for organizational culture to conceive efficiency and profit objectives [8] and the need to organize work and the resulting human relationships according to factors intrinsic to the tasks and factors extrinsic to the organization [47].

In relation to the definitions proposed in the literature for the term Organizational Development, the definition [29] stands out because it presents a vision that is consistent with that of the various authors and reflects a greater breadth in relation to the complexity that OD plays in organizations and the contexts in which they are inserted.

> [...] multidirectional, multicultural, and multicontextual macroprocess, motivated and influenced by factors internal to the organization or external to it, of integral development during its trajectory towards its mission and vision of the future, made up of continuous microprocesses of knowledge management and governance, as well as organizational learning for adaptation, change, and transformation [29, p. 60].

The definition of Freire et al. [29] in the face of the sustainability scenario and its complexity leads to co-producing solutions for sustainable organizational development.

Among the characteristics that require this rethinking are: the inclusion of environmental, social and corporate governance criteria in investment decisions; diversity, non-discriminatory policies and employee relations; the development and health of work teams; knowledge sharing; systems thinking for a flexible organizational model; the use of big data and analytical capacity for AI integration; adapting to change, redefining work and measuring results; the need to identify, assess and manage socio-environmental risks in projects; managers' commitment to competency-based people management; and performance evaluation in line with organizational goals.

To further the understanding of the SOD construct, a literature review of authors and studies related to SOD was conducted.

It is worth highlighting the definition of SOD in Tomiotto [90, p. 118], as a guiding assumption that represents a metamorphosis and a cycle in the business context to deal with contextual factors (internal motivations and external pressures).

The analysis of empirical studies led to an initial mapping of key themes and aspects for SOD, which, after being analyzed, resulted in five major themes: knowledge management, work satisfaction, human resource management, organizational culture, and relationship with stakeholders.

- Knowledge Management

In studies of Wendo et al. [98]; Turi et al. [95]; Gatarik [31]; Xu et al. [101], categorized as knowledge management, the following key aspects stand out: the influence on innovation in products, services and technology based on the support of organizational ethics, stakeholder interest, conservation, waste reduction, efficient organizational practices, fair working practices, the well-being of people and vulnerable groups; the role of the knowledge worker, empowerment, leadership, structure, strategy and culture; the commitment of organizational thinking and action based on mutual trust for cooperation, interest in finding good solutions, mutual respect and active listening.

- Work Satisfaction

In the research of Dragan et al. [19], Androniceanu [5], Fernandes et al. [24], Czakert et al. [18] and Lapina et al. [46], categorized as job satisfaction, among the key aspects identified are: positive attitudes to identify opportunities for sustainability, based on perception, motivation, learning, beliefs, and culture, among others; the ability to solve problems based on the reuse of acquired knowledge; the reduction of stressful aspects in the organizational environment; solutions for balancing work time, control, clarity of rules, teams, and leaders committed to well-being.

- Human Resource Management

The theme of human resource management, as characterized by the works of Mishra [54], Mutmainnah et al. [56], Criveanu [17], Reddy et al. [76] reveals key aspects such as: practices that encourage pro-environmental behavior; innovative behavior at work in which employees are interested in the values, goals, and objectives of the organization; the relevance of digital technologies that enable organizations to collect and analyze valuable information about the behavior and preferences of both internal and external audiences; the application of technologies to better understand human behavior, inspiring greater employee engagement and retention of talent.

- Organizational Culture

With regard to organizational culture, authors Negi et al. [63], Sarker and Ashrafi [79], Stiglbauer et al. [87], Knoll et al. [45], Stegeager and Thomassen [86], Rashkova et al. [74], Li and Lin [48], Naudé [60–62], Collin et al. [15] emphasize in their studies: the importance of incorporating the interests of regulators, competitors, consumers, shareholders, and the government into the organization's culture and performance, employees as internal customers, from training and development as an opportunity to enhance skills; companies becoming attractive and retaining professionals by demonstrating understanding and congruence with people's values and needs; awareness of beliefs related to risk, contributing to new practices, and proposing alternatives for organizational change; the internal reconfiguration of organizations in terms of beliefs, values, and modes of production; the mechanisms for developing mindfulness in organizations in order to distance habitual practices that generate social inequalities; the concentration of organizations' efforts on solving everyday problems and generating new possibilities; the analysis of knowledge and of nature of problems and feelings from different perspectives; the existence of dialogical spaces among employees as an intervention to promote innovation and the development of the organization.

- Relationship with stakeholders

The relationship with stakeholders characterizes the studies of Camargo Benavides and Ehrenhard [11], Shamsuzzoha et al. [81], Pardo and Ortiz [68], Lăzăroiu et al. [49], Ivascu [43], Vărzaru [97], Aggerholm et al. [2], Cioca et al. [13], Magrin et al. [50], Vărzaru et al. [97]. Key aspects identified in these works are: vertical integration in organizations and the coalition of companies beyond profit maximization, combining economic, social, and environmental sustainability; mutual cooperation between organizations and educational institutions for qualification and contribution to the organization and society; a dynamic management system characterized by the composition of boards, knowledge, information systems, strategies, and resources, supported by performance and evaluation indicators; the balance between traditional financial and non-financial indicators; recognition of stakeholders and analysis of the relationships established; the importance of sustainability reports that extend the reporting measures of financial reports.

The findings identified in the analyzed studies allow to derive a new definition of the SOD construct as an organizational macroprocess composed of strategies and practices that are in line with corporate governance principles and guidelines that meet collective contextual needs, considering factors both internal and external to the organization, with openness to transformation and future opportunities.

4 Stakeholder Equity

When addressing equity, it is important to recall the first article of the Universal Declaration of Human Rights, proclaimed by the General Assembly of the United Nations, which states that all human beings are born free and equal in dignity and rights. Endowed with reason and conscience, they should act towards one another in a spirit of brotherhood [59]. Another relevant reference to take into account is Article 5 of the Constitution of the Federative Republic of Brazil, which states that everyone is equal before the law, without distinction of any kind, and Brazilians and foreigners residing in the country are guaranteed the inviolability of the right to life, liberty, equality, security and property [9].

These provisions invite reflection on the concepts of diversity, equality, inclusion, and equity.

Diversity is understood from three perspectives, the first of which considers diversity in terms of the differences that exist between individuals belonging to social groups, with reference to a social identity such as nationality, ethnicity, gender, age, among others [94]. The second understanding is based on the specific differences between people, such as personality, beliefs, experiences, education, values, social status, and economic status [44]. And the third understanding of diversity goes beyond identifying differences to understanding differences and similarities, integrating knowledge and learning [89].

Equality begins with the understanding that all people are equal before the law and seeks to ensure equal treatment for all. On the other hand, since all individuals are different, legislation may not take into account all the specificities that make up the social context, thus contributing to inequalities. According to Miranda [53, p. 29], a society based on the formal principle of equality is par excellence a society characterized by the social inequalities it promotes.

Inclusion recognizes and values diversity, otherwise it would be exclusion, as Faleiros [22] states that by denying the guarantee and effectiveness of civil, political, social and environmental rights and gender, race, ethnicity and territorial equity, we practice exclusion. The act of inclusion presupposes attitudes of welcoming particularities, establishing dialogues, exchanges, and relationships that consider experiences, needs, interests and contributions, leading to people's engagement. In this sense, for there to be inclusion, there needs to be a context that promotes fairness, respect and belonging, identifying different ways of developing actions; as Mota et al. [55] points out, equity means questioning the way we carry out our processes.

In A Theory of Justice, Rawls [75], presents equity as the basis for articulating justice, emphasizing the just as a priority. The principles of this theory are: to ensure freedom based on the equal distribution of basic duties and rights, and to consider the maintenance of economic and social inequalities fair only if it is to the benefit of the less fortunate [82].

Equity tries to complete what justice does not [12], that is, it adapts the rules to the different conditions of people. It removes barriers to inequality so that everyone has the same opportunities for education, work, leisure, culture, and so on.

To promote stakeholder engagement, adaptation, and participation, it is necessary to go beyond treating everyone equally, regardless of need, to acting with equity.

The principle of stakeholder equity provokes reflection on the obligations that the organization and its stakeholders have towards each other [27], through a cooperative approach as described by Phillips [70]:

> Whenever a person or groups of persons voluntarily accept benefits from a mutually beneficial cooperative arrangement that requires sacrifice or contribution on the part of the participants, and in which there is the possibility of free choice, obligations of equity are created between the participants in the cooperative arrangement, in proportion to the benefits accepted Phillips [70, p. 116].

From this perspective, relations with stakeholders are based on a moral definition, in which norms regulate rules of coexistence and in order for them to be applied, they must consider behavior, as Torres [92] points out. This author emphasizes that the requirement presented in the norms must be guided by moral meaning, taking into account the motive, the end, the means, the results and the consequences.

To do this, the organization must be aware of itself, its purpose and what its needs are, as well as its stakeholders', be they those who are already part of it, those who want to be part of it directly, or those who are interested in its performance and results, as Torres [92] states.

In light of the reflections from the literature review, it is clear that the opportunity for SOD lies in creating a welcoming and fair environment for all, based on equity, to overcome organizational barriers to deal with the characteristics of each context and business models with a focus on the future. This understanding is corroborated by Mota et al. [55, p. 44], when pointing out that more ethical, fair, healthy, and sustainable relationships increase the possibility of attracting and retaining talent, making quality decisions, achieving results, and innovating.

Based on the studies analyzed, a synthesis of key aspects corresponding to the stakeholder equity relationship was developed. These key aspects include:

- Equity is questioning the way we carry out our processes, identifying other ways to develop actions [55, p. 44];
- Equity seeks to complete what justice does not, that is, it adapts the rules to people's different circumstances [12];
- Equity means treating all shareholders and other stakeholders fairly, taking into account their rights, duties, needs, interests, and expectations, as individuals or collectively [40];
- The principle of stakeholder equity provokes reflection on the obligations that the organization and its stakeholders have towards each other, from a cooperative perspective [27, 70];
- Both the organization and the stakeholders need to know themselves in terms of purpose and needs [92];
- Normative stakeholders act with the organization, establishing mutual obligations in a cooperative relationship. Derivative stakeholders do not act directly with the organization, although they have some impact [70];
- Fairness is also manifested in the policies and relationship strategies of individuals with the organization, as commitment and trust are sought, recognizing the contributions made to the organization, such as knowledge, skills, and experience [1, 6, 34, 35, 77, 85, 83];
- Recognizing the importance of a sense of belonging and connectedness comes from an inclusive perspective in which attention is paid to people, relationships, uniqueness, and differences, fostering construction and collaboration to achieve a common goal and connecting to create something new and greater [14, 26, 78, 84];
- Respect and appreciation for all participants in the process bring equity in relationships as a central point to be considered [65];
- Transformations and improvements in processes happen through people, and it is necessary to know their values, beliefs, and behaviors [30];
- […] more ethical, just, healthy and sustainable relationships [55, p. 44].

Based on this summary, the concept of equity described in Freire et at. [29, p. 65], as good corporate governance practice characterized by fair and isonomic treatment of all interested parties (stakeholders), taking into account their rights, duties, needs, interests, and expectations is emphasized. This definition has been adopted for this study.

5 Projects and Organizational Context as a Case Study

The purpose of the case analysis was to verify the applicability of the proposed guidelines by answering the following question: *What is the impact on sustainable organizational development in projects that take stakeholder equity into account?*

. To this end, the strategy of participant observation was adopted in projects developed at the financial institution analyzed, facilitated by the accessibility of one of the authors had to it. The financial institution is located in the southern region of Brazil and has approximately 10,000 employees.

The projects, identified as A, B, C and D, are analyzed in the light of stakeholder equity through objectives, premises, implementation and results achieved, leading to reflections and assumptions.

Project A aims to raise awareness among local communities and the financial institution's employees about the importance of accessibility for people with disabilities or reduced mobility, promoting equality and democratizing access to spaces and services.

The basic premises of the Project are:

- raise community awareness by developing skills aimed at humanization and sustainability to improve the quality of life, by showing films with accessibility resources, fostering understanding and comprehension of the needs of people with disabilities in all their diversity;
- recognize people with disabilities as part of society, bringing the topic up for discussion, inherent to the culture and behavior of companies and citizens;
- disseminate the various facets of accessibility that are often overlooked in communication, behavior, and public policies, requiring this issue to be clarified and expanded in discussions in society;
- involve communities and society through mobilization and concerted action. It is important that citizens are informed, contextualized and encouraged to take ownership of the issue, empowering them to engage in the search for equality and equity.

The results achieved include:

- 6000 participants from the Brazilian states of Rio Grande do Sul and Santa Catarina;
- 30 regions covering several municipalities;
- 235 local partnerships;
- involvement of 561 employees of the financial institution;
- 128 bank branches involved;
- 131 employees trained in Brazilian Sign Language (Libras), taking the opportunity to practise dialoguing in this language;
- 15% increase in employees' interest in learning Libras to better serve people with disabilities;
- 62% increase in employee participation in the Distance Learning Accessibility course offered by the financial institution.

The reflection on the culture of diversity promoted by Project A inspired the development of **Project B**, the aim of which is to define procedures that favor the planning cycle of training strategies for the diversity process in the financial institution.

The basic guiding premises of Project B are:

- identify stakeholders related to diversity;
- identify opportunities to raise awareness and understanding of the theme in the organizational culture;
- integrate colleagues who identify with and have worked in civil society on issues related to gender, people with disabilities, ethnicity, and sexual orientation, forming a team coordinated by the institution's Corporate University.

The results achieved include:

- awareness, sensitization and mobilization;
- needs assessment;
- mapping knowledge gaps;
- co-creation and co-production;
- information generation and dissemination;
- risk reduction and conflict prevention;
- improvement of the evaluation of the Corporate University and its actions;
- improvement of image and more confidence to act.

The experience of Project B involved identifying, diagnosing and developing specific strategies for working with stakeholders, leading to improved attention for young people who were experiencing their first experience of the job market with the financial institution. Young people showed a need for financial guidance and planning to manage their salaries in the face of the prospects they had for the future. In this context, **Project C** was developed with the aim of promoting a collaborative movement to provide financial education guidance to young people working at the financial institution.

The basic guiding premises of Project C are:

- that the young people experience financial education in order to change an already consolidated culture regarding the use of financial products and services;
- a process that contributes to the formation of responsible individuals and societies that consistently take actions to improve well-being.

The results achieved include:

- the involvement of employees in the analysis of opportunities, both internal and external to the organization, providing opportunities for interaction with young people through dialogue and collective reflection on financial education, as well as the applicability of the knowledge developed;
- the gradual involvement of employees in the strategies developed, as they identify with the proposals, volunteering or being encouraged by those already involved;
- the sharing of knowledge and expertise among employees, adding their experience and training to the preparation of contents in different formats;

- the identification of the young people's perceptions, which was very important for each edition of the strategies. The responses obtained made it possible to identify how the educational action contributed to the identification of dreams, planning, conscious consumption, savings, and other topics of individual interest to the respondents.

The selection process at the aforementioned financial institution is carried out through a public notice, which includes a broad competition and a vacancy reservation system for people with disabilities and self-identified black people, indigenous people and transgender people. In this context, **Project D** aims to integrate the new employees into the institution by welcoming them and giving them a context about the company, its capillarity, its positioning in the market, its organizational guidelines, its operational strategies, and its banking business.

The basic guiding premises of Project D are:

- market context through benchmarking with other banks;
- perception of the integration process by employees who joined the company through a previous selection process;
- perception of branch and regional superintendence managers;
- perception of administrative unit managers;
- integration program based on co-creation and co-production with professionals from different technical areas, covering administrative units and the branch network.

The results achieved include:

- a movement to integrate the administrative units and the branch network in order to offer a systemic vision to new colleagues;
- co-production of knowledge among the administrative units, with the aim of welcoming, guiding, and inspiring new employees to build their careers;
- evaluation of the experience of each class in a progressive and systematic way, identifying improvements in the interaction with the public and also among the participants, in the presentation of content, and in the guidance for managers in welcoming new employees.
- definition of a multidisciplinary team to manage people on standby for any needs that may arise during the face-to-face sessions.

By considering stakeholder equity, these projects have impacts on sustainable organizational development, such as: recognizing stakeholders in terms of their needs and expectations, diverse skills, degree of influence versus interest in relationship management, adapting the corporate education system to the changes and dynamics of society, collaborative learning, integration and appropriation of knowledge, reflecting in improvements in organizational performance, capacity for change, and adaptation to new and uncertain situations.

In order to achieve these impacts, the constituent elements of the SOD were systematized, considering stakeholder equity, as shown in Table 1.

Table 1 Constituent elements of the SOD, considering stakeholder equity

Constituent elements of the SOD—**Projects A** and **B**
• Recognizing that different people have different abilities and contribute to the development of organizations and society, thus demonstrating respect for the individual and acceptance, free from discrimination, prejudice, and stigma
• Recognizing the stakeholders in terms of their needs and expectations, uncovering skills and knowledge gaps; the need to pay attention to diversity awareness in organizational policies and plans, and learning styles and methodologies; a systematic and relational approach
• Mapping the stakeholders in the face of diversity makes it possible to identify all the parties interested in the issue, the level of influence x interest, and the level of management
• Involving stakeholders through a cognitive and affective way right from the conception of a project generates identification, enthusiasm, belonging and the necessary alliances
• Involving managers in strategies and actions contributes to understanding diversity and strengthening inclusive practices, as well as the adoption and integration of assistive resources and technologies to offer products and services
• Managing stakeholder relationships helps to prioritize educational activities to meet needs and guide planning, objectives, contents, and a balanced design of programs
• Considering the corporate training system as a means of sharing knowledge increases employees' awareness of their role in the organization and in the value chain; the dialogue and the development of competences promote organizational sustainability
• Documenting and institutionalizing the strategies developed with stakeholders strengthens: sensitization, awareness, and mobilization; the importance of identifying needs and knowledge gaps; the strategy of co-creation and co-production, generation, and dissemination of information; risk reduction and conflict prevention; confidence to act and the organization's reputation
• Transparency in the dissemination of information strengthens credibility and the common purpose, as it reduces anxiety, distrust, and resistance
Constituent elements of the SOD—**Projects C** and **D**
• The need for the corporate education system to adapt to the demands and changes resulting from the dynamics of society and the needs of the new generations for competence in daily practice and experience
• By contributing to the educational experience, stakeholders also experience learning to know, learning to do, learning to live together, and learning to be, promoting progress in "being competent for the common good" and in results for the community, self-motivation, and the application of skills and knowledge
• Collaborative learning goes through a process of maturing: respecting the participants and their life experiences; creating and maintaining a welcoming space for conversation and learning; developing expertise, feeling and thinking about lived experiences; providing an active role of self-direction and autonomy
• Integrating and appropriating knowledge facilitates interaction, understanding, vision and shared knowledge among stakeholders, meeting the stakeholders' expectations, achieving results and strengthening knowledge governance
• Taking into account the perception of internal and external stakeholders through various feedback channels contributes to learning, knowledge generation, organizational performance, capacity for change, and adaptation to new and uncertain situations

In addition to participant observation of the aforementioned projects, significant data and reports on equity were identified, including *Diversity, Equity and Inclusion Lighthouses* [99, 100]. The DE&I initiatives of the organizations analyzed in these reports pointed to five common success factors in the results achieved:

- Understanding the root of the problem: implies understanding the problem with a deep database; identifying the root causes; obtaining information from the target population; prioritizing and sequencing the problem areas;
- Meaningful definition of success: setting clear and quantifiable goals; articulating a clear case for change that connects employees to action;
- Responsibility and investment from corporate leaders: incorporating the goals of DE&I as a core business priority; holding leaders accountable for results achieved; CEO and senior leaders leading and shaping the desired change; ensuring resources for longevity in budget, expertise, and timeline;
- solution designed for the context: developing solutions that address root causes with scalability in mind; integrating changes into key processes and ways of working so that the impact is sustainable; equipping and encouraging employees to contribute.
- rigorous tracking and course correction: defining key performance indicators and milestones for periodic tracking; using data and feedback to correct course.

6 Guidelines and Methodology for SOD Based on Stakeholder Equity

Based on the aspects highlighted in the scientific literature and in the documentary analysis, the aim is to make progress in identifying procedures that foster the stakeholder equity relationship for sustainable organizational development.

From the specific perspective of the stakeholder equity relationship, guidelines were identified from both scientific and organizational knowledge according to the KAD framework. These guidelines enabled interpretations, convergences, and complementarities as transdisciplinary co-creation to contribute to solutions for organizational practices, aiming at SDO.

For the analysis process, the relevance of the guidelines and respective notes was observed, checking for possible repetitions, the possibility of combining them, the proximity between the contents, the compliance with equity, and the consideration of sustainable organizational development, as well as grammatical adjustments.

The resulting guidelines were systematized into 12 guidelines and 38 steps for a SOD method, based on stakeholder equity.

Next, the consistency of these guidelines and the method was checked through interviews with experts in the field of financial institutions working in the areas of management and governance, as well as experts from organizations in different segments of society with experience in managing processes aimed at sustainability.

Based on the observations, suggestions and points of attention recorded by the experts in the field, a table was prepared to analyze each statement of the guidelines and each statement of the corresponding step, in accordance with the respective governance principle [40], as shown in Table 2.

The meaning of each Governance principle and the respective guidelines identified are described below.

The **Integrity** principle, which focuses on the continuous improvement of the ethical culture, with attention to conflicts of interest, coherence between speech and action, as well as commitment to the organization and care for its stakeholders, society, and the environment, revealed two guidelines:

1. Recognize the diversity of stakeholders that make up the ecosystem of which the organization is a part, manifesting organizational ethics based on the principle of equity as a basic premise.
2. Establish a relationship of common good with awareness of diversity, equity, and inclusion in organizational policies and plans, learning styles and methodologies, linking efforts to values, mission, and business results.

The **Transparency** principle aims to provide stakeholders with true, timely, consistent, clear and relevant information. Adopting transparency goes beyond complying with the requirements of laws and regulations to provide information on financial, environmental, social and governance performance in support of business development and trust in relationships with all stakeholders.

3. Encourage openness to sharing and interfacing with the internal and external environment.
4. Understand the perceptions of people and groups to minimize polarizations and increase trust in the organization.

The **Equity** refers to the fair treatment of all stakeholders in terms of rights, responsibilities, needs, interests, and expectations, both individually and collectively. In this way, attention to a sense of justice, respect, diversity, inclusion, pluralism, and equal rights and opportunities means that the organization develops differentiated

Table 2 Fields in the results analysis table

<table>
<tr><td colspan="3">Principle of governance (1)</td></tr>
<tr><td colspan="3">Original guideline with number and explanation or corresponding step with number and explanation (2)</td></tr>
<tr><td>Level 1</td><td>Level 2</td><td>Interrelationship with scientific knowledge</td></tr>
<tr><td>Observations and suggestions from the experts (3)</td><td>Observations and suggestions from the experts (4)</td><td rowspan="2">Interrelation of observations and suggestions with scientific knowledge (7)</td></tr>
<tr><td>Points of attention from the experts (5)</td><td>Points of attention from the experts (6)</td></tr>
<tr><td colspan="3">Guideline or respective step after checking (8)</td></tr>
</table>

approaches in its relationships with stakeholders. Three guidelines are identified with this principle:

5. Understand what creates barriers to diversity, equity, and inclusion, and map the stakeholders in relation to diversity, identifying the level of interest and influence and the level of management to be developed.
6. Create policies, guidelines, and procedures for the workforce that are fair and that consider the well-being of vulnerable people and groups.
7. Recognize the mutual interest between the organization and stakeholders, adapting the corporate educational system to the demands and changes, based on the dynamics of society and the creation of sustainable values.

The **Accountability** principle is characterized by the exercise of leadership in a diligent and independent manner, taking responsibility for the consequences of actions or omissions. Similarly, actions must be reported in a clear, concise, understandable, and timely manner. This principle recognizes that decisions can lead to individual accountability as well as impacts on the organization, stakeholders and the environment. Based on the convergence between the principles derived from scientific knowledge and practical experience, two principles emerged:

8. Develop all leaders and managers to inspire learning and a collaborative environment, and provide feedback through open and respectful communication.
9. Consider that the board of directors and top management are influenced by the benefits of a culture of diversity, equity, and inclusion.
10. Commit the leaders to the results achieved.

The **Sustainability** principle is based on ensuring the economic and financial viability of an organization to reduce negative externalities in business and operations. At the same time, it seeks to increase positive externalities by considering the various capitals in business models, including financial, intellectual, human, social, natural, and reputational. Thus, this principle aims to strengthen the organization and its responsibility to society based on an understanding of the interdependence of the social, economic, and environmental dimensions. The analysis identified two guidelines:

11. Integrate policies and actions involving human rights, socio-environmental performance, and the fidelization of relations with communities.
12. Secure resources for initiatives and programs that contribute to stakeholder relationship management.

The steps for each guideline are systematized below as a method for SOD, as detailed in Table 3.

Table 3 Guidelines and method for SOD based on stakeholder equity

Corporate governance principle	Guideline	Method for SOD based on stakeholder equity
Integrity	1. **Recognize the diversity of stakeholders that make up the ecosystem of which the organization is a part, manifesting organizational ethics based on the principle of equity as a basic premise**	1. Identify the interests of stakeholders, taking into account the creation of sustainable values, such as: poverty reduction, solutions to social problems, corporate volunteering, partnerships with companies and civil society
		2. Adopt congruence with people's values and needs, including generational and cultural differences
		3. Adopt endomarketing to integrate the interests of stakeholders into the culture and performance of the organization
		4. Adopt digital technologies to collect and analyze information on the behavior and preferences of both internal and external audiences, taking into account the sustainability perspective
		5. Analyze the relationships established using technological tools to better understand the purpose of changes in human behavior
	2. **Establish a relationship of common good with awareness of diversity, equity, and inclusion in organizational policies and plans, learning styles and methodologies, linking efforts to values, mission, and business results**	6. Establish organizational policies and strategies, identifying processes that are not aligned and points to be developed
		7. Consider the corporate education system with employees, the value chain and the community to increase awareness and organizational sustainability
		8. Empower employees to have the authority and freedom to develop ideas

(continued)

Table 3 (continued)

Corporate governance principle	Guideline	Method for SOD based on stakeholder equity
		9. Develop critical and systemic thinking, as well as the search for solutions, supported by mutual respect and active listening
		10.Focus efforts on solving everyday problems and creating new opportunities
		11. Develop corporate citizens for sustainable business models and for the well-being of society in general
		12. Promote mutual cooperation between the organization and educational institutions for qualification and the exchange of practical cases and real experiences
Transparency	3. **Encourage openness to sharing and interfacing with the internal and external environment**	13. Create dialogic spaces for the perception of mental models (beliefs and values)
		14. Integrate the stakeholders approach as an organizational strategy in people management, based on dialogue, identification of expectations, partnership dynamics, and review of the organizational role
		15. Promote a collaborative environment to encourage effective and efficient knowledge transfer
	4. **Understand the perceptions of people and groups to minimize polarizations and increase trust in the organization**	16. Adopt a policy of transparency and promote it in the dissemination of information through all the company's communication channels
		17. Adopt a sustainability report to add social, environmental and climate vectors to the financial reports developed, based on standards such as GRI, SASB, TCFD, and Integrated Report

(continued)

Table 3 (continued)

Corporate governance principle	Guideline	Method for SOD based on stakeholder equity
		18. Monitor comments and statements from stakeholders
		19. Analyze feedback and use it to make necessary adjustments to the plan
Equity	5. **Understand what creates barriers to diversity, equity, and inclusion, and map the stakeholders in relation to diversity, identifying the level of interest and influence and the level of management to be developed**	20. Support customers, employees, suppliers in the value chain, and society, in socio-economic problems, adapting strategies to sustainable organizational development
		21. Work on psychological and social factors to identify new organizational opportunities, taking into account perception, motivation, learning, beliefs, and culture
Equity	6. **Create policies, guidelines, and procedures for the workforce that are fair and that consider the well-being of vulnerable people and groups**	22. Recognize that people have different backgrounds and experiences, and varied skills
		23. Promote the diversity of suppliers as part of the business ecosystem, helping to create better workplaces and supporting better performance and social advancement
		24. Develop the stakeholder relationship management
	7. **Recognize the mutual interest between the organization and stakeholders, adapting the corporate educational system to the demands and changes, based on the dynamics of society and the creation of sustainable values**	25. Implement mechanisms for individual and collective awareness of habitual practices that generate social inequalities
		26. Promote collaborative learning, taking into account the cognitive and affective backgrounds when involving stakeholders in projects
		27. Prioritize educational actions to meet needs and guide planning based on stakeholder relationship management

(continued)

Table 3 (continued)

Corporate governance principle	Guideline	Method for SOD based on stakeholder equity
Accountability	8. **Develop all leaders and managers to inspire learning and a collaborative environment, and provide feedback through open and respectful communication**	28. Train, develop, and retain employees by measuring progress in conjunction with performance management
		29. Develop preventive practices and solutions to balance working hours, mobility between home and work, clarity of agreements, and teams and leaders committed to well-being
		30. Define a performance evaluation system and its indicators linked to improvement, competitiveness, and value creation for society, including the suppliers in the value chain
Accountability	9. **Consider that the board of directors and top management are influenced by the benefits of a culture of diversity, equity, and inclusion**	31. Engage managers in inclusive leadership strategies and practices
	10. **Commit the leaders to the results achieved**	32. Integrate changes in key processes and ways of working, setting clear and quantifiable targets for change, and connecting with employees in action
		33. Define key performance indicators and link executive compensation to these indicators
		34. Implement financial and non-financial indicators (customers, internal organizational processes, learning processes, and organizational development) in an integrated manner with the SOD
		35. Document, institutionalize and report on the developed strategies

(continued)

Table 3 (continued)

Corporate governance principle	Guideline	Method for SOD based on stakeholder equity
		36. Develop an organizational remuneration and rewards system for healthy, stable and long-term relationships
Sustainability	11. **Integrate policies and actions involving human rights, socio-environmental performance, and the fidelization of relations with communities**	37. Promote pro-environmental understanding and behavior in organizational processes and procedures
	12. **Secure resources for initiatives and programs that contribute to stakeholder relationship management**	38. Make organizational practices efficient, so that waste reduction and conservation of natural resources can contribute to the budget for organizational citizenship initiatives and programs and commitment to the community

7 Final Considerations

The transdisciplinary process of co-production between academic knowledge and field or organizational knowledge characterized throughout this study resulted in the 12 guidelines and 38 steps that deepen the relationship between stakeholder equity and Sustainable Organizational Development.

This process of reflection brings theoretical and practical contributions as a result of this research.

The theoretical contributions include the relationship between the assumptions of Organizational Development and the complexity of the ability to adapt to change in organizations, resulting from literature reviews and reflections on OD and the context of sustainable development [20 21, 52, 58, 67, 73]; a new definition of the construct Sustainable Organizational Development and then the search for an understanding of the term SOD and its related aspects [2, 5, 11, 13, 15, 17–19, 24, 31, 43, 45, 46, 48–50, 54, 56, 60–63, 68, 74, 76, 79, 81, 86, 87, 95–98, 101]; the identification of key aspects of the stakeholder equity relationship, based on an examination of the concept of equity and its aspects [1, 6, 9, 12, 14, 22, 26, 27, 30, 34, 35, 40, 44, 53, 55, 59, 65, 70, 77, 78, 83–85, 89, 92, 94].

Practical contributions in terms of technologies, new strategies, and tools include the provision of guidelines and methodologies for Sustainable Organizational Development based on stakeholder equity, from the reflections that the various mentioned authors presented on diversity, inclusion, and equity, as well as the obstacles encountered and the initiatives that move towards an integrative perspective on the relationship of stakeholder equity; the socio-environmental impact, based on the interconnections among individuals, groups, and organizations in an integrative way for the sake of innovation, efficiency, responsiveness, and engagement of managers in the creation of value, generating a sense of belonging in the search for alternatives and solutions of common interest for something new and transformative. In this way, the benefits extend beyond the boundaries of the organization, reflecting on issues such as poverty reduction, social problem solving and corporate citizenship, and increasing legitimacy in the eyes of the stakeholders.

With regard to future research, it is suggested to:

- evaluate and validate the implementation of the "Guidelines and Methodology for Sustainable Organizational Development Based on Stakeholder Equity" in public and private organizational contexts in order to generate case studies that allow qualitative and quantitative analysis;
- evaluate the results of the implementation of the "Guidelines and Methodology for Sustainable Organizational Development Based on Stakeholder Equity";
- advance the scientific knowledge on the concepts of sustainable organizational development and stakeholder equity, as these are new concepts with much discussion and work to be done;
- analyze the renewal of corporate boards with the perspective of diversity, equity, and inclusion as a factor for improving commitments to socio-environmental performance and increasing performance;
- investigate the approach and perception of singularities and differences as a driving force for the process of organizational change towards something new and transformative.

References

1. Adams, J.S.: Inequity in social exchange. Adv. Exp. Soc. Psychol. **2**, 267–299 (1965)
2. Aggerholm, H.K., Andersen, S.E., Thomsen, C.: Conceptualising employer branding in sustainable organisations. Corp. Commun. Int. J. **16**(2), 105–123 (2011)
3. Alvares, L.M.A.R., Freire, P.S.: Knowledge Acquisition Design: Framework for Transdisciplinary Co-Production Research of Scientific and Technological Knowledge. Anthem Press, London (2022)
4. Alves-Mazzotti, A.J.: Relevância e aplicabilidade da pesquisa em educação. Cadernos de Pesquisa. São Paulo: Fundação Carlos Chagas/Cortez, no. 113, pp. 39–50 (2001)
5. Androniceanu, A.: Key issues related to the human resources management now a days. In: Proceedings of the 6th International Management Conference "approaches in organisational management", 15–16 Nov 2012
6. Assmar, E.M.L., Ferreira, M.C., Souto, S.O.: Justiça organizacional: Uma revisão crítica da literatura. Psicologia Reflexão e Crítica **18**, 443–453 (2005)
7. BID: Cívica, vol. 1, no 1 (2021). https://publications.iadb.org/pt/civica-vol-1-no-1-2021

8. Blake, R.R., Mouton, J.S.: A estruturação de uma empresa dinâmica através do desenvolvimento organizacional do tipo Grid. Trad. Meyer Stilman. São Paulo, Blücher, 126p (Série Desenvolvimento Organizacional) (1972)
9. Brasil: Constituição da República Federativa do Brasil (1988). Disponível em: https://www.planalto.gov.br/ccivil_03/constituicao/constituicao.htm
10. CEBDS: ESG: as três letras que estão mudando os investimentos (2019). Disponível em: https://cebds.org/esg-as-tres-letras-que-estao-mudando-comportamento-os-investimentos/
11. Camargo Benavides, A.F., Ehrenhard, M.: Rediscovering the cooperative enterprise: a systematic review of current topics and avenues for future research. VOLUNTAS **32**, 964–978 (2021)
12. Carvalho Filho, M.P.: Indenização por Equidade no Novo Código Civil, 2a edn. Atlas, São Paulo (2003)
13. Cioca, L.-I., et al.: Sustainable development model for the automotive industry. Sustainability **11**, 6447 (2019). https://www.mdpi.com/2071-1050/11/22/6447
14. Cisne, C.S.: Competitividade sistêmica: conhecimento como fator de produção de capital social para o desenvolvimento local. Dissertação (Mestrado), Programa de Pós-Graduação em Engenharia e Gestão do Conhecimento, Universidade Federal de Santa Catarina, Florianópolis (2015)
15. Collin, K., Paloniemi, S., Vähäsantanen, K.: Multiple forms of professional agency for (non)crafting of work practices in a hospital organization. Nord. J. Working Life Stud. **5**(3a) (2015)
16. Creswell J.W.: Projeto de Pesquisa: métodos qualitativo, quantitativo e misto. Artimed, Porto Alegre (2010)
17. Criveanu, M.M.: Investigando Intensidade Digital e E-Commerce como Drivers para Sustentabilidade e crescimento económico nos países da UE. Electronics **12**(10) (2023)
18. Czakert, J.P., et al.: Adaptation and Psychometric Cross-Cultural Validation of a Workplace Cyberbullying Questionnaire in Spain and Germany. Cyberpsychology Behav. Soc. Networking **24**(12) (2021)
19. Dragan, F., Luo, C., Ivascu, L., Ali, M.: Assessing the importance of psychosocial factors associated with sustainable organizational development during COVID-19. Front. Psychol. **12**, 647435 (2021). https://doi.org/10.3389/fpsyg.2021.647435
20. Elalfy, A., Palaschuk, N., El-Bassiouny, D., Wilson, J., Weber, O.: Scoping the evolution of corporate social responsibility (CSR) research in the sustainable development goals (SDGs) era. Sustainability **12**(14), 5544 (2020)
21. Falck, O., Heblich, S.: Corporate social responsibility: doing well by doing good. Bus. Horiz. **50**(3), 247–254 (2007)
22. Faleiros, V.P.: Inclusão social e cidadania. In: 32a Conferência Internacional de Serviço Social, pp. 1–15. DF, Anais, Brasília (2006). Disponível em: https://www.icsw.org/images/docs/Events/2006_Brazil/17_07_PDF/vicente_faleiros.pdf
23. Fernandes, et al.: ESG—Meio ambiente, sociedade e governança na educação. SENAR AR/PR, Curitiba (2023)
24. Fernandes, et al.: A case-base approach to workforces' satisfaction assessment. In: CONFENIS: Research and Practical Issues of Enterprise Information Systems, pp 191–206 (2016)
25. Ferrari, R.: Writing narrative style literature reviews. Med. Writ. **24**(4), 230–235 (2015)
26. Formanski, J.G.: A estrutura da rede social organizacional e sua influência no fluxo de conhecimento inovador. Tese (Doutorado), Programa de Pós-Graduação em Engenharia e Gestão do Conhecimento, Universidade Federal de Santa Catarina, Florianópolis (2018)
27. Freeman, E.R., et al.: Stakeholder Theory: the State of the Art. Cambridge Press, New York (2010)
28. Freeman, R.E.: Strategic management: a stakeholder approach. In: Lamb, R. (ed.) Advances in Strategic Management, vol. 1, pp. 31–60. JAI Press, Greenwich (1983)
29. Freire, P.S., et al.: Glossário: Governança Multinível do Conhecimento e da Aprendizagem e seus mecanismos de Universidade Corporativa em Rede e de Comunicação Dialógica, 152p. CRV, Curitiba (2021)

30. García, J., Durán, S., Parra, M., Martínez, H.: Inserção, integração e equidade no local de trabalho: Cenário empresarial pós-conflito na Colômbia. J. Soc. Sci. **24**(3), 36–49 (2018)
31. Gatarik, E.: The value of knowledge sharing in decision-making and organisational development: a model-theoretic systemic analysis of an intervention in an Austrian S.M.E. Econ Res Ekonomska Istra_Zivanja **32**(1), 148–167 (2019). https://doi.org/10.1080/1331677X.2018.1552175
32. Gil, A.C.: Métodos e técnicas de pesquisa social, 7th edn. Atlas, São Paulo (2019)
33. Godoy, A.S.: Introdução à pesquisa qualitativa e suas possibilidades. RAE Revista de Administração de Empresas, São Paulo **35**(2), 57–63 (1995)
34. Greenberg, J.: Equity and workplace status: a field experiment. J. Appl. Psychol. **73**(4), 606–613 (1988)
35. Greenberg, J., Ornstein, S.: High status job title compensation for under payment: a test of equity theory. J. Appl. Psychol. **68**(2), 285–297 (1983)
36. IBGC: Código das melhores práticas de governança corporativa. IBGC, São Paulo (2015)
37. IBGC: Governança Corporativa e Boas Práticas de Comunicação, São Paulo (2017)
38. IBGC: Diversidade e inclusão são energia para inovação, resultados sustentáveis e geração de valor aos negócios, São Paulo (2021)
39. IBGC: Diversidade e Inclusão. Revista RI, pp. 50–51 (2022)
40. IBGC: Código das melhores práticas de governança corporativa, 6a edição. IBGC, São Paulo (2023)
41. IPEA: Objetivos do Desenvolvimento do Milênio—Relatório Nacional de Acompanhamento. Ipea, Brasília (2004, 2005, 2007, 2010 e 2014)
42. International Finance Corporation: Executive Summary IFC 2005 annual report. IFC, Washington (2005). Disponível em: http://www.ifc.org
43. Ivascu, L.: Measuring the implications of sustainable manufacturing in the context of industry 4.0 Processes **8**, 585 (2020). https://www.mdpi.com/2227-9717/8/5/585
44. Kandola, R., Fullerton, J.: Diversity: more than just na empty slogan. Pers. Manag. **26**(11) (1994)
45. Knoll, M., Neves, P., Schyns, B., Meyer, B.: A multi-level approach to direct and indirect relationships between organizational voice climate, team manager openness, implicit voice theories, and silence. Appl. Psychol. Int. Rev. **70**(2), 606–642 (2021). https://doi.org/10.1111/apps.12242
46. Lapina, I., Briede, I., Aramina, D.: Interrelation of process management and employee stressors in organization. In: Proceedings of the 25th World Multi-Conference on Systemics, Cybernetics and Informatics (WMSCI) (2021)
47. Lawrence, P.R., Lorsch, J.W.: O desenvolvimento de organizações: diagnóstico e ação. Trad. Heyer Stilman. São Paulo, Blucher, 112p (Série. Desenvolvimento Organizacional) (1972)
48. Li, M., Lin, K.: A new paradigm of organizational transformation: enacting wholeness praxis in the oneness of problem and possibility. Syst. Pract. Action Res. **24**, 107–132 (2011)
49. Lăzăroiu, G., et al.: Sustainability management and performance in the urban corporate economy: a systematic literature review. Sustainability **12**, 7705 (2020). https://www.mdpi.com/2071-1050/12/18/7705
50. Magrin, M.E., Marini, E., Nicolotti, M.: Employability of disabled graduates: resources for a sustainable employment. Sustainability **11**, 1542 (2019). https://doi.org/10.3390/su11061542
51. Margherita, A., Secundo, G.: The stakeholder university as learning model of the extended entreprise. J. Manag. Dev. **30**(2), N175-186 (2011)
52. Van Marrewijk, M.: Conceitos e definições de RSC e sustentabilidade corporativa: entre agência e comunhão. J. Bus. Ethics **44**, 95–105 (2003). https://doi.org/10.1023/A:1023331212247
53. Miranda, M.G.: A organização escolar em ciclos e a questão da igualdade substantiva. Revista Brasileira de Educação **14**(40) (2009). Disponível em: http://www.scielo.br/pdf/rbedu/v14n40/v14n40a03.pdf
54. Mishra, P.: Green human resource management: a framework for sustainable organizational development in an emerging economy. Int. J. Organ. Anal. **25**(5), 762–788 (2017). https://doi.org/10.1108/IJOA-11-2016-1079

55. Mota, R., et al.: Diversidade, Equidade e Inclusão: tornar simples o que parece completo, 140p. Haikai, São Paulo, epub (2022)
56. Mutmainnah, D., et al.: The impact of directive leadership on innovative work behavior: the mediation role of continuance commitment. J. Indonesian Econ. Bus. **37**(3), 268–286 (2022)
57. Nações Unidas Brasil: Agenda 2030 para o desenvolvimento sustentável. Disponível em: Agenda 2030 para o Desenvolvimento Sustentável|As Nações Unidas no Brasil Acesso em 01 maio 2022 (2015)
58. Nações Unidas: Princípios para o investimento responsável (2007). Disponível em: https://www.unpri.org/download?ac=10969#:~:text=Os%20Princ%C3%ADpios%20para%20o%20Investimento%20Respons%C3%A1vel%20(PRI)%20foram%20criados%20por,para%20as%20pr%C3%A1ticas%20de%20investimento
59. Nações Unidas: Declaração Universal dos Direitos Humanos (1948). Disponível em: https://brasil.un.org/pt-br/91601-declaracao-universal-dos-direitos-humanos
60. Naudé, M.: Corporate governance, sustainable organisational development and employees as a valuable resource. Corp. Ownership Control **6**(4, Special Issues) (2009, Summer)
61. Naudé, M.: Increasing sustainable organisational development by using e-portfolios. Corp. Ownership Control **8**(1) (2010, Fall) (Continued—1)
62. Naudé, M.: Sustainable organizational development and reflection: a good combination? Corp. Ownership Control **9**(2) (2012) (Continued—3)
63. Negi, R., et al.: Effect of green marketing orientation dimensions on green innovation and organizational performance: a mediation-moderation analysis. Bus. Strategy Environ. (2023). https://doi.org/10.1002/bse.3429
64. Ollaik, L.G., Ziller, H.M.: Concepções de validade em pesquisas qualitativas. Educação e Pesquisa, [s.l.] **38**(1), 229–242 (2012)
65. Ostrom, E.: Crossing the great divide: coproduction, synergy, and development. World Dev., [s.l.] **24**(6), 1073–1087 (1996). Disponível em: https://doi.org/10.1016/0305-750X(96)00023-X
66. PNUD: Programa das Nações Unidas para o Desenvolvimento. Transformando nosso mundo: a Agenda 2030 para o desenvolvimento sustentável. PNUD, Brasília (2015). Disponível em: <http://www.br.undp.org/content/dam/brazil/docs/agenda2030/undp-br-Agenda2030-completo-pt-br-2016.pdf>
67. Pacto Global: Ambição pelos ODS: expandindo o impacto sobre os negócios na década de ação. Pacto Global das Nações Unidas, Nova York (2020)
68. Pardo, M.M., Ortiz, A.S.: Alcanzando el Éxito a través de la Sinergia entre las Tecnologías de la Información y la Cadena de Valor: El Caso de las PYME em el Cluster Minero de Antofagasta. J. Technol. Manag. Innov. **4**(1) (2009)
69. Patton, M.Q.: Paradigms and pragmatism. In: Fetterman, D.M. (ed.) Qualitative Approaches to Evaluation in Education: the Silent Scientifique Evolution Analysis. Praeger, New York (1988)
70. Phillips, R.: Stakeholder theory and organizational ethics. Berett-Koecler Publishers, São Francisco, CA (2003)
71. Pohl, C., Hadorn, G.H.: Principles for designing transdisciplinar research. GAIA Ecol. Perspect. Sci. Soc. **26**(3), 232–232 (2017)
72. Polk, M.: Transdisciplinary co-production: designing and testing a transdisciplinary research framework for societal problem solving. Futures **65**, 110–122 (2015). https://doi.org/10.1016/j.futures.2014.11.001
73. Porter, M.E., Kramer, M.R.: Creating shared value. Harv. Bus. Rev. **89**(1/2), 62–77 (2011)
74. Rashkova, Y., et al.: Addressing the societal challenges in organizations: the conceptualization of mindfulness capability for social justice. J. Bus. Ethics (2023)
75. Rawls, J.: A Theory of Justice, p. 475. Harvard University Press, Cambridge, Mass (1971)
76. Reddy, A.J.M., Rani, R., Chaudhary, V.: Technology for sustainable HRM: Na empirical research of health care sector. Int. J. Innov. Technol. Exploring Eng. (IJITEE) **9**(1) (2019). ISSN: 2278-3075

77. Robbins, D.W., et al.: Treinamento de resistência em pares agonista-antagonista: uma breve revisão. Revista de Pesquisa em Força e Condicionamento **24**(10), 2873–2882 (2010, outubro de).https://doi.org/10.1519/JSC.0b013e3181f00bfc
78. Salm, V.M.: A contribuição do ciclo do conhecimento para o desenvolvimento das redes de coprodução dos serviços públicos em meio ambiente. Tese (Doutorado), Programa de Pós-Graduação em Engenharia e Gestão do Conhecimento, Universidade Federal de Santa Catarina, Florianópolis (2020)
79. Sarker, M.A.R., Ashrafi, D.M.: The relationship between internal marketing and employee job satisfaction: a study from retail shops in Bangladesh. J. Bus. Retail Manag. Res. (JBRMR) **12**(3) (2018)
80. Schein, E.H., Bennis, W.G.: Personal and Organizational Change Through Group Methods. Wiley, New York (1966)
81. Shamsuzzoha, A., et al.: Development of value proposition to promote green innovation for sustainable organizational development. Cleaner Eng. Technol. **15** (2023)
82. Silveira, D.C.: Teoria da Justiça de John Rawls: entre o liberalismo e o comunitarismo. In: Trans/Form/Ação, pp. 169–190. Universidade Estadual Paulista, São Paulo (2007)
83. Siqueira, M.M.M., Gomide, S., Jr.: Vínculos do Indivíduo com o Trabalho e com a Organização. In: Zanelli, J.E., Borges-Andrade, J.E., Bastos, A.V.B. (eds.) Psicologia, organizações e trabalho no Brasil, pp. 316–350. Artmed, Porto Alegre, RS (2014)
84. Siqueira, M.M.M., Padovam, V.A.R.: Bases teóricas de bem-estar subjetivo, bem-estar psicológico e bem-estar no trabalho, vol. 24, no. 2, pp. 201–209. Teoria e pesquisa, Brasília, Psicologia (2008)
85. Spector, P.E.: Industrial and organizational psychology: research and practice, 6th edn. Wiley, Hoboken, NJ (2012)
86. Stegeager, N., Thomassen, A.O.: Sustainable organizational development. J. Workplace Learn. **33**(3), 149–154 (2021)
87. Stiglbauer, B., et al.: Work values across generations: development of the new work values scale (NWVS) and examination of generational differences. Front. Psychol. **13** (2022)
88. Stilingue, Global, P.: A evolução do ESG no Brasil. Environmental, Social, and Governance (2021)
89. Thomas, D., Ely, R.: Making differences matter: a new paradigm for managing diversity. Harv. Bus. Rev. **74**, 79–90 (1996)
90. Tomiotto, M.F.: Gestão da sustentabilidade organizacional: explorando suas bases teórico-empíricas a partir da Grounded Theory. Dissertação (Mestrado), Programa de Pós-Graduação em Administração, Universidade Estadual de Londrina, Londrina (2019)
91. Torraco, R.J.: Writing integrative literature reviews: guidelines and examples. Hum. Resour. Dev. Rev. **4**(3), 356–367 (2005)
92. Torres, L.H.: Teoria do Stakeholder: um estudo da aplicação do princípio de equidade do Stakeholder. Dissertação (Mestrado), Programa de Pós-Graduação em Administração na Universidade Federal do Estado do Rio Grande do Sul, Porto Alegre (2013)
93. Triviños, A.N.S.: Introdução à pesquisa em Ciências Sociais: a pesquisa qualitativa em educação. Atlas, São Paulo (1995)
94. Tung, R.L.: Managing cross-national and intra-national diversity. Hum. Resour. Manage. **22**, 461–477 (1993)
95. Turi, J.A., Sorooshian, S., Javed, Y.: Impact of the cognitive learning factors on sustainable organizational development. Heliyon **5**, e02398 (2019)
96. Vărzaru, A.A.: An Empirical framework for assessing the balanced scorecard impact on sustainable development in healthcare performance measurement. Int. J. Environ. Res. Public Health **19**(22), 15155 (2022)
97. Vărzaru, A.A., et al.: Assessing the contribution of managerial accounting in sustainable organizational development in the healthcare industry. Int. J. Environ. Res. Public Health **20**, 2895 (2023). https://doi.org/10.3390/ijerph20042895
98. Wendo, D.R., Rop, F.K., Mwanzu, A.M.: Knowledge management for innovativeness and sustainable organizational development: a case study of selected corporate institutions in Kenya. Libr. Philos. Pract. (e-journal) 2021. https://digitalcommons.unl.edu/libphilprac/5086/

99. World Economic Forum: Global Parity Alliance: Diversity, Equity and Inclusion Lighthouses (2024). Disponível em: https://www.weforum.org/publications/diversity-equity-and-inclusion-lighthouses-2024/
100. World Economic Forum: Global Parity Alliance: Diversity, Equity and Inclusion Lighthouses (2023). Disponível em: https://www.weforum.org/reports/global-parity-alliance-diversity-equity-and-inclusion-lighthouses-2023
101. Xu, J., Wu, H., Zhang, J.: Innovation research on symbiotic relationship of organization's tacit knowledge transfer network. Sustainability **14**, 3094 (2022). https://doi.org/10.3390/su14053094

Strategic Leadership and Value Creation: Evaluating the Impact of ESG Program Implementations in the Life Sciences Industry from Real World Evidence and Market Access Perspectives

Antonio Pesqueira, Andreia de bem Machado, and Marc François Richter

Abstract The life sciences industry is increasingly facing challenges related to environmental, social, and governance (ESG) issues, which impact its long-term sustainability. Real World Evidence and Market Access (RWEMA) functions play a crucial role in addressing these challenges by integrating real-world data to support ESG program implementations. The purpose of this study is to evaluate how ESG program implementations, guided by RWEMA, can enhance strategic leadership and value creation within the life sciences sector. A systematic literature review was conducted, focusing on real-world evidence and market access perspectives in the context of ESG integration. The research design included analyzing multiple documents and bibliographic sources to understand the impact of ESG practices on sustainability, compliance, and corporate value. The findings reveal that RWEMA has the potential to drive ESG integration, significantly improving brand reputation, financial stability, and adaptability in life sciences companies. The transformation of RWEMA from a supporting function to a strategic leader is critical in overcoming barriers to ESG adoption, thereby enhancing overall market credibility and sustainability. This research contributes to the understanding of ESG integration by providing an actionable framework for leveraging RWEMA as a strategic

A. Pesqueira (✉)
Technology and Architecture Research Department, Iscte—University Institute of Lisbon, Lisboa, Portugal
e-mail: Antonio_Pesqueira@iscte-iul.pt

A. de bem Machado
Graduate Program in Engineering and Knowledge Management, Federal University of Santa Catarina (UFSC), Florianopolis, Brazil
e-mail: andreia.bem@ufsc.br

M. F. Richter
Graduate Program in Environment and Sustainability (PPGAS), State University of Rio Grande do Sul (Uergs), São Francisco de Paula, Brazil
e-mail: marc-richter@uergs.edu.br

© The Author(s), under exclusive license to Springer Nature Switzerland AG 2025

A. d. b. Machado et al. (eds.), *Environmental, Social, Governance and Digital Transformation in Organizations*, Information Systems Engineering and Management 35, https://doi.org/10.1007/978-3-031-86079-9_8

driver. The insights presented here offer valuable guidance for life sciences companies aiming to align their operations with broader sustainability goals, ultimately ensuring long-term viability and stakeholder trust.

Keywords ESG integration · Real world evidence · Market access · Life sciences · Sustainability

1 Introduction

The integration of environmental, social, and governance (ESG) practices has become a key concern for industries worldwide, particularly within the life sciences sector. This industry's reliance on innovation, regulatory compliance, and stakeholder relationships makes it especially susceptible to the growing demand for sustainable and ethical practices. ESG principles are increasingly recognized not only as regulatory requirements but also as crucial drivers of long-term business value, risk mitigation, and market credibility [1, 2]. Addressing these challenges requires the implementation of effective strategies that integrate ESG considerations into all levels of organizational decision-making.

Key concepts in this context include ESG, which refers to a framework that incorporates environmental stewardship, social responsibility, and corporate governance to drive sustainable business practices [1]. Real World Evidence and Market Access (RWEMA) is another key concept, involving the use of real-world data to inform decisions regarding the accessibility and value of medical products [2]. RWEMA combines two elements, using real-world data to support and accelerate market access for new treatments, devices, or healthcare solutions. By integrating RWE (Real World Evidence) with MA (Market Access), companies can better demonstrate the value of their products in real-world conditions, thereby enhancing their chances of approval, adoption, and reimbursement within healthcare systems. The problem the present study aims to solve is the underutilization of RWEMA as a strategic tool for ESG integration. Despite its potential to bridge data-driven insights with sustainability initiatives, RWEMA's role in ESG leadership has not been fully recognized, leading to missed opportunities for enhancing corporate value, stakeholder trust, and long-term sustainability [3].

RWEMA functions have emerged as critical drivers of ESG integration in the life sciences industry. RWEMA utilizes real-world data from clinical practice and patient experience to support decisions related to access, pricing, and sustainability of medical products. By bridging the gap between research, commercial teams, and external stakeholders, RWEMA plays an essential role in transforming ESG principles from theoretical concepts into actionable practices that create tangible value. Despite its potential, the role of RWEMA in ESG leadership has often been overlooked, and its strategic significance underutilized. Recent literature (2020 onwards) emphasizes the growing importance of integrating ESG considerations into market access strategies. Studies have shown that ESG-driven initiatives contribute to enhanced corporate reputation, improved regulatory compliance, and increased stakeholder engagement [4]. Moreover, empirical evidence suggests that RWEMA, by leveraging real-world data, can significantly influence the effectiveness of ESG programs, thereby enhancing value creation and sustainability efforts [5, 6].

This study aims to explore the role of RWEMA as a strategic leader in implementing ESG programs within life sciences companies. Specifically, it seeks to evaluate how RWEMA's unique position can enhance ESG integration, ultimately contributing to corporate value creation, stakeholder trust, and long-term sustainability. The study also addresses how RWEMA's data-driven approach can provide empirical support for ESG decision-making, thereby establishing its position as a central player in sustainable business transformation.

The objectives of this research are threefold: (1) to assess the current challenges and opportunities associated with ESG integration in the life sciences sector, (2) to evaluate the role of RWEMA in enhancing ESG practices, and (3) to provide an actionable framework for leveraging RWEMA as a strategic driver of ESG initiatives. To achieve these objectives, the following research questions were defined: (a) What are the key challenges and opportunities related to ESG integration in the life sciences sector? (b) How can RWEMA enhance ESG practices within life sciences companies? (c) What actionable framework can be developed to leverage RWEMA as a strategic driver of ESG initiatives? To achieve these objectives, a systematic literature review was conducted, focusing on existing empirical evidence, market access strategies, and ESG practices in life sciences. The analyzed literature included peer-reviewed journal articles, industry reports, regulatory documents, and case studies published from 2020 onwards. This review encompassed both qualitative and quantitative studies, providing a comprehensive understanding of the interplay between ESG and RWEMA. The data used for the analysis involved bibliographic sources, real-world datasets from healthcare systems, and market analysis reports. This approach allowed for the identification of key areas where RWEMA can contribute to enhancing sustainability and governance within the industry.

The significance of this study lies in its ability to offer a new perspective on ESG integration, focusing on the often-underestimated role of RWEMA. By highlighting the strategic importance of RWEMA in driving ESG initiatives, this research not only contributes to the academic literature but also provides practical insights for life sciences companies seeking to improve their sustainability practices. The findings of this study are expected to assist industry leaders in making informed decisions that align their business operations with global sustainability goals, thus ensuring long-term competitiveness and resilience.

2 Methodology

2.1 Research Design

This study employed a qualitative research design, specifically adopting an interpretative approach. The choice of a qualitative methodology was driven by the need to explore and gain in-depth insights into the role of RWEMA in the integration of ESG practices within the life sciences industry. The study aimed to evaluate how RWEMA can enhance strategic leadership, value creation, and ESG integration through qualitative exploration of existing literature, documents, and real-world evidence.

The objectives of this research were threefold: (1) to assess the current challenges and opportunities associated with ESG integration in the life sciences sector, (2) to evaluate the role of RWEMA in enhancing ESG practices, and (3) to provide an actionable framework for leveraging RWEMA as a strategic driver of ESG initiatives. The following research questions were defined to guide the study: (a) What are the key challenges and opportunities related to ESG integration in the life sciences sector? (b) How can RWEMA enhance ESG practices within life sciences companies? (c) What actionable framework can be developed to leverage RWEMA as a strategic driver of ESG initiatives?

The chosen methodology effectively addressed these objectives by employing a bibliographic and documentary search to gather in-depth qualitative data from various sources, including peer-reviewed articles, industry reports, and regulatory documents [8, 9]. This approach allowed for a comprehensive exploration of the challenges and opportunities related to ESG integration, as outlined in the first objective. Thematic analysis [10] enabled the identification of key themes, providing insights into how RWEMA can enhance ESG practices, thus directly answering the second research question. Finally, the systematic review and synthesis of findings contributed to the development of an actionable framework, addressing the third objective and research question.

The theoretical conceptual framework underpinning this research draws from concepts of corporate sustainability, stakeholder theory, and ESG integration. It is guided by the principles of environmental stewardship, social responsibility, and corporate governance, as well as their application in healthcare and life sciences sectors.

2.2 Methodological Approach

The methodological approach taken for this study involved a bibliographic and documentary search rather than a bibliometric analysis. This interpretative qualitative approach was chosen to explore the existing literature in detail and to develop a nuanced understanding of how RWEMA can contribute to ESG practices, particularly in the context of the life sciences industry. According to Hart [9], a bibliographic search is essential for identifying gaps in the literature and understanding the current state of knowledge, which made it suitable for the exploratory nature of this research.

The bibliographic and documentary search provided the flexibility needed to incorporate multiple sources of information, such as peer-reviewed articles, industry reports, regulatory documents, case studies, and expert opinions [8]. This diverse set of sources enabled a more comprehensive understanding of the current practices and challenges related to ESG integration. The bibliographic search process involved identifying both recent and seminal works that were relevant to the study objectives, ensuring the inclusion of a broad spectrum of perspectives. Additionally, the use of regulatory documents allowed for the inclusion of practical guidelines and frameworks that are essential for understanding ESG requirements in different sectors [11].

The use of this method aligns well with the research objectives, allowing for a comprehensive examination of various factors contributing to ESG integration, including the challenges faced, potential solutions, and the evolving role of RWEMA. This approach allowed the study to explore ESG practices within the life sciences sector in an in-depth manner, ensuring that the findings were grounded in both theoretical and practical insights.

2.3 Data Collection

The data collection process involved systematically searching bibliographic and documentary sources related to ESG, RWEMA, and sustainability practices in healthcare and life sciences. Selection criteria for sources were based on relevance to the research objectives, recency (with a focus on publications from 2020 onwards), and credibility. Sources included peer-reviewed journal articles, books, industry reports, regulatory guidelines, and documentary records from credible databases such as Scopus, PubMed, and ScienceDirect.

The sampling for the data included documents that discussed ESG frameworks, real-world evidence applications, and market access strategies, as well as industry-specific challenges related to sustainability. Although quantitative sample size considerations do not apply in this qualitative context, the study sought to gather a broad and diverse set of sources to enrich the understanding of the subject. The demographic characteristics of the literature sources were also varied, involving different regions (such as Europe, North America, and Asia) and sectors (pharmaceuticals, biotech, and healthcare providers).

2.4 Data Analysis

The collected data were analyzed using a thematic analysis technique, as outlined by Braun and Clarke [10]. This technique was chosen because it is well-suited for identifying, analyzing, and reporting patterns (themes) within qualitative data, allowing for a rich and detailed understanding of the content. Thematic analysis involved systematically coding the content from the selected sources and identifying recurring themes and patterns related to ESG integration and the role of RWEMA. This approach enabled the study to synthesize a wide range of qualitative data, helping to capture both explicit and underlying themes. Thematic analysis allowed for the identification of key themes, debates, and gaps within the literature, providing insights into how RWEMA can enhance ESG integration and support corporate sustainability.

Key themes identified included the enhancement of corporate reputation, regulatory compliance, sustainability value creation, overcoming ESG adoption barriers, and providing empirical support for ESG decision-making. By employing thematic analysis, the study was able to comprehensively explore how RWEMA could play a strategic role in promoting ESG integration, contributing to better decision-making and more sustainable practices within the life sciences sector.

The analysis also included cross-referencing the findings against the defined research questions and objectives, ensuring consistency and alignment with the goals of the study.

2.5 Bias Removal Considerations

To address potential bias in data selection and interpretation, several measures were taken. The bibliographic search was conducted using well-established and credible databases, including Scopus, PubMed, and ScienceDirect, to ensure the quality and reliability of the information. Moreover, multiple sources were cross-referenced to validate the findings, thereby reducing the risk of confirmation bias. The interpretation of findings was also subjected to peer discussion and critical review to maintain a neutral perspective, aligning with the conventions of neutral scientific inquiry as outlined by Springer Nature and Elsevier. The goal was to apply the norms of objectivity and impartiality, ensuring that the analysis remained unbiased and scientifically rigorous.

2.6 Overview of Methods and Data Sources

Table 1 provides a concise overview of the methods and techniques used in this study, along with the data sources utilized at each step. This structured approach ensured a rigorous and systematic examination of ESG integration and RWEMA's role in

Table 1 Methods steps, techniques and data sources

Step	Method/Technique	Data source
Research design	Qualitative interpretative approach	N/A
Methodological approach	Bibliographic and documentary search	Peer-reviewed articles, industry reports, regulatory documents, case studies, expert opinions
Data collection	Systematic search	Scopus, PubMed, ScienceDirect, industry reports, books, regulatory guidelines
Selection criteria	Relevance, Recency, credibility	Publications from 2020 onwards, multiple credible databases
Data analysis	Thematic analysis technique	Selected literature and documents
Cross-referencing	Cross-checking against research questions and objectives	Research objectives, key questions, findings
Bias removal	Use of credible databases, peer discussion	Scopus, PubMed, ScienceDirect, expert peer review

Source The authors

the life sciences industry. The focus on credible data sources and systematic data analysis aimed to provide a comprehensive understanding while minimizing bias and maintaining scientific integrity.

3 Results

3.1 Enhancing Corporate Reputation and Stakeholder Trust

To cross-validate the findings against the research objectives and questions, it is evident that enhancing corporate reputation and stakeholder trust aligns with the objective of evaluating RWEMA's role in ESG integration. Specifically, this addresses research question (b): 'How can RWEMA enhance ESG practices within life sciences companies?' By demonstrating the impact of RWEMA-supported initiatives on corporate reputation, this study highlights how RWEMA contributes to building trust and credibility, which is a crucial element for ESG integration. RWEMA is a healthcare approach that focuses on using real-world data, including information from clinical practice and patient experience, to inform decisions related to access to medicines and medical devices. Unlike relying solely on data from controlled clinical trials, RWEMA seeks to incorporate real-world information on the efficacy, safety, and value of treatments. This can involve analyzing data from sources such as electronic medical records, health insurance databases, and prescription records to provide a more comprehensive view of a treatment's performance in real-world conditions. To develop a comprehensive understanding of how RWEMA could enhance corporate reputation and stakeholder trust, it was essential to examine the existing theoretical foundations and research findings. Corporate reputation is often built through transparency, ethical behavior, and the demonstration of value to stakeholders. Recent literature emphasizes the growing importance of ESG initiatives in fostering trust, particularly within the life sciences sector, which faces significant scrutiny from regulators and the public. Studies indicate that real-world evidence can serve as an effective tool in demonstrating accountability and value creation, thereby enhancing stakeholder confidence [4]. However, a gap remains in understanding how the integration of RWEMA specifically contributes to the strengthening of ESG efforts, especially in contexts requiring cross-functional collaboration. The key finding from this analysis is that RWEMA-supported ESG programs not only improve transparency but also create a robust foundation for stakeholder engagement by providing evidence-based justification for sustainability claims. The findings of this study highlight the transformative role of RWEMA in driving the integration of ESG practices within the life sciences industry. By systematically analyzing the existing literature, industry reports, and real-world datasets, several key themes have emerged, providing insights into how RWEMA can enhance ESG initiatives and create long-term value.

The results indicated that RWEMA played a significant role in enhancing corporate reputation and building stakeholder trust. ESG-driven initiatives supported by real-world evidence demonstrated the ability to improve transparency and accountability, which were crucial for maintaining trust among stakeholders, including patients, healthcare providers, investors, and regulators. By leveraging real-world data from various sources—such as healthcare systems, market reports, and industry documentation—RWEMA helped life sciences companies provide evidence-based justifications for their ESG programs. This comprehensive use of data sources not only strengthened corporate credibility but also contributed to deeper stakeholder engagement. The results provided a deeper understanding of how these data sources formed the main basis for evaluating the impact of ESG program implementations from Real World Evidence and Market Access perspectives.

The focus on value-based elements like Value Story Definition, Integrated Evidence, and Adaptive Value-Based Contracts within the Market Access Strategy sector highlights the critical role of governance in negotiating fair pricing and reimbursement strategies. These elements are crucial in ensuring that pharmaceutical products are not only effective and safe but also accessible and affordable, reflecting the social commitments of the ESG framework. The capture of complex relationships between RWEMA, and ESG dimensions, require detailed depiction of direct interactions and outcomes in enhancing its utilities and value management connections as seen in Fig. 1.

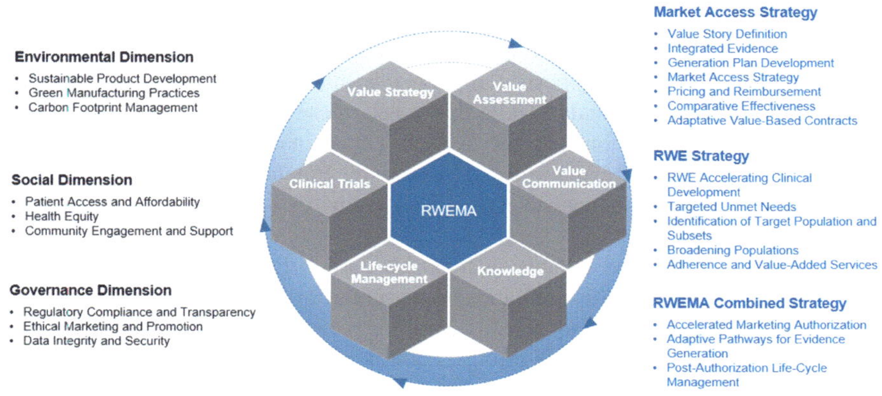

Fig. 1 Integrative framework of ESG, RWEMA strategies in the life sciences industry

3.2 Strengthening Regulatory Compliance and Risk Management

Cross-checking the results against the research objectives, strengthening regulatory compliance and risk management directly addresses the objective of assessing the challenges and opportunities associated with ESG integration in the life sciences sector. This also aligns with research question (a): 'What are the key challenges and opportunities related to ESG integration in the life sciences sector?' By providing real-time insights and enhancing adherence to ESG standards, RWEMA supports a proactive compliance strategy, contributing to risk mitigation, which is essential for successful ESG adoption.

Theoretical and empirical research has consistently highlighted the importance of regulatory compliance as a foundational component of ESG integration. RWEMA's role in enhancing compliance is particularly relevant in the life sciences sector, where regulatory standards are stringent and constantly evolving. Recent studies have found that real-world data, when leveraged effectively, helps organizations anticipate regulatory changes and proactively adapt their ESG practices [7]. This capability allowed life sciences companies to mitigate the risks associated with non-compliance and position themselves as leaders in the adoption of sustainable practices. Despite these findings, existing research often lacked a detailed examination of how RWEMA specifically contributed to effective risk management. The findings of this study build upon this gap by demonstrating that RWEMA provided critical real-time insights that supported companies in navigating regulatory landscapes and maintaining a proactive stance in compliance.

The integration of global goals and principles, such as the United Nations Global Compact (UNGC) and the Sustainable Development Goals (SDGs), provides a broad framework for aligning regulatory compliance with global sustainability efforts. Reporting frameworks like the Global Reporting Initiative (GRI), Task Force on

Climate-Related Financial Disclosures (TCFD), and Sustainability Accounting Standards Board (SASB) further emphasize the importance of transparent ESG reporting Bose et al. [12]. RWEMA supports compliance with these frameworks by enabling companies to gather and analyze data that meets these rigorous standards. Moreover, adherence to regulations like the Corporate Sustainability Reporting Directive (CSRD), EU Taxonomy Regulation, and Sustainable Finance Disclosure Regulation (SFDR) becomes more manageable through RWEMA's capabilities in generating data-driven insights.

In addition, ratings and indices such as the Carbon Disclosure Project (CDP), MSCI ESG Ratings, and Sustainalytics are essential in assessing the ESG performance of companies. RWEMA's role in compiling real-world evidence helps companies improve their scores in these ratings, thereby demonstrating their commitment to sustainability. Studies have shown that real-world evidence can play a significant role in informing market access decisions and supporting regulatory compliance for new therapies [13]. Another key finding is that RWEMA contributed to strengthening regulatory compliance and mitigating risks related to ESG issues. Life sciences companies operate in a highly regulated environment, and ESG compliance is becoming increasingly critical. For example, Agarwala et al. [14] emphasized the use of real-world evidence in supporting precision medicine, highlighting its importance in aligning clinical and regulatory needs. The results showed that RWEMA, by providing real-world insights, helped companies navigate complex regulatory landscapes, ensuring adherence to environmental, social, and governance standards. This proactive approach to compliance not only mitigated risks but also positioned companies as industry leaders in sustainable practices [15, 16].

3.3 RWEMA as a Strategic Leader in ESG Implementation

The key insights from the attached document reinforce the role of RWEMA as a transformative element in driving ESG integration within the life sciences industry. ESG programs are critical for life sciences companies, not only to enhance their adaptability and brand reputation but also to ensure mid- and long-term financial sustainability. The ability of RWEMA to lead these ESG efforts has been particularly evident during challenging events like the COVID-19 pandemic, which highlighted the necessity for strong ESG practices to maintain resilience [1].

RWEMA's unique position as a "bridging" function between internal stakeholders (such as R&D and commercial teams) and external partners makes it well-suited for taking on a leadership role in ESG initiatives. It functions as a credible scientific partner that prioritizes patient and customer-centric values, holding accountability, ensuring compliance, and effectively bridging strategic relationships. This role enables RWEMA to establish a foundation for cross-functional ESG initiatives that directly contribute to enhancing company reputation and market credibility [2].

Moreover, RWEMA's leadership in ESG extends beyond being a support function to becoming a strategic leader capable of overcoming long-standing barriers to

ESG integration. These barriers include a lack of standardized metrics, fragmented ESG strategies, and inadequate stakeholder collaboration. By transforming RWEMA into a strategic driver, life sciences companies can leverage real-world evidence to support ESG decision-making processes and build stronger relationships with regulators, patients, and investors. This proactive approach, supported by empirical data from RWEMA, fosters transparency, strengthens governance, and ensures that ESG claims are backed by actionable evidence, ultimately improving market access and compliance [3].

The incorporation of ESG metrics in business operations by RWEMA allows companies to align their activities with broader sustainability goals. For instance, ESG-driven value propositions enhance corporate performance by building investor trust, driving sustainable innovation, and contributing to long-term financial value. These activities not only align companies with the expectations of modern investors but also ensure adherence to global sustainability frameworks like the United Nations SDGs. This alignment not only enhances reputation but also secures the company's long-term viability by adapting to changing market expectations [17].

Overall, the transformation of RWEMA from an enabler to a strategic leader is a key determinant for the success of ESG initiatives within the life sciences sector. RWEMA's ability to provide empirical support and facilitate cross-functional collaboration empowers companies to overcome existing challenges in ESG implementation and positions them as leaders in sustainability and responsible governance [18, 19].

3.4 Driving Sustainability and Value Creation

The findings related to driving sustainability and value creation are closely linked to the research objectives of providing an actionable framework for leveraging RWEMA as a strategic driver of ESG initiatives. ESG a set of criteria that represents a major trend and an essential response by companies to address contemporary societal challenges. It emphasizes the integration of economic value generation with environmental, social, and corporate governance concerns. In practice, ESG reflects a company's responsibility and commitment to its market, consumers, suppliers, employees, and investors. It serves as a critical criterion for assessing corporate sustainability, extending beyond financial metrics to evaluate whether a company is a viable option for sustainable investment. Essentially, ESG aims to measure the capability and commitment of a company to generate positive financial, social, and environmental impacts. This addresses research question (c): 'What actionable framework can be developed to leverage RWEMA as a strategic driver of ESG initiatives?' By showing how RWEMA integrates real-world data to drive sustainability and create value beyond traditional financial metrics, the study builds a compelling case for utilizing RWEMA as a tool for value creation in the context of ESG integration. The concept of value creation in the context of ESG has evolved significantly in recent years. Traditional financial metrics are increasingly being complemented by non-financial measures, such as environmental impact and social contributions.

RWEMA contributes directly to sustainability by enabling life sciences companies to incorporate real-world data into decision-making, ultimately aligning their products and practices with global ESG standards. Studies have shown that companies utilizing real-world evidence are better positioned to develop sustainable innovations and respond to emerging social and environmental challenges [4]. Nonetheless, there is still a gap in understanding the role of RWEMA in capturing long-term sustainability benefits beyond traditional economic indicators. The current study's key finding is that RWEMA's ability to identify environmental and social impact areas allows for value creation that extends beyond financial performance, thus aligning corporate goals with broader societal needs. The study also reveals that RWEMA has a direct impact on driving sustainability and value creation within the life sciences sector. By integrating real-world data into decision-making processes, RWEMA supports the development of sustainable products and practices that align with global ESG standards. This approach enhances the long-term financial stability of companies by reducing costs associated with non-compliance, improving operational efficiencies, and attracting socially responsible investors. Moreover, the ability of RWEMA to identify and address key areas of environmental and social impact allows companies to create value beyond financial metrics, contributing to broader societal goals [20–22].

3.5 Overcoming Barriers to ESG Adoption

In cross-referencing the findings with the defined research questions, overcoming barriers to ESG adoption directly corresponds to the objective of assessing challenges and evaluating RWEMA's role in enhancing ESG practices. This addresses research questions (a) and (b) by demonstrating how RWEMA can help life sciences companies overcome challenges such as fragmented ESG strategies and cross-functional misalignment. By positioning RWEMA as a strategic leader, the study identifies key solutions for effectively integrating ESG initiatives. The barriers to ESG adoption in the healthcare industry include issues such as insufficient data integration, lack of standardized ESG metrics, and resistance to change. These challenges hinder the ability of healthcare organizations to fully embrace ESG initiatives that could lead to enhanced patient outcomes and sustainability. RWEMA's role in healthcare settings is crucial in addressing these barriers by providing a data-driven approach to link patient outcomes with ESG strategies, thereby fostering a culture of accountability and transparency.

Recent studies, such as Machado et al. [1], highlight the importance of digital transformation in enhancing ESG practices post-COVID-19, emphasizing that RWEMA's use of real-world data can support sustainable growth and bridge gaps in ESG implementation. Peixoto et al. [2] further underline the significance of standardized ESG metrics to assess sustainability effectively, which aligns with RWEMA's capability to provide comprehensive, data-driven insights that facilitate the measurement of ESG outcomes. Additionally, Richter and de Bem Machado [3] discuss the role of circular economy practices in overcoming sustainability challenges, suggesting that integrating RWEMA with circular economy initiatives can enhance environmental outcomes within life sciences companies. These perspectives provide further support for the transformative role of RWEMA in addressing and overcoming barriers to ESG adoption in the healthcare and life sciences sectors.

In the life sciences sectors like pharma and biotech, barriers are often more specific to the complexities of regulatory compliance and market demands. For hospitals, pharmacies, and drug manufacturing companies, compliance with regulations such as the Corporate Sustainability Reporting Directive (CSRD), EU Taxonomy Regulation, and Sustainable Finance Disclosure Regulation (SFDR) presents practical challenges. These challenges include navigating complex and sometimes conflicting regulatory requirements across different countries, managing the significant financial burden associated with meeting ESG compliance standards, and integrating these ESG goals with ongoing operational activities. Hospitals, for instance, face challenges in data collection and standardization required by the CSRD, while pharmacies struggle with the administrative workload of meeting SFDR disclosures. Drug manufacturers, particularly in biotech, need to align their product lifecycle with EU Taxonomy requirements, which can add substantial costs and operational constraints. RWEMA's role in these sectors is increasingly being recognized as a potential game changer. Studies suggest that leveraging real-world data to inform ESG strategies can improve the cohesiveness and impact of sustainability initiatives.

However, gaps remain regarding the specific mechanisms through which RWEMA can help overcome challenges such as cross-functional misalignment and inadequate stakeholder collaboration. This study builds on these insights by highlighting RWEMA's strategic role in transforming data into actionable insights, thereby enabling life sciences companies to effectively integrate ESG into their core business functions and overcome key challenges in adoption. The findings indicate that RWEMA's strategic role is essential in overcoming barriers to ESG adoption in the life sciences industry. The underutilization of RWEMA has often resulted in missed opportunities for effective ESG integration. The results show that by positioning RWEMA as a strategic leader, companies can overcome challenges such as lack of data-driven insights, fragmented ESG strategies, and limited cross-functional collaboration. This transformation is critical for enhancing the overall market credibility and sustainability of life sciences companies [23, 24].

3.6 Contribution to Empirical Support for ESG Decision-Making

The findings on empirical support for ESG decision-making align with the objective of providing an actionable framework for leveraging RWEMA as a strategic driver of ESG initiatives. This specifically addresses research question (c) by highlighting RWEMA's role in delivering data-driven insights that enhance the credibility of ESG programs. By using real-world evidence to support decision-making, RWEMA provides life sciences companies with a foundation for aligning ESG practices with stakeholder expectations, thereby contributing to long-term sustainability. Empirical support is crucial for the successful implementation of ESG initiatives, as it provides evidence that aligns corporate actions with stakeholder expectations. RWEMA's role in providing this empirical support is significant because it uses real-world data to demonstrate the practical outcomes of ESG programs. Recent studies have underscored the importance of using real-world evidence in supporting ESG claims and informing policy development, particularly in sectors like life sciences where data integrity is paramount [7]. A key gap in the literature concerns the scalability of RWEMA-driven ESG initiatives and their ability to provide consistent empirical support across different contexts. The findings of this study demonstrate that RWEMA can bridge this gap by delivering robust, data-driven insights that enhance the credibility and impact of ESG initiatives, thereby building a stronger case for sustained integration of ESG practices in corporate strategy. The results emphasize the importance of RWEMA in providing empirical support for ESG decision-making. By utilizing real-world evidence, life sciences companies are better equipped to demonstrate the effectiveness of their ESG programs and make data-driven decisions that align with sustainability goals. This empirical approach not only strengthens the credibility of ESG initiatives but also facilitates the development of targeted, actionable strategies that address specific sustainability challenges faced by the industry.

Overall, the findings of this study provide a comprehensive understanding of how RWEMA can be leveraged as a strategic driver of ESG integration in the life sciences sector. The evidence gathered highlights the potential of RWEMA to enhance corporate reputation, regulatory compliance, value creation, and sustainability, thereby ensuring long-term viability and competitiveness in a rapidly evolving industry (Table 2).

4 Discussion

4.1 Interpretation of Results

The findings of this study provide a deeper understanding of how RWEMA can play a pivotal role in integrating ESG practices within the life sciences industry. The results indicated that RWEMA significantly enhances corporate reputation, strengthens regulatory compliance, drives sustainability, and supports value creation. This is in line with Tunis et al. [13], who highlighted the importance of real-world evidence in informing market access decisions across different regions. RWEMA's role in enhancing compliance and credibility is also supported by the findings of Agarwala

Table 2 Key themes and findings overview

Key themes	Key findings	Implications for ESG integration
Enhancing corporate reputation	RWEMA-supported initiatives improve transparency, credibility, and stakeholder trust	Supports long-term corporate sustainability and market reputation
Strengthening regulatory compliance	Real-world data enables proactive compliance and risk mitigation in a regulated industry	Positions companies as leaders in regulatory adherence and sustainable practices
Driving sustainability and value creation	RWEMA drives value creation beyond financial metrics by identifying key ESG impact areas	Aligns corporate goals with broader societal and environmental needs
Overcoming barriers to ESG adoption	RWEMA helps overcome fragmented ESG strategies and enhances cross-functional collaboration	Enables more cohesive and effective ESG integration strategies
Contribution to empirical support	RWEMA provides data-driven insights to enhance ESG program credibility and effectiveness	Strengthens the foundation for sustainable business transformation

Source The authors

et al. [14], which emphasize the value of real-world evidence in supporting precision medicine and fostering stakeholder trust. By systematically utilizing real-world data, RWEMA contributes to improving transparency and accountability, which are essential elements in fostering stakeholder trust and building corporate credibility. This aligns with Ventola [16], who discussed the role of big data in improving pharmacovigilance and ensuring effective monitoring of adverse events. RWEMA's data-driven approach enhances the credibility of ESG initiatives by leveraging extensive datasets to provide actionable insights.

The research questions aimed to identify the challenges and opportunities associated with ESG integration, evaluate RWEMA's role in enhancing ESG practices, and provide an actionable framework for ESG leadership. The results directly addressed these questions by demonstrating RWEMA's potential in overcoming barriers to ESG adoption and transforming data-driven insights into actionable practices. By using thematic analysis, this study was able to uncover recurring themes that underline RWEMA's strategic contributions to ESG, including its role in risk management, reputation building, and facilitating stakeholder engagement. Furthermore, the findings align with the Global Goals and Principles, particularly the United Nations Global Compact (UNGC) and SDGs. By contributing to corporate sustainability and governance, RWEMA supports companies in aligning their practices with the UNGC principles, such as environmental stewardship and social responsibility. Additionally, RWEMA's strategic integration aids companies in making progress towards specific SDGs, including Goal 3 (Good Health and Well-being), Goal 9 (Industry, Innovation, and Infrastructure), and Goal 13 (Climate Action), by driving sustainable healthcare practices and fostering innovation in the life sciences sector.

The implications of these findings are substantial for the life sciences sector, suggesting that RWEMA's strategic integration can lead to more effective and credible ESG initiatives. The transformation of RWEMA from a support function to a strategic leader enhances the overall value proposition of ESG initiatives, providing companies with a competitive advantage and long-term sustainability. Moreover, this aligns with recent regulatory frameworks such as the Corporate Sustainability Reporting Directive (CSRD), EU Taxonomy Regulation, and Sustainable Finance Disclosure Regulation (SFDR). These regulations emphasize the need for transparent reporting, sustainability classification, and disclosure requirements for corporate activities. By integrating RWEMA strategically, companies can better comply with these regulatory standards, thus enhancing their ESG performance and ensuring that they meet evolving legal and market expectations. This approach not only supports regulatory compliance but also positions companies as industry leaders in sustainability and governance.

4.2 Comparison with Literature

The findings of this study align with and extend recent literature on ESG integration in the life sciences sector. Previous research has highlighted the growing importance of ESG considerations in building corporate reputation and fostering long-term business value [4]. The results of this study further validate these insights by demonstrating how RWEMA-supported ESG initiatives improve transparency and accountability, consistent with findings from recent studies [7]. Tunis et al. [13] also highlighted similar challenges and opportunities in the adoption of innovative therapies, emphasizing the importance of real-world evidence in decision-making processes to ensure regulatory compliance and market credibility.

Compared to earlier studies that have predominantly focused on the role of corporate governance in ESG integration, this study emphasizes the operational value of RWEMA in bridging the gap between theory and practice. For example, Collins et al. [15] discuss the limitations of relying solely on randomized clinical trials for assessing treatment efficacy, suggesting the value of real-world evidence. This study's findings reinforce that RWEMA, by leveraging real-world data, can provide empirical support that complements traditional methodologies, thereby making ESG programs more adaptable and effective in real-world scenarios.

Moreover, the results challenge the traditional view that market access functions are limited to supporting roles. This study demonstrates that RWEMA has a broader strategic significance, which extends beyond compliance and contributes to sustainability and corporate value creation. The empirical evidence provided by RWEMA plays a crucial role in supporting ESG claims and enhancing the credibility of corporate sustainability initiatives, as suggested by Agarwala et al. [14] and Ventola [16]. Additionally, Collins et al. [15] highlighted the limitations of randomized clinical trials and underscored the need for real-world data to complement clinical findings, thereby making ESG programs more adaptable and reflective of real-world conditions.

4.3 *Findings on ESG Regulations Across Europe, Asia, and the Americas*

ESG policies and regulations have a significant impact on the life sciences industry, with diverse approaches and standards in each region. In Europe, the European Union (EU) regulations on ESG are among the most stringent and comprehensive globally. The EU has integrated ESG considerations into various directives, including the Non-Financial Reporting Directive (NFRD) and the Sustainable Finance Disclosure Regulation (SFDR), which require companies to disclose environmental and social impacts and governance practices [4, 7, 25].

The SFDR requires financial market participants and financial advisors to disclose how they integrate ESG risks in their investment decisions and advisory processes. This regulation impacts the life sciences industry by influencing how investors assess companies in this sector, pushing them to adopt better ESG practices to attract investment. The NFRD requires large companies and groups to disclose information on how they operate and manage social and environmental challenges. This is pivotal for life sciences companies as it compels them to report on issues such as clinical trial ethics, supply chain sustainability, and environmental impacts of their operations [26, 27].

Expanding on the NFRD, the CSRD, which will phase in from 2023, will require all large companies and all listed companies (except listed micro-enterprises) to disclose detailed information on their sustainability actions. This includes how they handle social and employee matters, respect for human rights, anti-corruption and bribery, and environmental matters. This directive will ensure that life sciences companies are not only compliant with high-level ESG standards but also transparent about their compliance [25, 28, 29].

In Asia, ESG regulations vary widely but are rapidly evolving. China, for instance, has implemented mandatory ESG disclosures for listed companies and certain private firms, focusing heavily on environmental metrics due to the country's significant pollution challenges. Countries like Japan and South Korea also have robust ESG frameworks, emphasizing corporate governance and environmental conservation. These regulations often drive life sciences companies in Asia to adopt more sustainable practices, particularly in manufacturing and supply chain operations [4, 28].

In the United States, ESG regulations are less centralized than in Europe, with a combination of federal and state-level initiatives influencing the life sciences industry. The Securities and Exchange Commission (SEC) has been increasingly focused on

ESG matters, particularly in terms of disclosure requirements. The SEC has indicated plans to enhance transparency in how companies report their ESG practices, especially concerning climate-related disclosures. The Sustainability Accounting Standards Board (SASB) also provides a framework for life sciences companies to report on sustainability issues, offering guidelines that help public corporations disclose material, decision-useful information to investors [27].

Federal regulations such as those from the Environmental Protection Agency (EPA) and the Federal Drug Administration (FDA) also indirectly impact ESG practices within the life sciences sector. The FDA's guidelines on good manufacturing practices and ethical marketing tie directly into the governance and social pillars of ESG. Compliance with these regulations ensures not only product safety and efficacy but also enhances corporate governance standards [4, 29, 30].

Table 3 shows, the regulatory landscape across Europe, Asia, and the Americas influences how the life sciences industry integrates ESG considerations into its core operations. This diversity in regulatory approaches compels companies to remain adaptable and compliant with both global and region-specific standards, thereby enhancing their corporate sustainability profiles and demonstrating a commitment to ESG practices.

4.4 Implications

These following items summarize how the research findings answer the defined research questions and objectives.

1. **Challenges and Opportunities of ESG Integration in the Life Sciences Sector** The first objective of this study was to assess the current challenges and opportunities associated with ESG integration in the life sciences sector. The findings identified several challenges, such as regulatory fragmentation across different regions and the need for increased transparency in reporting. These challenges are compounded by the diverse regulatory frameworks across Europe, Asia, and the Americas, as outlined in the "Findings on ESG Regulations" section. The opportunities highlighted include leveraging RWEMA as a strategic tool to improve sustainability practices, enhance corporate reputation, and meet stringent regulatory requirements. RWEMA's integration provides life sciences companies with the ability to overcome barriers to ESG adoption, thereby turning challenges into opportunities for innovation and sustainable development.
2. **RWEMA's Role in Enhancing ESG Practices** The second research question focused on evaluating how RWEMA can enhance ESG practices within life sciences companies. The study demonstrates that RWEMA, by utilizing real-world evidence, plays a critical role in improving transparency, accountability, and compliance with ESG standards. This aligns with the literature emphasizing the value of real-world data in decision-making processes [14, 16]. The findings indicate that RWEMA helps address the social and environmental dimensions

Table 3 Overview of ESG implementations in healthcare, pharmaceuticals, biotechnology, and medical devices across regions

Sector	Region	Key ESG areas of intervention	Applied regulations and frameworks	RWEMA perspective
Healthcare	Europe	Waste management, patient data privacy, clinical trial ethics	CSRD, NFRD, EU Green Deal, SFDR	Enhances data transparency and supports compliance with strict EU standards
	Asia	Environmental impact reduction, health data management	China Mandatory ESG Disclosures, Japan ESG Framework, South Korea ESG Guidelines	Facilitates adoption of sustainable practices, particularly in reducing environmental footprint
	Americas	Health equity, patient safety, community engagement	SEC ESG Focus, EPA Regulations, FDA Good Manufacturing Practices	Supports detailed ESG reporting, aligned with state and federal requirements
Pharmaceuticals	Europe	Supply chain sustainability, clinical trial governance	CSRD, EU Taxonomy Regulation, SFDR	Provides real-world evidence to assess the sustainability of pharmaceutical production
	Asia	Manufacturing emissions, supply chain transparency	China Mandatory ESG Disclosures, Japan's Corporate Governance Code	Strengthens transparency in production processes and reduces environmental impacts
	Americas	Product safety, diversity and inclusion in workforce	SEC ESG Disclosure Plans, SASB Standards, FDA Regulations	RWEMA enhances compliance by providing credible data on product safety and governance
Biotechnology	Europe	Innovation in sustainable production, environmental impacts	CSRD, EU Green Deal, SFDR	Supports innovation in biotech through data-driven sustainability practices
	Asia	R&D focus on sustainable technologies, corporate governance	South Korea ESG Guidelines, China Green Bond Standards	RWEMA enables targeted ESG strategies focusing on sustainable biotech R&D

(continued)

Table 3 (continued)

Sector	Region	Key ESG areas of intervention	Applied regulations and frameworks	RWEMA perspective
	Americas	Ethical marketing, resource management	SEC ESG Reporting, SASB, EPA Guidelines	Provides empirical support for ethical governance and resource management in biotech firms
Medical Devices	Europe	Eco-design, circular economy, product lifecycle transparency	CSRD, EU Medical Device Regulation (MDR), EU Green Deal	RWEMA aids in compliance with eco-design requirements and promotes lifecycle transparency
	Asia	Waste reduction, product safety	China ESG Standards, Japan's Recycling Promotion Law	Facilitates implementation of ESG practices in reducing medical device waste
	Americas	Transparency in sourcing, workforce diversity	SEC ESG Standards, FDA Guidelines, California Consumer Privacy Act (CCPA)	Strengthens ESG practices by providing transparency in sourcing and workforce metrics

of ESG by supporting ethical practices, enhancing clinical trial governance, and providing empirical support for ESG claims. This role is particularly significant given the regulatory demands of the Corporate Sustainability Reporting Directive (CSRD) and Sustainable Finance Disclosure Regulation (SFDR), which require detailed sustainability disclosures.

3. **Actionable Framework for Leveraging RWEMA as a Strategic ESG Driver** The third objective was to provide an actionable framework for leveraging RWEMA as a strategic driver of ESG initiatives. The study suggests that RWEMA can transform from a supporting function to a strategic leader in ESG implementation. The thematic analysis identified key areas where RWEMA can contribute to ESG leadership, such as driving innovation in sustainable practices, improving regulatory compliance, and enhancing stakeholder trust. The framework proposed in this study emphasizes the strategic value of RWEMA in aligning corporate actions with the Sustainable Development Goals and the United Nations Global Compact (UNGC). By supporting companies in their sustainability journey, RWEMA ensures alignment with global sustainability principles, ultimately contributing to long-term value creation and competitive advantage.

The research findings successfully address the defined research questions by demonstrating how RWEMA can be a transformative force in the integration of ESG practices. By enhancing transparency, driving compliance, and facilitating stakeholder engagement, RWEMA proves to be an indispensable asset for life sciences companies aiming to align their operations with broader sustainability goals. In terms of policy implications, this study highlights the importance of regulatory frameworks that encourage the use of real-world evidence in ESG reporting. Policymakers could consider developing guidelines that incentivize the adoption of RWEMA-driven strategies, ensuring that real-world data becomes a cornerstone of ESG assessments and enhancing transparency and accountability in corporate sustainability practices.

5 Conclusion

5.1 Limitations

The study faced several limitations that impact the generalizability and interpretation of the findings. First, the methodology relied on a bibliographic and documentary search, which inherently limits the empirical nature of the results. Although a wide variety of sources were consulted, the absence of primary data collection means that the findings are based on existing literature rather than direct observations or interviews. This may limit the study's ability to capture nuanced, context-specific insights related to ESG implementation in the life sciences sector. Second, the use of thematic analysis, while effective in identifying key themes, is subject to researcher bias. Although efforts were made to minimize bias through peer review and cross-referencing, the interpretative nature of the analysis means that subjective influences may have impacted the results. Lastly, the focus on the life sciences sector means that the findings may not be entirely applicable to other industries, particularly those with different regulatory or operational contexts.

5.2 Future Research and Implications

Future research should aim to address the limitations of this study by incorporating primary data collection, such as interviews or surveys with key stakeholders in the life sciences industry. This would help to validate the findings of this research and provide richer, more context-specific insights into how RWEMA can drive ESG integration. Additionally, future studies could explore the role of RWEMA in specific ESG-related challenges, such as climate action, diversity and inclusion, or supply chain sustainability, to provide more targeted insights for practitioners and policymakers. From a theoretical perspective, future research could further investigate the evolving role of real-world evidence in ESG practices, particularly as regulatory frameworks continue to develop globally. Practically, the findings of this study highlight the need for life sciences companies to adopt a proactive approach in integrating RWEMA into their ESG strategies. Policymakers should consider developing guidelines that incentivize the use of real-world data in ESG reporting to enhance transparency and accountability.

5.3 Conclusion

Exploring the role of RWEMA in enhancing the integration of ESG practices within the life sciences sector. The research aimed to assess the challenges and opportunities of ESG integration, evaluate RWEMA's contributions to ESG practices, and provide a framework for leveraging RWEMA as a strategic ESG driver. The findings indicate that RWEMA plays a crucial role in improving corporate reputation, ensuring regulatory compliance, and supporting sustainability and value creation through data-driven approaches. By systematically utilizing real-world evidence, RWEMA enhances transparency, accountability, and stakeholder trust, positioning companies as leaders in ESG integration.

The study also demonstrated that RWEMA's role extends beyond compliance, contributing to strategic ESG leadership by driving innovation, supporting sustainable production practices, and enhancing stakeholder engagement. The research provides a valuable contribution to the understanding of how market access functions can evolve from a supportive to a strategic role in corporate sustainability. By aligning with global sustainability principles, such as the United Nations Global Compact (UNGC) and Sustainable Development Goals (SDGs), RWEMA proves to be an indispensable asset for companies aiming to align their operations with broader sustainability objectives.

In summary, this study highlights the transformative potential of RWEMA in the life sciences sector, providing a framework that companies can adopt to enhance their ESG initiatives. By addressing the challenges of ESG integration and capitalizing on the opportunities presented by real-world evidence, life sciences companies can achieve greater alignment with sustainability goals, improve regulatory compliance, and foster long-term value creation and competitiveness. The findings underscore the importance of integrating RWEMA into ESG strategies, thereby ensuring that life sciences companies can contribute meaningfully to global sustainability efforts and secure a sustainable future.

References

1. Machado, A.B., Richter, M.F., Peixoto, J.A.: Relations between digital transformation and sustainability post COVID-19: the pillars of ESG. In: Handbook of Research on Global Networking Post COVID-19, pp. 386–404. IGI Global (2022). https://doi.org/10.4018/978-1-7998-8856-7.ch021

2. Peixoto, J.A., Machado, A.B., Richter, M.F.: ESG indicator metrics used by organisations to assess the degree of sustainability in companies. Int. J. Adv. Eng. Res. Sci. **9**(9), 589–600 (2022). https://doi.org/10.22161/ijaers.99.64
3. Richter, M.F., de Bem Machado, A.: Circular economy and sustainable consumption behavior for green recovery. Multitemas, 183–201 (2023). https://doi.org/10.20435/multi.v28i70.3948
4. Liang, X., Lee, H., Jung, S.: ESG integration in the pharmaceutical industry: challenges and opportunities. Corp. Gov. **22**(3), 410–428 (2022)
5. Cifuentes-Faura, J.: European Union policies and their role in combating climate change over the years. Air Q., Atmos. Health, 1–8 (2022). https://doi.org/10.1007/s11869-022-01156-5
6. Dacks, R., et al.: The important role of environmental stewardship groups in supporting human health and well-being. Front. Sustain. Cities **3**, 710355 (2021)
7. Gebhardt, M., Kopyto, M., Birkel, H., & Hartmann, E.: Industry 4.0 technologies as enablers of collaboration in circular supply chains: a systematic literature review. Int. J. Prod. Res. **60**(23), 6967–6995 (2022). https://doi.org/10.1080/00207543.2021.1999521
8. Gough, D., Oliver, S., Thomas, J.: An Introduction to Systematic Reviews. SAGE Publications (2017)
9. Hart, C.: Doing a Literature Review: Releasing the Social Science Research Imagination. SAGE Publications (1998)
10. Braun, V., Clarke, V.: Using thematic analysis in psychology. Qual. Res. Psychol. **3**(2), 77–101 (2006). https://doi.org/10.1191/1478088706qp063oa
11. Booth, A., Sutton, A., Papaioannou, D.: Systematic Approaches to a Successful Literature Review. SAGE Publications (2016)
12. Bose, S.: Evolution of ESG reporting frameworks. In: Esty, D.C., Cort, T. (eds.) Values at Work. Palgrave Macmillan, Cham (2020). https://doi.org/10.1007/978-3-030-55613-6_2
13. Tunis, S., Hanna, E., Neumann, P.J., Toumi, M., Dabbous, O., Drummond, M., Chambers, J.D.: Variation in market access decisions for cell and gene therapies across the United States, Canada, and Europe. Health Policy **125**(12), 1550–1556 (2021). https://doi.org/10.1016/j.healthpol.2021.10.003
14. Agarwala, V., Khozin, S., Singal, G., O'Connell, C., Kuk, D., Li, G., Abernethy, A.P.: Real-world evidence in support of precision medicine: clinico-genomic cancer data as a case study. Health Aff. **37**(5), 765–772 (2018). https://doi.org/10.1377/hlthaff.2017.1579
15. Collins, R., Bowman, L., Landray, M., Peto, R.: The magic of randomization versus the myth of real-world evidence. N. Engl. J. Med. **382**(7), 674–678 (2020). https://doi.org/10.1056/nejmsb1901642
16. Ventola, C.L.: Big data and pharmacovigilance: data mining for adverse drug events and interactions. Pharm. Ther. **43**(6), 340 (2018)
17. Bari, N., Chimhundu, R., Chan, K.C.: Dynamic capabilities to achieve corporate sustainability: a roadmap to sustained competitive advantage. Sustainability **14**(3), 1531 (2022). https://doi.org/10.3390/su14031531
18. Delaney, E., et al.: The investigation of environmental sustainability within product design: a critical review. Des. Sci. **8**, e15 (2022). https://doi.org/10.1017/dsj.2022.11
19. McSweeney, D., Shelton, L.: Corporate financial disclosures and environmental, social, and governance concerns. Nat. Resour. Environ. **35**(2), 23–27 (2020)
20. Shabbir, M.S., Wisdom, O.: The relationship between corporate social responsibility, environmental investments and financial performance: evidence from manufacturing companies. Environ. Sci. Pollut. Res. **27**(32), 39946–39957 (2020). https://doi.org/10.1007/s11356-020-10217-0
21. Singhania, M., Saini, N.: Institutional framework of ESG disclosures: comparative analysis of developed and developing countries. J. Sustain. Finance Investment **13**(1), 516–559 (2023). https://doi.org/10.1080/20430795.2021.1964810
22. Su, X., Wang, S., Li, F.: The impact of digital transformation on ESG performance based on the mediating effect of dynamic capabilities. Sustainability **15**(18), 13506 (2023). https://doi.org/10.3390/su151813506

23. Ullah, H., et al.: Exploring the potential of metaverse technology in healthcare: applications, challenges, and future directions. IEEE Access (2023)
24. Usman, G., et al.: Effects of environmental pollution on wildlife and human health and novel mitigation strategies. World J. Adv. Res. Rev **19**(02), 1239–1251 (2023). https://doi.org/10.30574/wjarr.2023.19.2.1644
25. Jacob Rodrigues, M., Postolache, O., Cercas, F.: ESG practices and sustainability in the European context. J. Environ. Manage. **260**, 110089 (2020)
26. Kant, R., Anjali, S.: Supply chain sustainability in the life sciences sector: a review of current practices. Supply Chain Manag. **26**(2), 218–231 (2021)
27. Linton, J.D., Klassen, R., Jayaraman, V.: Sustainable supply chains: an introduction. J. Oper. Manag. **25**(6), 1075–1082 (2007). https://doi.org/10.1016/j.jom.2007.01.012
28. Kemp, R., Pearson, P.: Innovation for Sustainable Development: From Environmental to Sustainable Innovation. Edward Elgar Publishing (2007)
29. Mikalef, P., Pappas, I.O., Krogstie, J., Giannakos, M.N.: Big data and strategy: a review of the literature. J. Bus. Res. **98**, 574–590 (2019)
30. Pesqueira, L., Ringo, D., Van Tulder, R.: Governance and corporate responsibility in multinational enterprises. J. Int. Bus. Stud. **51**(1), 103–122 (2020)

Governance and Transparency: Ethical Leadership in the Information Age

Mário Parra da Silva and Ana Saraiva

Abstract The integration of Environmental, Social, and Governance (ESG) principles has emerged as a basis for the development of sustainability in organizations. With this in mind, governance and transparency also influence leadership in this information age. This chapter aims to elaborate on how governance and transparency help leaders act ethically in organizations. In today's world, where technology is advancing so quick, being transparent and having good governance models are crucial. We'll try to understand how governance and transparency affect ethical leadership in organizations in this modern age, by showing how different governance styles and levels of transparency impact ethical leadership, and try to give an answer to these three main questions: (1) How do organizational governance structures influence decision-making processes? (2) How do advances in technology and information facilitate transparency within organizational framework? (3) How do leaders adopt a culture of ethics, sustainability, and social responsibility within their organizations? To achieve that, our goal is to conduct a comparative and a quantitative analysis of the models used by organizations to assure ethical leadership and their impact on organizational behavior and performance. In conclusion, the aim is to provide a real framework illustrating how governance and transparency improve ethical leadership in the information age, thus providing practical insights for organizational improvement.

Keywords Governance · Organizational transparency · Ethical leadership · ESG principles · Technology in governance

M. P. da Silva · A. Saraiva (✉)
Portuguese Association of Business Ethics (APEE), Lisbon, Portugal
e-mail: anasaraiva@apee.pt

M. P. da Silva
e-mail: mps@stepup.pt

© The Author(s), under exclusive license to Springer Nature Switzerland AG 2025

A. d. b. Machado et al. (eds.), *Environmental, Social, Governance and Digital Transformation in Organizations*, Information Systems Engineering and Management 35, https://doi.org/10.1007/978-3-031-86079-9_9

1 Introduction

The integration of Environmental, Social, and Governance (ESG) criteria into corporate strategies has become increasingly essential as businesses navigate the complexities of modern markets, societal expectations, and environmental challenges. Over recent years, the drive adjacent to ESG and corporate sustainability has intensified, driven by a global push for responsible production and consumption models that align with the values of a more conscientious society [4]. This shift reflects a broader recognition that businesses have a pivotal role in addressing pressing issues such as climate change, social inequities, and governance failures. As stakeholders—including investors, consumers, and regulatory bodies—demand greater accountability and transparency, organizations are compelled to adopt and adapt ESG frameworks that not only enhance their operational sustainability but also strengthen their ethical and social responsibilities [11].

The implementation of ESG principles varies significantly across regions, industries, and markets, influenced by distinct cultural, economic, and regulatory contexts. However, among this diversity, best practices are shaping a new standard for corporate behavior worldwide. These best practices include a commitment to reducing environmental footprints, fostering diversity and inclusion within the workforce, ensuring transparent and ethical governance, and engaging in active dialogue with stakeholders. A leading example of such practices are the Principles for Responsible Investment (PRI), an initiative born from the collaboration between investors and international entities such as the United Nations Environment Program Finance Initiative (UNEP FI) and the UN Global Compact. PRI aims to guide the market toward responsible investment practices that prioritize long-term value creation and sustainability over short-term gains, reflecting a profound shift in how businesses are evaluated and valued [3, 4].

The PRI embodies the integration of ESG principles by encouraging companies to adopt strategies that align with broader societal goals, such as mitigating environmental impact, enhancing social welfare, and upholding robust governance standards. By adhering to these principles, organizations can better manage risks, seize opportunities, and build resilience against the unpredictable challenges of the global market. Moreover, the PRI serves as a framework for investors to assess the sustainability and ethical performance of their investments, fostering a culture of accountability and continuous improvement [12]. This alignment between investor expectations and corporate practices underscores the growing importance of transparency and ethical leadership in the information age, where access to information and the speed of communication have significantly altered the dynamics of corporate accountability [7].

In the information age, where technological advancements are rapidly reshaping industries, the role of governance and transparency has become even more critical. Technology enables unprecedented levels of information access and dissemination, making it easier for stakeholders to scrutinize corporate behavior and hold companies accountable for their actions. As a result, organizations that prioritize transparency and robust governance structures are better positioned to foster trust and credibility among their stakeholders [5]. Transparency in operations, decision-making, and reporting not only strengthens stakeholder relationships but also serves as a foundation for ethical leadership. Leaders who embrace transparency set a tone of openness and integrity, encouraging a culture where ethical considerations are integrated into every aspect of the organization's operations [13].

Ethical leadership, in turn, plays a vital role in advancing ESG objectives within organizations. Leaders who prioritize ethics and integrity are more likely to inspire their teams to align with the company's sustainability goals, fostering a culture of responsibility and accountability. This leadership approach extends beyond mere compliance with regulations; it involves actively promoting values-based decision-making and encouraging employees to consider the broader impact of their actions on society and the environment [10]. By embedding ESG principles into the organizational ethos, ethical leaders can drive meaningful change that resonates with stakeholders and supports long-term sustainable growth [9].

Furthermore, the incorporation of ESG principles into corporate governance frameworks reflects a broader transformation in how businesses are managed and evaluated. Traditional governance models, which often emphasized financial performance above all else, are increasingly being supplemented or even supplanted by models that consider a wider array of performance indicators, including environmental impact, social responsibility, and ethical governance [15]. This evolution in governance is not merely a response to external pressures but also a recognition that sustainable business practices are integral to maintaining competitive advantage and achieving long-term success. Organizations that effectively integrate ESG into their governance structures are better equipped to navigate the complexities of the modern business landscape, where success is defined not only by financial metrics but also by the ability to create positive social and environmental outcomes.

As businesses continue to adapt to the demands of the information age, the intersection of governance, transparency, and ethical leadership will remain a critical focal point. The ongoing evolution of ESG practices will require organizations to stay vigilant, continuously assessing and refining their approaches to governance and leadership [2]. By embracing transparency, fostering ethical leadership, and committing to sustainable practices, organizations can build resilient, responsible, and forward-thinking enterprises that contribute positively to society and the environment. The journey toward integrating ESG principles is not without challenges, but it offers a pathway to a more sustainable and equitable future for businesses and the communities they serve [1].

2 Chapter 1: Systematic Literature Review: Governance, Transparency, and Ethical Leadership in the Information Age

This systematic review aims to analyze the existing literature on how governance and transparency influence ethical leadership in organizations, particularly in the context of rapid technological advancements. The systematic literature review method is best known in the health sciences, where they usually includes statistical meta-analyses that aggregate available evidence from randomized controlled trials. Unlike traditional reviews, it uses 'systematic and explicit methods to identify, select, and critically appraise, relevant research, and to collect and analyse data from the studies that are included in the review' [6]. This methodological process was selected for this chapter due to its reliability and precision.

The review follows a structured approach involving the identification of relevant academic scientific papers. The process included:

1. The research strategy: GoogleScholar was chosen as the database and the keywords used for the research was the main ones of this chapter, such as "governance", "organizational transparency", "ethical leadership", "ESG principles", and "technology in governance".
2. Inclusion criteria: Only peer-reviewed journal articles, books, and scientific papers published since 2020. The focus was on studies that directly address the integration of governance and transparency in promoting ethical leadership.
3. Exclusion criteria: Papers not available in English or Portuguese, those lacking empirical data, or those not directly related to the research questions were excluded.
4. Data extraction: Key information was extracted from each selected article, including research objectives, methods, findings, and relevance to the review questions.

Synthesis: The extracted data were analyzed to identify common themes, gaps, and insights.

In the table above, we present the selection made after eliminating identical results (Table 1).

From the initial pool of articles, a total of 40 were selected for in-depth analysis. This first selection process involved investigating some of the abstracts to ensure the alignment with the research questions.

Based on the three research questions, we provide the findings of the literature review, namely (1) Influence of Organizational Governance Structures on Decision-Making Processes; (2) Role of Technology and Information in Facilitating Transparency; (3) Adoption of a Culture of Ethics, Sustainability, and Social Responsibility by the leaders. We provide the most relevant papers, divided by title, author and year, for each research question of this chapter (Table 2).

Table 1 Search equations

Search equation	Results	Selection
Governance AND organizational transparency	3340	6
Governance AND ethical leadership	9550	4
Governance AND ESG principles	8080	4
Governance AND technology in governance	18,500	6
Organizational transparency AND ethical leadership	9220	4
Organizational transparency AND ESG principles	1070	2
Organizational transparency AND technology in governance	14,400	3
Ethical leadership AND ESG principles	682	5
Ethical leadership AND technology in governance	9990	2
ESG principles AND technology in governance	1940	4
Total	*	40

3 Chapter 2: How the Governance and Transparency Influence the Leadership in This Information Age?

ESG Principles Importance

1. ***organizational governance structures influence decision-making processes***
2. ***advances in technology and information facilitate transparency within organizational framework***
3. ***how do leaders adopt a culture of ethics, sustainability, and social responsibility within their organizations?***

First, is important to understand what we mean by governance and by transparency. The latest definition provided by the World Economic Forum [14] describes governance within ESG as encompassing board composition, business ethics, and risk management to ensure that companies can meet their sustainability and ethical obligations effectively. Transparency, in this framework, emphasizes full disclosure and accountability across ESG metrics to build stakeholder trust and facilitate informed decision-making [14].

These two concepts significantly influence leadership in the information age by reshaping decision-making processes and organizational dynamics.

Governance extends beyond traditional state mechanisms, encompassing various modes of control and influence, which are increasingly driven by technology and digital data [7]. The emergence of "data governance" highlights the role of technology in defining authority and accountability within organizations. This shift requires leaders to adapt their strategies to leverage digital data effectively, ensuring informed decision-making and operational efficiency.

Transparency, bolstered by technological advancements, plays a crucial role in enhancing independent monitoring and accountability within organizations. It

Table 2 Findings

Research question	Articles	Authors	Year
1. Influence of organizational governance structures on decision-making processes	• Inter-organizational governance: a review, conceptualization and extension • From government to governance…to meta-governance: a systematic literature review. Public Management Review • Bibliometric analysis on corporate governance topics published in the journal of Corporate Governance: The International Journal of Business in Society • Corporate governance and sustainability: a review of the existing literature • An overview of recent developments in corporate governance	• Roehric • Gjaltema • Ellili • Pulejo • Almashhadani	• 2020 • 2020 • 2023 • 2022
2. Role of technology and information in facilitating transparency	• Governance by Data. Annual Review of Law and Social Science • Digital leadership in business organizations: an overview • Decentralized Network Governance: Blockchain Technology and the Future of Regulation • Transforming provider-customer relationships in digital servitization: A relational view on digitalization	• Johns • Araujo • Zwitter • Kamalaldin	• 2021 • 2021 • 2020 • 2020
3. Adoption of a culture of ethics, sustainability, and social responsibility by the leaders	• Corporate governance and sustainability: a review of the existing literature • Leadership in 2022: A perspective • Decentralized Network Governance: Blockchain Technology and the Future of Regulation • Corporate Governance and Ethical Leadership in Business Management • Ethical Leadership and its Effect on Organizational Integrity: A Multifaceted Examination • Integration of Environmental, Social, and Governance (ESG) criteria: their impacts on corporate sustainability performance	• Pulejo • Luedi • Zwitter • Malik • Azimov • Barbosa	• 2022 • 2022 • 2020 • 2024 • 2024 • 2023

strengthens alignment between agents and principals, protects shareholder rights, and imposes financial liabilities on corporate intermediaries [1]. This level of transparency ensures honesty, loyalty, and diligence, essential for ethical governance and leadership. Advances in technology and information have improved greater transparency within organizational frameworks, making it easier for stakeholders to access relevant information and hold leaders accountable [5].

In the digital era, leaders must navigate the complexities of digital transformation and inspire their teams to embrace new technologies despite inherent uncertainties. Digital leadership involves defining business strategies that leads to digital advancements, driving high business performance [2]. Leaders must foster a culture of continuous learning, integrity, and collaboration to adapt to the technological change. A strong digital leader will help define the digital business strategy, resulting in high business performance, which is crucial for companies aiming to be digitally advanced [8].

Ethical leadership is vital, as it integrates values-based decision-making and responsible behavior into the organizational strategy. Ethical leaders model integrity and transparency, setting standards that become ingrained in the corporate culture. This approach enhances employee morale and engagement, fostering a work environment where individuals feel valued and respected [10]. Ethical leadership involves demonstrating moral integrity, guiding employees towards ethical conduct, and, most of all, promoting sustainable practices [9].

The influence of governance and transparency on leadership also extends to promoting sustainability and social responsibility. Leaders are increasingly expected to incorporate ethical considerations into their business practices, ensuring that their organizations contribute positively to society and to the environment [11]. This involves not only compliance with ethical norms but also a proactive approach to ethical principles in decision-making processes, interpersonal relationships, and conflict resolution.

Governance mechanisms have adapted due to globalization, increased technical specialization, and functional differentiation, but they also need to adapt to the technical innovations of the digital domain [13]. Even when their full potential is still debated, the effects of technologies as governance instruments are increasingly tangible [15]. The impact of technology has fundamentally changed how those in charge of governance act, underscoring the importance of leadership that is responsive to these changes.

Overall, governance and transparency influence leadership by requiring ethical conduct, strategic adaptation to digital transformations, and a commitment to maintaining integrity and accountability within the organization. These elements are crucial for achieving sustained success and fostering a positive organizational culture in the information age. Effective leadership in this context requires a comprehensive understanding of the interplay between governance, transparency, and technological advancements, ensuring that organizations remain competitive and responsible. Strong governance ensures that organizations not only comply with regulatory standards but also instill ethical business practices that foster trust across stakeholders.

Transparency, a pillar of ESG, emphasizes the necessity of open, verifiable disclosure on ESG metrics, such as environmental impact and social equity, allowing stakeholders to evaluate an organization's genuine commitment to sustainability and ethical performance [14]. Moreover, as technology evolves, it introduces tools like advanced data analytics, blockchain for transparency, and AI-driven compliance solutions that can streamline ESG initiatives and reporting. This technological integration is essential for leadership, helping organizations to enhance data accuracy, strengthen risk management, and respond agilely to both regulatory and market demands in a landscape where ESG compliance is decisive.

4 Chapter 3: The Analysis

This study employed both quantitative and qualitative methodologies to investigate the state of governance, transparency, and ethical leadership across various organizations. We choose to distribute this survey to general managers, directors and to the main leaders from diverse types of organizations. The database for this study was provided by the APEE. The quantitative analysis involved a survey distributed to 890 respondents using QuestionPro Software, resulting in a completion rate of 32.73% (291 responses). However, after reviewing the responses, only 108 surveys were fully completed and valid for analysis. The dropouts (after starting were 222) and the survey took an average time to complete around 6 min. The qualitative analysis will not be presented in this chapter, due to compliance and data protection consideration. The figure presented above illustrates an image with the geographical distribution of responses by country (Fig. 1).

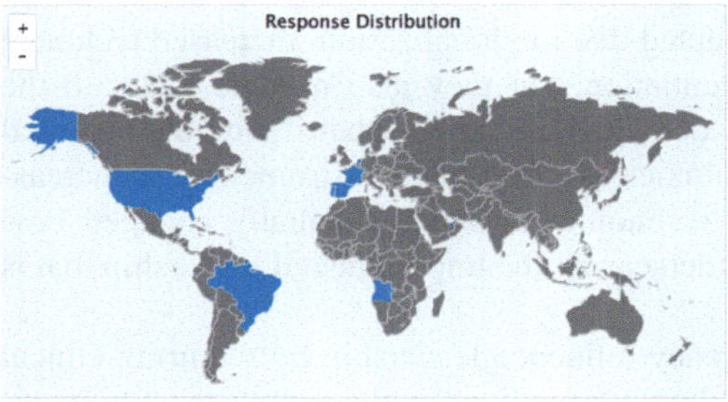

Countries	Responses
PT	96.67%
ES	1.21%
AO	0.61%
US	0.30%
AE	0.30%
BR	0.30%
FR	0.30%
NL	0.30%
Total	100.00%

Fig. 1 Response distribution of the survey worldwide, by country and percentage of total responses

We obtained responses from all over the world. Portugal was the main country, with 96.67% of the responses, followed by Spain with 1.21%.

The qualitative analysis supplemented the quantitative data with detailed insights from open-ended responses and comments provided by the participants. This analysis highlighted the variability in governance practices across different organizations and sectors, the challenges of implementing ethical leadership, and the significant role of technology in enhancing or hindering transparency. The authors choose to only present the quantitative analysis on this chapter. The quantitative analysis started with a survey (see annex), distributed in Portuguese and English to all kinds of organizations.

This survey was divided into three parts, related with the three main questions of this chapter:

Organizational governance structures (1); transparency with technology and information processes (2); leadership with a culture of ethics, sustainability and social responsibility (3).

4.1 Organizational Governance Structures

For the 1st question of the first part of the survey to the quantitative analysis we obtained the following results of the 108 different organizations that kindly provide us a response (Fig. 2):

1. How do you describe the organizational governance structure of your organization?

The survey results show a variety of governance structures in place, with the average response leaning slightly towards decentralized or mixed governance models. The mean response was 2.029, with a standard deviation of 0.813, indicating some variability among the organizations. The confidence interval at 95% suggests that this trend is consistent across the surveyed organizations (Fig. 3).

2. In which way the governance structure influences the decision making in your organization?

The influence of governance structure on decision-making varies, with 37.78% of organizations reporting a balance between centralized and decentralized decision-making. This suggests that many organizations are finding value in blending top-down decision-making with broader participation across various levels. Meanwhile, 26.67% rely primarily on decisions made by top management, and 35.56% involve different hierarchical levels in the decision-making process. The mean response here was 2.111, reflecting a slight preference for more inclusive decision-making practices (Fig. 4).

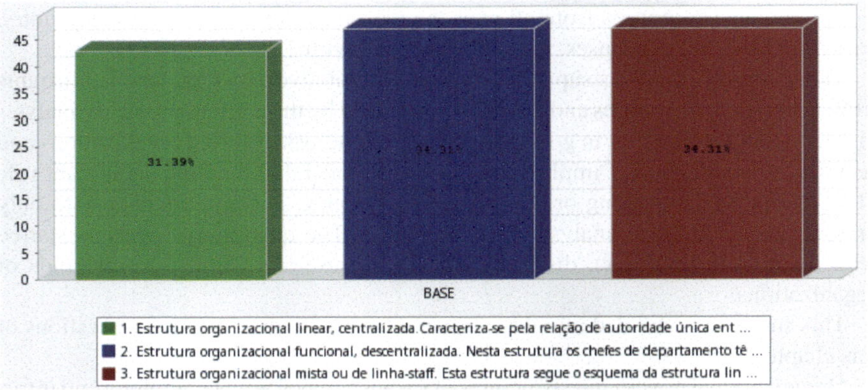

Total		137	100%
Mean: **2.029**	Confidence Interval @ 95%: **[1.893 - 2.165]**	Standard Deviation: **0.813**	Standard Error: **0.069**

Fig. 2 1st question

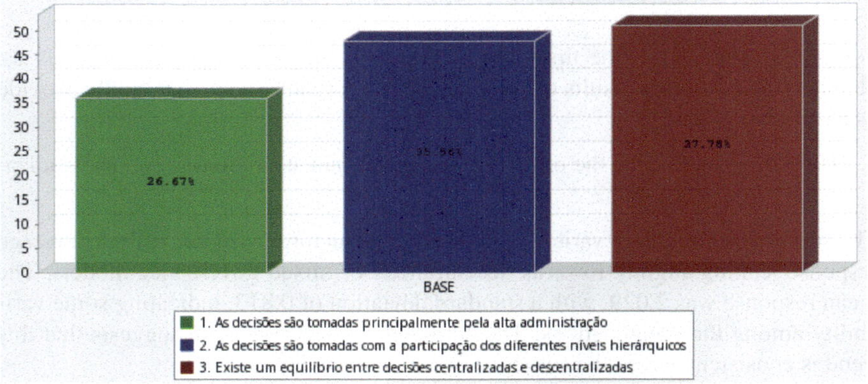

	Answer	**Count**	**Percent**
1.	As decisões são tomadas principalmente pela alta administração	36	26.67%
2.	As decisões são tomadas com a participação dos diferentes níveis hierárquicos	48	35.56%
3.	Existe um equilíbrio entre decisões centralizadas e descentralizadas	51	37.78%
	Total	135	100%
Mean: **2.111**	Confidence Interval @ 95%: **[1.976 - 2.246]**	Standard Deviation: **0.798**	Standard Error: **0.069**

Fig. 3 2nd question

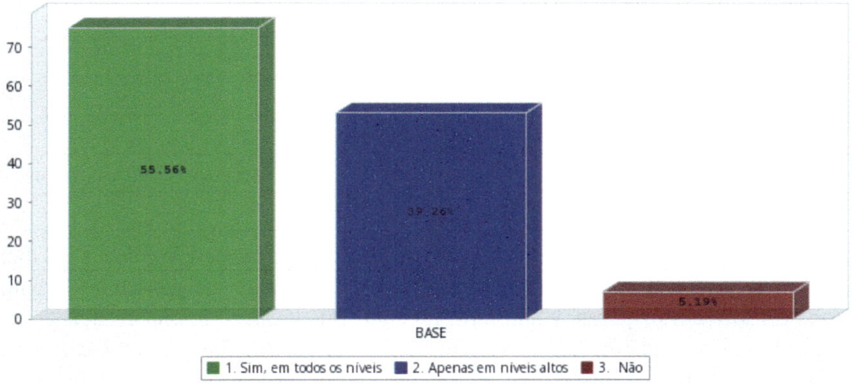

	Answer	Count	Percent
1.	Sim, em todos os níveis	75	55.56%
2.	Apenas em níveis altos	53	39.26%
3.	Não	7	5.19%
	Total	135	100%

Mean: **1.496**	Confidence Interval @ 95%: **[1.396 - 1.597]**	Standard Deviation: **0.597**	Standard Error: **0.051**

Fig. 4 3rd question

3. The employees are encouraged to participate in the decision-making process.

A majority of respondents (55.56%) indicated that employees at all levels are encouraged to participate in decision-making. This high level of inclusion supports the development of a more transparent and engaged organizational culture. However, 39.26% of organizations limit this participation to higher levels, which might restrict the flow of innovative ideas from lower levels. The mean response of 1.496, with a relatively low standard deviation, shows that most organizations are actively working to involve their employees in governance processes.

4.2 Transparency with Technology and Information Processes

See Fig. 5.

1. How do you classify the transparency level of your operations and decisions in your organization?

The majority of respondents (60.66%) classified the transparency level of their organization's operations and decisions as high. This suggests that many organizations prioritize transparency and have mechanisms in place to ensure that operations and decision-making processes are visible and accessible. A smaller portion (33.61%) rated their transparency as medium, indicating room for improvement, while only 5.74% reported low transparency. The mean response of 1.451 reflects a strong tendency towards high transparency, with a relatively low standard deviation, indicating consistency in responses (Fig. 6).

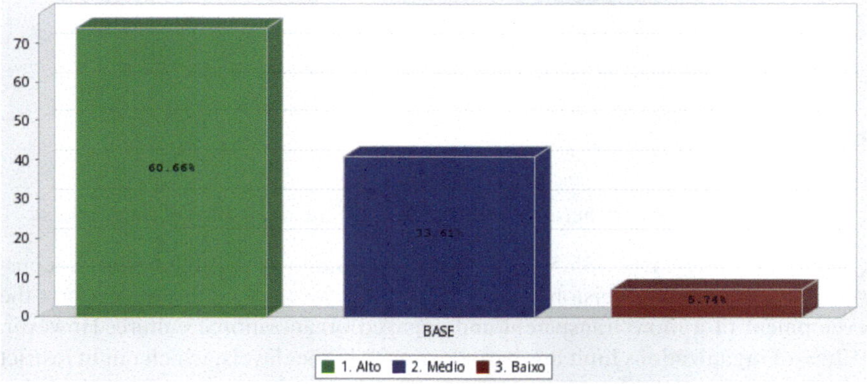

	Answer	Count	Percent
1.	Alto	74	60.66%
2.	Médio	41	33.61%
3.	Baixo	7	5.74%
	Total	122	100%
Mean: **1.451**	Confidence Interval @ 95%: **[1.344 - 1.558]**	Standard Deviation: **0.604**	Standard Error: **0.055**

Fig. 5 1st question

	Answer	Count	Percent
1.	Software(s) de gestão de informação interna	67	31.02%
2.	Plataformas de comunicação interna	77	35.65%
3.	Relatórios de Responsabilidade Social/ Sustentabilidade	48	22.22%
4.	Outro (especificar)	24	11.11%
	Total	216	100%

Mean: **2.134**	Confidence Interval @ 95%: **[2.003 - 2.265]**	Standard Deviation: **0.981**	Standard Error: **0.067**

Fig. 6 2nd question

2. What technology does your organization use in order to demonstrate transparency?

Organizations employ various technologies to enhance transparency, with internal information management software (31.02%) and internal communication platforms (35.65%) being the most commonly used tools. Additionally, 22.22% of respondents reported using sustainability and social responsibility reports to demonstrate transparency. The use of these technologies highlights a commitment to ensuring that information is readily available and accessible to stakeholders. The mean response of 2.134 suggests a balanced approach to leveraging multiple technological tools to achieve transparency (Fig. 7).

3. In which way the technology has on the transparency of the organization?

A significant majority (67.21%) of respondents believe that technology significantly facilitates transparency, enhancing the visibility and accessibility of information within the organization. However, 28.69% noted that while technology aims

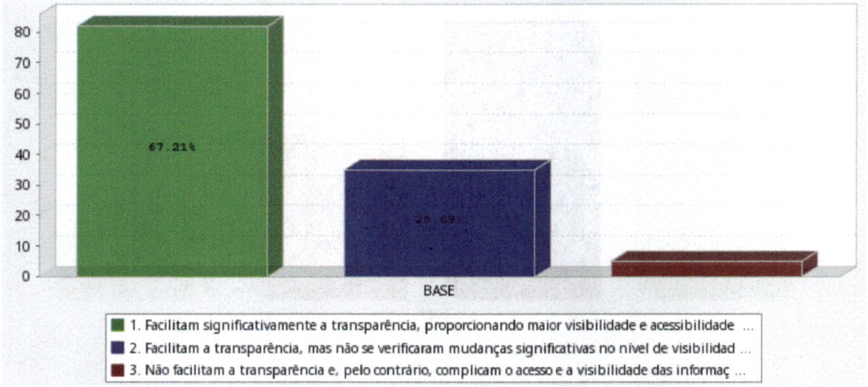

	Answer	Count	Percent
1.	Facilitam significativamente a transparência, proporcionando maior visibilidade e acessibilidade às informações	82	67.21%
2.	Facilitam a transparência, mas não se verificaram mudanças significativas no nível de visibilidade e acessibilidade às informações	35	28.69%
3.	Não facilitam a transparência e, pelo contrário, complicam o acesso e a visibilidade das informações	5	4.10%
	Total	122	100%

Mean: **1.369**	Confidence Interval @ 95%: **[1.269 - 1.469]**	Standard Deviation: **0.563**	Standard Error: **0.051**

Fig. 7 3rd question

transparency, it has not brought about significant changes in visibility and accessibility. Only a small percentage (4.10%) felt that technology complicates transparency, possibly due to challenges such as information overload or data management issues. The mean response of 1.369 indicates an agreement that technology positively impacts organizational transparency.

4.3 Leadership with a Culture of Ethics, Sustainability and Social Responsibility

See Fig. 8.

1. The leadership in your organization promotes a culture of ethics and sustainability?

A significant majority (76.47%) of organizations reported that their leadership promotes a culture of ethics and sustainability and has implemented a code of ethics or conduct. This demonstrates a strong organizational commitment to ethical

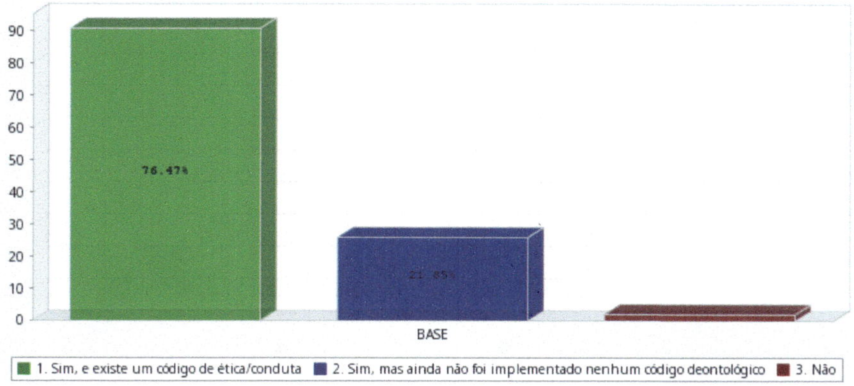

	Answer	Count	Percent
1.	Sim, e existe um código de ética/conduta	91	76.47%
2.	Sim, mas ainda não foi implementado nenhum código deontológico	26	21.85%
3.	Não	2	1.68%
	Total	119	100%

Mean: **1.252**	Confidence Interval @ 95%: **[1.167 - 1.337]**	Standard Deviation: **0.473**	Standard Error: **0.043**

Fig. 8 1st question

practices and sustainability. However, 21.85% of respondents indicated that while such a culture is promoted, a formal code of ethics has not yet been implemented, suggesting that these organizations may still be in the process of formalizing their ethical frameworks. Only 1.68% of respondents noted that their organization does not promote such a culture. The mean response of 1.252, with a low standard deviation, indicates consistent acknowledgment of the importance of ethical leadership across organizations (Fig. 9).

2. How are the sustainability principles integrated in the strategy and in the organizational structure of the organization?

The integration of sustainability principles varies across organizations, with 38.25% incorporating environmental and social policies into their strategies, and 34.56% implementing diversity and inclusion initiatives. Additionally, 20.74% of organizations have established a specific sustainability department, reflecting a structured approach to integrating sustainability into their operations. The mean response of 2.267, combined with a relatively low standard error (0.059), suggests that while there is a broad commitment to sustainability, the methods of integration differ among organizations (Fig. 10).

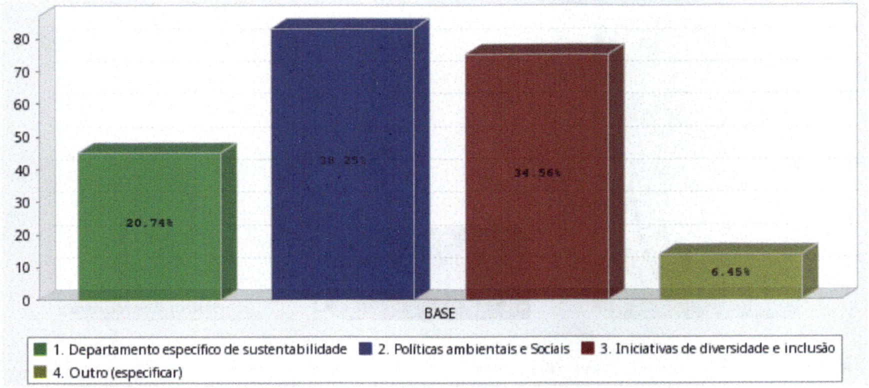

	Answer	Count	Percent
1.	Departamento específico de sustentabilidade	45	20.74%
2.	Políticas ambientais e Sociais	83	38.25%
3.	Iniciativas de diversidade e inclusão	75	34.56%
4.	Outro (especificar)	14	6.45%
	Total	217	100%

Mean: **2.267**	Confidence Interval @ 95%: **[2.153 - 2.382]**	Standard Deviation: **0.862**	Standard Error: **0.059**

Fig. 9 2nd question

3. In which way the organization guarantees the ethics, the sustainability and the social responsibility?

Organizations employ various methods to ensure ethics, sustainability, and social responsibility, with 40.74% offering training to employees, indicating a focus on building awareness and understanding of these principles. Internal and external audits (25.93%) and channels for reporting unethical behavior (26.75%) are also commonly used, highlighting the importance of accountability and transparency in maintaining ethical standards. The mean response of 1.992 reflects a balanced approach, with organizations utilizing multiple strategies to embed these values within their culture.

	Answer	Count	Percent
1.	Formação aos funcionários	99	40.74%
2.	Auditorias internas e externas	63	25.93%
3.	Canais para denunciar comportamentos antiéticos	65	26.75%
4.	Outro (especificar)	16	6.58%
	Total	243	100%
Mean: **1.992**	Confidence Interval @ 95%: **[1.870 - 2.114]**	Standard Deviation: **0.971**	Standard Error: **0.062**

Fig. 10 3rd question

5 Conclusion

The symbiotic relationship between corporate governance and ethical leadership lies at the heart of effective business management, underpinning organizational integrity, sustainable growth, and stakeholder trust. Corporate governance mechanisms provide the structural foundation for ensuring transparency, accountability, and responsible decision-making within organizations. Boards of directors, executive leadership, and governance policies establish the framework through which businesses are directed and controlled, setting the tone for ethical conduct and adherence to values-based principles. Without robust governance structures, ethical lapses and misconduct may occur, leading to reputational damage, legal liabilities, and erosion of stakeholder confidence [10].

Organizations led by ethical leaders often enjoy a reputation for integrity, attracting customers, partners, and talent who share similar values. This reputational capital is invaluable, providing a competitive edge that translates into improved financial performance [3].

Implementing robust ESG practices requires careful consideration, transparency, and ongoing evaluation to mitigate potential negative impacts and ensure sustainable results. Therefore, it would be important to establish standards and parameters that allow companies to understand and evaluate ESG criteria. In this sense, the International Organization for Standardization (ISO) could develop a global standardization on ESG that defines parameters, guidelines, and criteria with quality indicators, in line with the ISO 9001 standard already recognized worldwide [4].

The analysis made revealed several key trends in governance and transparency. For instance, 34.85% of respondents indicated that their organizations operated under a decentralized functional organizational structure, which allows for broader participation but presents challenges in maintaining consistent ethical standards. Additionally, 56.15% of respondents noted that employees at all levels are encouraged to participate in decision-making processes, which aligns with the principles of ethical leadership and transparency.

The survey results offer a comprehensive overview of how various organizational structures influence decision-making processes, transparency, and the adoption of ethical leadership practices. The data collected from 108 respondents provides valuable insights into the current state of governance in different organizations.

5.1 Key Findings

Organizational Structure

Linear Centralized Structure (31.39%): This structure is characterized by a clear hierarchy with centralized decision-making. While it ensures clear lines of authority, it may limit the involvement of lower-level employees in decision-making processes, which can hinder transparency and ethical leadership.

Functional Decentralized Structure (34.31%): This was the most common structure among respondents. It allows department heads autonomy, which can lead to more inclusive decision-making processes and greater accountability. This structure supports the adoption of ethical leadership by distributing decision-making power across different levels of the organization.

Mixed Line-Staff Structure (34.31%): Also popular among respondents, this structure combines elements of centralized and decentralized governance. It includes advisory bodies that provide technical recommendations while maintaining clear leadership roles within departments. This balance can foster a culture of transparency and ethical leadership.

Decision-Making Influence

Centralized Decision-Making (26.67%): A quarter of the respondents indicated that decisions are primarily made by top management, which can centralize power but may limit broader ethical considerations.

Inclusive Decision-Making (35.56%): The majority indicated that decision-making involves different hierarchical levels, promoting a more transparent and ethical environment.

Balanced Decision-Making (37.78%): A significant portion reported a balance between centralized and decentralized decision-making, which can enhance both efficiency and ethical standards within the organization.

Employee Participation

Encouragement Across All Levels (55.56%): Most respondents noted that employees at all levels are encouraged to participate in decision-making, which is crucial for fostering a transparent and ethical organizational culture.

High-Level Participation (39.26%): A notable percentage reported that only higher levels of the organization are involved, which may limit the broader application of ethical practices.

Transparency and Technology

High Transparency (60.66%): The majority of organizations reported a high level of transparency in operations and decisions, facilitated by internal management software, communication platforms, and sustainability reports.

Impact of Technology: A significant 67.21% of respondents stated that technological advancements have greatly facilitated transparency, improving the visibility and accessibility of information.

Ethical Leadership and Sustainability

Promotion of Ethical Culture (76.47%): A strong majority of organizations have implemented a code of ethics, indicating a commitment to ethical leadership.

Integration of Sustainability: Policies related to environmental and social governance are increasingly integrated into organizational strategies, with 38.25% of respondents affirming such practices.

In summary, the data suggests a notable diversity in organizational structures, decision-making practices, and transparency approaches across respondents, each playing a critical role in shaping ethical leadership and sustainability outcomes. Organizations with decentralized or mixed structures demonstrate a greater tendency towards inclusive decision-making and accountability, fostering ethical leadership by distributing decision-making authority. This inclusivity, paired with technological advancements, significantly enhances transparency, as most respondents indicated high levels of operational visibility facilitated by management software and communication tools.

Moreover, the emphasis on ethical culture, supported by codes of conduct and sustainability policies, indicates that a strong majority of organizations are actively embedding these values into their core strategies. By encouraging employee participation at all levels and integrating sustainability goals, these organizations not only uphold ethical standards but also strengthen their resilience and adaptability in an evolving regulatory landscape. This holistic approach positions them well to navigate complex challenges, balancing efficiency with a robust commitment to ethical and sustainable practices.

5.2 Implications

The survey highlights the importance of organizational structure in shaping the effectiveness of ethical leadership and transparency. Decentralized and mixed structures are particularly conducive to fostering an ethical culture, as they allow for greater employee participation and distributed decision-making. The integration of technology is also pivotal, as it enhances transparency and accountability across organizational levels. These findings underscore the need for organizations to carefully consider their governance structures and leverage technology to promote ethical practices and sustainability.

Annex 1: Survey in English

The questions in this questionnaire are divided into three sections, formulated based on three main guiding questions, and directed to the leadership of organizations to which it will be sent.

1. *How do organizational governance structures influence decision-making processes?*
2. *How do advances in technology and information facilitate transparency within the organizational framework?*
3. How do leaders adopt a culture of ethics, sustainability, and social responsibility within their organizations?

    Identification
    Name:
    Address:
    Sector:
    Number of employees:
    Respondent's Name:
    Position in the organization:
    Email:

1. **Governance Structure**

How do organizational governance structures influence decision-making processes?

Question 1

How would you describe your organization's governance structure?

1. Linear centralized structure: A clear hierarchy with centralized decision-making.

2. Functional decentralized structure: Each department reports to its respective head, allowing for greater autonomy in decision-making.

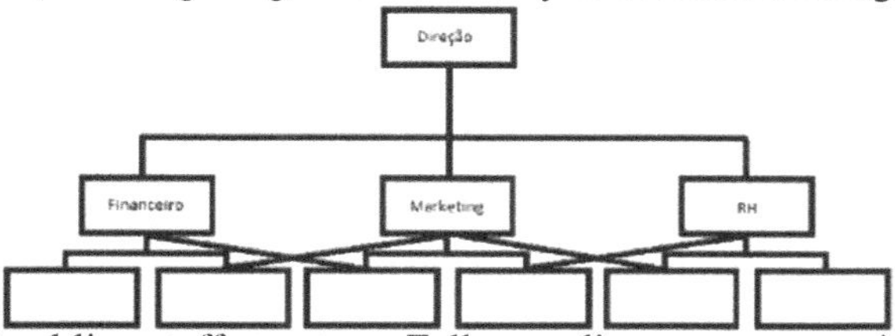

3. Mixed line-staff structure: Follows a linear structure with the addition of advisory bodies that provide specialized technical recommendations to lower levels without direct authority over them. Leadership remains with departmental heads.

Question 2

How does the governance structure influence decision-making in your organization?

() Decisions are mainly made by senior management.
() Decisions involve participation across hierarchical levels.
() There is a balance between centralized and decentralized decisions.

Question 3

Are employees encouraged to participate in the decision-making process?

() Yes, at all levels.
() Only at higher levels.
() No.

Open Response

Based on your answers in this section, please add any relevant comments or explanations for your choices.

2. **Transparency Facilitated by Technology**

How do advances in technology and information facilitate transparency within the organizational framework?

Question 1

How would you rate the level of transparency in your organization regarding operations and decisions?

() High.
() Medium.
() Low.

Question 2

Which technologies does your organization use to promote or demonstrate transparency? (Select multiple if applicable.)

() Internal information management software
() Internal communication platforms
() Social responsibility/sustainability reports
() Other (specify): ___________.

Question 3

In your opinion, how have technological advancements impacted transparency in organizations?

() They significantly enhance transparency, providing greater visibility and accessibility of information.

() They enhance transparency but have not significantly changed visibility and accessibility.

() They do not enhance transparency and complicate access and visibility.

Open Response

Based on your answers in this section, please add any relevant comments or explanations for your choices.

3. **Ethics, Sustainability, and Social Responsibility**

How do leaders adopt a culture of ethics, sustainability, and social responsibility within their organizations?

Question 1

Does the organization's leadership promote a culture of ethics and integrity?
() Yes, and there is a code of ethics/conduct in place.
() Yes, but no formal code has been implemented.
() No.

Question 2

How are sustainability principles integrated into the organizational strategy and structure? (Select multiple if applicable.)
() Dedicated sustainability department.
() Environmental policies.
() Diversity and inclusion initiatives.
() Other (specify): ___________.

Question 3

In what ways does the organization ensure ethics, sustainability, and social responsibility? (Select multiple if applicable.)
() Employee training.
() Internal and external audits.
() Channels for reporting unethical behavior.
() Other (specify): ___________.

Conclusion

Thank you for your participation. If you would like to receive a summary of your responses, please provide your email: ______.

Annex 2: Survey in Portuguese

(As questões para o questionário estão divididas em 3 partes, pois foram formuladas tendo em conta as 3 principais perguntas de partida e serão dirigidas à chefia das organizações para as quais este questionário será enviado).

1. *How do organizational governance structures influence decision-making processes?*
2. *How do advances in technology and information facilitate transparency within organizational framework?*
3. *How do leaders adopt a culture of ethics, sustainability, and social responsibility within their organizations?*

Identificação

Nome:

Morada:

Setor:

Número de funcionários:

Nome da pessoa que responde:

Cargo na organização:

Email:

1. **Estrutura de governação** *How do organizational governance structures influence decision-making processes?*

Pergunta 1

Como descreve a estrutura de governação da sua organização?

1. Estrutura organizacional linear, centralizada

2. Estrutura organizacional funcional, descentralizada. Cada setor responde à sua chefia direta.

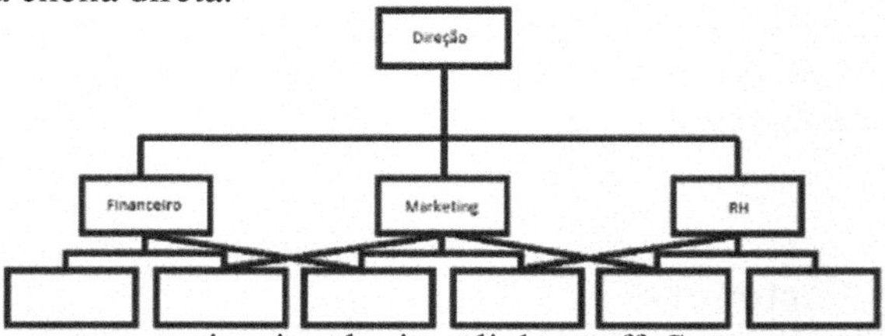

3. Estrutura organizacional mista, linha-staff. Segue o esquema da estrutura linear, mas distingue-se pela existência de órgãos de consultoria, que podem fazer recomendações técnicas e especializadas aos escalões inferiores, mas não os chefia. Quem os chefia são os chefes de cada departamento.

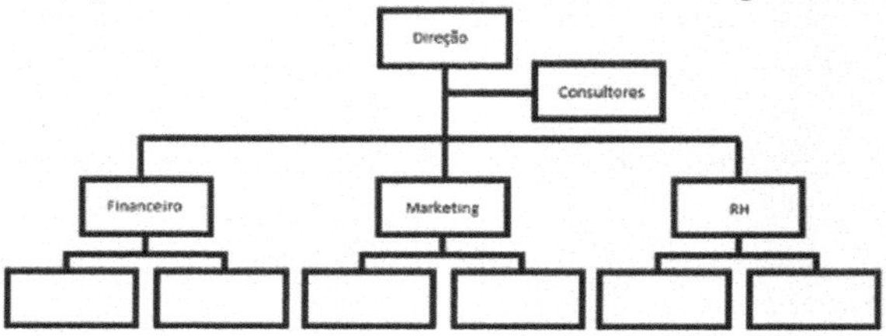

Pergunta 2

De que forma a estrutura de governação influencia a tomada de decisão na sua organização?

() As decisões são tomadas principalmente pela alta administração.
() As decisões são tomadas com a participação de diferentes níveis hierárquicos.
() Existe um equilíbrio entre decisões centralizadas e descentralizadas.

Pergunta 3

Os funcionários são incentivados a participar do processo de tomada de decisão?

() Sim, em todos os níveis.
() Apenas os funcionários em cargos altos de gestão.
() Não.

Pergunta de resposta livre/justificação de respostas

Face às respostas dadas nesta secção, poderá registar neste espaço todos os comentários que considerar relevantes para as suas escolhas.

2. **Transparência facilitada pelo avanço da tecnologia** *How do advances in technology and information facilitate transparency within organizational framework?*

Pergunta 1

Como classifica o nível de transparência da sua organização em relação às operações e decisões da mesma?

() Alto.
() médio.
() Baixo.

Pergunta 2

Que tecnologia a sua organização utiliza para promover/demonstrar transparência? (aqui pode selecionar várias).

() Softwares de gestão de informação interno.
() Plataformas de comunicação interna.
() Relatórios de Responsabilidade Social/ Sustentabilidade.
() Outro (especificar): ___________.

Pergunta 3

Na sua opinião, de que maneira os avanços tecnológicos têm impactado a transparência nas organizações?

() Facilitam significativamente a transparência, proporcionando maior visibilidade e acessibilidade às informações.

() Facilitam a transparência, mas não houve mudanças significativas no nível de visibilidade e acessibilidade às informações.

() Não facilitam a transparência e, pelo contrário, complicam o acesso e a visibilidade das informações.

Pergunta de resposta livre/justificação de respostas

Face às respostas dadas nesta secção, poderá registar neste espaço todos os comentários que considerar relevantes para as suas escolhas.

3. **Ética, sustentabilidade e responsabilidade social** *How do leaders adopt a culture of ethics, sustainability, and social responsibility within their organizations?*

Pergunta 1

A liderança da organização promove uma cultura de ética e integridade?
() Sim, e existe um código de ética/conduta.
() Sim, mas ainda não foi implementado nenhum código deontológico.
() Não.

Pergunta 2

Como é que os princípios de sustentabilidade são integrados na estratégia e na estrutura organizacional? (pode aqui selecionar várias opções também).
() Departamento específico de sustentabilidade.
() Políticas ambientais e Sociais.
() Iniciativas de diversidade e inclusão.
() Outro (especificar): ____________.

Pergunta 3

De que formas a organização garante a eticidade, a sustentabilidade e a responsabilidade social? (pode também aqui selecionar várias respostas).
() Formação aos funcionários.
() Auditorias internas e externas.
() Canais para denunciar comportamentos antiéticos.
() Outro (especificar): ____________.

Conclusão

Muito obrigado pela sua participação. Se desejar receber um resumo das suas respostas, por favor, forneça o seu endereço de e-mail:____________.

References

1. Almashhadani, H., Almashhadani, M.: An overview of recent developments in corporate governance. Int. J. Bus. Manag. Invention **11**(5), 39–44 (2022). https://doi.org/10.35629/8028-1105013944
2. Araujo, L., et al.: Digital leadership in business organizations: an overview. Int. J. Educ. Admin., Manag., Leadersh. **2**(1), 45–56 (2021). https://doi.org/10.51629/ijeamal.v2i1.18

3. Azimov, A.: Ethical leadership and its effect on organizational integrity: a multifaceted examination. SSRN. 1–5 (2024). https://doi.org/10.2139/ssrn.4777589
4. Barbosa, A., et al.: Integration of environmental, social, and governance (ESG) criteria: their impacts on corporate sustainability performance. Humanit. Soc. Sci. Commun. **10**(410) (2023). https://doi.org/10.1057/s41599-023-01919-0
5. Ellili, N.: Bibliometric analysis on corporate governance topics published in the journal of Corporate Governance: The International Journal of Business in Society. Corp. Gov. **23**(1), 262–286 (2023). https://doi.org/10.1108/CG-03-2022-0135
6. Gjaltema, J., Biesbroek, R., Termeer, K.: From government to governance…to meta-governance: a systematic literature review. Public Manag. Rev. **22**(12), 1760–1780 (2019). https://doi.org/10.1080/14719037.2019.1648697
7. Johns, F.: Governance by data. Annu. Rev. Law Soc. Sci. **17**, 53–71 (2021). https://doi.org/10.1146/annurev-lawsocsci-120920-085138
8. Kamalaldin, A., et al.: Transforming provider-customer relationships in digital servitization: a relational view on digitalization. Ind. Mark. Manage. **89**, 306–325 (2020). https://doi.org/10.1016/j.indmarman.2020.02.004
9. Luedi, M.: Leadership in 2022: a perspective. Best Pract. Res. Clin. Anaesthesiol. **36**(2), 229–235 (2022). https://doi.org/10.1016/j.bpa.2022.04.002
10. Malik, A.: Corporate governance and ethical leadership in business management. Manag. Sci. Res. Arch. **1**(03). 1–10 (2024). Available in: https://managementscienceresearcharchives.com/index.php/Journal/article/view/16
11. Naciti, V., Cesaroni, F., Pulejo, L.: Corporate governance and sustainability: a review of the existing literature. J. Manage. Gov. **26**, 55–74 (2022). https://doi.org/10.1007/s10997-020-09554-6
12. Pulejo, L.: Social reporting in European ethical banks: a comparative study. Int. J. Manag. Stud. Res. **3**, 196 (2022).
13. Roehrich, J., et al.: Inter-organizational governance: a review, conceptualisation and extension. Prod. Plan. Control **31**(6), 453–469 (2020). https://doi.org/10.1080/09537287.2019.1647364
14. World Economic Forum: Defining the ‘G’ in ESG Governance Factors at the Heart of Sustainable Business. We Forum. Harvard Business Review (2022). Available in: WEF_Defining_the_G_in_ESG_2022.pdf (weforum.org)
15. Zwitter, A., Hazenberg, J.: Decentralized network governance: blockchain technology and the future of regulation. Front. Blockchain **3**, 1–12 (2020). https://doi.org/10.3389/fbloc.2020.00012

Diffusion of Sustainable Business Ecosystems by Services Businesses

Liane Pinho Santos and João F. Proença

Abstract The purpose of this study is to investigate how business service can contribute to the diffusion of a sustainable business ecosystem. It was collected empirical data from thirty different actors that provide services to the automotive business and from industrial actors of the automotive businesses. Through service business, the automotive business players search for achieving sustainability goals. Embedded at interaction process between the industrial and the service business players emerge several effects, which are spread to their business relationships, occurring a deep sustainability diffusion process in business ecosystem. Service businesses players are responsible for: decreasing the difficulties on business processes of automotive business industry; developing green production and green activities over the business ecosystem; promoting technological development; increasing trust and commitment in business relationships due to the interaction processes, namely related with innovation; developing a green atmosphere in business relationship; and developing a more sustainable business ecosystem.

Keywords Sustainability · Automotive business · Service business · Sustainable service business ecosystems

L. Pinho Santos (✉) · J. F. Proença
School of Economics and Management, University of Porto, Porto, Portugal
e-mail: lianepinhosantos@hotmail.com

J. F. Proença
e-mail: jproenca@fep.up.pt

J. F. Proença
ADVANCE/ISEG Research, ISEG, University of Lisbon, Lisbon, Portugal

© The Author(s), under exclusive license to Springer Nature Switzerland AG 2025

A. d. b. Machado et al. (eds.), *Environmental, Social, Governance and Digital Transformation in Organizations*, Information Systems Engineering and Management 35, https://doi.org/10.1007/978-3-031-86079-9_10

1 Introduction

The importance of service business has increased over time due to their connection to different sectors and their contribution to business' growth. Business services are business that serve business ecosystems [29]. In this way, service business are considered as internal and external growth promoters for businesses [10].

Service businesses helps to provide solutions for sustainability challenges and/or are responsible for an important part of research on sustainability [15, 57]. Sustainability incorporate social, economic and environmental concerns, based on the triple bottom line of sustainability [16]. Based on that, businesses are focusing on making all their processes more sustainable in order to reduce their environmental, social and economic impact and do not compromise future generations [52]. Sustainability integrated in business means a set of strategies in which companies achieve economic success while ensuring social needs and environmental preservation [56, 64]. By doing this, sustainability concerns are integrated into business processes, and then its effect spreads across business ecosystems [42, 68].

Businesses are changing the way business is developed due to sustainability [42, 46]. However, the way businesses have reacted to sustainability are a topic under investigation. It remains a research topic with several gaps in business ecosystems. It is known that companies have changed the way they relate to each other, either in individual business relationships, or in the business ecosystems where they are embedded. Business relationships are characterized by links between actors, who develop activities by exchanging their resources [25]. In business ecosystems, the actors are interconnected by their business relationships to a common goal [31]. Business ecosystem can be defined as "constellations of partners for focal firms" [42, pp. 385].

The automotive industry is a rich and a good example of an industrial network embedded in a business ecosystem, which includes a diversified group of interesting and heterogeneous industrial and service businesses that in recent years have been under pressure to become more sustainable [9, 13, 19, 62]. There are governmental pressures that influence the implementation of sustainability [19, 45, 62], with temporal deadlines, since it is necessary to comply global and environmental goals [55]. Thus, taking in account the theoretical background of business and service ecosystems, we select to research the automotive industrial network that are slowly responded to the implementation of sustainability [62]. Our research intends: (i) To discover the main factors and the underlying processes that emerge in developing sustainable business service ecosystems; And (ii) To show why and how service business firms have an essential role in the diffusion and implementation of sustainability.

We decide to collect our data at the Portuguese automotive industrial network considered a rich and appropriate case to understand the phenomenon in analysis, and following [55] research suggestions that investigated the challenges for developing a sustainable return supply chain at the Portuguese automotive industrial network. It is known that companies still find it difficult to be fully sustainable, since they cannot recycle the totality of the products [27, 34, 41, 53] and service processes [19, 23]. Moreover, the literature demonstrates difficulty in implementing sustainability due to lack of knowledge and training [32, 55], which called for partnerships between industrial and service business for knowledge development [47, 55]. In addition, services are necessary to manage both the automotive production and the activities developed by the companies [11]; service business or green services providers are essential for sustainable business performance [3, 33, 18]; and sustainable services give companies a competitive advantage by reducing their environmental impact and creating a sustainable service climate [39]. One of the goals of service companies is to promote resource conservation, recycling of materials used, energy efficiency [33]. The need to reduce the environmental impact of automotive industry makes green service companies and green service delivery linked to corporate strategy and business processes [23]. With this, the effects are not only for the company, but also for the ecosystem where they are inserted [9, 33]. In addition, service businesses are providing more than specific services, they are able to cause long-term change in business necessary to accomplish governmental guidelines [68]. As said before, the lack of knowledge of automotive business is one of the main obstacles that companies faced. For that reason, partnerships between industrial and services companies help to obtain the knowledge for the implementation of sustainability need for innovation and governmental targets achievement [55]. Service businesses have been important for increasing the knowledge needed for sustainable development and for implementing sustainability in companies. A good example of this type of services are supplied by Knowledge-Intensive Business Services (KIBS) [10]. KIBS are based on professional and/or technical knowledge in the field [30]. KIBS provide knowledge, external information, and facilitates innovation, which adds internal value, employment, and increased labour productivity [10]. Innovation can be seen as a result and as a process that emerges in a service ecosystem [12, 30], and, developing and receiving the knowledge is necessary to overcome all the difficulties related to the automotive industry [55].

This article is divided into several parts. In Sect. 1, we find the introduction to the study. In Sect. 2, the literature review on sustainability, sustainable business processes and service business ecosystem are presented. In Sect. 3, we present the methodology of our study and the data collection process and analyses. In Sect. 4, we present our results. Section 5 discusses the findings that are confronted with literature. Finally, in Sect. 6, we present the main conclusions, namely the paper contribution, limitations, and suggestions for future research.

2 Literature Review

2.1 Business Services, Sustainability, and Sustainable Business Processes

Business services have been studied over the years, considering their relationship with industry. Gadrey et al. [22] state that industry grows through the help of services, which demonstrates a source of competitiveness. Services were initially defined with a specific contract, aiming to change the reality of the industry [22]. And, on the other hand, business services are business that serve business ecosystem [29]. So, business services firms are finding solutions to problems, which may not mean a physical resource or good [70]. Services can be contracted to increase industry capabilities, change structure, work models and knowledge [50]. Services have shown an increasing importance for industry development, finding solutions or as a form of innovation in the industry [20].

Sustainability has been an innovation' promoter [40], and on the other hand, service businesses have been essential in creating innovation for the industry [35, 51].

Rodriguez refers to the "urgent need" to include in services business, sustainability in their business strategies [51, pp. 2].

Sustainability is part of the business process, just as all strategies and decision-making should be based on sustainability. However, there is still a lack of knowledge about how companies can implement sustainability in their processes [51]. Sustainability causes business development as well as new business relationships [52]. Business relationship are established and developed by actors through their shared activities using their resources [25]. The actors had an important role, they are those who carry out the activities and/or control the resources used. The content and structure of business relationships and the dynamic between business relationships could be explained by the ARA (Actors-Resources-Activities) Model developed by IMP Group [47]. The actors are those who carry out the activities and/or control the resources. The actors can be individuals or companies that develop business relationships with others in the exchange process and/or aiming to mobilize resources [25, 46]. Different activities are conducted and coordinated within an organization, which are linked to other activities of other organizations [25, 46]. The importance of sustainability integrated in business is increasing and it is influencing the establishment of new business relationships or the development of established and older business relationships [54]. For instance, [54] show that business actors intending to achieve sustainability in business processes, develop sustainable business relationships with their counterparts, and if the (new) partner is not sustainable, the business relationship will decrease business and it can even terminate. In situations when business relationship last for a long period of time, it means an increase of stability. For that reason, longer and stable business relationships reduce the instability related to activities adaptations for sustainability [54]. Longer and stable business relationships allow actors to build trust, which represents a great commitment and collaborative participation in case of changes on business processes.

Sustainability, when integrated into business processes can create a sustainable atmosphere between companies, considering the climate of business relationships [7]. In this way, we can understand and manage business relationships. A sustainable atmosphere means more commitment from companies in terms of sustainability, more information shared and more trust on both sides [25, 49].

The atmosphere is a dynamic process, in which, the perception of the actors involved is influenced, regarding the climate surrounding the business relationships—sustainable climate. This atmosphere emerges from the sharing and exchange that exists from the interaction between the business ecosystem [25, 49]. In this way, if the substance of the business relationships change, the dynamic process of the relationship changes, affecting the sustainable atmosphere [25, 49].

Sustainability can be seen as an outcome of business interaction and interdependencies in business ecosystem [36, 52]. A sustainable business ecosystem meaning that sustainability is spread at the business ecosystem. Business ecosystem can be explained as mutual interdependent business relationships [8, 24], in each, the actors can be analysed as a part of a interdepend system environment [1]. Each actor acts differently, considering that they can act in overlapping business relationships [42]. The business ecosystem considers business relationships, and the relationships established by actors based on a purpose. In this way, they establish relationships to respond to a necessary solution ensuring structure to occur [42]. All business relationship are inter-linked and change over time [63]. A change in one part of the system leads to a reciprocal change in another [8].

2.2 Sustainable Service Business Ecosystem

One of the aspects to consider is the development or the process of becoming sustainable. Thus, in addition to making all internal processes sustainable, in business ecosystems it is intended that their partners also develop and implement the same type of process. This happen because business relationships are based on several elements as actors, activities exchanged and resources shared between partners [5, 38, 56]. Some companies have shown difficulty in achieving sustainable goals or objectives without help, support, or cooperation from partners [5, 56]. Moreover, companies in industrial networks establish partnerships with service business to overcome the existing difficulties in the implementation of sustainability. Thus, business services may find solutions to sustainability problems. Services can be contracted to increase industry capabilities, change structure, work models, knowledge [29]. However, these business solutions are integrated in business ecosystem, and since the service is integrated into the ecosystem, it is known that any change in activities may affect the performance of all other partners [48].

A network of business service can be defined as a "complex interdependent relationships between actors who are either traditional service providers and that provide services to other firms, or that are traditional industrial companies which provide tailored solutions to their clients by integrating a service into the core product that they offer" [50, pp. 952].

Thus, service business can be defined as the provision of services within business relationships. In this way, the parties interact and have access to resources to solve their own and their partner's problems [38, 50].

One of the greatest difficulties for the sustainability implementation on industrial networks is knowledge. Knowledge is one of the mediators for developing sustainability in business. It is necessary a lot of knowledge to develop sustainability [52], and service businesses are needed to increase the knowledge of companies in industrial networks [50]. The growing importance of these companies is due to the higher global competitiveness, the qualification of employees, the rapid need for technological change and the continuous need for technological development and innovation. Thus, partnerships with service companies are needed in order to increase the productivity of companies [30]. Service businesses are associated with the creation, development, and diffusion of knowledge within companies, which diffuses into business ecosystems. In this way, these services lead to technological and non-technological innovation of the partners, being useful for all the stakeholders [30]. The service companies contracted for the diffusion of knowledge, leads to increased competitive advantage and financial growth [30, 61].

A good example of this type of services are supplied by KIBS and research and development service businesses [4, 10, 30, 67]. KIBS are based on professional and/or technical knowledge in the field [30]. KIBS provide knowledge, external information, and facilitates innovation, which adds internal value, employment, and increased labour productivity [10]. Research and development service business helps business growing due to the need of create product and process innovations continually. This type of service business is a main contributor for innovation [21]. Innovation can be seen as a result and as a process that emerges in a service ecosystem [12, 30], and developing and receiving the knowledge is necessary to overcome all the difficulties related to sustainable business practices at the automotive industry [55].

Moreover, sustainability has been a promoter of change in the automotive industrial network [62, 66] and to the automotive service business ecosystem [55]. The automotive industrial network is considered one of the main industries responsible for global temperature increase and greenhouse gas emissions [17, 65]. Thus, the automotive industrial network has come under government pressure to achieve sustainable goals—such as carbon neutrality set by the European union by 2050 [58, 2]. Carbon neutrality requires changes in automotive production and aftermarket, which causes changes in activities and resources. With this, innovation, knowledge and technological development are necessary to implement sustainability, which can be provided by service business [55].

With the increasing concern with global sustainability, a set of guidelines and policies that promote sustainable practices have emerged [56]. Regulatory policies, in whatever form, are the enablers for companies to adopt and implement sustainability. In doing so, they lead companies to reduce their environmental impact in all associated processes [56].

With the increasing pressure of environmental problems and threats, the business ecosystems have developed and implemented sustainable practices in their activities. This is proving to be a source of competitive advantage [5]. An effective way to promote sustainable practices is through the definition of common objectives and procedures by all stakeholders in the business ecosystems [5, 56]. Relationships prove to be very important for the development of sustainability, besides being essential for example, for global industrial networks [52]. There are several aspects necessary to facilitate sustainable business practices, but one of the most important is the transfer of knowledge between partners. This knowledge can be transferred by training all the stakeholders, training in capacity building, providing technological support, assisting in product development, advising on product quality, stimulating the partner to increase their focus on the products, commitment is required, and the partners have their own education and training programs [5, 56].

In order to become a fully sustainable partner you need to evaluate and give feedback, develop evaluation programs, give feedback on partner performance, set targets for environmental improvement, audit your partners, encourage the creation of a working group in order to solve environmental issues, information sharing and communication between all stakeholders [5]. Stable and long-lasting business relationships facilitates the sustainability implementation [54]. On the other hand, time and commitment are necessary to implement sustainability in business processes, and this effect spreads in business ecosystems [64]. In addition, investment in research and development increases the demand from other companies in research for innovation [4].

Service businesses tend to collaborate with other businesses and even a larger network, since they are the basis of innovation for many industrial firms. In this way, these partnerships prove mutually beneficial, as it leads to building growth for the service businesses themselves, as well as for the industrial firms [30]. The knowledge about the industrial business and of service business can complement each other and lead to the development of solutions, innovations and new products [4]. The benefits are spread in the business ecosystems creating a "innovation ecosystem", especially when the actors develop mutual goals for sustainability achievement [43].

Knowledge provided by service businesses change the way business ecosystems act. There are changes in activities and production. Service business helps industrial business to have sustainable production and activities. The interlinks between industrial and service business firms leads to a greener production and activities [37, 59]. At the automotive industry, green production means that all automotive production processes are considering the reduction of environmental impact, using sustainable materials and technologies during the industrial process [37, 59]. In addition, green production depends on knowledge, which in turn is related to service business firms and ecosystems as said before. Green production refers to all the steps of production, the manufacturing, the product used and the end-of-life [3, 59]. With green production, the activities became greener, that is, they are preventing pollution, waste reduction by their recycling, using recycle or biodegrade material. Also, the energy need for the production are from renewable sources [3, 37]. The concern with environmental pollution promotes the technological development to reduce the pollution during the production and after. That is, reduction the pollution create by the car [3].

Green production contributes to economic, social, and environmental performance [59]. Trough green production and green activities caused by service businesses; the automotive business develop a more sustainable business performance. Other relevant aspect are the business relationships established among actors, and adapted from greener activities and resources [54]. When the partners shared their commitment for sustainable practices, the sustainable business performance is achieved [14]. Thus, and as it was said before, sustainability has been an innovation' promoter [40]. Green production and activities are enhancer for product innovation, which affects sustainable business performance [59, 60].

In sum, green production and green resources support sustainable business performance [3, 18] and then industrial and service business ecosystem. Sustainable business performance is supported by business relationships, and by actors organizing their business relationships, their activities and resources used. Thus, sustainable business performance improves sustainable competitive advantages and then sustainability practices spread in business ecosystems in an easier way [28, 61].

Through the literature review it was possible to understand sustainability and how companies establish business relationships to facilitate and implement sustainability in all business processes. However, it was possible to identify also a knowledge gap about the factors and processes underlying the development of a sustainable business service ecosystem, as well as the role of service companies in the diffusion and implementation of sustainability. Thus, the research intends to collect information to respond to this gap.

3 Methodology

3.1 Research Design

The purpose of this study is to investigate and understand how and why business service contribute to the diffusion of a sustainable business ecosystem and the underlying processes and practices.

Considering the research question, we decided to do an exploratory study within the automotive industrial network, considered a rich and adequate research unit of analysis to achieve our investigation objectives. Our research deeply investigates the role of service businesses for sustainability implementation, collecting and analysing relevant data for the investigation and understand the business ecosystem where industrial companies and service business firms are operating and interacting.

In this research, we need to understand and explore the context of the automotive business network and the established business relationships with service companies to solve sustainability issues. We decided to do the empirical research at the

Portuguese automotive industrial network, which is respecting and applying the European governmental guidelines related to sustainability. The Portuguese automotive industrial network constitutes a business ecosystem based in two main activities: the production of car components and the final vehicles assembling. With documental data we verified that multiple and diverse business are collaborating with automotive industrial network. Indeed, in Portugal there are four final vehicle producers and several suppliers' companies responsible for different components as plastic, door panels, seats, paints, textiles, suspensions, tires, among other components needed for the final vehicle. In parallel, we observed that are several services organisations that are providing services for the automotive industrial companies, namely universities and research institutions, consulting firms, engineering firms, and/or product development, etc.

3.2 Research Data Collection

By exploring the Portuguese Automotive industry, we verify a business ecosystem with multiple companies—final assemblers, components producers, and services companies. To answer the research goals, we focus on services companies that are providing services for the automotive businesses, as well as on automotive business companies to understand both sides of the business relationship.

Several companies operating at the automotive business ecosystem are providing diverse services to automotive companies, as for instance, automation and engineering services, training, product and processes development, project management and development, prototype engineering and others. These service business firms or entities include for profit and non-profit organizations, as well as some public institutions namely related with some public and state universities, researcher centers, consulting firms, etc. Thus, these services companies are providing services for the final assemblers vehicle companies and/or for components companies.

We collected empirical data from 30 interviewees from various service providers and from different automotive industrial companies.

Respondents from the service business were selected considering the following criteria: (i) The interviews should be done at service business entities commercializing, developing, and/or operating at the Portuguese automotive business; And (ii) The entities must be involved in sustainability projects related to automotive business.

The criteria for selecting the interviewees from the automotive industrial businesses were: (i) The company should be based in Portugal and/or should have operations embedded at the automotive Portuguese automotive ecosystem; (ii) The company must have a research and/or development department; And (iii) The company must have public sustainability reports.

With these criteria, it was possible to collected data from a very interesting and well diversified group of interviewees belonging to 19 services providers organisations and to 11 automotive industrial business companies, see Table 1.

Table 1 Identification of interviewees from services organisations and from automotive industrial companies

	Businesses	Activities of companies	Interviewee function
Services businesses	A	Training center	Managing director
	B	Engineering and product development	Product development manager
	C	University	Research department director
	D	Research institute	Research department director
	E	Research institute	Research department director
	F	Research and development center	Research and development manager
	G	University	Research department director
	H	Research and development center	Research and development manager
	I	Research and development center	Research and development manager
	J	Research institute	Research department director
	K	Consulting and training	Managing director
	L	Product research center	Product development manager
	M	Engineering services	Managing director
	N	Automation services	Owner and main director
	O	Engineering and project management	First engineer
	P	Project management and development	Sustainability director
	Q	Development of products and processes	Product development manager
	R	Software application development	Innovation manager
	S	Prototype engineering	Head chief of engineering department
Automotive businesses	1	Plastic supplier	Quality engineer
	2	Vehicle electronics supplier	Innovation manager

(continued)

Table 1 (continued)

	Businesses	Activities of companies	Interviewee function
	3	Supplier of plastic and metal components	Marketing and communication coordinator
	4	Technical parts for vehicle interiors firm	Human resources director
	5	Automotive filters	Quality engineer
	6	Textile supplier	Product development manager
	7	Multimedia supplier	Internal control coordinator responsible
	8	Rubber and tyre supplier	Marketing manager
	9	Interior and cockpit supplier	Managing director
	10	Navigation systems	General director
	11	Glass and glass panels supplier	Logistics director

We used a script with open-ended questions and with various themes to guide the interviews.

The interview script for the service business providers is presented as follows:

- How would you classify your business relationship with the companies in the automotive industrial network?
- How does it contribute to a sustainable business relationship?
- Can you identify your importance in the business ecosystem? ("role")
- What differentiates your company from other service providing companies?
- What is the added value of providing services to the automotive businesses?
- What services are you providing to the automotive businesses?
- How do the services help you comply with European directives?
- How does the knowledge you provided contribute to sustainability in the automotive businesses? (If applicable)
- How does the knowledge you have provided contributed to a more sustainable business ecosystem? (If applicable)
- How does the knowledge provided contribute to innovation? (If applicable)
- How does the knowledge provided contribute to technological development? (If applicable)
- As a service company, what are the benefits of being part of the business ecosystem?
- What are the main difficulties or obstacles that you faced when developing your interaction with the automotive industrial production actors?
- How was the interaction process developed with those actors? Can you give me some examples, including of cooperation and/or of conflict episodes?

- Any other issue that you consider relevant?

For the interviewees from the automotive industrial network, we used the following main topics to guide the conversations, as follows:

- Role of service business for the automotive business;
- Benefits for establishing the business relationship with service firms;
- Main reason for establishing the business relationship—sustainability, business process difficulties', European guidelines, others?
- Changes in automotive business due to services provided;
- Services business facilitates sustainable business processes?
- Role of knowledge for sustainable business ecosystem/innovation/technological development;
- Benefits of being part of the business ecosystem;
- Other relevant.

The service businesses and the automotive industrial companies involved in this research are represented with letters and numbers, respectively, see Table 1 and Fig. 1. Business relationships established between service business and automotive companies are represented by arrows in Fig. 1. The automotive businesses companies are symbolized with squares and services business with circles. The letter and number inside each square and/or circle represent the interviewee referred in Table 1. Figure 1 shows the several business relationships were researched. In addition, we also collect data from some services businesses that are not providing services for the automotive business companies interviewed. However, all the data collected is related with companies belonging to the automotive business ecosystem.

3.3 Research Data Analysis

After data collection, all the interviews data was transcribed and coded considering Bardin methodology for content analysis [6]. Through this method, it was possible to create categories and sub-categories related with data. Each category had different meaning. We also made data triangulation, considering the multiple data source and the interviews process. The content analysis allowed us to create a data structure, which facilitated the data analysis processes, see Fig. 2.

The coding process was based on meaning categories. After data exploring and full reading, it was created the data structure: we identify the "Difficulties on business processes" which were related with the "Business relationship between business services actors and industrial production actors", namely the lack of technological development and innovation, the lack of knowledge and the accomplishment of European Union guidelines. After this, the data was grouped by similar themes and meaning categories. This organization helps to creates a sequential process of the data, facilitating the understanding of the results. With this, it was created a meaningful data structure. The categories that emerge from the data were "Business

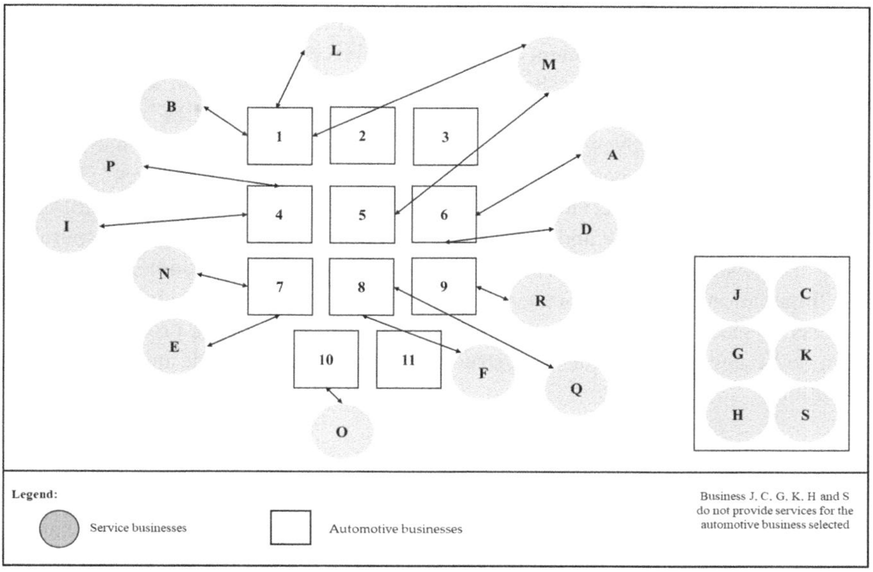

Fig. 1 Data collection: business relationships researched between service business and automotive businesses (J, C, G, K; H and S are providing services for automotive companies that were not involved in the data collection, despite belonging to the automotive business ecosystem)

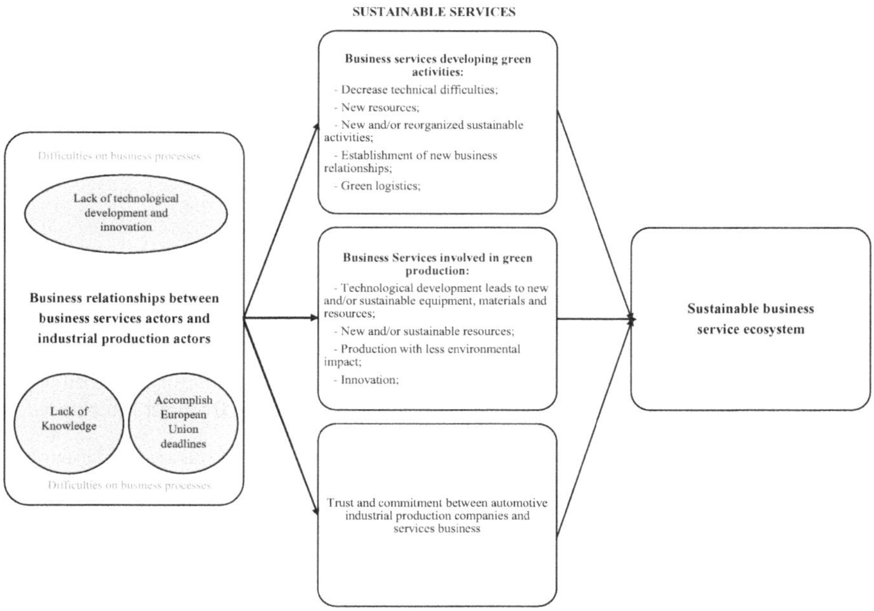

Fig. 2 Data analysis process (categories and sub-categories related with the data)

services developing green activities", "Business services involved in green production" and "Trust and commitment between automotive industrial production companies and services business". The first two categories had several sub-categories. The sub-categories of "Business services developing green activities" are "Decrease technical difficulties", "New resources", "New and/or reorganized sustainable activities", "Establishment of new business relationships", and "Green logistics". The sub-categories of "Business services involved in green production" are "Technological development leads to new and/or sustainable equipment, materials and resources", "New and/or sustainable resources", "Production with less environmental impact", and "Innovation". The three categories materialise in final category "Sustainable business service ecosystem" as shown in Fig. 2. After the coding process, interpretation and categorization, the data structure was presented to the interviewees for validation.

4 Results

After the content analysis, it was possible to understand how service companies contribute for diffusion of a sustainable business service ecosystem. We have results from both sides of the business relationships (service providers and producing companies), which gives an encompassing vision of the phenomenon under research. Bellow, we present our findings considering the categories and sub-categories presented before at Fig. 2.

4.1 Business Relationships Between Business Services Actors and Industrial Production Actors

Industrial companies of the Portuguese automotive industrial network felt governmental pressures, which have influenced their activities. As known in the field, by 2050 it is necessary to achieve carbon neutrality and being the automotive industry one of the main causes of carbon dioxide emissions, there is a need to change processes in this industry and business, which called for establishing new business relationships with service firms. This is illustrated by a quote from an interview:

> The automotive industry must comply with European standards. For this reason, sustainability is already embedded in the companies, however, not all processes are sustainable. (…) This is one of the main reasons for establishing the business relationship. To make the processes more sustainable to comply with European regulations. (Business Q)

Nowadays, automotive industry business is changing their processes to be fully sustainable. In this way, companies have focused on changing resources, materials, and logistics issues to reduce environmental impact.

> The pressure made by the European guidelines led us to look for solutions together with the service companies. Considering the time limits and the changes required to carry out several processes while maintaining parallel production, (…) that was our solution. (Business 5)

> Some of the companies we supplied material to had already asked us for changes in resources, which led to a reorganization of the activities, as well as in logistics. However, with the increasing need for quick responses we needed outside help. The service companies were initially contracted only for direct services, but now we work together for the demands of the automotive companies that are suppliers and need more and more sustainable resources. (Business 3)

As mentioned before, automotive business mentioned some difficulties in business processes such as lack of knowledge, lack of technological development and innovation, and deadline issues [55]. Thus, partnerships or the establishment of business relationships with service companies are the most used solutions to overcome these difficulties in search of sustainability of companies in the automotive industry. The service companies are hired directly by the automotive companies.

> These business relationships between services and industry has been intensified due to the need to overcome technical and technological issues that are delaying the implementation of full sustainability in automotive industry. The research and development service businesses presents solutions to the automotive companies. (Business F)

4.2 *Sustainable Services*

The partnerships between the automotive industry and the service business demonstrate a mutually beneficial business relationship. We found that the service businesses were able to provide answers to the automotive industrial companies' requests, as well as to overcome the difficulties experienced for the implementation of sustainability.

It was possible to analyse three categories at the data collected in the interviews, as majors' benefits of establishing business relationships with service business: Business services developing green activities; Business services involved in green production; and Trust and commitment between automotive industrial production companies and service business.

4.2.1 Business Services Developing Green Activities

The concern with sustainability led the automotive business industrial companies' increase their concern about the activities developed. Thus, service businesses were fundamental to make the activities more sustainable. These service business firms helped the industrial companies through the reduction of technical difficulties, supporting the use of new resources, and changes in logistics – green logistics. As a result of interaction process with the service business firms, new and/or reorganized sustainable activities and new business relationships were established and developed.

> Knowledge provided by the research and development service business was essential to reduce all the technical difficulties of the automotive industry (Business P)

> The lack of technical knowledge to overcome the major difficulties were our biggest hindrances. The technical difficulties were only overcome by the solutions presented by the service companies. (Business 11)

It was found that through the services provided, the automotive industry was able to understand the resources used, considering the product biological and technical specificities. In other words, the service businesses by studying the resources used, make it possible, the total recycling.

> One of the most current trends in the automotive industry is total car recycling. Some models have already been presented that are made of recyclable materials and can be recycled after disposal. This is one of the main limitations for the automotive industry. However, by researching the product (technical and biological), we can ensure that it can be recycled and kept as the same product, or used in another part of the car. (Business D)

> Plastic recycling has been going on for a long time, as has the whole process of integrating recyclable plastics into the automobile (…). However, maintaining the quality of the plastic, as well as a second, third or more lives for the plastic, was one of our biggest obstacles. As we recycled the material, we could use it a second time, but a third time was unthinkable, turning it into waste for the car. The loss of quality in the recycling process was our biggest obstacle, but it was made easier by the services, since we were able to give other lives to the plastic and/or transform the same product into something different to be used in the car. (Business 1)

Automotive businesses industrial companies have mentioned the impossibility of fully recycling the products as a major difficulty. Having the knowledge of how it is possible to recycle and maintain quality after use, led to a more sustainable activity. Some examples are direct: tire rubber recycling, which after being treated creates tires without the use of new materials; on the other hand, they mentioned indirect examples, i.e., the fabric used to give volume and comfort to the seats, is transformed into the carpets under the car's pedals. In this way, they manage to recycle the totality of the products and/or transform the products and, in this way, guarantee sustainability since the service provided overcame the technical difficulties.

Still on the difficulties in sustainability implementation, we found the emergence of new resources for the automotive industry, as was the case of replacing the leather needed for the seats for coconut fibre and latex seats. With these new resources, the environmental impact in the automotive industry was reduced, meeting environmental objectives. Other resources that have emerged are due to the reduction of the use of fuel engines, and to the electric motors increase, so companies have had to modify their activities to create and develop the electric motor. Direct current motors are no longer manufactured, and nowadays alternative current motors are used. Thus, the accessories also changed. In addition, with the motor electrification', there was also the need of energy efficiency while simultaneously improving autonomy, i.e., increasing the number of kilometres driven without electric charging. In the construction of the electric car, the question of the environmental impact of the batteries (mainly lithium) and their impact always arises.

> Sustainability is only guaranteed when the car is produced without environmental impact, its use means no pollution and its destruction does not cause any kind of waste or environmental impact. For this reason, it is so important, that in all phases of car construction attention is paid to this. The use and/or creation of resources from renewable sources is essential to ensure less environmental impact and sustainability. (Business E)

Business services are integrated in business ecosystem. For that reason, they observed changes in business relationships with automotive business, as well, in some cases the establishment of new business relationships:

> With the new activities developed towards sustainability, the automotive industrial business companies refer that some of their previous partners and suppliers no longer make sense, since the materials are no longer used. On the other hand, suppliers have not been able to change their more sustainable way of sourcing, as they continue to produce without paying attention to their environmental impact. (Business S)

> For all the influence of sustainability, new companies have started to emerge with things we have established new business relationships to stay current with change. At the same time, in our established business relationships, we promote the same request, sustainability in all processes, helping in its implementation or accepting help from service companies and applying it to our suppliers. (Business 4)

Despite not be referred as difficulty, automotive industrial businesses companies demonstrate a concern with the logistic. Rather than reducing the environmental impact of the companies' materials and resources, it would be desirable for suppliers to be located close to the vehicle production-plant. For that reason, service business firms were contacted to try to reduce the impact of logistics as well—green logistics. Automotive industrial businesses companies refer the need of a technological solution that would bring the businesses closer together, since it is not always possible for suppliers to be physically close.

4.2.2 Business Services Involved in Green Production

The biggest change in the automotive industry is leading to greener production, that is, more sustainable production. Services businesses were able to support the supply of new equipment, materials, and resources with less environmental impact.

Furthermore, the technological development was pointed out as one of the main solutions for the implementation of sustainability in the automotive industry. According to the interviewees, the technological development services are the answer for what is pointed to today, what is needed. And service business has demonstrated to be essential for this technological development.

> Technological development is a continuous need, more than just for production, activities, and others. Technology is needed to make processes more sustainable as well. We see an increase in demand, mainly from companies that do not have their own research capacity and, thus, hire us to develop new technology, equipment for something new or to provide a solution to a difficulty faced. (Business J)

Also, the technological development is being the focus of the establishment of the business relationships between industry business and services business, i.e., companies in the automotive industry hire services business to provide solutions to the difficulties presented. The technological development are needed to reach the technical, technological and product difficulties.

> (…) in these cases, most of the services are one-off. It's something requested, done, answered. We work on it again, in case of a new need. However, more and more it's something continuous, like a work plan. The automotive design teams get together and say what they need and give us guidelines and we develop what's necessary taking into account the deadlines requested. (Business S)

In this way, service business firms have been essential to introduce new equipment to build materials, new and/or sustainable resources, accessories, and recycle the materials used in the automotive business. Furthermore, through research, new and/or sustainable materials and resources began to be introduced in the automotive industry. Such as biodegradable materials, materials whose origin does not come from non-renewable sources. These materials leaded to the development of different resources, which reduced the environmental impact of automobile manufacturing.

According to the interviewees, their services led to the biggest change in the automobile, which is sustainability during the production process and after-sales, i.e., the car is also sustainable.

> It is no use saying that it is a sustainable industry if the resources used to create more sustainable cars come from non-renewable sources. (Business E)

With more sustainable resources, the existing materials in automotive can be fully recycled, becoming the same products again, with no reduction in quality. In the same way, in case of resource' loss of quality, are used to create another product within the automotive. There are known cases of automotive brands that have created, for example, office bags with materials that were previously used in cars. In other words, although the materials are not directly used in the automobile, they are not disposable. This is in line with the circular economy, cradle-to-cradle, and environmental sustainability.

> The companies came to us with concrete difficulties. For example, difficulty in creating a sustainable material, difficulty in recycling the product at the end-of-life, ensuring a durability of the recycled material like the product with a non-renewable origin. For some, we already had solutions and we were supplying direct services. For others is required some research and product development to achieve the request. (Business P)
>
> (…) being part of the process of building a fully sustainable vehicle was unthinkable a few years ago. The truth is that it is a reality and there are already several fully recycled models with little environmental impact in their production and use. (Business 2)

With more sustainable materials and resources, with the help of technological development which development equipment, automotive business is being able to reduce the environmental impact in production. In this way, with the activities mentioned above, the car is getting a lower environmental impact, which is necessary to comply with European legislation. And, as said before service business firms have been determinant in this process.

The legislation for the reduction of greenhouse gas emissions and for carbon neutrality, caused additional pressure, which some business could not respond immediately to the needs.

> (...) changes may seem slow, but the car has been changing since concern for sustainability began to rise. Electric cars are the solution today. However, artificial intelligence and automation has shown great technological evolutions for the vehicle, demonstrating great differences in a few years. (Business O)

More than the technological development is required innovation. In other words, according to the interviewees, the work of service businesses is divided into two paths: technological development for what is pointed to today, what is needed; and innovation, that is, creating something that is not actually needed and that may create an answer to something that at the moment is not thought possible to achieve and / or even thought of.

> The need to innovate and find solutions beyond what is expected or expected for today, ends up fostering a more intense and mutually beneficial work for us and the service companies. As much as it is a contractual relationship, the companies work with us and together we create solutions and grow side by side. The service companies have led us to stand out and that is beyond description. (Business 7)

The need for innovation is essential for the development of a sustainable environment. Technical, technological and product/resource innovation is essential to maintain the sustainability in all the business processes. For example, if sustainability is not implemented in automotive industry, the automotive manufacturing will maintain its production based on non-renewable resources; in the aftermarket, the cars consume for its action non-renewable resources contributing to the emission of greenhouse gases. Thus, innovation is not only urgent, but also the answer to changing the automotive industrial network as it is known.

> (...) innovation has been the key to the success of the relationships established. Our services have been continuously requested, since we develop innovative products, which highlights the different companies in their sectors. Innovation, in addition to create a future need, increases the reputation of the automotive companies. We have observed an increase in the companies' positioning globally, which has been very beneficial. (...) establishing this kind of relationship, besides the expected innovation, also helps the environmental, social, and economic sustainability of the auto companies. (Business I)

4.2.3 Trust and Commitment Between Automotive Industrial Production Companies and Services Business

The establishment of business relationships between service business firms and automotive industrial business companies, considering all previous said, are needed to achieve sustainability. It was observed an increase in trust and commitment when the automotive business gave the service business opportunity to demonstrate something new and/or innovative, and at the end of the day the relationship ended up being strengthened. In this way, trust facilitated that the automotive industrial business companies and the service business firms started to share knowledge, which resulted in project creation, in getting richer and quicker outputs, and in developing more directed solutions to the needs of the automotive industrial network.

> When automotive business companies gives us the freedom to demonstrate what we have, or works with us to develop something new, we get faster and/or more innovative results. (Business H)

Thus, one of the greatest and most important benefits outcomes of these business relationships was the increase of trust between parts.

> (…) especially when automotive industry companies gave us freedom to present innovation or other solutions that were not so immediate (…). (Business F)

On the other hand, service business gained reputation because of their ability to innovate. In this way, the business relationships between automotive industrial companies and service business firms were mutually beneficial. Furthermore, there was an increase in value for both parties due to the indirect benefits and mutual benefits obtained.

Other greater benefits for services businesses are national and international recognition and, as a result, collaboration on other projects with other services business, as research centers and/or universities. For automotive businesses, this recognition means being the pioneers in such innovation or create the innovation for the customer, creating their need.

> Our relationship gets much stronger. The existing trust, together with this commitment that is to differentiate ourselves and achieve something innovative is something really synergistic. Since the first time, we have never stopped working together and the truth is that we have mutually won, both in terms of reputation and international recognition of what we have achieved (…). (Business G)

It was possible to observe that the increase trust and commitment strengthens the competitive position for both parties, which also led to the creation of a sustainable atmosphere and a sustainable climate.

4.3 Sustainable Business Ecosystem

The development of new and/or reorganized sustainable activities and green production, along with technological development and innovation have brought great benefits to both the automotive and service business. Thus, with sustainability being implemented in the automotive industry due to services, it has created not only more sustainable business relationships, but also a sustainable atmosphere or green atmosphere.

> With the results achieved with the help of the service businesses, it was created a sustainable climate, not only within the car companies, which spread in the system among all other relationships. (Business L)

The implementation of sustainability led to a power increase at the business ecosystem for both service businesses and industrial players. Furthermore, sustainability was spread throughout the business relationships, leading to an increase in the closeness of the relationship, commitment, cooperation, and trust.

It was observed that, in the case of a desire for bilateral growth, mainly based on the need for innovation, the business relationships became even closer with a greater commitment. With this, automotive and service business, shared their knowledge. The relationship is not just the contract as initially. In this way, the business relationship was strengthened, and sustainability was spread throughout the business ecosystem.

> The gains from sustainability go beyond our direct relationships. (…) The effects are spread across various relationships, leading to a contagion of sustainability into various processes. (Business 9)

Green atmosphere means that sustainability is present in business relationships. Considering, all the business relationship established in automotive industrial network, the diffusion of sustainable atmosphere happens. It was possible to verify that the effects spread through the different business of the business ecosystem. And the contribution of services business was essential for internal sustainability of automotive business. At the same time, through business relationships, the diffusion of sustainability through the business ecosystem occurred.

> (…) sustainability is an "opening door", that is, all relationships are based on sustainability (…). Business B)

> Technological development and innovation bring added an informed strength, which accelerated the spread of sustainability. (Business A)

The creation of a sustainable business ecosystem is a continuous process. It is necessary to establish business relationships considering the objectives to be achieved and/or the difficulties faced. That said, service companies are seen as essential for the implementation of sustainability in all processes in the automotive industry. Having all the processes sustainable, the effects are easily spread across all other business relationships leading to the creation of a sustainable climate, which in turn creates a sustainable atmosphere, and a sustainable business ecosystem. All steps are important, and business relationships need to be sustainable, so that the atmosphere and the ecosystem are, consequently, sustainable.

5 Discussion

The aim of the research is to understand how service businesses are contributing to a more sustainable business ecosystem. Thus, this section presents our findings discussing how service businesses processes are developed in what concerns actors, resources, and activities in order to get a more sustainable ecosystem. Given the results, our findings show that services business has an essential role in the diffusion and implementation of sustainability. The discussion was divided based on the research questions that formed the basis of this study.

5.1 Main Factors and Underlying Processes in Development of Sustainable Business Service Ecosystem

The relationships established between the service business firms and the industrial business companies are based on relationships settled to overcome the difficulties that the implementation of sustainability has caused. Thus, the relationships between the parties were limited in time. Considering the development process, it is necessary to explain and discuss the interactions and the dynamics that exist between these businesses. As well as to discuss on how the processes between actors, resources and activities are developed.

One of the main characteristics of the business relationships is the time of the relationship, that is, time can define the nature of the interaction as a sequential process. The companies established a business relationship with a concrete objective, based on difficulties that arose by the implementation of sustainability. So, they enlist a service. These difficulties are solved by the service companies, facilitating the implementation of sustainability. After this, in some cases, after verifying mutual benefits, the relationship is no longer based on time, nor on being a sequential process, as both parties work with greater commitment to the common good and achieve mutual benefits, namely related to reputation as our results showed. Through this, they establish a long-lasting and stable business relationship.

These relationships are based on interaction, and on commercial benefits for the partners. The industry was getting the services they need, and the service firms were rewarded for the services provided. However, because of the greater commitment due to innovation, the business relationships become strategic. Then, industrial businesses companies and service business firms shared their knowledge, and for that reason, both improved their positioning in the business ecosystem, as well as, the relationships got greater value, and the effects of sustainability were spread more easily. Reinforcing [39] about the need for good management of business relationships with service companies, as this helps to achieve sustainability.

Our study confirms that business relationships are dynamic, not static. In which both parties adapt over time, as the business relationships evolves. This dynamic process is influenced by time and by the increased commitment in the business relationship. In this way, the benefits of sustainability search lead industrial companies and services firms to unite, increasing their commitment, thus facilitating the diffusion of sustainability in the business ecosystem.

One of the ways that facilitates the development of sustainability is the exchange process between actors related to the resources and activities involved in the business ecosystem. Industrial actors played the initial role in establishing the business relationships with the service business firms; however, it was in the resources and in the activities involved in the relationships that there emerged the biggest changes. It was verified that business relationships were maintained, and even strengthened due to research in sustainability. Moreover, the need to establish new business relationships with other actors was due to the impossibility of guaranteeing a green production. It was necessary not only to guarantee the reduction of environmental impact during production, but also in the automotive industry itself. In this way, resources are probably the most important element of the three, since they are conditioning the diffusion of sustainability. The need for new and/or more sustainable resources, as well as, the integration of recycled resources and/or resources that have been studied and know what kind of resources they will be, is essential for the implementation of sustainability. This is in accordance with [23] which refers that green services preserve or create systems focused on resources that were previously considered disposable. Green services assist in the reduction of negative effects on the environment by facilitating an implementation of sustainability.

Thus, with the introduction of new business relationships as well as new resources came the need to develop new activities. Thus, time on business relationships is fundamental to facilitate the diffusion of sustainability. In more stable relationships, where the partner's work is known, it becomes easier to anticipate results and reduce possible conflicts that may exist. Therefore, the efficiency of the activities is due to the time of the business relationships. Supporting [44] that refers that longer relationship are based in higher levels of trust which represents a conflict reduction.

5.2 The Role of Service Business Firms in the Diffusion and Implementation of Sustainability

Our research shows that service companies are essential for the implementation of sustainability in the automotive industry. Without service providers, technical and technological difficulties prevented a more sustainable performance in the automotive industrial network. The automotive business showed difficulties, as lack of knowledge, which prevented a more sustainable logistics, production, and after-sales. The services businesses, in this way, reduced these limitations by increasing knowledge, providing support related to the equipment, new and/or more sustainable resources, and materials with less environmental impact are examples of the vital role of service companies in the implementation of sustainability. Supporting [30] that refers that business services contributes for changes and promote the development of new products and services. The knowledge of service business firms contributes for business innovativeness and competitiveness [30, 61].

On the other hand, the diffusion of sustainability through the business ecosystem is due to the strengthening of these established business relationship based on the increased trust and commitment. Innovation goes beyond the previously contracted service. Through the need for innovation, as answer to some problems that have not yet been solved and/or the need to develop new things, the industrial companies and service business decided to share their knowledge for common goal. This commitment reveals an increase in trust, with the goal of strengthening their positioning in the industrial network, namely their power position, reputation, and the associated economic benefits. Through this greater trust and commitment, the effects of sustainability are spread across all stakeholders, spreading throughout the business ecosystem. Without this greater commitment on the relationship, the diffusion of sustainability would be slower and, in some cases, even non-existent. Since the competitiveness that helps to maintain the need to stand out, was not as high. Completing [69] who refers that businesses in cases of co-innovation "securing a stronger position in the power dynamics between suppliers and buyers, innovation collaboration, market share expansion, cost reduction through bulk purchasing, knowledge and workforce sharing, and securing larger contracts" [69, pp. 554].

The diffusion is made by the business relationships themselves, affecting the business ecosystem. As refer in [42] the globalization, and environmental awareness are affecting the way businesses react. We can confirm that the need to accomplish governmental guidelines and sustainability cause changes in the business ecosystem that are diffusing through business relationship, affecting actors, resources, and their activities.

6 Conclusion, Contribution, Limitations, and Future Research Suggestions

Our research demonstrates that business services contribute for business ecosystem sustainability diffusion. Through services business, automotive business achieves sustainability goals, and their affects are spread to their other business relationship, occurring a diffusion in the business ecosystem.

We found out that the service companies are responsible for: (i) decreasing the difficulties on business processes of automotive business industry; (ii) developing green production and green activities over the business ecosystem; (iii) promoting technological development; (iv) increasing trust and commitment in business relationships due to the interaction processes, namely related with innovation; (v) developing a green atmosphere in business relationship; and, finally (vi) developing a more sustainable business ecosystem.

Our study contributes theoretically to increase knowledge about sustainability theory, business relationships, businesses ecosystems, and service theory. The business relationships established between service business and automotive industrial companies are the first step to overcome automotive industry difficulties in order to get a more sustainable ecosystem. With service business firms' help, sustainability is implemented at automotive industrial business. Our research shows that the need of innovation leads to a higher commitment between actors, which result in more trustful business relationships sharing knowledge. Through interaction and business both industrial players and service businesses players reinforce their competitive positioning in the automotive industrial network, as well as power positions, reputation, and the associated economic benefits.

This paper also contributes to improve the knowledge of managers to deal with sustainability difficulties. In this situation, the business relationships with services businesses are essential. With that, a green activities and production are achieved, as well, innovation and technological development helps in automotive sustainability.

Finally, our research has some limitations. First, this an exploratory study developed over the automotive industrial network that is buying business services to overpass several difficulties related with sustainability. Further research about the role of services for sustainability should be developed in other industrial areas that should confirm and enlarge our findings. Moreover, future research can get more and different data from the automotive business, which would enrich our understanding of the entire business ecosystem, namely deepening the sustainability issues.

Acknowledgements João F. Proença gratefully acknowledges financial support from FCT—Fundação para a Ciência e Tecnologia (Portugal), national funding through research grant UIDB/04521/2020 & UID06522.

References

1. Aarikka-Stenroos, L., Jaakkola, E., Harrison, D., Mäkitalo-Keinonen, T.: How to manage innovation processes in extensive networks: a longitudinal study. Ind. Mark. Manag. **67**, 88–105 (2017)
2. Agency, International Energy: Portugal's energy policies set a clear pathway towards 2050 carbon neutrality (2021)
3. Al-Shami, S., Rashid, N.: A holistic model of dynamic capabilities and environment management system towards eco-product innovation and sustainability in automobile firms. J. Bus. Ind. Mark. **37**, 402–416 (2022)
4. Audretsch, D.B., Belitski, M.: The role of R&D and knowledge spillovers in innovation and productivity. Eur. Econ. Rev. **123**, 103391 (2020)
5. Bai, C., Sarkis, J.: Green supplier development: analytical evaluation using rough set theory. J. Clean. Prod. **18**, 1200–1210 (2010)
6. Bardin, L.: Análise de conteúdo. Edições 70. Lisboa, Portugal (2011)
7. Bayne, L., Purchase, S., Soutar G.N.: Network change processes for environmental practices. J. Bus. Ind. Mark. 1832–1845 (2021)
8. Breslin, D., Kask, J., Schlaile, M., Abatecola, G.: Developing a coevolutionary account of innovation ecosystems. Ind. Mark. Manag. **98**, 59–68 (2021)
9. Brodie, R.J., Fehrer, J.A., Jaakkola, E., Conduit, J.: Actor engagement in networks: defining the conceptual domain. J. Serv. Res. **22**, 173–188 (2019)
10. Bumberová, V., Milichovský, F.: Sustainability development of knowledge-intensive business services: strategic actions and business performance. Sustainability **11**, 5136 (2019)
11. Chan, T.-Y., Wong, C., Lai, K.-H., Lun, Y.H., Ng, C., Ngai, E.: Green service: construct development and measurement validation. Prod. Oper. Manag. **25** (2015)
12. Chandler, J.D., Danatzis, I., Wernicke, C., Akaka, M.A., Reynolds, D.: How does innovation emerge in a service ecosystem? J. Serv. Res. **22**, 75–89 (2018)

13. Damert, M., Baumgartner, R.J.: External pressures or internal governance—what determines the extent of corporate responses to climate change? Corp. Soc. Responsib. Environ. Manag. **25**, 473–488 (2018)
14. de Abreu, M.C.S., Ferreira, F.N.H., Proença, J.F., Ceglia, D.: Collaboration in achieving sustainable solutions in the textile industry. J. Bus. Ind. Mark. 1614–1626 (2020)
15. Desmarchelier, B., Djellal, F., Gallouj, F.: Environmental policies and eco-innovations by service firms: an agent-based model. Technol. Forecast. Soc. Chang. **80**, 1395–1408 (2013)
16. Farooque, M., Zhang, A., Thürer, M., Ting, Q., Huisingh, D.: Circular supply chain management: a definition and structured literature review. J. Clean. Prod. **228**, 882–900 (2019)
17. Fawzy, S., Osman, A.I., Doran, J., Rooney, D.W.: Strategies for mitigation of climate change: a review. Environ. Chem. Lett. **18**, 2069–2094 (2020)
18. Fernando, Y., Jabbour, C.J.C., Wah, W.-X.: Pursuing green growth in technology firms through the connections between environmental innovation and sustainable business performance: does service capability matter? Resour. Conserv. Recycl. **141**, 8–20 (2019)
19. Field, J.M., Fotheringham, D., Subramony, M., Gustafsson, A., Ostrom, A.L., Lemon, K.N., Huang, M.-H., McColl-Kennedy, J.R.: Service research priorities: designing sustainable service ecosystems. J. Serv. Res. **24**, 462–479 (2021)
20. Fischer, T., Gebauer, H., Gregory, M., Ren, G., Fleisch, E.: Exploitation or exploration in service business development? J. Serv. Manag. **21**, 591–624 (2010)
21. Fisscher, O.A.M., Visscher, K., Pearson, A., Weisenfeld, U.: Research and development as a competence creating business in a business. Creativity Innov. Manag. **10**, 251–258 (2001)
22. Gadrey, J., Gallouj, F., Weinstein, O.: New modes of innovation. Int. J. Serv. Ind. Manag. **6**, 4–16 (1995)
23. Guyader, H., Ottosson, M., Frankelius, P., Witell, L.: Identifying the resource integration processes of green service. J. Serv. Manag. **31**, 839–859 (2019)
24. Hakala, H., O'Shea, G., Farny, S., Luoto, S.: Re-storying the business, innovation and entrepreneurial ecosystem concepts: the model-narrative review method. Int. J. Manag. Rev. **22**, 10–32 (2020)
25. Håkansson, H., Snehota, I.: Developing Relationships in Business Networks. Routledge, London (1995)
26. Hallin, A., Karrbom-Gustavsson, T., Dobers, P.: Transition towards and of sustainability—understanding sustainability as performative. Bus. Strateg. Environ. **30**, 1948–1957 (2021)
27. Hansen, E.G., Schmitt, J.C.: Orchestrating cradle-to-cradle innovation across the value chain: overcoming barriers through innovation communities, collaboration mechanisms, and intermediation. J. Ind. Ecol. **25**, 627–647 (2021)
28. Haseeb, M., Hussain, H.I., Kot, S., Androniceanu, A., Jermsittiparsert, K.: Role of social and technological challenges in achieving a sustainable competitive advantage and sustainable business performance. Sustainability **11**, 3811 (2019)
29. Henning, M.: Regional labour flows between manufacturing and business services: reciprocal integration and uneven geography. Eur. Urban Reg. Stud. **27**, 290–302 (2019)
30. Hipp, C., Gallego, J., Rubalcaba, L.: Shaping innovation in European knowledge-intensive business services. Serv. Bus. **9**, 41–55 (2015)
31. Iansiti, M., Levien, R.: The Keystone Advantage: What the New Dynamics of Business Ecosystems Mean for Strategy, Innovation, and sustainability. Harvard Business Press (2004)
32. Kavalić, M., Nikolić, M., Radosav, D., Stanisavljev, S., Pečujlija, M.: Influencing factors on knowledge management for organizational sustainability. Sustainability **13**, 1497 (2021)
33. Khan, M., Ajmal, M.M., Gunasekaran, A., AlMarzouqi, A.H., AlNuaimi, B.K.: Measures of greenness: an empirical study in service supply chains in the UAE. Int. J. Prod. Econ. **241**, 108257 (2021)
34. Kumar, S., Putnam, V.: Cradle to cradle: reverse logistics strategies and opportunities across three industry sectors. Int. J. Prod. Econ. **115**, 305–315 (2008)
35. Lacity, M., Willcocks, L.: Conflict resolution in business services outsourcing relationships. J. Strateg. Inf. Syst. **26**, 80–100 (2017)

36. Lacoste, S.: Sustainable value co-creation in business networks. Ind. Mark. Manage. **52**, 151–162 (2016)
37. Lopes, L.J., Pires, S.R.I.: Green supply chain management in the automotive industry: a study in Brazil. Bus. Strateg. Environ. **29**, 2755–2769 (2020)
38. Lyons, P., Brennan, L.: Assessing Value from business-to-business services relationships: temporality tangibility, temperament, and trade-offs. J. Serv. Res. **22**, 27–43 (2019)
39. Malik, M.M., Abdallah, S., Hussain, M.: Assessing supplier environmental performance: applying analytical hierarchical process in the United Arab Emirates healthcare chain. Renew. Sustain. Energy Rev. **55**, 1313–1321 (2016)
40. Matinaro, V., Liu, Y.: Towards increased innovativeness and sustainability through organizational culture: a case study of a Finnish construction business. J. Clean. Prod. **142**, 3184–3193 (2017)
41. McDonough, W., Braungart., M.: Cradle to Cradle: Remaking the Way We Make Things. North Point Press (2002)
42. Möller, K., Nenonen, S., Storbacka, K.: Networks, ecosystems, fields, market systems? Making sense of the business environment. Ind. Mark. Manag. **90**, 380–399 (2020)
43. Oskam, I., Bossink, B., de Man, A.-P.: Valuing value in innovation ecosystems: how cross-sector actors overcome tensions in collaborative sustainable business model development. Bus. Soc. **60**, 1059–1091 (2021)
44. Ozpolat, K., Dresner, M.: A dark side of long-term VMI relationships: supply chain trust. Res. Logistics Prod. **8**
45. Pichler, M., Krenmayr, N., Schneider, E., Brand, U.: EU industrial policy: between modernization and transformation of the automotive industry. Environ. Innov. Soc. Trans. **38**, 140–152 (2021)
46. Proença, J.F., Torres, A.C., Marta, B., Silva, D.S., Fuly, G., Pinto, H.L.: Sustainability in the coffee supply chain and purchasing policies: a case study research. Sustainability **14**, 459 (2022)
47. Proença, T., Proença, J.F., Costa, C.: Enabling factors for developing a social services network. Serv. Ind. J. **38**, 321–342 (2018)
48. Pullman, M.E., Thompson, G.: Strategies for integrating capacity with demand in service networks. J. Serv. Res. **5**, 169–183 (2003)
49. Purchase, S., Olaru, D.: Substance in business-to-business relationships. J. Bus. Bus. Mark. **11**, 23–52 (2004)
50. Ramos, C., Roseira, C., Brito, C., Henneberg, S.C., Naudé, P.: Business service networks and their process of emergence: the case of the Health Cluster Portugal. Ind. Mark. Manag. **42**, 950–968 (2013)
51. Rodríguez, R., Svensson, G., Wood, G.: Determining corporate direction in sustainable development: a multi-dimensional framework in B2B. J. Bus. Ind. Mark. **36**, 1–17 (2021)
52. Sabatini, A., O'Toole, T., Gregori, G.: Integrating sustainability in business network initiation: the case of an Italian pasta maker. J. Bus. Ind. Mark. **36**, 1894–1908 (2021)
53. Sadhukhan, J., Christensen, M.: An in-depth life cycle assessment (LCA) of lithium-ion battery for climate impact mitigation strategies. Energies **14**, 5555 (2021)
54. Santos, L.P., Proença, J.F.: Sustainable business practices in business relationships: a case study research on the Portuguese automotive industrial network. In: 37th Annual IMP Conference, Cork, Ireland, 26–27 Aug 2021 (2021)
55. Santos, L.P., Proença, J.F.: Developing Return Supply Chain: a research on the automotive supply chain. Sustainability (2022)
56. Sarkis, J: Handbook on the Sustainable Supply Chain. Edward Elgar Pub (2019)
57. Schaltegger, S., Burritt, R.: Measuring and managing sustainability performance of supply chains. Supply Chain Manag Int. J. **19**, 232–241 (2014)
58. Sikora, A.: European green deal—legal and financial challenges of the climate change. ERA Forum **21**, 681–697 (2021)
59. Silva, G.M., Gomes, P.J., Sarkis, J.: The role of innovation in the implementation of green supply chain management practices. Bus. Strateg. Environ. **28**, 819–832 (2019)

60. Skålén, P., Gummerus, J.: Conceptualizing services and service innovation: a practice theory study of the swedish music market. J. Serv. Res. **26**, 10946705211062693 (2022)
61. Storey, C., Kahn, K.B.: The role of knowledge management strategies and task knowledge in stimulating service innovation. J. Serv. Res. **13**, 397–410 (2010)
62. Szász, L., Csíki, O., Rácz, B.-G.: Sustainability management in the global automotive industry: a theoretical model and survey study. Int. J. Prod. Econ. **235**, 108085 (2021)
63. Tate, W.L., Ellram, L.M.: Sustainable supply chains and social networks: an overview. In: Handbook on the Sustainable Supply Chain (2019)
64. Tura, N., Keränen, J., Patala, S.: The darker side of sustainability: tensions from sustainable business practices in business networks. Ind. Mark. Manag. **77**, 221–231 (2019)
65. Vasylieva, T., Lyulyov, O., Bilan, Y., Streimikiene, D.: Sustainable economic development and greenhouse gas emissions: the dynamic impact of renewable energy consumption, GDP, and corruption. Energies **12**, 3289 (2019)
66. Vaz, C.R., Rauen, T.R.S., Lezana, Á.G.R.: Sustainability and innovation in the automotive sector: a structured content analysis. Sustainability **9**, 880 (2017)
67. Vildåsen, S.S., Havenvid, M.I.: The role of interaction for corporate sustainability. IMP J. **12**, 148–170 (2018)
68. Vink, J., Koskela-Huotari, K., Tronvoll, B., Edvardsson, B., Wetter-Edman, K.: Service ecosystem design: propositions, process model, and future research agenda. J. Serv. Res. **24**, 168–186 (2020)
69. Wang, X., Zeshui, X., Qin, Y., Skare, M.: Service networks for sustainable business: a dynamic evolution analysis over half a century. J. Bus. Res. **136**, 543–557 (2021)
70. Wolfson, A., Tavor, D.: Sustainability via servicing: from individual action to institutional action. J. Serv. Sci. (JSS) **11**, 1–10 (2018)

Smart Cities, People at the Center: A Collaborative Mapping with AI for a Sustainable Future

Bruno Santos Cezario, Otacilio José Moreira, Oswaldo Borges Peres, Patricia Bilotta, Carlos Alberto Pereira Soares, and André Luis Azevedo Guedes

Abstract The evolving landscape of cities faces paradigm shifts driven by the increasing concern for sustainability. In this context, collaborative mapping using Generative Artificial Intelligence (Gen-AI) focusing on citizens for sustainable and resilient development plays a crucial role, emerging as a revolutionary method to enhance sustainable performance analyses. This study explores the influence of Gen-AI in the implementation of collaborative mapping with citizens, aiming to fill the existing gap in understanding the impacts of this technology on the transparency and reliability of information related to the sustainability of territories. The methodology was exploratory, using bibliographic and documentary research with qualitative and quantitative data for systematic analysis, following the Preferred Reporting Items for Systematic Reviews and Meta-Analyses (PRISMA). VOSviewer software was utilized to review publications and bibliometric analyses. Sustainable development, artificial intelligence, smart cities, and sustainable cities served as foundational

B. S. Cezario (✉) · O. B. Peres · P. Bilotta · A. L. A. Guedes
Postgraduate Program in Local Development (PPGDL), Centro Universitário Augusto Motta—UNISUAM, Rio de Janeiro, RJ, Brazil
e-mail: brunoscezario@souunisuam.com.br

O. B. Peres
e-mail: oswaldoperes@souunisuam.com.br

P. Bilotta
e-mail: pbbilotta@gmail.com

A. L. A. Guedes
e-mail: andre.guedes@souunisuam.com.br

O. J. Moreira
Postgraduate Program in Administration (PPGAd), Federal University of Niterói—UFF, Niterói, RJ, Brazil
e-mail: otaciliomoreira@id.uff.br

C. A. P. Soares
Postgraduate Program in Civil Engineering (PPGEC), Fluminense Federal University, Niteroi, RJ, Brazil
e-mail: capsoares@id.uff.br

© The Author(s), under exclusive license to Springer Nature Switzerland AG 2025

A. d. b. Machado et al. (eds.), *Environmental, Social, Governance and Digital Transformation in Organizations*, Information Systems Engineering and Management 35, https://doi.org/10.1007/978-3-031-86079-9_11

data search and analysis parameters. The results indicate that citizen participation supported by artificial intelligence can generate beneficial effects for reinforcing best sustainable practices favoring territorial development, aligning with the 2030 Agenda and the United Nations Sustainable Development Goals (SDGs).

Keywords Sustainable development · Artificial intelligence · Resilient development · Sustainable cities

1 Introduction

Currently, there is much discussion about the impact of technologies on society, and there is a global need for cities and communities to become more sustainable to address this issue, aligning with the sustainable goals outlined in the 2030 Agenda. This proposal will require innovative and transformative solutions, including artificial intelligence [1]. Artificial intelligence has played a crucial role in decision-making, involving the best governance practices for cities and corporations worldwide. However, there is no clear understanding of the impacts of Generative Artificial Intelligence as a moderator in the transformation of "smart cities" [2] in decision-making to address environmental, social, and governance challenges in collaboration with the local population, guiding society toward a sustainable evolution.

One of the main objectives of smart and sustainable cities is to combine innovation with technology and address social and urban problems sustainably while providing comfort for their citizens, all while respecting and considering the environment [3]. Previous studies have emphasized that digital transformation is crucial in changing social and economic behaviors, directly impacting inclusive and sustainable development [4]. It is also a reality that the urban population has considerably increased worldwide in recent times, necessitating the integration of technologies with a sustainable bias to enhance the quality of life in society.

The population's collaboration in transforming cities is vital, as sustainable development consists of a set of guidelines and goals that should place the citizen at the focal point of the process [5]. It is essential in the pursuit of socio-environmental harmony, given that the challenges for concise local development and sustainable urban growth involve various concerns at the urban scale, such as excessive consumption of often scarce natural resources, social equity, and climate threats [6]. It would be impossible to create a smart city without the support of society in terms of participation and the implementation of the transformative process. For the development of a smart city, it is imperative to consider the information provided by citizens to address local issues and respond to their needs, aiming to serve all society ecosystems directly and tangibly [7].

The use of Generative Artificial Intelligence (Gen-AI) in digital transformation has proven to be highly useful due to its performance and flexibility, with a wide range of possible applications in data processing [8]. In this context, society can leverage this technology to generate and capture data that influence decision-making and can lead to sustainable and resilient development. This digital technology can be employed to address urban urgencies by mapping data [9], contributing to tackling the paradigms of sustainable urban issues, such as the increasing demand for natural resources, the accumulation of waste, and the growing environmental impact caused by carbon emissions into the atmosphere.

Establishing the foundations for a smart city must align with environmental, social, and governance aspects, which are consolidated in the ESG (Environmental, Social, and Governance) framework. These factors are drivers of sustainability for a smart city, integrating technologies with sustainable development goals [10]. These socio-environmental aspects require the cooperation of all social actors, such as private companies, government, academia, and civil society, who must work together to integrate environmental, social, and management principles into their actions, thereby energizing local development. These concepts are generally used in the corporate sector; however, their application is focused on sustainable practices to mitigate environmental impacts and direct efforts toward building a fair and responsible world, committing to the best processes.

It is also known, according to studies, that the use of artificial intelligence presents an opportunity that comes with intrinsic risks due to the real potential of its application. The importance of understanding how machine learning algorithms evaluate their decision-making has been observed, aiming to promote knowledge about artificial intelligence systems to ensure integrity, transparency, and proper management [11]. Another important aspect of directing this research is to adjust technologies so that they can focus on people at the core of sustainable development, mitigating the impacts of climate change [9]. In this scenario, this article proposes to investigate collaborative mapping using artificial intelligence, focusing on citizens for sustainable and resilient development. In the context of this research, the study conducted is directly aligned with the goals established in the Sustainable Development Goals of the UN 2030 Agenda. It directly connects to the global transformation outlined in the Pact for the Future created by the UN in 2024, aiming for a more equitable and prosperous world for all humanity.

2 Methodology

The research was exploratory, bibliographical, and documentary, using secondary qualitative data. Initially, an extensive bibliographic search was carried out on the Web of Science and Scopus using the keywords "sustainable development," "artificial intelligence," "smart cities," "intelligent cities," and "sustainable cities". The recommendations of the Preferred Reporting Items for Systematic Reviews and Meta-Analysis (PRISMA) [30] were adopted for the bibliographic research. The following filters were applied: scientific and review articles in English published in the last five years, peer-reviewed, and open access.

The database obtained is aligned with the following sustainable goals of the UN 2030 Agenda (Fig. 1): zero hunger (SDG 2); good health and well-being (SDG 3);

Fig. 1 Bibliographic research on sustainable development goals

quality education (SDG 4); clean water and sanitation (SDG 6); affordable and clean energy (SDG 7); industry, innovation, and infrastructure (SDG 9); sustainable cities and communities (SDG 11); responsible consumption and production (SDG 12); climate action (SDG 13); life on land (SDG 15).

The bibliographic data of the references found were exported as a "ris" metadata file and processed with the help of Vosviewer 1.6.20 software [12]. This system conducts a quantitative bibliometric analysis to monitor published research, providing publication patterns regarding a specific field of literature. The software enables the construction and visualization of bibliometric networks or knowledge clusters, which are groupings and interconnections among the articles, allowing for the visualization of occurrences and facilitating the categorization of documents, whether by keywords or authors. Thus, a Systematic Bibliometric Literature Review (SBLR) is obtained, characterized by the screening and selecting knowledge references to ensure the effectiveness and accuracy of the interpreted data [13].

It was possible to map, analyze, and identify a network of co-occurrence relationships among the keywords from the selected set of articles using this methodology, enabling the creation of a conceptual arrangement of a knowledge domain. It means that when keywords are repeated in the articles comprising the analysis data, the concepts assigned to those words are related. Figure 2 shows how Vosviewer software creates networks of keywords among the selected articles, data is cross-referenced, and the research is divided into four main clusters, represented by colored circles. Each grouping (cluster) brings together keywords that share similarities and interactions among the articles. The connection between them occurs through structural associations, and the stronger this link, the greater the interaction between the themes. Figure 2 illustrates the co-occurrences of the keywords found in all the articles returned after the search vector query. It can be observed that they are grouped into a knowledge network, subdivided as follows: network 1 in green, network 2 in red, and network 3 in blue, respectively.

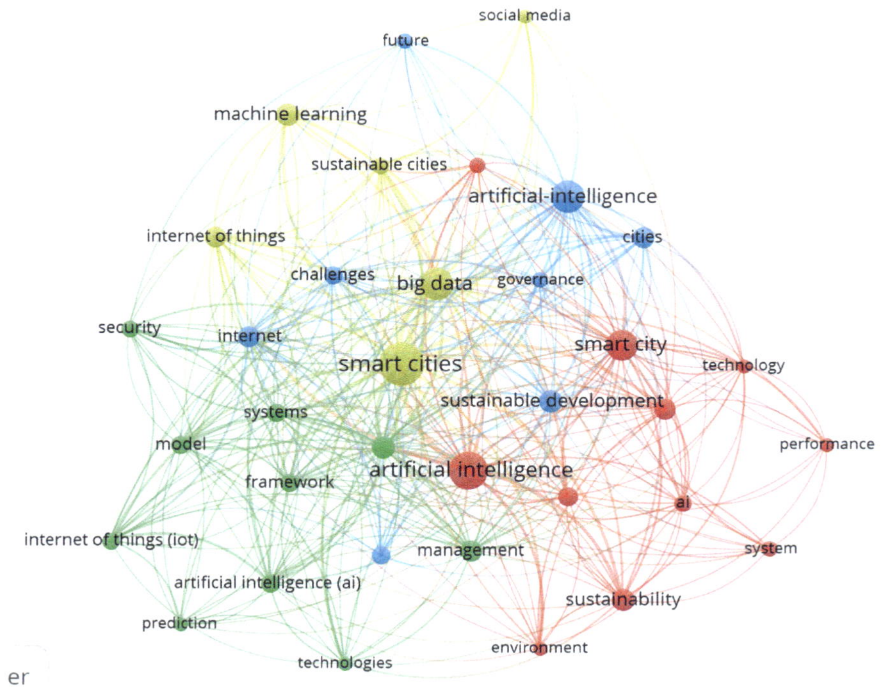

Fig. 2 Network of clusters

It is noted that the keywords used in the initial research are among the knowledge networks guided by the bibliometric study aimed at identifying indicators to dynamize and advance scientific knowledge [13], indicating that the choice of keywords was effective in this research, as shown in Fig. 3.

However, still using quantitative methods to implement the eligibility criteria, emphasizing the logic of selection, evaluation, and synthesis of the study [14], a search was conducted in the abstracts of the 120 articles using the keywords "sustainable development", "artificial intelligence", "smart cities", and "sustainable cities" to identify articles relevant to the theme. At this stage, 103 selected items remained. To ensure that all possible articles for the study were included, a second round of mapping was performed on the bibliographic materials containing at least two of the mentioned keywords, resulting in 54 articles. Considering the use of Vosviewer software and the systematic review, 6 references from other sources discussing the methodologies used and relevant to this theme were added, incorporating these references into the bibliographic base of this investigation. Finally, applying qualitative eligibility criteria, the abstracts and conclusions of these articles were read in full, leading to the selection of 31 references, while 29 articles were discarded due to lack of relevance to the theme (Fig. 4).

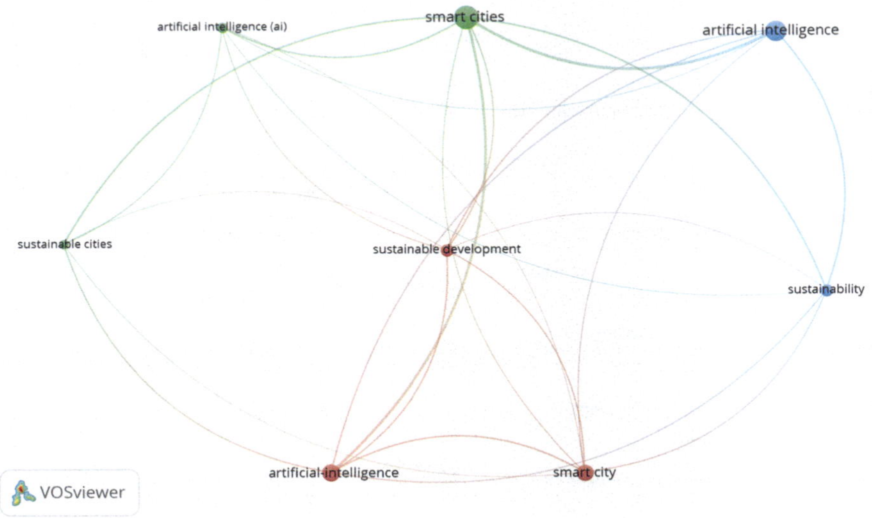

Fig. 3 Keyword cluster network

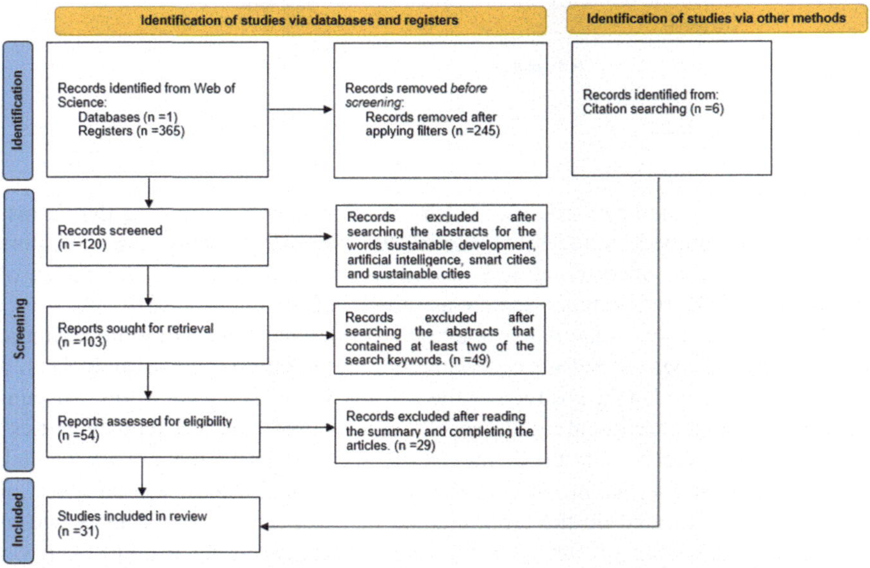

Fig. 4 Prisma flowchart. *Source* Author

3 Artificial Intelligence and the Sustainable Participatory Vision

When we talk about smart and sustainable cities, we envision society acting together for the greater good; we visualize happy people and public services functioning perfectly, all with the use of technology in harmony with society, etc. However, we know that the path ahead is long and winding, as there are many challenges along the way, one of the main ones being the transformation of a city into a smart cities in a sustainable manner with a focus on the environment. For this, we must always refer to the Sustainable Development Goals (SDGs) of the United Nations (UN), outlined in the 2030 Agenda, as shown in Fig. 5.

These objectives should serve as a guiding framework for transforming urban centers assisted by technology. However, technology can be used in the service of sustainable development in managing systems and collecting data with greater accuracy, saving costs and natural resources. Computational sustainability, which is an essential goal for smart cities, utilizes artificial intelligence, machine learning, and digital resources to promote the development of cities with greater responsibility towards environmental perspectives, reducing ecological impact and fostering sustainable development. Combining these technologies enables citizens to make informed decisions, actively participate in their communities, and adopt sustainable lifestyles [15].

Artificial intelligence and sustainability are recurring points in business meetings, as technology and the environment increasingly guide management and governance

Fig. 5 The 17 sustainable development goals (SDGs). *Source* UN [16]

actions for territories. In parallel, technological growth proclaims a need for innovation in city governance for sustainable economic development [17]. On the other hand, environmental issues demand greater responsibility and an improvement in sustainable awareness, moving towards controlling waste and consuming natural resources.

It is evident that artificial intelligence and sustainability are extremely important subjects for decision-making focused on local development, and the shift to smart cities is a movement that various locations are following in search of efficiency and optimizing ways to achieve sustainable development [18]. Conventional capital patterns are gradually transforming, with a greater focus on investments that highlight environmental responsibility within corporations. At this moment, the field of digital transition still presents a limited analysis regarding sustainability factors.

Integrating artificial intelligence and sustainability encompasses various structural concepts such as reducing waste, greater energy efficiency with integrated responses, and increasing the capacity to integrate analyzed data. However, citizens must take responsibility for mitigating adverse environmental impacts by collaboratively participating in decision-making. In this sense, for a sustainable evolution in the conceptual and theoretical field, where the local population commits to common purposes and plays a fundamental participatory role in defining and sustaining public administration [4], it is possible to outline potential levers that can be used to engage stakeholders in paths and strategies based on sustainability.

4 Artificial Intelligence and a Sustainable Collaborative Vision

Social responsibility has become an essential pillar for the evolution of smart cities, regardless of their size or population. Some studies generally report hope that Open Government Data (OGD) will drive economic growth and promote social development [19]. For this reason, implementing social policies by public authorities is one of the ways to apply equity in society, as they are considered one of the primary means of growth and achieving objectives for the development of territories. This collaborative social vision has gained strength with the support of the United Nations in its framework of Sustainable Development Goals (SDGs) that make up the 2030 Agenda (UN).

The governance of cities requires a structure that supports addressing the challenges of social issues, assisting in decision-making that promotes sustainable development and improves the quality of life for the population [20]. In this context, using artificial intelligence and other technologies helps contribute to the best social practices and can significantly benefit the population. It is known that generative AI (Gen-AI) is a rapidly evolving field of study that has served as operational support in addressing socio-environmental challenges related to urban growth. With the advancement of social public policies, societal entities constantly need accountability regarding socio-environmental issues; in this context, companies have utilized technologies to obtain data and practice conduct aimed at integrating the social aspect into their business, seeking a greater perspective for growth.

This impact of artificial intelligence for social good refers to the use of technologies and skills powered by technology to promote the well-being of the population in their territories [11]. The UN's 2030 Agenda and sustainable development proposals have shifted the community's focus towards social issues, encouraging the generation of public policies aligned with the socio-environmental axis reflecting citizens' perspectives. It is no longer sufficient to merely comply with the country's existing laws, as investors, business partners, and the community demand that public management pursue corporate social responsibility practices beyond what the law mandates. Thus, implementing technologies with a social vision is a long way to go. However, with the advancement of artificial intelligence, various studies are exploring the possibility of using it to solve social issues, a concept known as AI for social good, contributing to more informed decision-making [11].

5 Overview of Artificial Intelligence in Logistics and Supply Chain

Galaz et al. [21] state that using AI in Logistics and Supply Chain involves leveraging advanced algorithms and data analysis to create more efficient, sustainable, and resilient operations. AI can optimize decision-making processes by predicting demand, automating inventory management, and improving supply chain visibility. According to the authors, in alignment with environmental, social, and governance (ESG) principles, AI enables organizations to minimize their environmental impact by reducing waste, optimizing resource use, and improving energy efficiency.

According to Hao and Demir [22], the integration of AI in logistics helps mitigate systemic risks by anticipating disruptions and ensuring the continuity of the supply chain through real-time data analysis. Furthermore, AI-driven insights can align with ESG goals, as the authors state, enabling ethical sourcing, monitoring working conditions, and reducing the carbon footprint of transportation and logistics operations.

Industry 4.0 technologies, powered by AI, have the potential to transform the supply chain landscape by increasing operational efficiency and sustainability. Saxena et al. [23] highlight how AI and other advanced technologies support ESG goals by enabling more intelligent resource allocation and promoting transparency throughout the supply chain. Galaz et al. [21] demonstrate how practical applications of AI can address specific ESG-related issues, such as reducing emissions or improving working conditions, further reinforcing AI's role as an essential driver for sustainable logistics and supply chain management.

6 AI-Driven Sustainability Benefits in Reverse Logistics

Artificial Intelligence (AI) is crucial in waste reduction, resource conservation, and process efficiency improvement in logistics and supply chain operations. According to Hao and Demir [22], by enabling real-time analysis and process automation, AI optimizes decision-making and identifies opportunities to minimize resource consumption and reduce waste, positively impacting both operational costs and the environmental aspects of supply chains. In the context of reverse logistics, Saxena et al. [23] and Radclyffe et al. [24] agree that AI can be applied to minimize the carbon footprint through optimized routes for collecting returned products, reducing distances traveled, and efficiently planning transportation, thus promoting sustainability. For the authors, AI-driven reverse logistics operations allow companies to recover, reuse, or recycle materials more effectively, avoiding improper disposal while generating significant cost savings and new revenue streams. For example, returned products can be refurbished and resold, or their components can be reused, increasing efficiency and adding value to the operation.

7 Digital Transition as a Favor of Technological Innovation

The disadvantage of the digital transition involves a change in the mentality of companies that start to use technological processes to meet their objectives and enhance their performance, which is directly linked to innovation. For Rosário and Dias [13], it is a proactive approach that anticipates the next broad opportunity and implements relevant measures to exploit for the benefit of the organization and its key stakeholders.

Rosário and Dias [13] complement this by clarifying that the process of improving industrial sustainability through intelligent management systems that can perform multiple functions aims to improve energy and resource efficiency and reduce waste.

In this sense, Galaz et al. [21] consider that digitalization, artificial intelligence, and automation seem to be increasingly taking shape in sectors that are fundamental to sustainability, taking into account technological advances and social and environmental pressures. Society's demands for better management of scarce natural resources bring with them the need to understand the impacts of climate change, stimulating research and development.

The authors complement by noting that, from a technological point of view, predictive advances resulting from the increase in computational capacity and new developments in the robotics industries have paved the way for new approaches aimed at productivity and decision-making.

8 Sustainable development Through Corporate Socio-Environmental Practices

With the advancement of human thinking aimed at socio-environmental practices in companies, notions and criteria for environmental, social, and governance practices have been incorporated into corporations. These terms are designated by the acronym ESG—Environmental, Social, and Governance—which is based on the principle of implementing good practices for sustainable business development. This term was first introduced in 2004 at the United Nations Global Compact.

ESG practices have strengthened the purpose, organizational culture, and strategy of both the private and public sectors, improving their financial indicators, attracting and retaining talent, and valuing the business conduct adopted. This set of practices is focused on environmental preservation, social responsibility, and corporate transparency. The ESG policy is directed along three main axes: the social axis, which strengthens practices of diversity, equity, and inclusion, promoting respect for plurality in corporate business processes; the sustainable axis, which integrates governance strategy and corporate engagement with the environment, promoting the adoption of practices aligned with the UN Agenda 2030; and finally, the governance axis, which integrates a business perspective for risk reduction and encompasses the adoption of governance policies aligned with strategic approaches towards sustainability.

The incorporation of ESG practices in corporations is vital for business growth, as it contributes to an increase in the value of companies, in addition to providing corporate ratings by ESG rating agencies [25], based on the integration of socio-environmental policies practiced by the corporations. Furthermore, companies that adopt these ESG practices are well-regarded by their consumers, improving the company's image in society. Given the investment from various sectors in policies and practices that engage the community, these practices have fostered sustainable development for territories. By consolidating ESG practices, the environment, social issues, and governance promote sustainable concerns, outlining a direction for territorial development.

9 Sustainable development and Technological Collaboration

Since the advent of society, humans have been connecting with technology, which presents challenges in adaptability and transformative potential. This technology-society connection has made our world more dynamic, sometimes even making it difficult for some people to adapt. Given that territories play a fundamental role in achieving the UN's Sustainable Development Goals, it is essential to understand the opportunities and obstacles associated with using intelligent technologies to promote urban sustainability [6]. Technological tools have assisted society in various ways, aiding in the progress and integration of governance processes, as seen in the case of artificial intelligence in collecting and processing information at an unprecedented capacity.

In this context, a connection between technology and society makes it difficult to separate these actors to distinguish ends from means. Therefore, it is important to overcome technological obstacles in specific contexts that are less developed in sustainable transformation [7].

Consequently, it is essential to leverage technological control to make the most of these tools and use them effectively in managing and developing specific public policies aimed at sustainable development. By harnessing digital technologies, we can promote the evolution of cities on their journey to becoming intelligent and sustainable, particularly concerning community engagement and participatory governance.

Innovative examples should be followed for the evolution of cities, creating economically sustainable environments through interactions between technological and socioeconomic factors. It emphasizes adopting principles aligned with those defined by the Sustainable Development Goals as parameters for evaluating the progress of smart cities [7]. However, technological tools are always a sensitive point when considering their integration into society, as caution must be exercised in their use. It is necessary to maximize the benefits of their qualities, including the ability to integrate people and services assisting in public policies for the sustainable development of territories.

10 Innovation Ecosystem and Sustainable development

In a growing context involving challenges in sustainable local development and a concise connection with the environmental matrix, technologies are seen as a factor that enables significant advancements in the systematic improvement of territories aimed at sustainability. Recent studies reveal that for sustainable urban development, smart cities utilize technological systems to improve society's quality of life.

It is safe to say that the assertive use of solutions for sustainable development must involve all actors, whether public or private entities, in a way that integrates them into the process to solve urban problems and indicate possible solutions. Therefore, universities, governments, businesses, and civil society must be aligned to integrate a unified vision in mapping resolutions for problems involving sustainable evolution. The interconnection among socioeconomic actors has a strategic vision to encourage innovation in smart cities and transform territories, improving the quality of life in urban areas while addressing the challenges of rapid urban growth and reducing the impacts of climate change [26].

However, this relationship within the innovation ecosystem has introduced new actors, such as artificial intelligence, which has played a prominent role in transforming smart cities [2]. It enhances the creation process, refining knowledge and considering approaches for sustainable development. The innovation ecosystem and its actors are aligned with the SDGs, creating a synergy favoring territorial development, as shown in Table 1.

Table 1 The relationship between the innovation ecosystem and the SDGs

Innovation ecosystem	SDGs
Companies	
University	
Civil society	
Artificial intelligence	
Government	

11 Citizen Participation in the Transformation of Sustainable Cities

Much is said about the actions that need to be taken to transform cities into smart and sustainable cities; however, little is mentioned about how to do this, with citizen participation at the center of the issue. It is also known that there are studies indicating which priorities are important from society's perspective for increasing the intelligence of cities and that some of the indicators are linked to management and applications involving technology.

Identifying the problems faced by citizens so that they can become agents of change and positive transformation in large cities is one of the first steps in the evolutionary path toward quality urban development [5]. Urban transformation must align with spaces for people and their needs; only then can territories transform integrally and effectively. It is the direction that must be taken for a more sustainable city, defining goals and outlining strategies that transform the territory correctly. For this, technological applications are essential, as they serve as an integrative vehicle between cities and the population.

It would be impossible to create a sustainable city without the support of society in participating and implementing the transformative process. Smart cities emerge as a solution to the challenges posed by urbanization, digital transformation, and the growing social demand for more effective and sustainable urban services, aiming to enhance the quality of life [27]. It is important to observe the information provided by citizens to address local issues and respond to their needs to develop a smart city. Structural principles must be created to meet the demands of smart cities in a direct and attainable way.

Another important point for the success of a sustainable city lies in innovation as a characteristic, and this innovation should be provided through the technologies or technological applications used without losing sight of the needs of the people. It is known that imperfections in smart cities are often caused by organizational culture, difficulties in behavioral change, lack of professional grounding, and lack of focus on technology that goes against the priorities of citizens and the expected effects. It is essential to utilize technological innovation to better adapt to the individual needs of each group of individuals, which is crucial for managing complex urban infrastructures, predicting social demands, and avoiding potential crises. It impacts all dimensions of urban life, enhancing value delivery to citizens [15].

Following this reasoning, it is important to collaboratively integrate citizens in the mapping and indicate their priorities, showing the public authorities the real need for investments. To achieve this goal, smart cities need a partnership between the public and private sectors to implement information technology platforms that allow for collecting and analyzing large volumes of data [8]. We also know from previous studies that indicators of urban planning and environmental sustainability are among the main factors related to urban life in cities [28]. The active participation of people in the innovation process allows for the development of new solutions to their demands in real-life contexts, where they test these innovations, being integrated into the co-creation process and redesigning the networks that make up the socioeconomic fabric.

Thus, the applications should have intense popular participation so that the entire society can benefit from the thorough analysis of the provided data, enabling us to encourage data-driven decision-making by public authorities based on reliable information. Technologies such as artificial intelligence and the Internet of Things can transform challenges into opportunities, presenting viable solutions to problems identified by society and reinforcing the role of citizens in the decision-making process [29]. The collaborative action of society plays a crucial role in sustainable local development, participating in the urban evolution process by generating information that, when correctly analyzed, can influence the creation of public policies for local improvement within the territories.

Reinforcing the idea of societal participation through the citizen, it is understood that cities, on the other hand, are not merely abstract mechanisms aimed at generating sustainability; they are spaces where real individuals live, work, study, and develop. This social collaboration transforms urban governance and improves service delivery, promoting inclusion, efficiency, and a citizen-centered perspective [15] to meet people's needs. Consequently, sustainable actions need to be relevant and meaningful in this context.

12 Conclusion

Local and sustainable development should be an objective pursued at all costs by those responsible for managing localities. This mindset should be increasingly disseminated throughout society and constantly demanded from citizens. Thus, there must be adequate technological tools for collaboration and control, measuring the quality of life and services provided to society. The articles used in this research support a constant inquiry in city management regarding improving urban socio-environmental conditions with the citizen as the central focus of policies and demands concerning a fairer and more equal society based on innovation and sustainable technologies.

There should be integration among all of society to maximize decisions that address sustainable development, tangible through concise data; for this purpose, artificial intelligence proves to be very valuable in controlling, analyzing, and integrating data, as per the literature. With the rapid advancement of technologies, smart cities utilize these innovations to enhance the quality of life for their inhabitants. Artificial intelligence serves as a data analysis tool, making the urban environment more personalized, efficient, and automated, and is also essential in managing urban infrastructures. This digital technology transforms city governance by incorporating digital technology into various aspects of urban life, adjusting service delivery to citizens.

Evidently, the collaborative idea suggested in this article among the actors of society aligns with the 2030 Agenda and is grounded in the research referenced in the systematic and bibliometric methodological approach, as seen in the production of knowledge clusters and the references found. It reinforces the notion that the connection between ESG practices, focusing on the citizen, and artificial intelligence forms a tripod that should be considered in the evolution of smart and sustainable cities.

The results of this research show that it has limitations, as its bibliographic framework relied on articles while excluding other types of documents that could highlight additional aspects of this work. On the other hand, it is evident in the transcription of this document that artificial intelligence plays a substantial role in guiding indicators that enable resilient urban evolution. Efforts should be directed toward citizen participation in all social spheres, including businesses, the public sector, universities, and civil society. It would create truly effective participatory collaboration in favor of local, sustainable development based on innovation and creative solutions that help society achieve the sustainable goals outlined in the UN 2030 Agenda.

References

1. Leal Filho, W., Mbah, F.M., Dinis, M.A.P., Trevisan, L.V., Lange, D., Mishra, A., Rebelatto, B., Hassen, T.B., Aina, Y.A.: The role of artificial intelligence in the implementation of the UN sustainable development goal 11: fostering sustainable cities and communities. Cities **150**, 105021 (2024). ISSN 0264-2751, https://doi.org/10.1016/j.cities.2024.105021
2. Yigitcanlar, T., Corchado, J.M.: Inteligência artificial verde: rumo a uma tecnologia eficiente, sustentável e equitativa para cidades e futuros inteligentes. Sustentabilidade **13**, 8952 (2021). https://doi.org/10.3390/su13168952
3. Cezario, B.S., Guedes, A.L.A.: Technological applications for smart cities: mapping solutions. In: Anwar, S., Ullah, A., Rocha, Á., Sousa, M.J. (eds.) Proceedings of International Conference on Information Technology and Applications. Lecture Notes in Networks and Systems, vol. 614. Springer, Singapore (2023). https://doi.org/10.1007/978-981-19-9331-2_48
4. Caputo, F., Magliocca, P., Canestrino, R., Rescigno, E.: Repensando o papel da tecnologia para o engajamento dos cidadãos e o desenvolvimento sustentável em cidades inteligentes. Sustentabilidade **15**, 10400 (2023). https://doi.org/10.3390/su151310400
5. The Pact for the Future: Seventy-ninth session Agenda item 123 Strengthening of the United Nations system (2024). Available in: www.un.org/pga/wp-content/uploads/sites/109/2024/09/The-Pact-for-the-Future-final.pdf
6. Liang, D., Yunhong, L.: Frontiers of policy and governance research in a smart city and artificial intelligence: an advanced review based on natural language processing. Front. Sustain. Cities **5** (2023). ISSN 2624-9634, https://doi.org/10.3389/frsc.2023.1199041
7. Parra-Domínguez, J., Gil-Egido, A., Rodríguez-González, S.: SDGs as one of the drivers of smart city development: the indicator selection process. Smart Cities **5**, 1025–1038 (2022). https://doi.org/10.3390/smartcities5030051
8. Hilal, A.M., Alfurhood, B.S., Al-Wesabi, F.N., Hamza, M.A., Duhayyim, M.A., et al.: Artificial intelligence-based sentiment analysis for health crisis management in smart cities. Comput. Mater. Continua **71**(1), 143–157 (2022). https://doi.org/10.32604/cmc.2022.021502
9. Costa, D.G., et al.: A Survey of Emergencies Management Systems in Smart Cities. IEEE Access **10**, 61843–61872 (2022). https://doi.org/10.1109/ACCESS.2022.3180033

10. De Giovanni, P.: Sustentabilidade do Metaverso: Uma Transição para a Indústria 5.0. Sustentabilidade, vol. 15, 6079 (2023). https://doi.org/10.3390/su15076079
11. Mhlanga, D.: Inteligência Artificial Centrada no Homem: A Abordagem Superlativa para Alcançar Objetivos de Desenvolvimento Sustentável na Quarta Revolução Industrial. Sustentabilidade **14**, 7804 (2022). https://doi.org/10.3390/su14137804
12. Perianes-Rodriguez, A., Waltman, L., van Eck, N.J.: Constructing bibliometric networks: a comparison between full and fractional counting. J. Informetrics **10**(4), 1178–1195 (2016). ISSN 1751-1577, https://doi.org/10.1016/j.joi.2016.10.006
13. Rosário, A.T., Dias, J.C.: Sustentabilidade e Transição Digital: Uma Revisão de Literatura. Sustainability **14**, 4072 (2022). https://doi.org/10.3390/su14074072
14. The PRISMA: Statement: an updated guideline for reporting systematic reviews 2020. BMJ Results **71** (2021). https://doi.org/10.1136/bmj.n71
15. Alves, V., Fdez-Riverola, F., Ribeiro, J., Neves, J., Vicente, H.: Encouraging eco-innovative urban development. Algorithms **17**, 192 (2024). https://doi.org/10.3390/a17050192
16. UN, United Nations: 2030 Agenda: Sustainable Development Goal (2015). Available in: https://brasil.un.org/pt-br/sdgs
17. Zhou, Y.: The application trend of digital finance and technological innovation in the development of green economy. J. Environ. Public Health **8**, 1064558 (2022). https://doi.org/10.1155/2022/1064558
18. Blanco, R.L., García, M.C., Alonso, R., Prieto, J., Corchado, J.: Pollutant time series analysis for improving air-quality in smart cities. Int. J. Interac. Multimedia Artif. Intell. **8**, 98 (2023). https://doi.org/10.9781/ijimai.2023.08.005
19. Gao, Y., Janssen, M., Zhang, C.: Understanding the evolution of open government data research: towards open data sustainability and smartness. Int. Rev. Adm. Sci. **89**(1), 59–75 (2023). https://doi.org/10.1177/00208523211009955
20. Renaud, J., Karam, R., Salomon, M., Couturier, R.: Deep learning and gradient boosting for urban environmental noise monitoring in smart cities. Expert Syst. Appl. **218**, 119568 (2023). ISSN 0957-4174, https://doi.org/10.1016/j.eswa.2023.119568
21. Galaz, V., Centeno, M.A., Allahan, P.W., Causevic, A., Patterson, T., Brass, I., Baum, S., Farber, D., Fischer, J., Garcia, D., McPhearson, T., Jimenez, D., King, B., Larcey, P., Levy, K.: Artificial intelligence, systemic risks, and sustainability. Technol. Soc. **67**, 101741 (2021). https://doi.org/10.1016/j.techsoc.2021.101741
22. Hao, X., Demir, E.: Artificial intelligence in supply chain decision-making: an environmental, social, and governance triggering and technological inhibiting protocol. J. Model. Manag. **19**(2), 605–629 (2024). https://doi.org/10.1108/JM2-01-2023-0009
23. Saxena, A., Singh, R., Gehlot, A., Akram, S.V., Twala, B., Singh, A., Montero, E.C., Priyadarshi, N.: Technologies empowered environmental, social, and governance (ESG): an industry 4.0 landscape. Sustainability **15**(309) (2023). https://doi.org/10.3390/su15010309
24. Radclyffe, C., Ribeiro, M., Wortham, R.H.: The assessment list for trustworthy artificial intelligence: a review and recommendations. Front. Artif. Intell. **6** (2023). https://doi.org/10.3389/frai.2023.1020592
25. Hughes, A., Urban, M.A., Wójcik, D.: Classificações ESG alternativas: como a inovação tecnológica está remodelando o investimento sustentável. Sustentabilidade **13**, 3551 (2021). https://doi.org/10.3390/su13063551
26. Sánchez, O., Castañeda, K., Vidal-Méndez, S., Carrasco-Beltrán, D., Lozano-Ramírez, N.E.: Exploring the influence of linear infrastructure projects 4.0 technologies to promote sustainable development in smart cities. Results Eng. **23**, 102824 (2024). ISSN 2590-1230, https://doi.org/10.1016/j.rineng.2024.102824
27. Liu, Z., He, Y., Demian, P., Osmani, M.: Immersive technology and building information modeling (BIM) for sustainable smart cities. Buildings **14**, 1765 (2024). https://doi.org/10.3390/buildings14061765
28. Năstasă, A., Dumitra, T.C., Grigorescu, A.: Artificial intelligence and sustainable development during the pandemic: an overview of the scientific debates. Heliyon **10**(9), e30412 (2024). https://doi.org/10.1016/j.heliyon.2024.e30412

29. Wang, K., Zhao, Y., Gangadhari, R.K., Li, Z.: Analyzing the adoption challenges of the internet of things (IoT) and artificial intelligence (AI) for smart cities in China. Sustainability **13**, 10983 (2021). https://doi.org/10.3390/su131910983
30. Moher, D., Liberati, A., Tetzlaff, J., Altman, D.G.: PRISMA Group. Preferred reporting items for systematic reviews and meta-analyses: the PRISMA statement. PLoS Med. **6**(7), e1000097 (2009). https://doi.org/10.1371/journal.pmed.1000097

The Positive Impact of ESG in the Business World

Luís Manuel Roberto

Abstract Conduct, value-add for society and long-term perspective, are currently distinguish responsible business from the others, and ESG compels the business world to be more transparent in the way they approach sustainability, being a vital factor in strengthening of their reputation. ESG criteria have a direct effect on the companies live, by helping them to evaluate their performance concerning **environmental , social sustainability, and governance practices**, also allowing to identify and manage potential operational risks related to their activity. ESG is becoming a standard practice for many investors and asset managers who are looking for the alignment of financial objectives with sustainability targets. They believe that companies with strong ESG performance are better positioned to long-term success, making them also more competitive. What positive impacts can ESG bring to the business? Can ESG positively contribute to the innovation and the development of new market opportunities? What about people management? Can we expect the increase of people engagement and productivity, through the ESG implementation? Could the processes changes as result of ESG criteria, impact the business financials results? Can artificial intelligence support the ESG deployment?

Keywords Sustainability · ESG · Business · Reputation · Innovation · Financial · People · Risk · AI

1 Introduction

Nowadays, Environment, Social and Governance (ESG) factors are a common language used by companies that have sustainability integrated into their business strategy.

The purpose of the companies with the generation of value for society, makes the distinction between responsible business and all others, and ESG compels companies

---

L. M. Roberto (✉)
Lisbon, Portugal
e-mail: luismanuelroberto@gmail.com

© The Author(s), under exclusive license to Springer Nature Switzerland AG 2025
A. d. b. Machado et al. (eds.), *Environmental, Social, Governance and Digital Transformation in Organizations*, Information Systems Engineering and Management 35, https://doi.org/10.1007/978-3-031-86079-9_12

to act with integrity and transparency in the way they manage sustainability, making in vital indicators for the business success in the medium and long term.

This chapter aims to provide researchers with the main impacts, as result of implementation of ESG factors in companies, as follow:

- Brand reputation and value
- Innovation and market opportunities
- Financial Investment
- People Management
- Risk Management

2 Background

> Sustainable development is development that meets the needs of the present without compromising the ability of future generations to meet their own needs.
>
> — Gro Harlem Brundtland

Before we begin any approach to ESG factors, it is important to understand the evolution of sustainability, and how companies seek to develop their strategies.

In the last century, companies' main objective was to generate profits and pay dividends to their shareholders.

The connection with the community was non-existent and the concept of stakeholders was practically null.

To better understand the beginnings of sustainability, we must go back to the 1980s and learn about the journey of Gro Harlem Brundtland, a Norwegian politician, diplomat and doctor, international leader in sustainable development and public health.

She was a member of the Norwegian Labor Party since her youth and became her country's first female head of government in 1981 and is currently the United Nations (UN) Special Envoy for Climate Change.

Brundtland, was responsible for coordinating the Brundtland report (as it became known), entitled "**Our Common Future**", a document that announced for the first time the concept of sustainable development, a topic under discussion since the 1970s.

The Brundtland report highlighted for the first time the interconnection between the **economic**, **social and environmental** aspects of development, highlighting their interdependence, promoting an integrated approach to tackling global challenges such as poverty, inequality and environmental degradation.

The report also included data on global warming, acid rain and the destruction of the ozone layer, suggesting to the UN General Assembly the need to monitor these new impacts, and the establishment of goals to be followed by nations around the world, to prevent the advancement of environmental destruction and climate imbalance.

This is how the Brundtland report became the driving force for sustainable development.

In 2015, the Member States of the United Nations created the 2030 Agenda for Sustainable Development, translated into the 17 Sustainable Development Goals (SDGs), providing a framework for global action to address the issues facing humanity and the planet, with the goal of achieving a more equitable, prosperous and sustainable future for all.

In September 2000, the UN launched the Millennium Development Goals, committing to fight poverty and hunger, gender inequality, environmental degradation and HIV/AIDS, to improve access to education, health care health and drinking water by 2015.

One of the main goals was to reduce by half the 36% of the world's population living in poverty. Fifteen years later, the target was exceeded, with a strong contribution from India and China, through their economic growth.

At the same time, there was a growing involvement of the business world in the life of society, with the voluntary integration of social and environmental concerns into business strategy, and relationships with its stakeholders, leading many leaders to believe that sustainability will be fully interconnected not only with the core business of the company, but also with the entire value chain, driven by a sustainability strategy.

Sustainability becomes part of companies' vocabulary and corporate agenda.

The concern with a more sustainable future and the promotion of practices and actions with the aim of providing sustainability, became a transmission of trust and security for investors, and active cooperation with government bodies.

Business must play an essential role in improving society, including advancing economic opportunity and promoting fair business practices.

Responsible business goes far beyond environmental and social issues.

Security, diversity, human rights, transparency, the independence of corporate bodies, the fight against corruption, ethics, interaction with the media, and more generally the governance model, finance and sustainable investments, are factors which, when introduced into company management, contribute to reducing financial risks and preserving markets.

Sustainability must also be seen by companies as an opportunity for growth, alongside financial stability, people management and customer satisfaction. Innovation, through digital and automation, can also contribute to the emergence of new business opportunities, when combined with maximizing the SDGs.

According to a study carried out by the Business and Sustainable Development Commission, it is expected that with the implementation of the SDGs by 2030, \$12 trillion dollars in business value can be generated, with the development of mobility systems, solutions for healthcare, and energy efficiency, occupying pole position.

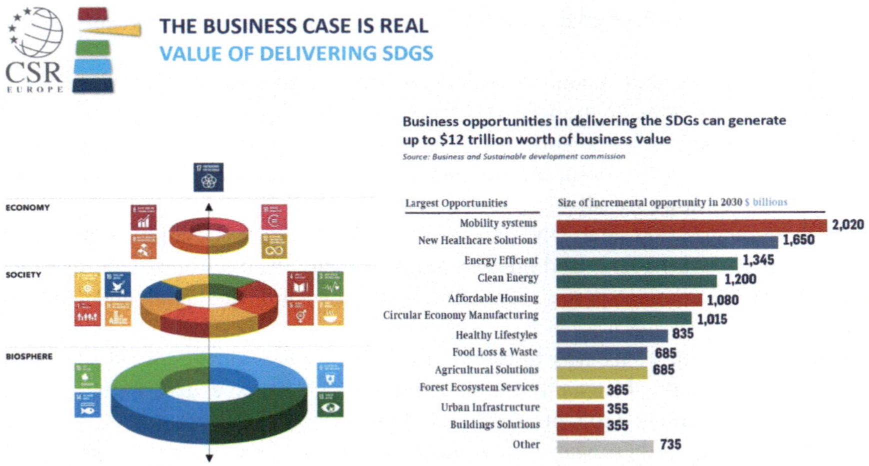

The 2024 Sustainable Development Report is written by a group of independent experts from the SDG Transformation Center, an initiative of SDSN—Sustainable Development Solutions Network, which highlighted that only 16% of the SDG targets are on track to be achieved by 2030, with the remaining 84% showing limited progress or a reversal in implementation.

The pace of SDG progress varies significantly across country groups. Nordic countries continue to lead on SDG achievement, with the BRICS countries making significant progress while poor and vulnerable nations lag far behind. As in previous years, European countries—notably the Nordic countries—top the 2024 SDG Index. Finland is ranked first, followed by Sweden (#2), Denmark (#3), Germany (#4), and France (#5).

Yet even these countries face significant challenges in meeting several SDGs. Since 2015, average SDG progress in the BRICS (Brazil, the Russian Federation, India, China, and South Africa) and BRICS + countries (Egypt, Ethiopia, Iran, Saudi Arabia, and the United Arab Emirates) has outpaced the world average, while East and South Asia has emerged as the region that has made the greatest progress toward the SDGs.

In contrast, the gap between the world's average SDG performance and the performance of the poorest and most vulnerable countries, including the Small Island Developing States (SIDS), has widened since 2015.

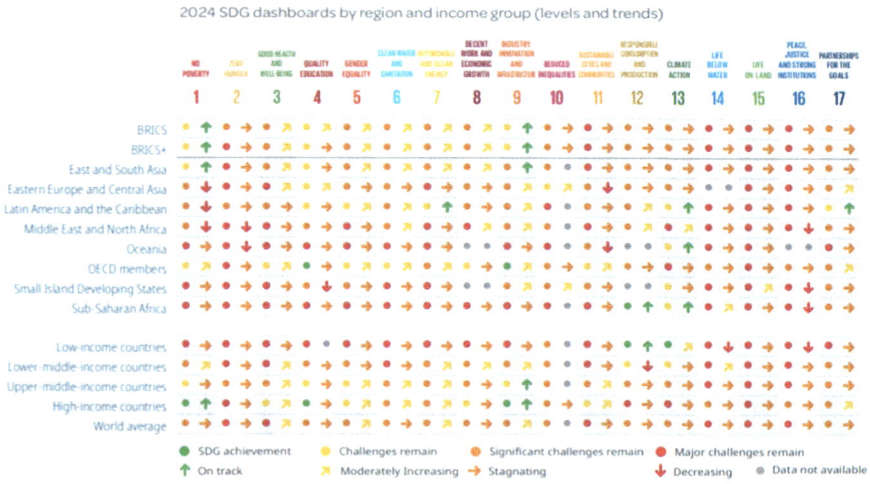

Note: Excluding OECD specific indicators. Population-weighted averages
Source: Authors

It is a fact that countries and companies are moving forward at different speeds, and not all of them are in the same position when it comes to sustainability.

There is, however, a broad consensus on the importance of companies for a sustainable future, and sustainability is no longer a matter of individual whish, but a need for the entire ecosystem.

3 ESG by Definition

ESG factors must be recognized as a process of continuous improvement, and not just as a legal obligation

The history of ESG has evolved from religious values-based investing to a globally recognized framework for integrating environmental, social, and governance factors (ESG) into financial decision-making.

In 2004, Kofi Annan, then Secretary-General of the United Nations, invited major institutional investors to participate in a study titled "Who Cares Wins". The resulting report established the term ESG and emphasized integrating these factors into capital markets. This was a pivotal moment for ESG's formalization.

Another report, "Freshfields Bruckhaus Deringer" (2005), outlined how ESG factors could be legally incorporated into investment analysis under existing fiduciary responsibilities.

Two years later, United Nations Principles for Responsible Investment (UNPRI) were launched, an investor initiative in partnership with UNEP and the UN Global Compact. The PRI asked institutional investors to integrate ESG factors into investment decisions. The initial signatories represented $4 trillion in assets under management (AUM).

The Global Reporting Initiative (GRI) became a prominent standard for companies to disclose their sustainability practices, influencing investors to demand greater transparency on ESG issues.

Since then, ESG investing has continued to evolve and gain momentum globally, with investors increasingly recognizing the importance of considering non-financial factors alongside traditional financial metrics, when assessing the viability and impact of long term of investments.

The current form is deeply intertwined with sustainable development goals and the global response to climate change, supported by influential figures and institutions worldwide.

Photo by Akshta Pandey, Medium Caption: SDG through the lens of ESG

Currently, ESG factors are standard practice for many investors, asset managers and companies looking to align financial objectives with environmental, social and governance considerations, and that why ESG is considered as "social license" to operate.

Sustainable practices, and company externalities such as carbon emissions, the impact of companies on employees, and human rights violations, have become unavoidable challenges for their management and operation, contributing to their assessment, as financial information were concerned.

ESG factors have a positive impact on the companies lives by helping to evaluate their performance in terms of environmental and social sustainability and governance practices.

By analyzing their materiality matrix, companies are being able to identify and manage potential risks related to their activity, allowing them to mitigate negative impacts on the way they are operating and managing their activities.

Investors increasingly consider that companies with strong ESG performance are better positioned for long-term success and are more likely to mitigate risks related to sustainability management and operations.

What are the criteria typically used to evaluate a company's operations, and their impact on society and the environment?

ESG evaluate the sustainability and societal impact of companies, as follow:

1. **Environmental (E)**: This examines how a company manages the environmental impact. The main areas are:
   - Climate change and carbon emissions
   - Energy efficiency and renewable energy usage
   - Pollution, waste management, and water usage
   - Biodiversity and deforestation
   - Compliance with environmental regulations
2. **Social (S)**: This assesses how a company manages relationships with employees, suppliers, customers, and communities. The mains areas, are:
   - Employee working conditions, diversity, and inclusion
   - Labor standards and human rights
   - Customer satisfaction and product safety
   - Community engagement and philanthropy
   - Data privacy and protection
3. **Governance (G)**: This evaluates how a company is managed and how it ensures accountability and transparency. The main areas are:
   - Board diversity, independence, and structure
   - Executive compensation and incentives
   - Shareholder rights and voting procedures
   - Ethical business practices and anti-corruption policies
   - Risk management and compliance with laws and regulations

4 The Positive Impact

4.1 Brand Reputation and Value

> It takes 20 years to build a reputation and five minutes to ruin it. If you think about that, you'll do things differently.
>
> — Warren Buffett

Reputation is crucial to the success of companies across industries.

In an increasingly competitive world, building a strong reputation requires consistent efforts, ethical behavior, and strategic actions over time.

Deliver consistently high-quality products or services is fundamental to build trust, but is not enough, when we have more and more demanding consumers, concerned about the relationship between companies and society, in terms of environmental and social responsibility.

According to a 2020 survey by Forbes, 88% of consumers state that they are more loyal to companies that support social or environmental issues.

The integration of sustainability and ESG criteria into business operations has a profound impact on brand reputation and value.

Based on a survey by Edelman in 2022, 91% of business leaders believe that their company's ESG performance is directly tied to their reputation, Companies that prioritize sustainability, social responsibility, and strong governance build positive public perception, foster stakeholder trust, and are better positioned for long-term success.

As ESG continues to grow in importance, companies with robust ESG commitments will experience stronger market positioning and enhances reputation, as responsible corporate citizens, resonating with a broad audience, including consumers, employees, investors, and regulators.

A strong ESG commitment builds trust and can enhance brand loyalty, especially with conscious consumers who prioritize sustainability, fair labor practices, and community engagement.

ESG-friendly companies can differentiate themselves from their competitors by showing they care about more than just profit. This creates a positive image and helps brands stand out in the market, gaining a competitive edge.

Companies with strong ESG practices tend to receive positive coverage from the media and from industry analysts, contributing to a stronger public image and enhancing their brand reputation over time.

A well-established ESG framework helps brands withstand crises better. When a company is known for consistently ethical behavior, it is more likely to recover from scandals or setbacks without significantly damaging its reputation or stock price.

48% of companies that experienced a negative ESG-related incident saw a significant reputational hit, with 32% reporting financial losses as a direct consequence, per a RepRisk report from 2020.

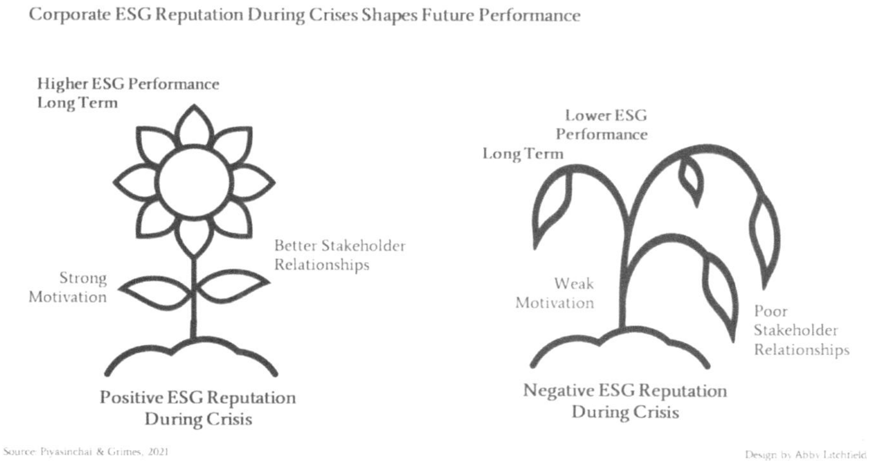

According to "**Reputational Spirals**" working paper written by Nareuporn Piyasinchai, during the 2008–2009 financial crisis, two thirds of the companies received little or no criticism on ESG issues (an average of less than 20 negative news stories per year). Happily, these companies generally showed learning and innovation in the sustainability arena, both during the crisis and beyond. Their ESG scores increased by approximately 4% in the two years post-crisis and continued that positive trend.

4.2 Innovation and Market opportunities

> Without change, there is no innovation, creativity, or incentive for improvement. Those who initiate change will have a better opportunity to manage the change that is inevitable.
>
> — William Pollard

Environmental, Social, and Governance (ESG) criteria play a vital role in driving innovation and unlocking market opportunities for companies.

As businesses incorporate ESG principles into their strategies, they can create new products, services, and business models that not only meet growing consumer and regulatory demands, but also position them ahead of direct competitors.

This can include developing renewable energy solutions, biodegradable materials, or energy-efficient technologies, which are increasingly in demand of governments prioritize sustainability.

For example, companies in the automotive industry are innovating by producing electric vehicles (EVs) and hybrid models, aligning with the global goal for reduced carbon emissions.

As part of European Green Deal, Fit for 55, presented in July and December 2021, refers to the EU's target of reducing net greenhouse gas emissions by at least 55% by 2030, in line with 2030 Agenda (**European Council-Council of European Union**).

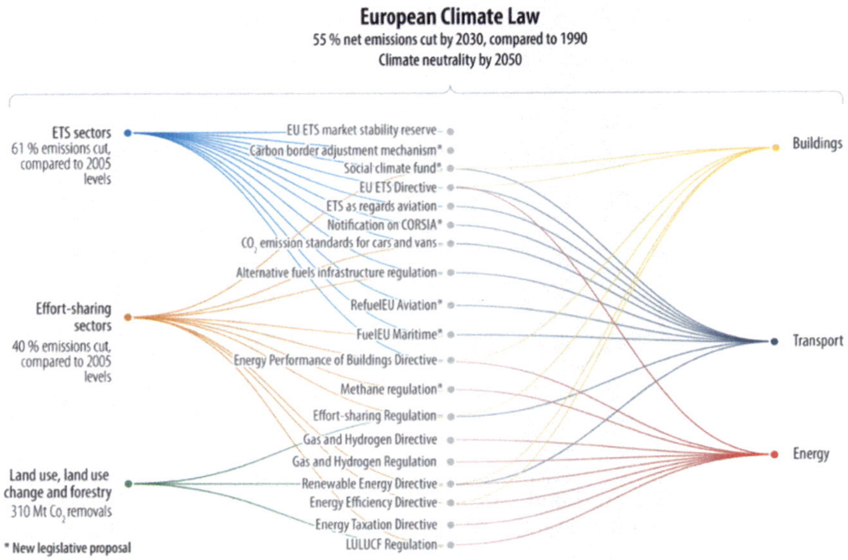

Legislative initiatives in the 'fit for 55' package

ESG principles inspire businesses to adopt circular economy models, where products are designed to be recycled, reused, or repaired, minimizing waste.

This shift creates new market opportunities, particularly in sectors like fashion, electronics, and manufacturing, as consumers and businesses alike seek out sustainable consumption options.

Companies in industries such as manufacturing or data centers are optimizing their energy usage through sustainable energy practices, which helps them save on **operational costs** while contributing to environmental goals.

Companies that adhere to ESG standards can more easily enter new markets with stringent environmental and labor regulations. For example, many European and Asian markets have tough sustainability regulations, and companies with strong ESG practices are better equipped to comply, giving them a competitive advantage when expanding globally.

Businesses are innovating in areas such as sustainable packaging by reducing plastic usage or creating compostable and recyclable alternatives. This trend opens new markets and caters to eco-conscious consumers while also reducing regulatory risks linked to plastic waste.

4.3 Financial Investment

Successful investing is about managing risk, not avoiding it.

— Benjamin Graham

ESG criteria have a profound impact on business economics.

By addressing environmental, social and governance factors, companies can improve their financial performance and gain a competitive advantage in the marketplace.

The integration of sustainable practices into business strategy, establishing alignment with ESG criteria, becomes increasingly crucial for the economic success of companies.

In the last years, the rise of sustainable investing has made companies with strong ESG credentials more attractive to investors, venture capitalists, and institutional investors who prioritize environmental and social considerations alongside financial returns.

According to PwC survey in 2021, 79% of investors agree that ESG risks and opportunities play an important role in investment decisions.

A Harvard Business School working paper (2020) found that high ESG-scoring companies saw an average annual return on equity (ROE) of 9–15% higher compared to companies with low ESG scores. Banks and financials agencies are starting look at companies ESG ratings when they have to approve investments, and open access to **capital**.

UNCTAD estimates that the value of sustainability-themed investment products in global financial markets totaled US$5.2 trillion in 2021, a 63% increase from 2020.

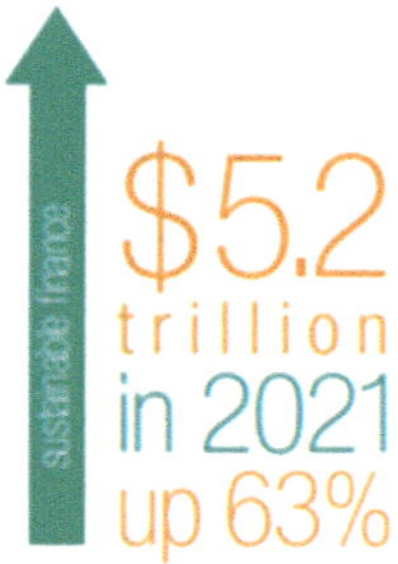

UNITED NATIONS CONFERENCE ON TRADE AND DEVELOPMENT
UNCTAD

ESG practices can lead to improved financial performance by minimizing operational risks and costs, and by improving efficiency, such as through energy-saving measures, adding to the company's **financial value** and long-term sustainability.

Companies are often better prepared to handle risks related to environmental and social changes, and regulatory pressures, reflecting into **financial stability**.

Companies know that aligning with consumers' growing preferences for environmentally and socially responsible brands, companies can expand their customer base and drive **sales growth**. Strong ESG practices signal to investors that the company is forward-thinking, socially responsible, and less likely to face risks that could impact financial performance.

Companies with a focus on sustainability can raise funds through green bonds or sustainability-linked loans, which can finance projects related to energy efficiency, or renewable energy infrastructure. These funding mechanisms provide companies with additional capital to innovate and expand into sustainable business models.

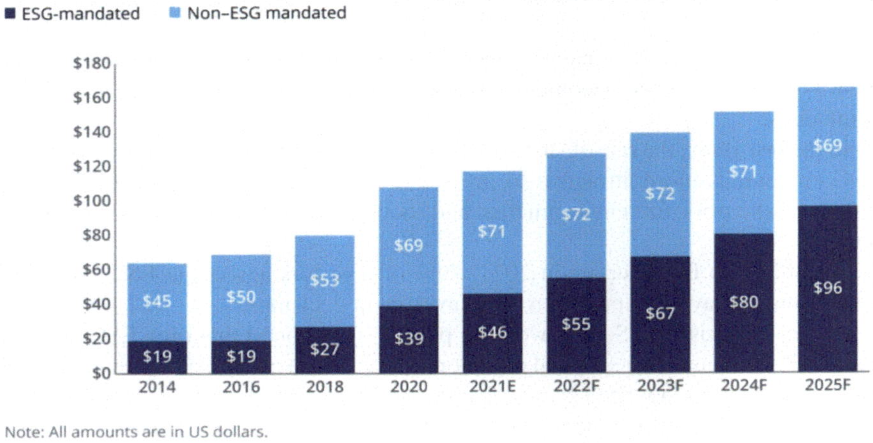

As reported by the New York Times, assets in ESG funds have risen by 38% in the past year. ESG-focused institutional investment is predicted to soar 84% to US$33.9 trillion in 2026, making up 21.5% of assets under management. It represents a dramatic and continuing shift in the asset and wealth management (AWM) industry according to PwC's Asset and Wealth Management Revolution 2022 report.

The case study of Tesla demonstrates that companies with strong ESG performance can benefit from higher revenue growth, better financial stability, and a lower cost of capital. Furthermore, investors and stakeholders are increasingly recognizing the role of ESG in risk management and long-term profitability, which also underscores the growing importance of sustainability in corporate strategy.

Tesla's business model is built on creating sustainable energy solutions, from electric vehicles to solar energy systems.

Tesla's strong ESG focus younger ones, consider Tesla a prime choice due to its pioneering role in renewable has resonated with investors, contributing to the company's valuation surging past $1 trillion in 2021, making it one of the most valuable companies in the world. ESG-driven investors, especially energy.

By focusing on sustainability, Tesla has gained a significant competitive advantage in the electric vehicle market, which is expected to grow exponentially as governments and consumers transition away from fossil fuels.

4.4 *People Management*

> People will forget what you said, people will forget what you did, but people will never forget how you made them feel.
>
> — Maya Angelou

The impact of Environmental, Social, and Governance (ESG) on people management has been widely studied, with numerous facts and statistics showcasing how companies that prioritize ESG, particularly the social pillar, improve workforce well-being and overall business performance.

By aligning People Management with ESG, companies can attract and retain the best talent, contribute to the motivation and well-being of their employees, and ensure a long-term sustainable success.

Employee satisfaction is closely linked to ESG performance, with companies that perform well on ESG seeing higher levels of employee retention and engagement.

A study by Gartner in 2020 found that 85% of employees believe that their company's sustainability efforts make them feel more fulfilled in their work, highlighting the importance of ESG for employee satisfaction and retention.

According to Harvard Business Review, 2019, companies that actively focus on social responsibility and employee well-being have 20% higher employee engagement levels than those that don't.

Mental health and well-being have become central to ESG-related people management, especially post pandemic crisis, COVID.

Edelman's Trust Barometer, in 2021, demonstrated that 71% of employees stated they would lose trust in their employer if they did not prioritize health and safety during the pandemic.

Generations Y and Z currently make up a significant portion of the workforce, who aspire to work in companies with a strong commitment to social aspects.

The social dimension of ESG drives companies to introduce changes in areas like diversity, equity, and inclusion (DEI).

Companies that foster diverse and inclusive workforces benefit from a broader range of ideas and perspectives, leading to innovation and creativity in problem-solving and product development.

According to a 2020 Cone Communications study, 83% of employees would be more loyal to a company that helps them contribute to social or environmental issues.

Adopting HR practices to attract and retain diverse talent can gain a competitive advantage in recruitment and improve organizational performance.

ESG pushes companies to focus on employee well-being and mental health through the implementation of initiatives such as flexible work arrangements, mental health programs, or healthier work environments which contributes to the productivity and retention while fostering a positive company culture.

A Deloitte survey in 2021 showed that 61% of employees expect their companies to support mental well-being, part of the "Social" pillar in ESG.

The pandemic, regulatory changes, and growing investor focus on social responsibility have accelerated the adoption of ESG frameworks in managing human capital, making it a key aspect of business strategy moving forward.

As ESG continues to evolve, impact on people management will become even more critical to driving sustainable growth and create a sustainability culture.

4.5 Risk Management

> The biggest risk is not taking any risk. In a world that is changing quickly, the only strategy that is guaranteed to fail is not taking risks.
>
> — Mark Zuckerberg

The European Commission and Governments around the world are implementing regulations related to environmental protection, labor practices, and corporate governance.

By proactively adopting sustainable practices, the companies can ensure compliance with existing regulations, minimizing the risk of their operations while also becoming more competitive.

Environmental, Social and Governance (ESG) frameworks encourage companies to broaden their understanding of risk, moving beyond traditional financial and operational risks to include environmental, social, and governance-related threats.

Companies that integrate "Environmental" factors into their risk management can better prepare for disruptions such as extreme weather events, supply chain interruptions, and regulatory changes. At a time when climate changes are impacting life on the planet, businesses can anticipate the effects of extreme weather events or regulatory shifts related to carbon emissions, be better prepare and avoid disruptions.

Companies that ignore "Social" risks, such as human rights violations, poor working conditions, or lack of diversity and inclusion, face potential lawsuits, reputational damage, and loss of consumer trust. ESG frameworks can help identify these risks early, allowing businesses to implement corrective or improving measures.

Bad corporate "Governance" governance can lead to ethical breaches, fraud, and regulatory penalties. Strong governance practices, ensure transparency, accountability, and ethical decision-making, reducing the likelihood of governance failures.

According to Deloitte's 2021 Global Risk Management Survey, 55% of companies view sustainability-related risks, such as climate change, as critical to long-term business strategy and reputation,

Based on a survey by KPMG in 2022. 75% of CEOs say that a strong ESG performance helps protect their company from future regulatory and reputational risks.

In general terms, ESG factors help companies proactively identify risks that can impact operations and the supply chains.

Through effective monitoring, companies can mitigate risks, capitalize on opportunities, enhance transparency, and create long-term value

Ensures that ESG commitments are not just words, but are translated into meaningful action and outcomes.

5 Artificial Intelligence in Support of ESG

Artificial Intelligence (AI) can significantly enhance and support the deployment of Environmental, Social and Governance (ESG) strategies by enabling organizations to collect data, automate processes, and manage related risks more effectively.

AI through IoT devices can track environmental data such as energy usage, emissions, waste, and water consumption in real-time. This allows companies to gather accurate data, essential for ESG reporting and compliance.

AI can play a significant role in advancing sustainability, helping companies to previously analyze the impact of their sustainability strategy, and improve practices in several areas of their operations such as Energy Management, Logistics Efficiency, Waste Reduction, Recycling Optimization, Emissions Monitoring, Product Life Cycle, Risk Assessment and Sustainable Investments.

By using AI in these areas, companies can make more informed decisions based on data compilation, optimize resource utilization and contribute to global sustainability goals.

It is important to consider the ethical implications, data privacy and responsible development of AI to ensure this technology is implemented in a way that aligns with sustainable principles.

ESG data often comes from diverse sources, such as social media, financial reports, news, and AI can process large volumes of data more quickly and efficiently than humans, enabling organizations to analyze trends, identify risks, and make data-driven decisions.

AI and several integrated tools can analyze public sentiment around social issues, diversity and inclusion, and corporate reputation by scanning public forums, providing insights into how a company's ESG initiatives are perceived by stakeholders.

Can facilitate the automation the collection and aggregation of ESG metrics, streamlining the reporting process and ensuring consistency and accuracy.

This reduces the labor-intensive work of manually compiling ESG reports, and operational costs.

Organizations can automatically check for compliance with regulatory frameworks and international ESG standards, such as the Global Reporting Initiative or the UN's Sustainable Development Goals), minimizing the risk of human error.

AI can process future trends and potential ESG risks based on historical data, helping companies to prepare reports that are not just reflective of past performance, but also future vison, giving them an edge in strategic planning.

A survey made by Deloitte found that 79% of executives believe AI will help in monitoring ESG performance, and 77% believe it can improve the quality of ESG data, reflecting the potential of AI to automate data collection and analysis, leading to more accurate and reliable ESG metrics.

AI can optimize supply chain operations to reduce environmental impact by identifying inefficiencies, minimizing waste, and reducing transportation emissions. This helps businesses lower their carbon footprint and improve sustainability throughout their supply chain.

AI's capabilities in processing large datasets, automating complex tasks, and making predictive analyses make it a powerful tool for supporting the deployment of ESG initiatives.

From improving data accuracy in reporting to optimizing resource use, monitoring social and environmental risks, and enhancing governance practices, AI can help organizations more effectively to meet their ESG goals, while driving long-term sustainability and business success.

6 Sustainability and ESG, Future Trends

The world will continue to witness the growing transformation of the business world, because of new environmental, social and (good) governance challenges.

Companies will continue to focus on sustainability practices, responding to the 2030 Agenda—Sustainable Development Goals, and European Green Deal, implementing appropriate structuring measures to integrate sustainability into business practices, keeping the eyes on consumers, investors and regulators.

Climate Actions: More and more companies hope to be able to reduce the carbon tax and adopt renewable energy sources in the fight against climate change.

The recent agreement signed at COP28 to gradually phase out fossil fuels by 2050 brings new challenge to all activities related to the energy transition, either through new legislation or through a set of measures that allow the phasing out, with an impact on the entire industry.

Social: Companies will continue to pay attention to social issues, giving priority to diversity and inclusion, with a focus on fair work policies and the well-being of their employees.

ESG criteria emphasize the well-being of employees, covering physical, mental and emotional health, providing a safe and healthy work environment and promoting work-life balance and promoting diversity at all levels. Create inclusive cultures, address pay equity and implement transparent hiring practices.

Companies aligned with ESG criteria recognize more and more the importance of diversity and inclusion through the equal opportunities, eliminating discrimination, employee training, professional development and mentoring programs to improve their skills.

Human Resources: Companies' commitment to sustainability and the need to create organizational structures, suitable for the implementation and development of new business processes and sustainable models, will increase the demand for resources with sustainability and ESG skills, simultaneously enhancing the development of new career opportunities.

Longevity and human responsibility are a new theme that will become part of the companies' sustainability agenda, in the People pillar.

How to manage the change, how to ensure that senior employees will continue to be useful, and how to prepare them for longevity, will be topics under consideration in the coming years.

Reporting: From January 2024, the Corporate Sustainability Reporting Directive (CSRD) reinforce companies' sustainability reporting obligations, representing a significant change in the way they communicate their sustainability practices with different stakeholders, reflecting the growing understanding that sustainability, ESG factors and company finances are interconnected.

SMEs, although they will be obliged to the new reporting model from 2026 onwards, they will be faced with this requirement immediately by their supply chain, as suppliers of goods and/or services to companies eligible under the directive.

Technology: Technology and innovation, through AI artificial intelligence, will play a significant role in improving ESG practices, namely in the data collection, helping companies to previously analyze the impact of their sustainability strategies.

The impact of companies on sustainability is profound, and everywhere, the business world is expected to respond to the challenge and call of the United Nations—Forward Faster, for the priority areas of intervention—gender equality—climate actions—reducing inequalities—reducing supply chain disruption—finance and investment.

The adoption of sustainability practices and the mitigation of associated management risks, via ESG factors will not only contribute positively to the environment and society, but also to the long-term success of the companies themselves.

7 Conclusions

> Companies that do not start now the implementation of their sustainability and ESG strategy, are unlikely to be successful in the near future

ESG regulations are expanding quickly across geographies.

Governments in various countries, are introducing directives that requires obligation to meet sustainability standards.

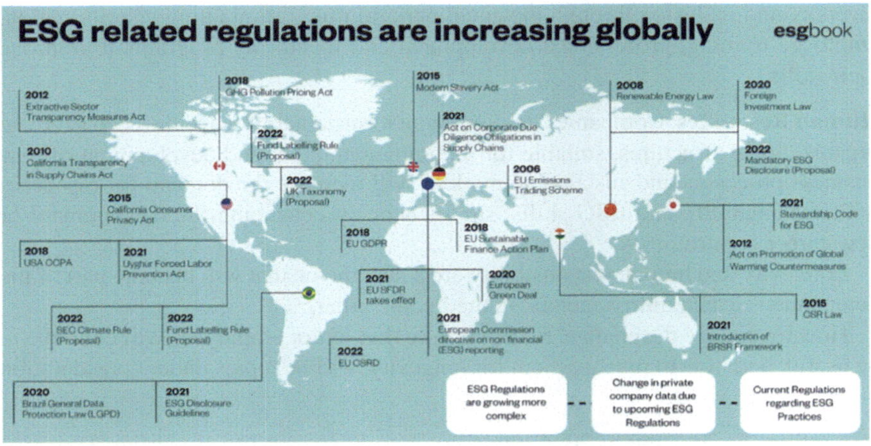

Reporting is the key element to ensure companies are providing transparent data about their environmental and social impacts.

The CSRD—Corporate Sustainability Reporting Directive has been approved by EU in January 2023, and places sustainability reports on an equal footing with financial information.

The Directive strengthens the rules regarding the social and environmental information that companies must communicate, allowing investors to access the information they need to assess the impact of companies on people and the environment, assess the risks and financial opportunities arising from climate change and other sustainability issues.

It aims to improve the flow of information about sustainability in the business world, making companies' sustainability reports more understandable.

Operational compliance is becoming an area of attention, with regulators looking for supply chain ethical practices. Themes like human right and environmental standards are part due diligence law in some European countries.

Companies that will be not able to adopt new models to achieve the carbon emission reduction as targeted, may face legal penalties.

Stay informed and proactively aligned with legislation and directives is essential to ensure a long-term success.

Responsible companies will make a difference by strengthening their codes of ethics and conduct, as well as formalizing clear and transparent internal management rules applied to recruitment, development and training, work-life balance, health and safety, diversity and inclusion.

It is hoped that companies will be able to progress on sustainable goals and committed to supporting systemic changes to achieve a more sustainable Europe.

As part of this entire ESG effort, companies need to create appropriate sustainability structures to their size, responsible for changing their business models and ensuring the development of new processes, and with that, new opportunities to manage sustainability departments and areas, begin to be obvious.

Companies are establishing sophisticated new roles with responsibility for managing their sustainability in what has now become a competitive market.

For these reasons, the implementation of ESG practices must be considered as an "ongoing journey" which requires a set of competencies and a sustainability culture.

Commitment

Long journeys often involve numerous stages, challenges, and setbacks. Commitment keeps going when initial motivation fades and helps to maintain consistent effort, allowing to break through obstacles and progress step by step.

Commitment keeps the focus on the end goal, and is about choosing to stay on a path, not just when it's easy or exciting, but also when it's challenging. That's the case of ESG factors.

Research published in the Journal of Business Venturing shows that entrepreneurs and business leaders who remain committed to their long-term journeys, despite challenges and obstacles, are 34% more likely to achieve sustained growth.

Leadership

Leadership is the driving force behind the successful implementation of ESG. Leadership is essential for ESG integration into the business strategy, to ensure that it's not just a side initiative but embedded into every business decision.

Strong leadership ensures accountability for ESG performance across all levels of the companies and play a key role in creating a culture of sustainability, ensuring that ESG becomes part of the company's identity, inspiring staff to innovate and participate in sustainability projects and initiatives.

ESG factors involves balancing the expectations of diverse stakeholders, including investors, customers, employees, regulators, and communities.

Leaders act as key communicator and influencer, ensuring that stakeholders understand the company's ESG commitments and progress.

Monitoring

By tracking ESG commitments, companies can keep stakeholders informed about progress and gaps, making sure they are following the defined plans.

Without monitoring, it's impossible to know if ESG initiatives are effective.

Tracking performance metrics allows companies to measure the real-world impact of their ESG strategies and provides quantifiable data to make decisions for improvement, innovation or even highlight opportunities, where social initiatives could be expanded to create more positive community impacts.

Continuous monitoring helps to identify ESG-related risks, manage these risks and mitigate potential negative impacts.

Culture of Sustainability

Sustainability culture refers to the mindset, values, and collective behaviors within a company or organization that actively promote sustainability.

It reflects how much sustainability is integrated into the business, decision-making processes and long-term planning.

Developing a culture of sustainability involves promoting awareness and commitment to ESG factors, at all levels of the company.

Therefore, it is up to leadership to establish the initiatives and sustainability goals that the company proposes to develop, involving employees in daily actions or strategic projects.

Encourage innovation and collaboration between teams to improve sustainability efforts across all departments, including creating incentives for sustainable behaviors.

Taking responsibility for the company's impact on future generations, and communicating the company's sustainability performance internally and externally, means sharing good practices among employees, partners and suppliers, and making them aware of the company's concerns and positioning in relation to the challenges of sustainability.

Internally, communication is vital for integrating the sustainability and ESG strategy into the business.

How to integrate it, why we do it, and what changes result from it, need to be communicated effectively throughout the organization.

It is important to consider that the adoption of sustainability practices, will be only achieved when all employees internalize the basic principles and make this commitment.

Training in sustainability should be seen as an investment in companies' human capital, thus ensuring that new leaders have the skills and abilities for sophisticated new roles.

Qualifying and developing people's skills is a multiplying and mobilizing factor for the new sustainability strategy.

However, not everything it's good news when it comes to the speed at which we are working to achieve the goals.

In fact, since 2019 there has been a stagnation in sustainability progress in Europe, because of the pandemic crisis, and more recently the invasion of Ukraine.

The lack of cooperation between the central government and the private sector is a fact, which requires to make up for lost time and quickly move to action, where cooperation and interaction between companies, governments, academia and NGOs is necessary.

The level of commitment to people, whether through the adoption of sustainability practices or new management models, is decisive for the sustainability of companies.

Thinking about the path to sustainability and ESG factors, and understanding the impact on the business, people and the planet, is a continuous improving, cooperation and learning process for all businesses.

Bibliography

1. Luís Roberto, A Caminho da Sustentabilidade, 7 Passos para a ação book (2024)
2. Brundtland Report (Our Common Future) (1987)
3. PwC survey (2021)
4. Harvard Business School working paper (2020)
5. Cone Communications study (2020)
6. Edelman Survey (2022)
7. Deloitte's Global Risk Management Survey (2021)
8. KPMG survey (2022)
9. Forbes survey (2020)
10. RepRisk report (2020)
11. SDSN, Sustainable Development Report (2024)
12. CSR Europe: SDG Master Class (2018)
13. Network for Business Sustainability (2021)
14. Nareuporn Piyasinchai: Reputational Spirals (2021)

Opportunities and Challenges of Sustainability Reports for Corporate Governance and Transparency in Portugal

Paula Figueiredo, Irene Ciccarino, Ana Filipa Joaquim, Cátia Arnaut, and Ana Diogo

Abstract In the past decade, sustainability has emerged as a core focus within the European research and development community, spurred by the European Commission's introduction of new regulations through the Corporate Sustainability Reporting Directive (CSRD). Under the recently introduced European Sustainability Reporting Standards (ESRS), a significant number of companies, including small and medium-sized enterprises (SMEs), are mandated to produce regular and comprehensive information about their impacts and commitments toward sustainable development. This study provides an overview of the 100 biggest companies in the district of Leiria and the municipality of Ourém, Portugal, to evaluate how prepared they are to deliver their sustainability reports next year. This chapter discusses this significant sustainability step, providing a framework to analyse the mandatory reports. Therefore, it addresses the lack of consensus on report guidance in the literature, undermining the knowledge about disclosure phenomena overall. Researchers have been focusing on sustainable, responsible investment disclosure and corporate governance, and they are gradually including their impacts on sustainability. The CSRD can potentially overcome these conceptual issues, promoting positive social and environmental effects. In this sense,

P. Figueiredo (✉)
ECEO, Universidade Lusófona, Lisboa, Portugal
e-mail: p40297@ulusofona.pt

I. Ciccarino
Business Research Unit of University Institute of Lisbon (BRU-ISCTE-IUL), Lisboa, Portugal
e-mail: irene.ciccarino@iscte-iul.pt

A. F. Joaquim
ISLA Santarém – Instituto Politécnico and IPRI NOVA University, Santarém, Portugal
e-mail: afilipajoaquim@gmail.com

C. Arnaut
FCSH-NOVA University, Lisboa, Portugal

A. Diogo
ISLA—Polytechnic University of Management and Technology, Vila Nova de Gaia, Portugal
e-mail: ana.diogo@islagaia.pt

© The Author(s), under exclusive license to Springer Nature Switzerland AG 2025
A. d. b. Machado et al. (eds.), *Environmental, Social, Governance and Digital Transformation in Organizations*, Information Systems Engineering and Management 35, https://doi.org/10.1007/978-3-031-86079-9_13

a systematic perspective was adopted to verify if prior reports have aspects related to the three categories of ESRS. The assumption is that the degree of previous information existence could enable compliance with the ESRD. The resultant description can reveal latent aspects of each company context, allowing inferences about their degree of compliance and the amount of effort needed to ensure it. At the time of this survey in the second half of 2024, it was found that less than half of the companies were aligned with the CSRD. Therefore, this paper highlights the lack of readiness among companies in the district of Leiria and the municipality of Ourém to fully comply with the European Sustainability Reporting Standards (ESRS). This suggests a significant gap in corporate governance structures and the public disclosure of sustainability data. There is a need for greater clarity and guidance so that companies can avoid disclosure issues and practices such as greenwashing. In addition, the results point to the urgent need for training and support to help companies adapt to these new regulatory requirements.

Keywords Sustainability · ESG · European union's corporate sustainability reporting directive · CSRD · Portugal

1 Introduction

The European Union's Corporate Sustainability Reporting Directive (CSRD) adopted in December 2022 and operative from January 2023, represents a significant shift in corporate transparency within the European Union [1]. It is a regulatory framework to enhance corporate transparency and accountability regarding sustainability matters. Enacted by the European Commission, this directive expands on the predecessor Non-Financial Reporting Directive (NFRD), intending to harmonise sustainability reporting standards across the EU [1–3].

The CSRD requires businesses to disclose non-financial information in their annual reports, ensuring stakeholders are informed about the company's impact on society and the environment. It aims to provide standardised, comparable, and reliable sustainability data to facilitate informed decision-making by investors, consumers and policymakers.

The CSRD came into force at the beginning of 2024, significantly changing the ESG reporting landscape [1–3]. It is estimated to affect over 50,000 companies. Even non-EU companies with subsidiaries or branches in the EU's internal market will become subject to reporting requirements by 2028 [4].

The CSRD is important in advancing corporate sustainability practices and fostering a transition towards a more sustainable economy. By compelling companies to disclose comprehensive ESG information, the directive facilitates better risk management, improved stakeholder engagement, and increased accountability. Investors can make more informed investment decisions, considering financial performance and environmental and social impacts. Additionally, consumers can choose products and services from companies aligning with their values, driving market demand for sustainability [5].

This was a necessary enforcement mechanism to prevent "greenwashing," "social washing", and "ESG washing", where companies engage in superficial or misleading sustainability reporting to improve their public image without genuinely addressing environmental or social issues [5, 6]. Despite its potential benefits, the implementation of the CSRD poses several challenges. One major challenge is the complexity of sustainability reporting, particularly for smaller companies lacking resources and expertise in this area. Compliance costs could be substantial, especially for companies operating across multiple jurisdictions with varying reporting requirements [7].

This study aims to evaluate the readiness for compliance with the CSRD of the one hundred largest companies in the district of Leiria, Portugal. The research offers an analytical framework to assess existing corporate sustainability reports and gauge alignment with the European Sustainability Reporting Standards (ESRS). By identifying the degree of preparedness among these companies, the study addresses the current gap in consensus on sustainability report guidance [8]. It offers insights into the challenges and best practices for achieving compliance [9, 10].

This paper contributes to the broader discussion on corporate sustainability by examining Portuguese companies as they navigate these new regulatory requirements. The findings serve as a basis for potential consultancy or training programs to support businesses in effectively disclosing their sustainability efforts and avoiding practices like greenwashing.

The major contributions of this chapter are as follows:

- Provides an analytical framework for assessing compliance with sustainability standards.
- Highlights best practices and identifies key challenges for Portuguese companies in meeting CSRD requirements.
- Enhances understanding of corporate sustainability disclosure, aiding both researchers and policymakers.
- Contributes to discussions on overcoming "greenwashing", "social washing", and "ESG washing" by promoting genuine corporate transparency in ESG reporting.

2 Sustainable Responsible Investment

Sustainable and responsible investment has been an increasing trend in literature and business [11, 12]. Corporate sustainability (CS) and the evolution of social and environmental disclosure (SED) have occurred since the 1970s. Sustainability is a multidimensional concept that is difficult to tackle, reinforcing the importance of CSRD as a powerful component to enhance performance and compliance toward sustainable development [6].

Corporate social responsibility (CSR) and Environment Social Governance (ESG) are concepts often used interchangeably, causing much confusion [12]. Corporate social responsibility (CSR) is a guiding principle that defines the relationship between a company and its stakeholders. It emphasises the responsibility of an organisation to contribute positively to societal well-being. By fulfilling its obligations to the broader community, a company fosters a mutually beneficial relationship, promoting goodwill and trust. However, many CSR practices were primarily centred on voluntary initiatives companies undertake to address their social and environmental impacts, often as part of their philanthropic activities or reputation management strategies [13].

While the awareness of sustainability issues grew and stakeholders increasingly demanded more transparency and accountability from companies, the prominence of CSR changed toward ESG [14]. Environment social governance (ESG) is more of an indicator than a principle. In the aftermath of the 2008 financial crisis, the market shifted toward ESG-guided investments [11]. Since the 90s, ESG metrics have been mainly applied to filter out nonethical companies, also being a common proxy for sustainability performance [15]. The ESG indicators as a control tool have evolved to encompass environmental stewardship, social responsibility, and corporate governance practices. Its objective is to elucidate the business model, delineate implemented sustainability measures and their outcomes, and scrutinise risks stemming from business relationships, products, or services that could jeopardise the company's sustainability [14].

Research on sustainable and responsible investment has focused on disclosure and corporate governance, and it is gradually shifting to include its impacts on sustainability in general. Literature still needs more consensus on report guidance, undermining knowledge about disclosure phenomena overall [8]. By focusing on the report, the CSRD has the potential to overcome these conceptual issues, promoting positive social and environmental impacts.

3 CSRD Historical Perspective

Over the past decade, corporate reporting has significantly shifted to investors and shareholders, expanding beyond traditional financial metrics to include sustainability reporting. The Barnier Directive (Non-Financial Reporting Directive—NFRD Directive—2014/95/EU) enshrined in European law the requirement for large public-interest entities to produce and disseminate a non-financial statement. It has connected the ESG to execution for the first time. This statement is mandated to cover various aspects such as environmental impact, social responsibility, employee welfare, human rights adherence, and efforts against corruption to provide stakeholders with a comprehensive understanding of the company's operations, performance, outcomes, and influence [16, 17].

Indeed, contemporary management perspectives advocate for a holistic approach that encompasses both financial and non-financial dimensions to safeguard the generation of corporate value. Reflecting this trend, on December 14, 2022, the European Commission adopted the Corporate Sustainability Reporting Directive (CSRD) to enhance sustainability reporting practices to facilitate the transition towards a fully sustainable and inclusive economic and financial system.

Consequently, sustainability reporting has emerged as a pertinent concern for companies. This directive replaced the NFRD Directive [1, 3]. The replacement of the term "non-financial information" for "sustainability information" is also a significant shift to consolidate theory and practice under a concept that emphasises sustainability over the economic dimension [18].

The CSRD focuses on (1) Listed companies on regulated markets (excluding micro-enterprises); (2) Large companies meeting two out of three size criteria in two consecutive financial years on its balance sheet date: (a) more than 250 employees, (b) more than €50 million net turnover; (c) more than €25 million in total assets; (3) Non-EU companies with subsidiaries exceeding the aforementioned thresholds within the EU [1, 2, 4].

The implementation of CSRD will occur gradually with non-EU companies with subsidiaries or branches in the EU's internal market, which will become subject to reporting requirements by 2028. The first report under CSRD will be in 2025 to large companies and companies already under the NFRD Directive. The tendency is that the number of companies affected will increase considerably, from the previous 11,700 (i.e., NFRD Directive companies) to an estimated 50,000. Companies in the European Union, with another 3000 or more U.S.-based companies estimated to be affected. Ultimately, the scope of the CSRD may be shaped by how each E.U. member state transposes it into law [4].

To overcome the lack of consensus on standardised methodologies and verification mechanisms [3], the EU Commission has published sustainability reporting standards to supplement the CSRD. It aims to provide clarity and consistency in how companies disclose their sustainability performance. This helps meet stakeholders' expectations and, promotes greater accountability and facilitates informed decision-making by investors, consumers, and policymakers [2]. Additionally, evidence suggests that standardised reporting enables benchmarking and comparing companies' sustainability efforts, driving continuous improvement in corporate sustainability practices across the European Union [3]. The implementation of the CSRD is slated to occur in three stages, commencing from the financial year starting January 1, 2024 [1].

Mandating ESG reporting can reduce ESG decoupling by increasing transparency and accountability in corporate practices. For instance, in Malaysia, listed companies must disclose sustainability statements, which help investors understand the company's approach to ESG issues, thereby aligning corporate actions with reported ESG commitments [19]. There is positive evidence that corporate ESG performance significantly enhances green innovation. It can also be a mechanism of influence by alleviating corporate financing constraints, improving human capital levels, and reducing management myopia, which can, therefore, increase innovation. In this sense, insights underscore the importance of integrating ESG considerations into corporate strategies to foster sustainable development and reduce ESG decoupling by holding companies accountable to their reported sustainability goals [20].

This positive evidence supports the EU Parliament's proposal for a structured reporting approach to mitigate the risk of companies presenting a misleading image of their ESG performance [3, 21]. Agencies concerned with other kinds of risks also significantly consider ESG risks [22]. For all these features, there is a great expectation of CSRD as a positive impulse to achieving ONU'S sustainable development goals and also encouraging best practices across industries [3]. This expectation has been discussed in other efforts to set sustainability standards [21].

However, ensuring the accuracy and reliability of reported data presents challenges due to the variability in ESG ratings and methodologies used by different providers. This variability can affect consistency and comparability [2, 20, 22].

4 Research Design

This study provides an overview of the one hundred biggest companies in the district of Leiria and the municipality of Ourém in Portugal to evaluate their preparedness to deliver their sustainability report next year. The local journal "Região de Leiria" annually publishes a list of the one hundred biggest companies in the district [23], which was used to select the sample as they will likely meet the requirement to submit the sustainability report next year. The list was created in 2023 and published on July 17th, 2024.

A systematic perspective was adopted to verify if prior reports have aspects related to the three categories of ESRS. The assumption is that the degree of prior information existence could enable compliance with the ESRD. Thus, the ESRD elements in each category are the unit of context in a distinguishable and comparable dataset, with sustainable reports being the unit of analysis.

The characteristics of these reports allow for content analysis [24]. This research design examines the available data, evaluating its quality and purpose. It also allows an abductive approach to infer contextual phenomena [25]. The resultant description can reveal latent aspects of each company context, allowing inferences about their degree of compliance and the amount of effort they will need to ensure it [26]. Prior studies related to sustainability applied a similar process [9, 10].

The study commenced with a review of the literature to provide a theoretical analysis of the European Sustainability Reporting Standards (ESRS). The explanation presented in the next section aims to make ESRS easy to understand for users. As a result, a theoretical framework is proposed. It was used to analyse the company's reports and can also aid in replicating this study. The framework supplies orientation questions to help with thematic coding. The codes, a priori defined, were refined and validated by the team, composed of five researchers [25, 26].

Data were analysed using pattern matching from July 20th to September 29th. Each researcher independently performed data analysis, ranking the attributes of the theoretical framework on a 5-point Likert scale. The results are a synthesis of systematic patterns found in both the literature and the collected data. Conclusions are based on systematic recurrence until theoretical saturation is reached, enabling an explanation of the themes raised in the theoretical framework presented in the previous section. The findings presented in this chapter result from critical consensus among the team [24].

5 Report Features

As far as is known, the literature lacks consensus on standardised methodologies and verification mechanisms [3, 22]. On 31 July 2023, the European Commission published the Commission Delegated Regulation (EU) 2023/2772, supplementing CSRD regarding sustainability reporting standards. It establishes the European Sustainability Reporting Standards (ESRS). This reporting framework encompasses a broad spectrum of quantifiable and qualitative information, covering targets, impacts, risks, opportunities, and strategies related to sustainability. It was designed to provide detailed guidelines for mandatory reporting under the CSRD, ensuring standardisation across the EU for disclosing sustainability-related information [3].

Materiality analyses the dimensions of the considered impact. Traditionally, the focus of sustainability reporting has been on the impacts of companies on the external environment and society, known as "single materiality." "Double materiality" is a concept within sustainability reporting that acknowledges the dual impacts of environmental, social, and governance (ESG) issues on companies' financial performance and broader societal and environmental well-being. From a financial perspective, ESG factors can present both risks and opportunities for companies. For example, climate change may pose physical risks to infrastructure and supply chains, while regulatory changes related to environmental protection may impact operational costs. On the other hand, investments in renewable energy or sustainable practices can enhance long-term competitiveness and profitability [21].

Therefore, materiality discusses whether the information chosen to be communicated is relevant to a particular topic and to whom it is relevant. Sometimes, the significance is unintentional or deliberately unsuitable from a stakeholder's perspective. Therefore, materiality defines the content and indicators highlighted in the sustainability strategy and report [18, 27, 28].

While this transition has also been noted in the market, sustainability reports analysis has underscored inconsistencies in defining materiality and reporting information. Challenges stem from prioritising company-centric views of materiality instead of considering the stakeholder's perspective. Companies are urged to clarify their methods for identifying relevant stakeholders and to explain how stakeholder feedback influences decision-making processes [18].

On the other hand, there is no consensus about the best focus on materiality. For instance, important authorities, such as the International Sustainability Standard Board (ISSB—IFRS standard), have defended the focus on single materiality to clarify the report's requirements and improve their comparability. Abhayawansa [13] explains that single materiality is most useful when focusing on how a company's activities affect its financial performance. This approach emphasises accountability for financial and non-financial capital and suits organisations looking to incorporate sustainability into their core business strategy. It ensures that their reporting aligns with financial outcomes and long-term value creation.

Conversely, those who advocate double materiality highlight that it is beneficial when considering both the financial impact of sustainability on the company and the company's impact on society and the environment. It is ideal for organisations that seek to address a broader range of stakeholders, including investors, customers, and the community, by providing a comprehensive view of their social and environmental performance and financial sustainability [13].

The ESRS seems to address both perspectives while emphasising double materiality. It also defines materiality as context-specific, meaning companies should assess materiality based on their own impacts and circumstances rather than applying a universal checklist. While the ISSB and ESRS approaches diverge on materiality (i.e., single vs. double), efforts are being made to align global standards such as ISSB and European standards (ESRS) for ease of global reporting.

The ESRS has three categories: (1) Cross-cutting Standards, (2) Topical Standards, and (3) Sector-specific Standards [3].

Cross-cutting standards are general standards that apply universally to all businesses, regardless of the industry they operate in. They encompass broad topics relevant to every company, serving as a set of general rules for businesses to follow. These standards aim to help companies report essential information regarding their impact on people and the planet. By ensuring businesses provide details about their leadership, future plans, and management of risks and opportunities related to sustainability, cross-cutting standards facilitate a better understanding of a company's global impact.

Companies are mandated to report in four main areas: how the company is run (governance), its strategy for addressing sustainability issues, its approach to identifying and managing risks and opportunities (impact, risk, and opportunity management), and its progress towards sustainability goals (metrics and targets).

The governance area looks at how a company ensures it acts responsibly towards the planet and people, including the establishment of rules and teams to oversee its actions. Strategy involves how a company plans to address environmental and societal impacts, ensuring its plans contribute positively to the future. The focus on impact, risk, and opportunity management is about assessing the significant effects a company might have on the world and planning to address these effects. Finally, metrics and targets involve companies sharing their progress in achieving goals to become more environmentally and socially responsible, detailing the steps they are taking and the progress made.

The Topical Standards focus on specific sustainability topics such as the environment, social issues, and governance. They provide detailed guidelines addressing specific areas of concern within these broad categories. Companies are required to share detailed information about how their actions impact important areas like the environment or people's rights. These standards add specific details to the general rules that all companies must follow, ensuring comprehensive coverage of important sustainability aspects. By adhering to Topical Standards, companies can demonstrate how they are addressing significant challenges.

Finally, the Sector-specific Standards are tailored to specific industries. They deal with impacts, risks, and opportunities that are most relevant and likely to be material for businesses within a particular sector. Sector-specific standards are in development and will be rolled out progressively.

The ESRS focuses on four reporting areas: Governance, Strategy, Impact, Risk, and Opportunity Management, and Metrics and Targets, designed to offer a comprehensive view of a company's sustainability approach from governance oversight to specific performance metrics. Companies are encouraged to disclose how sustainability governance is integrated into their overall corporate governance structures, including board-level oversight of Environmental, Social, and Governance (ESG) issues to ensure sustainability is managed and monitored at the highest company levels.

Additionally, it is important to note that under the CSRD, companies must subject their sustainability reports to external assurance, meaning an independent audit is needed to verify the sustainability data's accuracy, enhancing the reports' credibility and reliability. This assurance requirement distinguishes the CSRD/ESRS from previous voluntary frameworks, making sustainability reporting more robust.

When reporting on greenhouse gas emissions (Scopes 1, 2, and 3), emphasising the challenge of Scope 3 emissions, which cover the entire value chain, is crucial. This is due to the complexity of tracking emissions outside a company's direct control, such as from suppliers or end-users, making data collection and accuracy difficult. The social aspects of reporting, such as collective bargaining, non-employee workers, and social dialogue, are essential but should be more strongly emphasised in relation to human capital management, a significant focus under the ESRS. Companies will also need to report on critical social issues like the gender pay gap, working conditions, and diversity and inclusion efforts, highlighting a company's commitment to fair treatment, equality, and social responsibility, reflecting the broader scope of sustainability.

For the ESRS to be comprehensive, it has become complex. Companies, especially small and medium-sized enterprises (SMEs), may encounter challenges complying with ESRS. Although the CSRD and ESRS will initially apply to large companies, SMEs might find the complexity and data requirements of the ESRS overwhelming when they eventually become subject to these standards. There are also concerns regarding implementation costs and the comparability of reports across jurisdictions, particularly for companies operating in both EU and non-EU countries, as they may need to reconcile the ESRS with ISSB standards.

This paper summarises the ESRS regulation using a theoretical framework to discuss opportunities, challenges, and the need for clarification. It is presented in Table 1.

Due to the lack of reports, a questionnaire based on three pillars was created and made available through Google Forms to the workers of the sample considered: Leadership Behavior, Sustainability—The Triple Bottom Line (TBL), and Corporate Social Responsibility.

In accordance with the CSRD and ESRS guidelines, reports must be public. Hence, other sources of data collection are not used. The theoretical framework can be

Table 1 Theoretical framework

No.	Binary code	Descriptive question
1	Stakeholders (STK)	Do they inform who their stakeholders are?
2	Relationship (REL)	Do they describe how they deal with stakeholders? (i.e. positive or negative issues)
3	Relevance (RES)	Do they relate stakeholder participation as an indicator or result of their actions?
4	Participatory process (PP)	Do they explicitly state how they involve stakeholders?
5	Participatory techniques (PT)	Do they indicate participatory tools or techniques?
6	Participatory Result (PR)	Do they present any results of stakeholder participation?
7	Training and education action (TE)	Do they report any training and education action (TE) to employees or local communities?
8	Disclosure and transparency (INFO)	Do they have systems for disclosure of information and transparency practices?
9	Opinions collection (OC)	Do they systematically collect stakeholders' opinions?
10	Partnerships (PART)	Do they report partnerships?
N	*Examples code*	*Descriptive question*
11	Ex. PT	How do they foster stakeholders' participation?
12	Ex. PP	How long have they been doing the participatory efforts?

Source Authors

summarized under the ESG framework's three pillars—Environmental Performance, Social Performance, and Governance Performance with three reference variables that highlights the main concern according to the literature (Table 2).

The questionnaire was exclusively based on closed-answer questions. All questions, except for the entity's characterisation, were based on a five-point Likert scale, ranging from 1 (completely disagree) to 5 (completely agree).

Table 2 Quality assessment variables

Pillar	Main variables considered		
Environmental Performance	Greenhouse gas emissions	Energy and resource consumption	Pollution
Social performance	Employment	Job health and security	Human rights
Governance performance	Governing structure	Sustainability policies	Risk management

Source Authors

However, the responses collected were limited, with only 48 responses gathered over the three-month period (July, August, and September 2024), despite the first two months being holiday months. In September, a new call was made, but with limited results. The sample comprised 39.6% of respondents linked to management and 60.4% not linked to management.

5.1 Analysis

This study involved analysing public information available on the institutional websites of the one hundred largest companies in the district of Leiria and the municipality of Ourém. An initial analysis revealed that the list included one company from the primary sector of Portuguese economic activity—agriculture, fishing, and mining. Forty-nine companies belong to the secondary sector, characterised by industrial activities, civil construction, and water and energy supply to the population. Fifty companies are part of the tertiary sector and provide services.

Subsequently, the companies with an internet presence were identified through their institutional websites, reducing the sample to sixty-nine. The rest do not have an internet presence, and there are no official reports about them on open-source websites.

In the sampled organizations, only 23% of listed companies have information about their environmental performance, with energy and resource consumption being the most reported variables at 25% and 24% respectively.

For social performance, evidence drops to 21% on average, with job health and security being the most referred to variable at 27%.

In terms of governance performance, there is an average of 24% availability of evidence within the sampled universe, with Sustainability Policies being the most prominently reported variable at 34%.

Our analysis shows that only 4% of the organisations provide evidence in all domains, while 58% have no evidence in any of the ESG domains.

Regarding the respondents' workplace survey, 35.4% work for companies with 151 to 250 employees, followed by 25% working for entities with 251 or more employees. In terms of job seniority, the majority (31.3%) have been with the company for 31 years or more, followed by 20.8% who have been with the company for five years or less.

Although the questionnaire was conducted anonymously, it was found that the responses regarding the first axis—leadership behaviour, were mostly based on a value of four (I agree), with response rates between 35.4% and 54.2%, suggesting that the workers perceive the leadership behaviour as adequate.

The sustainability axis—The Triple Bottom Line (TBL)—is divided into social, environmental, and economic dimensions. In the social dimension, respondents mainly selected "*Agree*" (level 4). However, for the statements "I am proud to be part of this organisation" and "*I would recommend it as an excellent workplace," 21 and 18 respondents agreed, respectively*". In the environmental dimension, only the statement "*I adequately complete the tasks proposed to me in an environmentally friendly way*" received twenty-one responses, standing out from the others.

In the economic dimension, particularly concerning the retention of key employees, thirteen respondents indicated that the company where they work has a worse performance (level 2) compared to other organisations of the same type of activity.

Finally, on the axis related to Corporate Social Responsibility, twenty-six respondents agreed (level 4) with the statement, "*Our organisation participates in activities that aim to protect and improve the quality of the environment*."

6 Discussion

Based on our analysis, most of the one hundred largest companies in the district of Leiria showed significant gaps in compliance with the European Sustainability Reporting Standards (ESRS). While some organisations have made progress in aligning with ESG principles, particularly in the secondary sector, many are still unprepared to fully comply with the Corporate Sustainability Reporting Directive (CSRD). The analysis suggests the importance of the mandatory reinforcement given by CSRD, corroborating the theory [19, 20].

A key issue observed is that the companies in the sample focused on ESG only when it was legally mandatory. For example, only 23% of the organisations in our sample have ISO certification and even fewer display evidence of their ESG efforts on publicly accessible platforms. The gap between internal compliance and public disclosure raises concerns about the transparency and accountability of corporate sustainability practices, as evidence from other studies also suggests [19, 21].

The data collected from the companies' sustainability reports and the worker questionnaire reveal several areas of misalignment. The lack of structured information about environmental, social, and governance (ESG) performance is particularly pronounced in the tertiary sector (26.5% of the sample). In contrast, companies in the secondary sector (42.6% of the sample), such as those in manufacturing and construction, provide more detailed information on sustainability efforts, although still insufficient. Even though standards and procedures may be in place and audited, there is no evidence of ESG compliance in the organisations' online presence. This suggests partial compliance with the ESG framework, but a gap remains regarding corporate sustainability reporting being made public on the organisations' websites.

Moreover, the challenge of double materiality—addressing the financial impact of sustainability issues on companies and the environmental and societal effects of companies' operations—has not been consistently addressed. Many organisations prioritise financial materiality, often neglecting broader environmental and social impacts that are crucial under the CSRD.

In the secondary sector, most companies that present ESG indicators are involved in product manufacturing and transformation, followed by those in the construction sector. The primary environmental indicators found were related to sustainability policies and pollution levels. However, only a small percentage, about 4% (four) of companies provide detailed information on these aspects.

Most companies provide information about their sustainable policies in the tertiary sector, which is mostly linked to wholesale trade and followed by transport companies. However, the information provided must be more accurate and complete, especially regarding concrete measures and data. The data collected from the sample workers are considered quite balanced, but their limited participation does not allow for conclusive findings in this study.

Finally, the shift from voluntary corporate social responsibility (CSR) initiatives toward mandatory environmental, social, and governance (ESG) reporting marks a significant challenge for companies, particularly SMEs. As these new regulatory frameworks evolve, companies need to transition from a narrow focus on financial metrics to a broader integration of sustainability into their core business strategies.

Due to the study's delimitation and the difficulties in obtaining data, the results cannot be generalised. Data availability also presented some constraints, such as a relatively limited exploration of double materiality or the absence of best practices to share. However, they raise an important alert to Portuguese companies and highlight evidence of the challenges in implementing a sustainability disclosure standard. It also emphasises the need for more attention and guidance to avoid disclosure problems or harmful practices such as greenwashing, whether intentional or accidental.

7 Conclusion and Future Scope

The research provided an analytical framework to assess existing corporate sustainability reports and gauge alignment with the European Sustainability Reporting Standards (ESRS). In this sense, the paper contributes to the broader discussion on corporate sustainability, allowing discussion and replication.

At the time of this survey in the second half of 2024, it was found that less than half of the 100 largest companies in the district of Leiria and the municipality of Ourém were aligned with the CSRD. This paper highlights the lack of readiness among companies in the district of Leiria and the municipality of Ourém to fully comply with the European Sustainability Reporting Standards (ESRS). This suggests a significant gap in corporate governance structures and the public disclosure of sustainability data.

The implementation of the CSRD represents a major shift in how companies report on ESG issues. While some organisations are prepared to meet these new requirements, many face challenges related to data collection, materiality assessments, and the integration of sustainability into their overall corporate strategies. This is particularly true for SMEs, which may lack the resources and expertise to comply with these complex standards.

For future research, it would be beneficial to expand the sample size to include companies from other regions of Portugal, providing a broader understanding of

CSRD readiness at the national level. The comparison of the progress of Portuguese companies with their European counterparts to identify best practices and key challenges across different jurisdictions would also be relevant. This would provide a more accurate view of the more integrated ESG domains in the organisational practices and the ones that still need to be developed into standardised organisational procedures.

Further exploring the impacts of sustainable and responsible investment (SRI) on corporate behaviour and decision-making, given the rising importance of ESG metrics in investment decisions. This perspective would also benefit from research on how double materiality is being adopted in practice and identifying specific barriers companies face when addressing both financial and non-financial sustainability impacts.

Finally, extending the analysis to include more quantitative data, such as employee perspectives on ESG compliance, will offer a more comprehensive understanding of how these directives are implemented at the operational level. This will also allow for deeper insights into corporate transparency, helping to prevent practices like greenwashing and promoting genuine accountability in sustainability reporting. This study also recommends an urgent effort toward training and support to help companies adapt to these new regulatory requirements.

References

1. Directive (EU) 2022/2464 of the European Parliament and of the Council of 14 December 2022 (2022)
2. Commission Delegated Regulation (EU) 2023/2772 of 31 July 2023 supplementing Directive 2013/34/EU of the European Parliament and of the Council as regards sustainability reporting standards (2023)
3. Tőzsér, D., Lakner, Z., Sudibyo, NA., Boros, A.: Disclosure compliance with different ESG reporting guidelines: the sustainability ranking of selected European and Hungarian Banks in the socio-economic crisis period. Adm. Sci. **14**(3) Article 58 (2024). https://doi.org/10.3390/admsci14030058
4. Deloitte, DC.: Deloitte heads-up CSRD corporate sustainability reporting directive Frequently Asked Questions. Deloitte Consult. (2023) Retrieved 04/05/2024 from https://dart.deloitte.com/USDART/home/publications/deloitte/heads-up/2023/csrd-corporate-sustainability-reporting-directive-faqs
5. Esteves, C., Ciccarino, I.: Greenwashing, a negative externality of sustainability? Revista de Comunicação Dialógica **9** (2023). https://doi.org/10.12957/rcd.2023.74397
6. Aboud, A., Saleh, A., Eliwa, Y.: Does mandating ESG reporting reduce ESG decoupling? Evidence from the European union's directive 2014/95. Bus. Strateg. Environ. Strateg. Environ. **33**(2), 1305–1320 (2024). https://doi.org/10.1002/bse.3543
7. Allgeier, S., Feldmann, R.: CSRD sustainability reporting for non-listed SMEs: European regulators remain challenged. Eur. Co. Financ. Law. Rev. **20**(3), 438–446 (2023). https://doi.org/10.1515/ecfr-2023-0019
8. Khamisu, M.S., Paluri, A.R.: What's past is prologue: reminiscing research on environment social governance (ESG) disclosures. Int. J. Discl. Gov. Discl. Gov. (2024). https://doi.org/10.1057/s41310-024-00231-w

9. Ciccarino, I.D.M., Fagundes, C., Fernandes, M.E.T.: Sustainable forest management and stakeholders participation in Portuguese FSC's certified companies. In: Rethinking Management and Economics in the New 20's. Springer Proceedings in Business and Economics (2023) https://doi.org/10.1007/978-981-19-8485-3
10. Ciccarino, I.D.M., Rodrigues, S.C.S.F.: Social entrepreneurship: what people are looking for when they talk about it?. In: Handbook of Research on Approaches to Alternative Entrepreneurship Opportunities. IGI Global (2020)
11. El Ghoul, S., Karoui, A., Patel, S., Ramani, S.: The green and brown performances of mutual fund portfolios. J. Clean. Prod. **384**, Article 135267. (2023). https://doi.org/10.1016/j.jclepro.2022.135267
12. Gillan, S.L., Koch, A., Starks, L.T.: Firms and social responsibility: a review of ESG and CSR research in corporate finance. J. Corp. Financ. **66**, 101889 (2021). https://doi.org/10.1016/j.jcorpfin.2021.101889
13. Tunio, R.A., Jamali, R.H., Mirani, A.A., Das, G., Laghari, M.A., Xiao, J.: The relationship between corporate social responsibility disclosures and financial performance: a mediating role of employee productivity. Environ. Sci. Pollut. Res. Pollut. Res. **28**(9), 10661–10677 (2021). https://doi.org/10.1007/s11356-020-11247-4
14. Li, T.-T., Wang, K., Sueyoshi, T., Wang, D.D.: ESG: research progress and future prospects. Sustainability **13**(21) (2021). https://doi.org/10.3390/su132111663
15. Widyawati, L.: A systematic literature review of socially responsible investment and environmental social governance metrics. Bus. Strateg. Environ. Strateg. Environ. **29**(2), 619–637 (2019). https://doi.org/10.1002/bse.2393
16. Glaveli, N., Alexiou, M., Maragos, A., Daskalopoulou, A., Voulgari, V.: Assessing the maturity of sustainable business model and strategy reporting under the CSRD shadow. J. Risk. Financ. Manag. **16**(10), Article 445. (2023). https://doi.org/10.3390/jrfm16100445
17. Directive (EU): 2014/95 of the European Parliament and of the Council of 22 October 2014 (2014)
18. Miettinen, M.: Are materiality determination practices evolving in the wake of increasing legislation on sustainability reporting? Findings from EU pharmaceutical companies' reports. Int. J. Law. Manag. **66**(3), 363–392 (2024). https://doi.org/10.1108/IJLMA-09-2023-0221
19. World Economic Forum., Schwab Foundation for Social Entrepreneurship.: Unlocking the social economy towards an inclusive and resilient society (2022)
20. Lian, Y., Li, Y., Cao, H.: How does corporate ESG performance affect sustainable development: a green innovation perspective. Front. Environ. Sci. **11**, 1170582 (2023). https://doi.org/10.3389/fenvs.2023.1170582
21. Abhayawansa, S.: Swimming against the tide: back to single materiality for sustainability reporting. Sustain. Account. Manag. Policy. J. **13**(6), 1361–1385 (2022). https://doi.org/10.1108/sampj-07-2022-0378
22. Sustainalytics ESG Risk Ratings.: Retrieved 01/06/2024 (2024) from https://www.sustainalytics.com/esg-data
23. Http: https://www.regiaodeleiria.pt/seccao/revistas/100-maiores-empresas/100-maiores-empresas-2023/
24. Flick, U.: The SAGE Handbook of Qualitative Data Analysis, SAGE Publications Inc., Ed. SAGE Publications Inc. (2014)
25. Krippendorff, K.: Content Analysis: An Introduction to Its Methodology, 2nd ed. Sage Publications, Inc. (2004)
26. Bardin, L.: Análise de Conteúdo. Editora Persona (1977)
27. Claro, P., Esteves, N.: Sustainability-oriented strategy and sustainable development goals. J. Clean. Prod. **39**(4), 613–630 (2021). https://doi.org/10.1108/mip-08-2020-0365
28. Torelli, R., Balluchi, F., Furlotti, K.: The materiality assessment and stakeholder engagement: a content analysis of sustainability reports. Corp. Soc. Responsib. Environ. Manag. Responsib. Environ. Manag. **27**(2), 470–484 (2019). https://doi.org/10.1002/csr.1813

Assessing the Digital Competencies of Graduates for Sustainable Development: Bridging the Gap Between Academia and Industry

Fernanda Pereira, Marta Amaral, Luis Bruno, Isabel Brito, João Paulo Barros, André Silva, and Carlos Delgado

Abstract Organizations increasingly seek graduates with strong digital competencies to drive innovation and sustainability initiatives aligned with the United Nations' Sustainable Development Goals (SDGs) and Environmental, Social, and Governance (ESG) principles. The presented study aims to assess organizations' perceptions of new graduates' digital skills and compare them to the students' perceptions. The study results can be used to redesign academic programs to better contribute to the promotion of SDGs and ESG. A mixed-methods approach is employed,

F. Pereira (✉) · I. Brito · J. P. Barros
Polythecnic Institute of Beja, Beja, Portugal
e-mail: fernanda.pereira@ipbeja.pt

I. Brito
e-mail: isabel.sofia@ipbeja.pt

J. P. Barros
e-mail: joao.barros@ipbeja.pt

M. Amaral
CiTUR—Centre for Tourism Research, Development and Innovation, Polytechnic Institute of Beja, Beja, Portugal
e-mail: marta.amaral@ipbeja.pt

L. Bruno
Polytechnic Institute of Beja, Polytechnic Institute of Beja/LabSI2, Beja, Portugal
e-mail: lbruno@ipbeja.pt

I. Brito · J. P. Barros
Center of Technology and Systems (UNINOVA-CTS) and Associated Lab of Intelligent Systems (LASI), Caparica, Portugal

A. Silva
Astrolábio, Leça Do Balio, Portugal
e-mail: inovacao@astrolabio.com.pt

C. Delgado
PD Consulting, Beja, Portugal
e-mail: carlos.delgado@pdconsulting.pt

© The Author(s), under exclusive license to Springer Nature Switzerland AG 2025

A. d. b. Machado et al. (eds.), *Environmental, Social, Governance and Digital Transformation in Organizations*, Information Systems Engineering and Management 35, https://doi.org/10.1007/978-3-031-86079-9_14

combining quantitative surveys, to assess students' competencies, and focus groups with organization representatives regarding recent graduates' digital competencies from the Portuguese Dynamic Digital Competence Framework. This includes Information Literacy, Communication and Citizenship, Content Creation, Security and Privacy, and Solutions Development. The focus groups delve into experiences with onboarding and training new hires, identifying gaps between curricula and organizations' demands. The findings inform curricula revisions and pedagogical approaches to better equip graduates with the digital skills and sustainability mindset required by modern organizations. This research recommends academic-industry collaboration to design programs promoting global sustainability goals aligned with industry needs. By bridging academic offerings and industry needs, universities can develop a workforce prepared to leverage technology for positive societal and environmental impact.

Keywords SDG · ESG · Digital literacy · Digital transition · Students' competencies

1 Introduction

The InovEAP project "Innovation in Teaching and Learning 4 Post-Covid@IPBeja", developed in the Polytechnic Institute of Beja in Portugal, encompassed a set of initiatives and strategic investments for the development and promotion of new capabilities and skills in students, teachers, and institutional organisation with the general aim of overcoming the main gaps and challenges the COVID-19 pandemic situation caused in the Polytechnic Institute of Beja and higher education in general, as well as in the job market. The InovEAP project contributed to modernising pedagogical practices and better preparing the institution to meet current and future societal needs while addressing two main priorities: (i) Digital Transition and (ii) Equality and New Skills. Moreover, its primary goal is to promote innovation in learning, understood as the interconnection between the complex set of practices, methods, and configurations that are part of the institution's efforts to improve teaching and provide better learning for its students. In other words, innovation in education is as much about innovation in the classroom and the work between teachers and students as sustainable and lasting innovations at the institutional level [1].

One of the project's tasks is the survey and diagnosis of needs and opportunities for improvement linked to the digital transition in articulation with business, scientific, and technological entities. The aim is to determine the higher education institutions' current capabilities in the digitalisation of education and to know what digital skills are needed in the job market, namely in the production sector, public administration, and others. This diagnostic will allow higher education institutions to consider new teaching methods and practices to better prepare students as future professionals for the real needs of the job market (current and future).

To this end, the aim is to promote initiatives such as carrying out benchmarking actions, which include study visits and meetings, to identify and collect good practices and methodologies for both teaching and learning and work that takes advantage of digital capabilities, online and in person questionnaires, focus groups, brainstorming, among other participatory dynamics. All of these elements are internal and external to the higher education institution.

In addition to the internal approach to higher education institutions, entities from the public or private sector were consulted, namely companies and public organisations, as well as scientific and technological entities belonging to the Research and Innovation ecosystem. The latter were identified in the protocols presented in this application and other entities in which the higher education institution intends to establish new partnerships, relationships, or protocols.

This chapter is structured into 6 sections with the following content: The Introduction section describes the context of the chapter; (ii) Background introduces the main concepts relevant to the study; (iii) related work brings us the most current work in this area; (iv) we define the methodology to be used, which will consist of using surveys on students' proficiency levels and focus groups to assess companies' needs and satisfaction and discussion results.

2 Background

The "National Digital Skills Initiative e.2030, Portugal INCoDe.2030" is an integrated public policy initiative launched in 2017 to enhance digital skills in Portugal. Digital skills are crucial for fully exercising citizenship and play a key role in improving employability by meeting the demands of an increasingly digital labor market, including digital transformation. Moreover, "the digital economy will be one of the drivers of economic activity in the coming decades, It is essential to ensure that the population will increasingly have the necessary and adequate qualifications to succeed in this digital transition, which implies a strong digital commitment to education, inclusion, digital literacy, information literacy and the promotion of the development of the digital skills necessary for the full exercise of citizenship".[1] Equipping society with digital competencies is key to ensuring everyone can contribute to and benefit from a more inclusive, equitable, and sustainable world, i.e. accelerate progress towards Sustainable Development Goal (SDG) 4, 8, 9 and 10.[2] SDG4 to "ensure inclusive and equitable quality education and promote lifelong learning opportunities for all". SDG8 to "promote sustained, inclusive and sustainable economic growth, full and productive employment and decent work for all". SDG9 to "build resilient infrastructure, promote inclusive and sustainable industrialization and foster innovation". SDG10 to "reduce inequalities within and among countries by promoting social, economic, and political inclusion, and addressing

[1] https://www.incode2030.gov.pt/en/incode-2030-en/.

[2] https://sdgs.un.org/goals.

disparities in income and opportunities". Therefore, digital competencies must be integrated into academic curricula to prepare students for the demands of the twenty-first century to better prepare the population for an uncertain future, recognizing that existing inequalities necessitate diverse approaches to education and training.

The Portugal INCoDe.2030 initiative addresses the concept of digital competencies broadly. Digital competencies encompass the use of digital technologies to create solutions for various problems, integrating interdisciplinary knowledge, data analysis, artificial intelligence, and advanced instrumentation. They also involve communication networks, mobile systems, and the development of cyber-physical systems, extending beyond ICT to include electronics, automation, and robotics. These competencies can be developed to variable levels of proficiency based on qualifications and goals, with measures designed to promote them across all levels of society. In this context, the INCoDe.2030 program launches the Dynamic Framework of Reference for Digital Competencies (in Portuguese: "Quadro Dinâmico de Referência para as Competências Digitais"—QDRCD[3]), an instrument for assessing the society's digital competencies. Based on the European Digital Competence Framework for Citizens 2.1,[4] the QDRCD has three main objectives: support the definition of policies and strategies; design education programs; and verify and certify skills, either by self-diagnosis or by certified entities.

QDRCD has 4 dimensions: (1). Five areas/fields of competence; (2). Competences for each of the five areas/fields; (3). Four proficiency levels for each competence; and (4). Examples of use present manifestations of actions in a real context of competencies. Table 1 presents the competencies and their area/field.

Instead of eight proficiency levels of the European Digital Competencies Framework for Citizens, QDRCD has four levels, composed of (i) Task complexity, (ii) Autonomy, and (iii) Cognitive Domain (see Table 2).

3 Related Work

The increasing demand for graduates with proficient digital competencies is fueled by the need for organizations to drive innovation and meet sustainability goals. These competencies empower companies to benefit from digital transformation to advance the United Nations' Sustainable Development Goals (SDGs) and integrate environmental, social, and governance (ESG) principles into their operations.

Recent studies emphasize that organizations are increasingly seeking graduates with strong digital competencies to drive innovation and sustainability efforts aligned with the United Nations' Sustainable Development Goals (SDGs). Digital literacy and skills are widely recognized as critical for advancing the SDGs, particularly in areas such as quality education (SDG 4), decent work (SDG 8), and innovation (SDG

[3] https://www.incode2030.gov.pt/en/qdrcd-en/.

[4] https://op.europa.eu/en/publication-detail/-/publication/50c53c01-abeb-11ec-83e1-01aa75ed71a1/language-en.

Table 1 QDRCD competencies

Areas/fields of competence	Competences
Information literacy	Searching and filtering information Information assessment Information storage and retrieval
Communication and citizenship	Interaction through digital technologies Sharing information and content Citizenship through digital technologies Collaboration through digital technologies Code of conduct in a digital environment Digital identity management
Content creation	Digital content production Integration and re-elaboration Copyright and licenses
Security and privacy	Device protection Personal data protection Health protection Environment protection
Development of solutions	Technical troubleshooting Identification of technological needs and responses Innovation and creative use of technology Identification of gaps in digital competence

Source https://www.incode2030.gov.pt/en/qdrcd-en/

Table 2 QDRCD levels

Level	Task complexity	Autonomy	Cognitive domain
1	Simple tasks	With some autonomy or support	Remind/Understand
2	Tasks and routines associated with concrete problems	Autonomy	Apply
3	Different tasks and routines associated with concrete problems using different strategies	Autonomy	Apply
4	Solve complex problems	Autonomy	Create

Source https://www.incode2030.gov.pt/en/qdrcd-en/

9). Moreover, several studies underscore the importance of digital social innovations, where emerging technologies help address societal challenges, supporting key SDGs like reducing inequalities (SDG 10) and improving infrastructure (SDG 9) [2].

So, this convergence of digital transformation and social innovation is regarded as a crucial means of fostering inclusive growth, enhancing industry productivity, and promoting responsible consumption practices [2].

[3] conducted an in-depth literature search that demonstrated the importance of digital learning, digital literacy and digital ethics in developing sustainable organizations and competent students in the context of digital transformation in business and entrepreneurship. The study advances the existing literature on entrepreneurship education and digital learning, emphasizing that integrating entrepreneurship education with digital learning orientation can foster entrepreneurial skills among students. This organized knowledge can help educators and policymakers to better include these concerns in entrepreneurship education initiatives.

In a further study, [4] explored the relationship between sustainability and the teaching of digital competencies in the university setting. The primary aim was to understand how sustainability is integrated into higher education through teaching practices involving information and communication technologies (ICT), with a focus on developing digital competencies. The findings revealed key areas of sustainability linked to digital teaching competence, including inclusion, educational quality, and lifelong learning.

Moreover, Amballor and Naik [5] propose that in a global context characterized by cultural, demographic, and political diversity, accelerated progress toward sustainable development necessitates the enhanced utilization of digital technologies. Education for sustainability has therefore become a key mechanism for cultivating graduates who can understand, value, implement, and advocate for the achievement of the Sustainable Development Goals (SDGs). In this context, digital technologies play a vital role in enhancing the delivery of sustainability education. The authors conclude that integrating technology into the learning process to create sustainable development has numerous advantages. Firstly, it can facilitate a transdisciplinary approach to addressing societal challenges. Secondly, it can equip students and teachers to shift their attention from the 'what' of thinking to the 'how'. Furthermore, cooperation between stakeholders and the education system can be highly beneficial, particularly in the design of programmes to understand, appreciate and implement the integration of digital technologies and teaching in the classroom.

Another study [6] assumes that the concept of digital literacy is an indispensable element of sustainable education and social concerns, which, together with the environment, the economy, social justice and human rights, form the basis of the concept of sustainability. The authors posit that an enhancement in digital proficiency is perceived to have a favorable impact on the utilization of ICT and its correlation with sustainable development. The findings of this study, which compared four countries (Portugal, Spain, Brazil and Colombia) in terms of literacy, digital competencies and digital resources and experiences according to the PISA 2018 report, indicate that there is a need to support the promotion of digital competence, which is a crucial element in the sustainable, educational and social development of a community.

In their research, [7] highlight the ways in which skills are evolving in line with technological advancement and the emergence of new responsibilities associated with sustainable development. The advent of new social challenges has given rise to the necessity for a new generation of skills that can be deployed in a manner that will yield beneficial outcomes for societal well-being. The authors posit that there are well-known approaches to describing competences based on the assumption that they are applicable to any sector. However, they argue that this is not the case. Furthermore, the authors posit that it is crucial to consider the quantity and quality of competencies within the context of the industry, identifying specific capabilities for enhancing them with the aid of digital technologies. The findings demonstrate that industrial transformation can advance sustainability when it is founded upon a profound comprehension of technology and industry. The outcome of this research is the formulation of a conceptual framework for competencies in digital transformation across a range of industries. This represents a foundational step towards the advancement of existing conceptualizations of digital competencies for transforming industry in a manner that supports sustainability.

The study conducted by [8] illustrates the significance of literacy and the imperative of promoting sustainability in a digital society. However, this digitization of the educational process presents a number of challenges. It necessitates the availability of software and hardware, as well as digital literacy, which is the result of a multifaceted process of literacies. It also requires teachers and students to alter their perspectives and practices in order to acquire new teaching and learning skills, with the aim of combating the digital divide and fostering the widest possible social inclusion for the promotion of a sustainable society.

Considering the purpose of this study to redesign academic programs to better contribute to the promotion of SDGs and ESG, researches reinforce the need to invest and reskill their people to stay competitive in the fast advancement of technology, which is altering the nature of skills and abilities required in the workplace, which also needs to make a mental shift among the employees who can make that technology useful [9].

Higher education institutions are increasingly prioritizing the acquisition of digital competencies by their students. These skills are being integrated with entrepreneurship education in order to prepare graduates for the role of leading sustainability initiatives in the workforce. The integration of digital learning and blended education models into business and entrepreneurship curricula is becoming increasingly prominent, enabling graduates to demonstrate proficiency in the use of digital tools for sustainable innovation [3, 10].

For instance, Bagur-Femenias et al. [11] conducted research with the objective of identifying methods for adapting teaching and learning methodologies to align with the characteristics of the emerging generation of digital natives, with the aim of fostering their commitment to sustainable development. The results demonstrate a significant relationship between the skills of digital natives, individual dynamic capabilities and an enhanced approach to managing situations with regard to relations with key stakeholders and promoting innovation and commitment to the social and environmental issues demanded by society. Consequently, the findings suggest the incorporation of individual dynamic capabilities and innovation courses in education for sustainable development, with the objective of accelerating the implementation of the sustainable development goals. Furthermore, these findings can inform the development of university programmes and courses.

Moreover, Amballor and Naik [5] argue that in a global context characterized by diverse cultures, demographics, and political ideologies, accelerated progress towards sustainable development hinges on the enhanced utilization of digital technologies. It is evident that there is a discrepancy between those who are aware of the SDGs and those who possess proficiency in technology. There is a necessity for integration between these two distinct groups to facilitate the mapping of digital technologies with the appropriate SDGs. They propose that teachers in the higher education system require enhanced exposure to comprehend and implement this integration.

The training of citizens in digital competencies represents a strategic objective for countries' education systems, with a view to the realization of sustainable development goals. In light of the findings presented by Razinkina et al. [12], it can be concluded that digital transformations have given rise to a significant discourse surrounding the necessity to re-examine the content of higher education, particularly with regard to the training and development of students' digital competencies, both at the level of universal competencies and professional competencies. Moreover, the advancement of digital competencies is a fundamental requirement for the realization of the majority of the Sustainable Development Goals (SDGs), particularly SDG 4, which aims to "Ensure inclusive and equitable quality education and promote lifelong learning opportunities for all."

The main objective of [13] research is to develop a training course model designed to enhance the digital literacy of undergraduate students, with due consideration of the objectives of sustainable development and the implementation of a competency-based approach. The authors conclude that the digital literacy models proposed within the framework of international projects, research teams and authored developments demonstrate the complex nature of this phenomenon. The formation of digital competences is considered an essential task in the context of sustainable development, given the multidimensionality of the tasks involved. The research suggests a model for digital skills training based on three training modules: 'Ethics and digital security', 'Self-development in the digital environment', and 'Self-organisation and teamwork'.

[14] state that several factors promote the use of digital technologies in education: (1) Digital skills and literacy: Students and faculty must acquire multiple skills, such as creativity, problem-solving, and critical thinking, which are crucial for success and effective time and content management; (2) Digital technology and infrastructure investment: European countries have invested significant resources in integrating digital technologies into education [15]; (3) Digital technology transformation and evolution: Digital literacy is shaped by the availability of technology, with higher levels seen in "high-tech" countries; (4) Urban–rural learning contexts: Low-tech countries face challenges in adapting their education systems to global demands and digital-savvy generations. [16] add that integrating digital literacy into curricula empowers students and facilitates flexible learning.

4 Methodology

The research methodology employed in this study aimed to assess students' competencies and, simultaneously, to evaluate organizations' perceptions of new graduates' digital competences and compare them to the students' own perceptions with the intention of exploring experiences with onboarding, training new hires identifying gaps between curricula and organizations' demands.

A mixed-methods approach was used to achieve the proposed objectives, combining quantitative data collection with a survey applied to a sample of higher education students in a selected study case institution; and analysis with qualitative insights based on focus group methods with employers.

Student's survey was structured in six parts. Survey's first section aimed to gather data to gain insights into student's demographics, such like age, gender, level and year of studies. The other sections were oriented by five dimensions (competences) recommended in the Portuguese Dynamic Digital Competence Framework, i.e., Information Literacy, Communication and Citizenship, Content Creation, Security and Privacy, Solutions Development (see characterization in background chapter).

For each competence, and based on the options highlighted in each area of competence, a correspondence was made between the selected usage examples and the proficiency levels, measured by a scale of five points which: 0—Not qualified; 1—Basic; 2—Intermediate; 3—Advanced; 4—Highly specialized. After that, and to identify the level of global proficiency in the area of competence, the average proficiency level of the different inherent skills.

Empirical survey study was approved by the data protection officer and the ethics committee.

To capture organizations' perceptions of new graduates' digital competences, focus group was conducted. The focus group method represents a source of data collection that intensifies access to information formations about a phenomenon, either due to the possibility of generating new conceptions or analysis and problematization of an idea in depth [17]. Focus groups are therefore particularly useful when the researcher wants to investigate people's thoughts because the interaction between participants can elicit data and ideas that might not be uncovered in one-on-one questioning. To optimize the data obtained from focus groups, careful attention must be paid to the composition and number of groups, as well as the selection and training of the moderator, and the development of the question route [18].

Attending to theorical justification for the use of focus group method, several focus group sessions were organized with the aim of reaching the indicator achievement of 90 entities (as was foreseen in the project that supports the research) and these sessions were ordered in accordance with the affinity between scientific areas.

Once again and based on the competencies identified in the study on digital technologies for employability (non-ICT) in the future—Training Paths and Pilot Project, prepared by CPED within the scope of INCoDe. 2030 (September 2022), a guide interview was created which served as the basis for discussion in each of the 5 groups of competencies (see background chapter). The focus group meetings were guided based on the domains and competencies mentioned above and were always prepared and carried out by the same group of people in order to have coherence in the conclusions to be obtained.

After the meetings were held, synopsis was prepared and subsequently a summary table that allowed us to more easily visualize the results obtained for each area of knowledge, thus enabling easier diagnosis.

4.1 Survey Data Collection

The quantitative study was based on data from the survey conducted in a higher education institution selected as study case, with 342 students enrolled in academic year 2022–2023.

In a registered total population of 3106 selected students (including all students from several years of short cycles; bachelors; post-graduations and master's studies), a stratified probability sample was established, and it was calculated with a level of 95% confidence and a margin of error of 5%. As a result, a total sample of 342 students was calculated.

The data were collected during February 2023 using the Google Forms platform in the classrooms. Contacts were previously made with all course coordinators with the purpose of planning in each faculty, which professors would control survey's application by type, level and years of the courses in a restricted time. Student's adherence to the process was clearly a limitation and it was necessary to repeat the process more than once.

Of the 50 courses that structure the training offer of the Higher Education Institution studied, we managed to reach the target sample in 32; in 8 courses we managed to obtain data, but it was not possible to reach the target sample; and in 10 courses it was not possible to obtain any data. After cleaning the database, there were 295 valid responses resulting in 86.3% response rate.

In terms of demographic characterization of students 'sample, 50% of the ones are between 17–21 years followed by 28% between 22–25; 56.3% are female and 43.7% male. Regarding the distribution of students' answers by study cycle, 16.7% answers were from short cycles; 76.3% were from bachelor's degrees; 0.7% were from post-graduate degrees; and 6.4% were from master's degrees.

4.2 Focus Group Data Collection

Invitations for each focus group (FG) were sent in advance and on all occasions, the presence of the organization/employer was confirmed by telephone. Given some negative responses, for various reasons, contact was made with other more companies, in order to obtain a representative and reliable sample. The predicted indicator of the project involved contacting 90 companies.

Companies were grouped into 6 anticipated categories considering economic areas: (1) Food Science and Technology; Agronomy; (2) Attorneyship; Social Work; (3) Business Management; (4) Computer Science; Environment; (5) Education; (6) Tourism, Heath and Sport.

To ensure representativeness, the team contacted a total of 146 companies/entities, an average of 24 per group. The distribution of the number of invitations among the 6 focus groups was as follows: (FG1) Food Science and Technology; Agronomy = 24 invitations; (FG2) Attorneyship; Social Work = 26 invitations; (FG3) Business Management = 23 invitations; (FG4) Computer Science; Environment = 24 invitations; (FG5) Education = 26 invitations; (FG6) Tourism, Heath and Sport = 23 invitations.

All the process was planned and conducted between 15th February until 28th April 2023. Sessions were held online to better reconcile the representatives' schedules and the difficulties inherent to geographic distance. This was still a limitation, and the team decided to extend the invitation to a greater number of companies in the area of influence of the higher education institution selected study case.

As a result, we had the presence of 34 companies. Sample data characteristics included: (FG1)—6 companies; (FG2)—7 companies; (FG3)—9 companies; (FG4)—4 companies; (FG5)—2 companies; and (FG6)—6 companies.

The focus groups ranged from agricultural companies, elderly centers, associations (e.g. accounting, agriculture; wine; business), to computer science companies; tourism accommodations; municipalities organizations; daycare centers, all from the region.

5 Results and Discussion

This section presents the quantitative results obtained from the survey applied to students and the insights gathered from the focus group with representatives of the organizations/entities.

5.1 Survey

The data gathered from the students' responses to the representative sample were processed according to different variables and dimensions to get the maximum expressiveness of these results and a complete understanding of the competencies. First, an overview of the students' global competencies is presented and distributed by each one of the five areas of competency. This allows us to understand the areas of competencies where the students present less and more proficiency. Another important insight generated was the students' global competencies to the five areas of competencies and by each study cycle level. This made it possible to analyze the effect of the study cycle on the student's proficiency. Another variable analyzed was

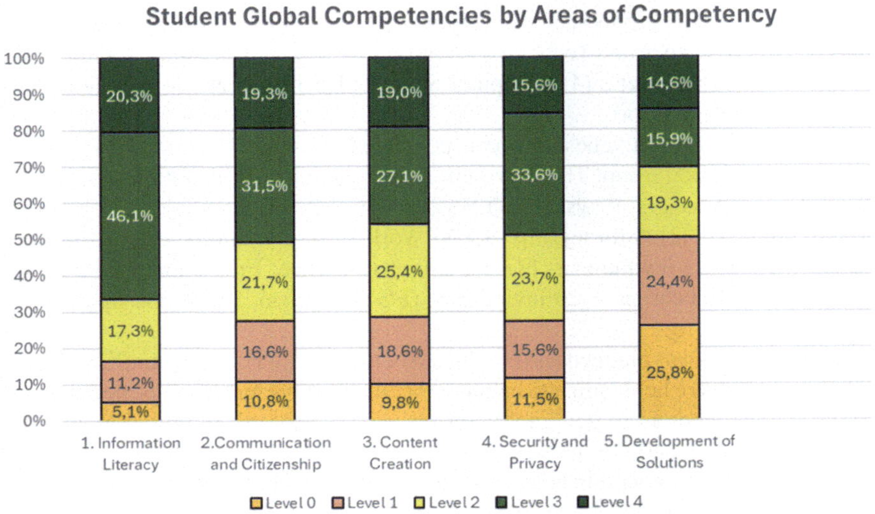

Fig. 1 Student global competencies by areas of competency. **Source* own elaboration

the student's curricular year to understand the effect on their competencies while attending the program. In this case, the low and advanced student competencies groups were analyzed by curricular year. Finally, the effect of gender on digital competencies was analyzed in this study.

Student Global Competencies by Area of Competencies

Overall competencies were obtained by averaging the different levels of digital proficiency for each competency included in the various areas of competence. Figure 1 exhibits the results for each competency field aggregated by the proportions of the 5 proficiency levels: 0 (Not qualified), 1 (Basic), 2 (Intermediate), 3 (Advanced), and 4 (Highly specialized).

This first area of competence (Information Literacy) encompasses the skills of "Searching for and filtering information", "Evaluating information" and "Storing and retrieving information". These tasks are related to obtaining, processing, and storing information through digital tools. From the results, 66.4% (196) of the students surveyed obtained a proficiency rating of advanced (46.1% in level 3) or highly specialized (20.3% in level 4), this being the area with the highest overall proficiency level. Similarly, this is the area of competence with the lowest proportion of non-proficient students (5.1%) and with basic skills (11.2%). Among the different components of this area of competence, the competence "Searching for and filtering information" stands out as the one in which students feel the least capable, with 20% of students not being capable or having basic competencies.

The second area of competence (Communication and Citizenship) encompasses the "Interaction through digital technologies", "Sharing information and content",

"Citizenship through digital technologies", "Collaboration through digital technologies" and "Code of conduct in a digital environment" competencies. It is related to interaction, collaboration, and conduct in digital media. 50.8% of the students have advanced (31.5% in level 3) or highly specialized (19.3% in level 4) skills in this area. Of the 6 different competencies that make up this area of competence, the "Code of conduct in a digital environment" and "Citizenship through digital technologies" competencies stand out as the ones in which there are the fewest advanced competencies, with 40% and 34.2% of students, respectively, having no or only basic competences.

Regarding the third area of competence (Content Creation), this is the second area with the highest proportion of untrained students (28.4%) encompassing the non-proficient students (9.81% in level 0) and with basic skills (18.6% in level 1). Among the components that make up this area of competence, the following stand out competency "Integration and reworking" as the most critical, with 23.1% of students not having the necessary skills. On the other hand, the competency "Copyright and licenses" stands out as the competency in which students are most proficient, with 62% of students having advanced or highly specialized proficiency.

In the fourth area of competence (Security and Privacy), 11.5% of students have no competence at all, and 15.6% have basic competencies. On the other hand, 49.2% have advanced (33.6% in level 3) or highly specialized skills (15.6% in level 4). It should be noted that the "Protection of personal data" and "Protection of the environment" competencies are the most critical, as 26.1% and 28.5% of students, respectively, are non-proficient or have basic skills.

In the fifth and final area of competence (Developing Solutions), it should be noted that this is the area where there is the lowest level of competence, with 25.8% of students having no competence in this area and 50.2% having low competence (not trained or basic competence). In terms of advanced skills, only 30.5% of students have advanced or highly specialized skills.

Summarizing is possible to verify that the area of competence "Information Literacy" presents the lowest proportion of students with low competencies and the highest proportion with advanced competencies. This suggests that in general students have suited skills to develop proficient tasks of "Searching for and filtering information", "Evaluating information" and "Storing and retrieving information. On the other hand, students present the highest proportion of low competencies and the lowest proportion of advanced competencies in the "Developing Solutions" area of competencies. This area requires specialized skills to resolve complex problems using and developing digital technologies and media or research outputs. The results suggest the academy should improve and stimulate the more advanced digital skills in the "Developing Solutions" area to empower students to solve advanced problems for companies and society.

Student Competencies by Areas of Competency and Study Cycle Degree

In terms of overall proficiency, considering the study cycle degree attended by students the results are exhibited in Fig. 2. The aim was to understand the effect of the cycle degree attended by students on their digital competencies. It is supposed

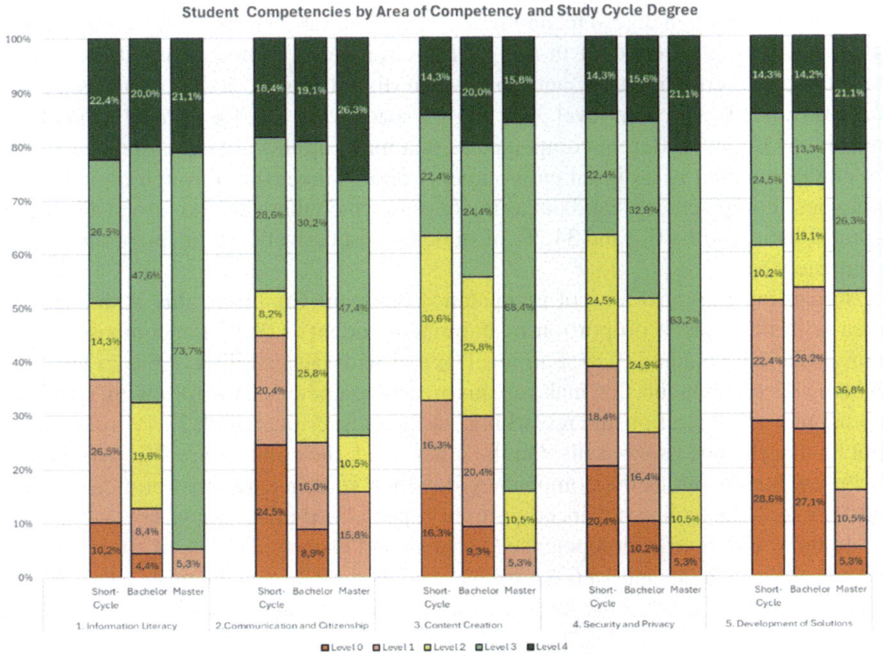

Fig. 2 Student competencies by area of competency and study cycle degree. **Source* own elaboration

that students should present progress in digital competencies through the students' academic careers for different cycle degrees because they will be exposed gradually to more complex digital tasks and more specialized digital tools. The chart of Fig. 2 groups each of the five areas of competencies by three study cycle degrees (Short-Cycle, Bachelor, and Master). Due to the short number of programs and the student sample from the postgraduate degree, the results are not presented in this paper.

In all areas of competence, except for the "Development of Solutions" area, students attending Short-Cycle's degree have the lowest levels of digital competence. Effectively, it presents the highest concentration of students with low competencies (no or basic competencies). On the other hand, the Master's degree presents the highest proportion of students for all five areas of competencies with the more advanced competencies (advanced and highly specialized).

In "Information Literacy", 10.2% of Short-Cycle degree students have no competencies, and 26.5% have basic competencies totaling 36.7%. This last value is about 3 times greater than the low competencies from Bachelor students (12.8%) and 7 times greater than from Master students (5.3%). In this area, the Master students have the more advanced competencies with 94.8 (73.7% in level 3 and 21.1% in level 4) against 67.6% of Bachelor students (47.6% in level 3 and 20.0% in level 4).

The results suggest that the students improve their digital skills in this area through the order of degrees they attend.

In the "Communication and Citizenship" area the Short-Cycle students present the highest low competencies (44.9% against 24.9% and 15.9% for Bachelor and Master, respectively) from the three degrees. Master students have the highest more advanced (advanced and highly specialized) competencies (73.7%) than Bachelor (49.3%) and Short-Cycle (47.0%).

In "Content Creation", 32.6% of the students have the lowest competencies (16.3% without competencies and 16.3% with basic competencies). This value is not very different from Bachelor's (29.7%) but it is higher than Master's students (5.3%). Related to more advanced competencies, Master's students continue with the highest (84.2%) competencies with a great difference to Bachelor (44.4%) and Short-Cycle (36.7%).

Regarding "Security and Privacy" area the trend maintains presenting Short-Cycle's students with the highest low competencies (38.8% against 26.6% and 5.3% for Bachelor and Master, respectively). Relatively to the more advanced competencies, Master presents the highest value (84.3%) with great difference to the Bachelor (48.5%) and Short-Cycle (36.7%).

In the area "Solution Development", the undergraduate programs present the greatest low (no or basic competence) competencies recorded, with 51.0% for Short-Cycles and 53.3% for Bachelor students. The Master students present 15.8% low competencies which is about 3 times lower than undergraduate programs. The results show Master students present the more advanced competencies (47.4%), against the Bachelor (275.5%) and Short-Cycle (38.7%). In this case, not following the trend in the previous areas, the Short-Cycle presents better results than the Bachelor students. This can be caused by the greater weight in the student sample of information technology programs in Short-Cycles than in Bachelors.

To conclude, the master's degree students naturally present the highest digital competencies in all five areas of competencies. This shows students can improve their digital competencies when exposed to more learning time through a sequence of various study cycle degrees attended. This can be reinforced when comparing the bachelor and short-cycle students. These last students have poorer competencies than bachelor students, except for "Solution Development" area. This also naturally presents in general the lowest (more lower competencies and less advanced competencies) for the three study cycle degrees. Despite these results, it is important to develop a fine analysis at the level of study programs to understand which from them have the most and least digital skills.

Student Competencies by Curricular Year and Degree

This study also analyzed the effect of the student curricular year for the five areas of competency and three-cycle degrees on their proficiency level. Based on the collected data, it is important to understand the evolution of the following realities in the different cycles of study degrees: (1) proportion of students with no skills per year of attendance; (2) proportion of students with low skills (no skills and basic proficiency) per year of attendance; (3) proportion of students with advanced skills (advanced and

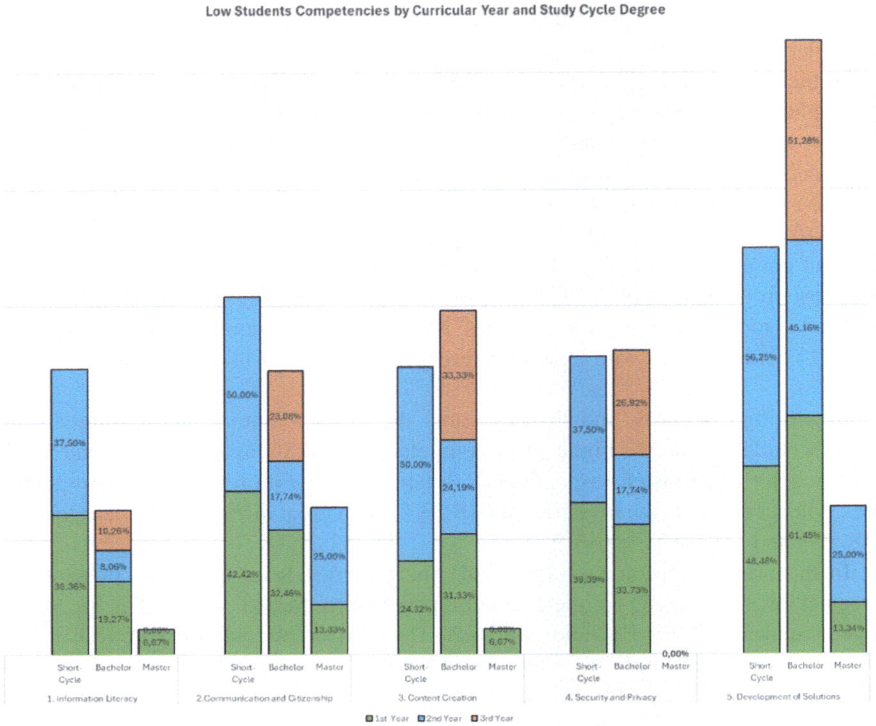

Fig. 3 Low students' competencies by curricular year and cycle level. **Source* own elaboration

highly specialized proficiency) per year of attendance. Figure 3 shows the results of low competencies per curricular year by degree and Fig. 4 exhibits the advanced competencies for the same variables.

Concerning the low competencies (no or low competence) by curricular year, the short-cycle degree results don't present in general an expressive difference between the first and second curricular years for five areas except for "Content Creation". For "Information Literacy" (36.6% vs. 37.50%), "Communication and Citizenship" (42.00% vs. 50.00%), "Content Creation" (24.32% vs. 50%), and Development of Solution" (48.48% vs. 56.25%) areas, strangely students for this degree have worse results in the second year than in the first year.

Regarding the bachelor's degree, the results show the students' low competencies improve (less proportion of students) for all five areas from the first to second-year in "Information Literacy" (19.27% vs. 8.06%), "Communication and Citizenship" (32.46% vs. 17.74%), "Content Creation" (31.33% vs. 24.19%), "Security and Privacy" (33.73% vs. 17.74%) and "Development of Solution" (61.45% vs. 45.16%) areas. However, results from the second to third year, have the opposite direction. The results get worse for all five areas: "Information Literacy" (8.06% vs. 19.27%), "Communication and Citizenship" (17.74% vs. 23.08%), "Content

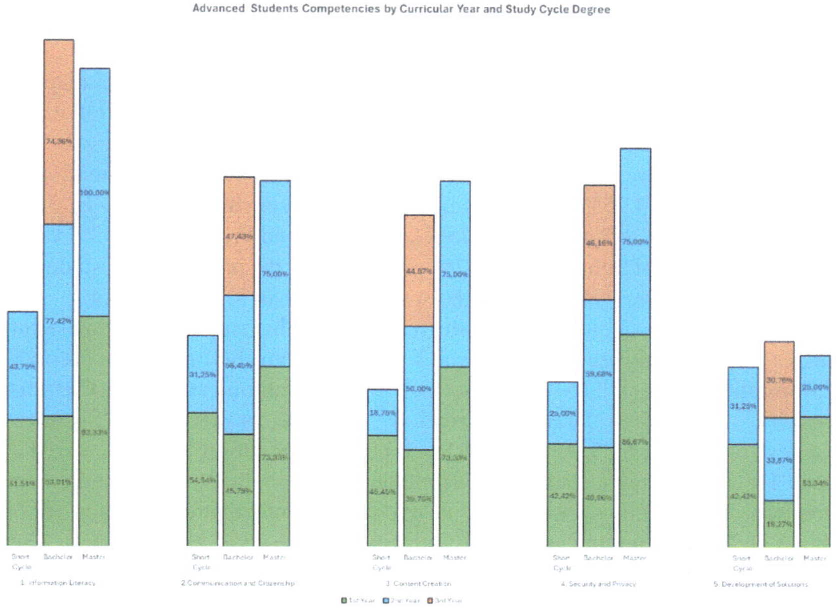

Fig. 4 Advanced student competencies by curricular year and cycle level. **Source* own elaboration

Creation" (24.19% vs. 33.33%), "Security and Privacy" (17.74% vs. 26.92%) and "Development of Solution" (45.16% to 51.28%).

The master's students have the best results for low competencies in all areas for the first and second year compared to short-cycle and bachelor students with great distance. But the results from the first year to the second year aren't consistent for all the areas. In the "Communication and Citizenship" (13.33% vs. 25%) and "Development of Solution" (13.34% to 25.00%) students get worse results from the first to second year. In the other three areas, the low competencies decrease (i.e. improve competencies) between the first and second year for ("Information Literacy" (6.67% vs. 0.00%), "Content Creation" (6.67% vs. 0.00%) and "Security and Privacy" (0.00% vs. 0.00%).

In conclusion, the proportion of short-cycle students with lower digital skills is not very different from the first to second year. It suggests the program curriculums didn't improve students' learning and skills in some areas of digital competencies. For the bachelor students what is more evident is the worse results from the second to third year although both are better than for the first year. The difference between these two second and third years is perhaps due to students being faced with more complex problems related to digital tasks in the third year they have a better awareness of their gaps and digital proficiency capabilities. The master's proportions of students with low competencies between the two years are predictable. A lower proportion of students or even none for the second year compared to the first year. This suggests these students enter these programs already with the skills from the undergraduate programs and then improve them in the 1st year of the master's degree.

Concerning the advanced competencies (advanced and highly specialized proficiency) identified by curricular year, Fig. 4 exhibits the results. After it will be analyzed and discussed these results.

For the short-cycle degree, in all areas of competencies ("Information Literacy" (51.51% vs. 43.75%), "Communication and Citizenship" (54.54% vs. 31.25%), "Content Creation" (45.45% vs. 18.75%), "Security and Privacy" (42.42% vs. 25%), and Development of Solution" (42.42% vs. 31.25%)), the proportion of students with advanced competencies is lower in the second year than in the first year.

Regarding the bachelor's degree, the results show that students' advanced competencies improve (higher results) for all five areas from the first to the second-year "Information Literacy" (51.51% vs. 43.75%), "Communication and Citizenship" (54.54% vs. 31.25%), "Content Creation" (45.45% vs. 18.75%), "Security and Privacy" (42.42% vs. 25.00%) and "Development of Solution" (42.42% vs. 31.25%) areas. However, results, even if little from the second to third year, have the opposite direction. The results get worse for all five areas: "Information Literacy" (77.42% vs. 74.36%), "Communication and Citizenship" (56.45% vs. 47.43%), "Content Creation" (50.00% vs. 44.87%), "Security and Privacy" (59.68% vs. 46.16%) and "Development of Solution" (33.87% to 30.76%).

The master's students have the best results for advanced competencies in all areas for the first and second years compared to short-cycle and bachelor students with great distance. The results from the first year to the second year aren't consistent for all areas. In the "Security and Privacy" (88.67% vs. 75.00%) and "Development of Solution" (53.34% to 25.00%) students get worse results from the first to second year. In the other three areas, the advanced competencies are better between the first and second-year students for ("Information Literacy" (93.33% vs. 100.00%), "Communication and Citizenship" (73.33% vs. 75.00%), and "Content Creation" (73.33% vs. 75.00%).

In summary for the short-cycle degree, the proportion of students with advanced digital skills is lower for the second than for the first year. There could be two reasons for this: (i) the second-year students have a better awareness of their gaps and digital proficiency capabilities because they were exposed to more challenges with digital tools from the first year or (ii) the students in the second year didn't develop their digital competencies in the previous academic year due to the pandemic Covid-19 where they didn't have many presential classes. For the bachelor students what is more evident is the worse advanced skills results from the second to third year although all are better than for the first year. These differences may have the same explanation described for the similar case of first to second year in short-cycle degree. For the master's cases, the proportions of students with advanced competencies between the two years are predictable and better than the similar years of the other two types of degrees. However, the results are not consistent for all areas of competencies from the first year to the second year. Special attention should be given to student learning in the areas of "Security and Privacy" and "Development of Solution" that deal with problems of greater digital complexity.

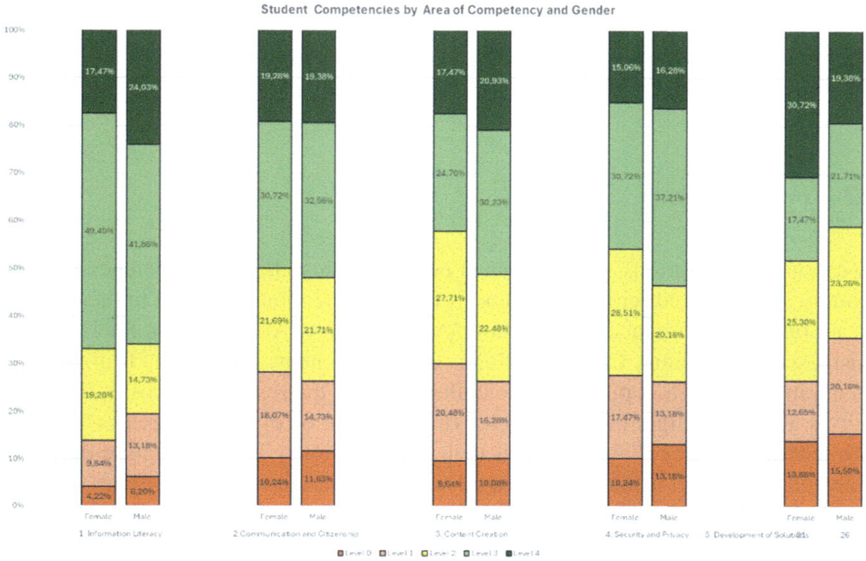

Fig. 5 Student competencies by area of competency and gender. **Source* own elaboration

Student Competencies by Gender

Figure 5 shows the results obtained concerning the level of digital skills taking gender into account. The effect of the gender variable was studied to understand whether our students' men and women have similar digital competencies that allow equal job opportunities for both and assure inclusion and a sustainable society.

Regarding the proficiency level "not qualified" (Level 0), female students present a shorter proportion than males for all areas of competencies with small differences: ("Information Literacy" (4.22% vs. 6.20%), "Communication and Citizenship" (10.24% vs. 11.63%), "Content Creation" (9.64% vs. 10.08%), "Security and Privacy" (10.24% vs. 13.18%), and Development of Solution" (13.86% vs. 15.50%)). Also, for students with lower competencies (not qualified and with a basic level of proficiency (Level 1)), in all areas of competence, except for the "Security and Privacy" area, the proportion of female students is lower than the proportion of female students.

However, considering the students who have higher competencies (advanced (Level 3) or highly specialized (Level 4) proficiency), female students present better results in "Information Literacy" (66.87% vs. 65.89%) and Development of Solution" (48.19% vs. 41.09%). On the other hand, male students present higher results (not expressively) in the areas "Communication and Citizenship" (50.00% vs. 51.94%), "Content Creation" (42.17% vs. 51.16%), and "Security and Privacy" (45.78% vs. 53.49%)).

The results show that these male and female students have similar digital competencies with small differences between them for the group of reduced skills level of proficiency and to the higher competencies students' group. This makes us believe that learning offered in the study programs in digital skills is reaching both genders equally contributing to a more inclusive, digitally skilled, and sustainable society.

5.2 *Focus Group*

To capture organizations' perceptions of new graduates' digital competencies, focus group was conducted with the presence of 34 organizations grouped in these activities' areas: Food Science and Technology, Agronomy, Attorneyship, Social Work, Business Management, Computer Science, Environment, Education, Tourism, and Heath and Sport. The opinions were collected from the representatives of the institutions about the digital competencies of the actual or old employees or students in internships who collaborate with their organizations. This qualitative data was analyzed according to the defined methodology that supports one report with the most relevant comments and suggestions for each activity area and area of competence. Also, it was asked for representatives of organizations to identify the relevance of the five different areas of competencies for each study-cycle program based on digital skills needs from their organizations to provide higher quality services.

To understand the feedback provided by representatives of the organizations the results of qualitative data were distributed in three groups of competencies ("SQ—Suitably Qualified"; TI—To Improve; and TA—To Acquire") for each activity's areas and each area of competency as exhibited in Table 3. When analyzing this data, it was possible to see that our professionals/students have the most appropriate competencies in 57.3% of their cases, must improve the actual competencies in 24.6%, and need to acquire effective competencies in 18%.

These results were discussed in detail related to the strong points and gaps identified in the competencies of the professionals/students for each study cycle. For each competence, the representatives exposed the main elements considered relevant to

Table 3 Summary focus group results

	Agronomy	Food Science	Attorneyship	Social Work	Business	Tourism	Computer Science	Health	Education	Sport	Environment
Area 1 - Information literacy											
1. Searching and filtering information	SQ	SQ	SQ	TI	SQ	SQ	SQ	SQ	SQ	TI	TI
2. Evaluation of information	SQ	SQ	SQ	TI	SQ	SQ	SQ	SQ	SQ	TI	TI
3. Information storage and retrieval	SQ	SQ	SQ	SQ	SQ	SQ	SQ	SQ	SQ	SQ	SQ
Area 2 - . Communication and Citizenship											
1 Interaction through digital technologies	SQ	SQ	SQ	SQ	SQ	SQ	SQ	SQ	SQ	SQ	SQ
2. Sharing information and content	SQ	SQ	SQ	SQ	SQ	SQ	SQ	SQ	SQ	SQ	SQ
3.Citizenship through digital technologies	SQ	SQ	SQ	SQ	SQ	SQ	SQ	SQ	SQ	SQ	SQ
4. Collaboration through digital technologies	TI	TI	TI	TI	TI	TI	SQ	TI	TI	TI	TI
5. Code of conduct in a digital environment	TO	TO	TO	TO	TO	TO	SQ	TO	TO	TO	TO
6. Digital identity management	TO	TO	TO	TO	TO	TO	SQ	TO	TO	TO	TO
Area 3 - Content creation											
1. Digital content production	TI	TI	TO	TO	TI	TI	SQ	TI	TI	TI	TI
2.Integration and re-elaboration	TI	TI	TI	TI	TI	TI	TI	TI	TI	TI	TI
3. Copyright and licenses	TI	TI	TI	TI	TI	TI	TI	TI	TI	TI	TI
4. Security and Privacy											
1. Device protection	SQ	SQ	SQ	SQ	SQ	SQ	SQ	SQ	SQ	SQ	SQ
2. Protection of personal data	TI	TI	TI	TI	TI	TI	TI	TI	TI	TI	TI
3. Health protection	TI	TI	TI	TI	TI	TI	TI	TI	TI	TI	TI
4. Environmental protection	TI	TI	TI	TI	TI	TI	TI	TI	TI	TI	TI
Area 5 - Development of solutions											
1. Technical troubleshooting	TO	TO	TO	TO	TO	TO	TI	TO	TO	TO	TO
2. Identification of technological needs and responses	TO	TO	TO	TO	TO	TO	TI	TO	TO	TO	TO
3. Innovation and creative use of technology	TO	TO	TO	TO	TO	TO	TI	TO	TO	TO	TO
4. Identification of gaps in digital competence	SQ	SQ	SQ	SQ	SQ	SQ	SQ	SQ	SQ	SQ	SQ

improving digital processes in organizations, for example, digital tools, procedures, attitudes, tasks, law, and other significant.

To measure the relevance of the digital competencies to the market each of the five different areas of competencies for the respective study-cycle program of the three degrees was rated on a scale of 0 to 5 (where 0 is the worst relevance and 5 is the best relevance).

The study cycle programs object of this study were in a posterior phase divided into 10 groups ("Engineering, manufacturing and construction", "Information and Communication Technologies", "Agriculture, forestry, fisheries and veterinary", "Business, administration and law", "Education", "Natural sciences", "Mathematics and statistics", "Arts and humanities", "Social sciences", "Journalism and information and Services" and "Health and welfare") according to the ISCEF classification that allows to better measuring the academic performance involves among other factors the intellectual level, motivation, and skills [19].

To assist teachers and those responsible for decision-making, a structured model was developed into quadrants to position the 10 ISCEF groups, considering the average proficiency identified in each area of competence among students from different courses of these groups, as well as the relevance of each competency area identified from employers, as shown in Fig. 6. Each quadrant classifies the ISCEF group for two variables: diagnosed competencies and search of competencies by market. The quadrant "4." is the best case (existence of students' competencies and search of that by the market). On the other hand, the quadrant "1." is the worst case.

Based on this structured model of competencies, one study was developed for each of the five areas of competencies. Figure 7 shows the results of this model in the "Information Literacy" area. One can see that the organizations have high expectations in this area of digital competencies (7 of the 10 areas are classified with the highest value (level 5)). The diagnosed competencies are between 2.49 and 3.20

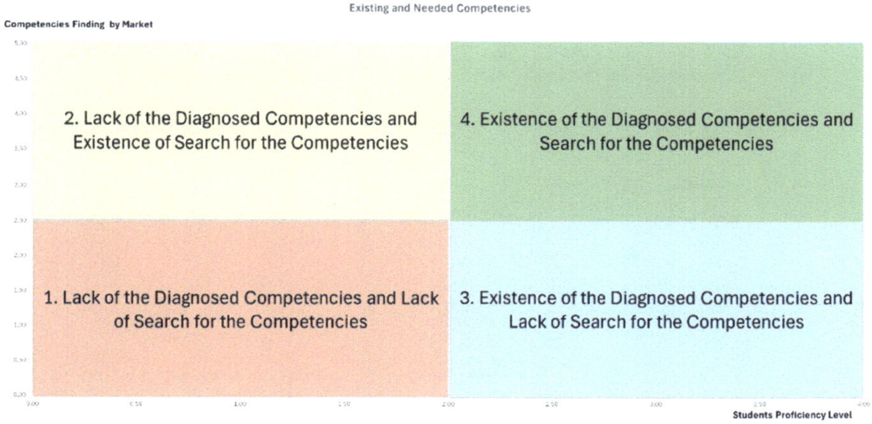

Fig. 6 Existence of competencies versus diagnosed competencies. **Source* own elaboration

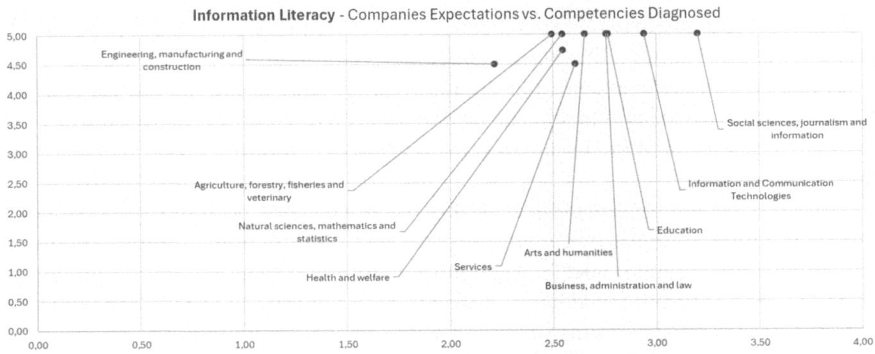

Fig. 7 Information literacy—company expectations versus competences diagnosed. **Source* own elaboration

which means that all the results are in the fourth quadrant and are positive in this area of competence for all the ISCEF groups of study-cycles.

Figure 8 exhibits the results of this structured model in the "Communication and Citizenship" area. One can see that the organizations just have high expectations in the area "Information and Communication Technologies" and most areas (i.i six) are classified a little bit above the middle of the scale (level 3.0). This suggests the employees in general don't consider this area very relevant to the market for these ISCEF groups. The diagnosed competencies are between 2.09 and 2.93, which means even without high results are in the fourth quadrant and are positive in this area of competence for all the ISCEF groups of study-cycles programs.

Figure 9 exhibits the results of this structured model in the "Content Creation" area. One can see that the organizations have the same expectations for all the ISCEF groups that were rated with the same value (level 3). middle of the scale (level 3.0). This suggests the employees in general don't consider this area very relevant to

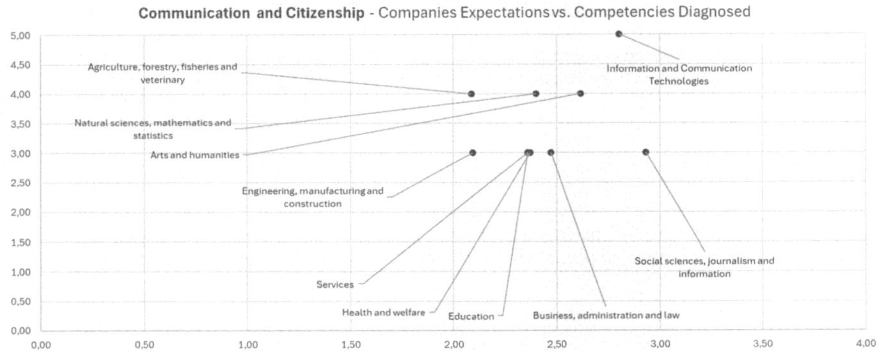

Fig. 8 Communication and citizenship—companies expectations versus competences diagnosed. **Source* own elaboration

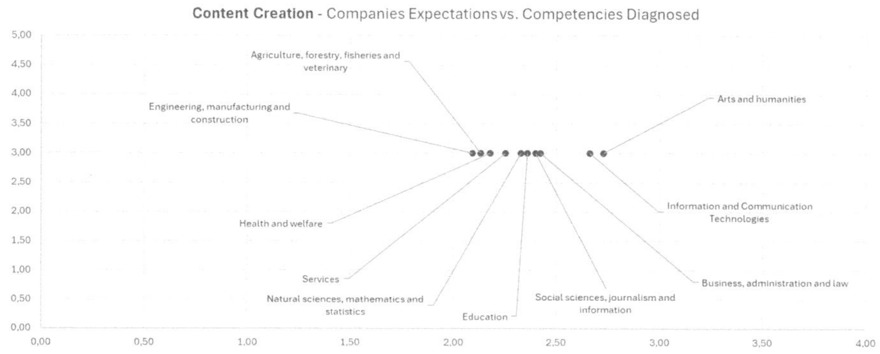

Fig. 9 Content creation—companies expectations versus competences diagnosed. **Source* own elaboration

the market for these ISCEF groups. The diagnosed competencies for this area are between 2.09 and 2.66, which means results are in the fourth quadrant and are positive in this area of competence for all the ISCEF groups of study-cycles programs.

Figure 10 exhibits the results of this structured model in the "Security and Privacy" area. One can see that the organizations have the same expectations for all the ISCEF groups, except for the "Engineering" area. The most rated value was level 4.0. This means organizations consider the digital competencies of this area than the "Content Creation" area. The diagnosed competencies for this area are between 2.15 and 2.93, which means results are in the fourth quadrant and are positive in this area of competence for all the ISCEF groups of study cycles programs.

Figure 11 exhibits the results of this structured model in the "Development of Solutions" area. Eight of the ten ISCEF groups were rated to 3.0 level, "Arts" to 3.5 level, and "ICT" to 4.0 level. These seem to be low values for the added value that this area of competence adds to the organization. The diagnosed competencies for

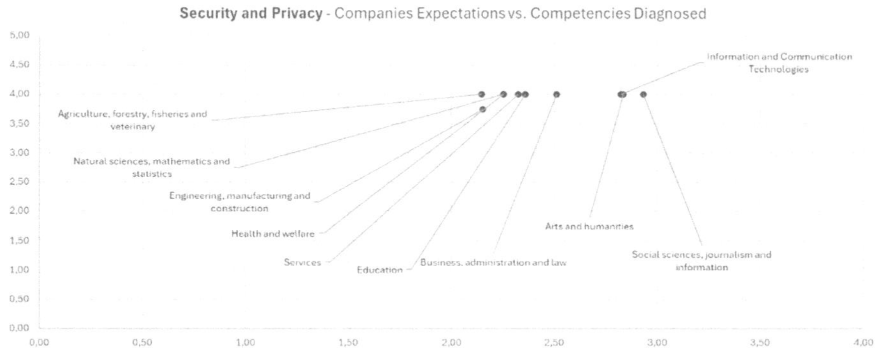

Fig. 10 Security and privacy—companies expectations versus competencies diagnosed. **Source* own elaboration

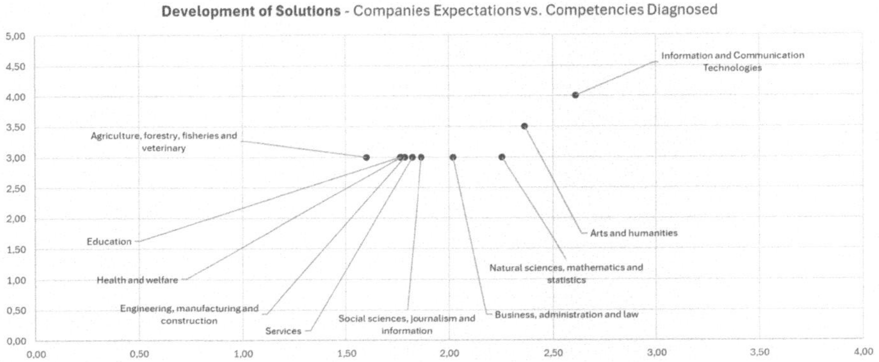

Fig. 11 Development of solutions—companies expectations versus competencies diagnosed. **Source* own elaboration

this area are between 1.6 and 2.61. This is the most critical area of all with the lowest values. Six of the ten groups are rated below the threshold level (i.e. basic level (2.0)), which means results are in the second quadrant (lack of the diagnosed competencies and existence of search for the competencies) in this area of competence for all the ISCEF groups of study-cycles. Naturally the students of "ICT" presented the best result for this area.

In short, the organizations consider the more relevant digital competencies for the five areas of competencies in this order: "1st—Information Literacy", "2nd—Security and Privacy", "3rd—Communication and Citizenship", "4th—Development of Solutions" and "5th—Content Creation". Related to the diagnosed competencies, for all five areas except for "Development of solutions" the students presented positive values (above level 2) which means they have at least basic digital competencies.

6 Conclusions and Future Work

The study carries out a global analysis of the digital skills of students at the higher education institution as well as the adjustment between students' digital skills considering the digital skills needs of employers in the region.

It is noted that the area of competence of Information Literacy is the most consolidated, denoting an average proficiency level of 2.7, that is, advanced, denoting that 66.4% of students have, at least, advanced level skills. It should also be noted that only 5% of students are not qualified in this skill.

Conversely, the Solutions Development area of competence is the most critical and, on average, the proficiency level of students is 1.7, that is, intermediate. It should be noted that 25.7% of students do not have any skills and 24.4% of students have basic level skills.

The results show that these male and female students have similar digital competencies with small differences between them for the group of reduced skills level of proficiency and to the higher competencies students' group. This makes us believe that learning offered in the study programs in digital skills is reaching both genders equally contributing to a more inclusive, digitally skilled, and sustainable society.

Regarding the adjustment between students' skills and the needs of current and future employers, there are many courses/areas in which there is a mismatch between existing skills and the needs of employers. It is important to help students develop digital skills, as well as adapting curriculum plans to help develop these skills.

Considering the above, it will be necessary to restructure the curricular plans so that, in every year, of the different study cycles, there is a curricular unit for the development of digital skills, ensuring that students are obliged to choose specific training actions adjusted their skills development needs. In this way, there is an efficient allocation of resources as students from different years and courses can attend the same training activities.

In terms of research, it would be advisable to carry out this type of assessment regularly to assess the evolution and adjustment of students' digital skills.

Funding Information This work was co-financed by POCH Programa Operational Capital Humano, Portugal 2020, União Europeia Fundo Social Europeu and by the Portuguese Agency FCT—Fundação para a Ciência e Tecnologia, in the framework of project CTS/00066.

References

1. Kim, J., Maloney, E.: Innovation learning and the future of higher education. Johns Hopkins University Press (2020)
2. Dionisio, M., de Souza Junior, S.J., Paula, F., Pellanda, P.: The role of digital social innovations to address SDGs: a systematic review. Environ. Dev. Sustain. **26**, 5709–5734 (2023). https://doi.org/10.1007/s10668-023-03038-x
3. Satar, M., Alharthi, S., Omeish, F., Alshibani, S., Saqib, N.: Digital learning orientation and entrepreneurial competencies in graduates: is blended learning sustainable? Sustainability **16**(17), 7794 (2024). https://doi.org/10.3390/su16177794
4. Colás-Bravo, P., Conde-Jiménez, J., Reyes-de-Cózar, S.: Sustainability and digital teaching competence in higher education. Sustainability **13**(22), 12354 (2021) https://doi.org/10.3390/su132212354
5. Amballoor R.G., Naik, S.B.: Digital technologies and education for sustainable development. In: Sharma, R., Shishodia, A., Gupta, A. (eds.) Fostering Sustainable Development in the Age of Technologies. Emerald Publishing Limited, Leeds, pp. 225–237 (2023) https://doi.org/10.1108/978-1-83753-060-120231016
6. Gutiérrez-Ángel, N., García-Sánchez, J.-N., Mercader-Rubio, I., García-Martín, J., Brito-Costa, S.: Digital competence, validation and differential patterns between Spanish and Portuguese areas as assessed from the latest PISA report as a pathway to sustainable education and social concerns. Sustainability **14**, 12721 (2022). https://doi.org/10.3390/su141912721
7. Dneprovskaya, N., Kang, S., Inessa Shevtsova, I.: Evolution of the competencies to embrace digital technology for sustainable development. In: Extended Abstracts of the 2022 CHI Conference on Human Factors in Computing Systems (CHI EA '22). Association for Computing Machinery, New York, NY, USA, Article 461, pp 1–8 (2022) https://doi.org/10.1145/3491101.3519730

8. Sá, M.J., Santos, A.I., Serpa, S., Ferreira, M.C.: Digitainability—digital competences post-COVID-19 for a sustainable society. Sustainability **13**, 9564 (2021). https://doi.org/10.3390/su13179564
9. Shweta Patel, S., Mishra, P.: New skills for sustainability in the socially challenging era of digital transformation: opportunities and challenges. Handb. Res. Digit. Solut. Soc. Econ. Needs, 305–312 (2023). https://doi.org/10.4018/978-1-6684-4102-2.ch015
10. Saxena, P., Seetharaman, A., Shawarikar, G.G.: Factors that influence sustainable innovation in organizations: a systematic literature review. Sustainability **16**(12), 4978 (2024). https://doi.org/10.3390/su16124978
11. Bagur-Femenías, L., Buil-Fabrega, M., Juan, PA.: Teaching digital natives to acquire competences for sustainable development. Int. J. Sustain. High. Educ. **21**(6), 1053–1069. https://www.proquest.com/scholarly-journals/teaching-digital-natives-acquire-competences/docview/2440237324/se-2 (2020) https://doi.org/10.1108/IJSHE-09-2019-0284
12. Razinkina, E., Zima, E., Pankova, L.: Formation of digital competencies in higher education as the basis for realization of sustainable development goals. In: E3S Web of Conferences, vol. 296, 08012 p. (2021). https://doi.org/10.1051/e3sconf/202129608012
13. Shelyugina, O.A., Komarova, O.S., Chernyaeva, I.V.: Formation of digital literacy of undergraduate students in the context of sustainable development goals. J. High. Educ. Theory Pract. **22**(4) (2022). https://doi.org/10.33423/jhetp.v22i4.5125
14. Odusanya, O.E.S., Ademosu, E.A.: Higher education in digital phenomenon. TWIST **19**(3), 894–899 (2024)
15. European Commission, Joint Research Centre, Carretero, S., Vuorikari, R., Punie, Y.: DigComp 2.1 – The digital competence framework for citizens with eight proficiency levels and examples of use. Publications Office. (2017). https://doi.org/10.2760/38842
16. Ramzijah, R., Nazariah, N., Fattah, A.: Literasi Digital di Universitas Jabal Ghafur Pidie: Membangun Generasi Muda yang Melek Teknologi. Konferensi Nasional Literasi Digital dan Kerelawanan **2**, 1–6 (2024)
17. Backes, D., Colomé, J., Erdmann, R., Lunardi, V.: Grupo focal como técnica de coleta e análise de dados em pesquisas qualitativas o mundo da saúde, São Paulo **35**(4), 438–442 (2011) http://www.saocamilo-sp.br/pdf/mundo_saude/88/10_GrupoFocal.pdf
18. D´Amato, P.: Focus group methodology Part 1: considerations for design. Int. J. Ther. Rehabil. **15**(2) (2013). https://doi.org/10.12968/ijtr.2008.15.2.28189
19. UNESCO Institute for Statistics (2015) International Standard Classification of Education (ISCED) ISBN 978-92-9189-179-5 Ref UIS/2015/INS/6 https://doi.org/10.15220/978-92-9189-179-5-en

Sustainable Healthcare Investments: ESG Strategies for Hospital Infrastructure Renovation

Michele Dolcini, Andrea Brambilla, and Stefano Capolongo

Abstract The healthcare sector has increasingly demonstrated its attention and adaptability to ESG (Environment, Social, Governance) challenges. Hospital infrastructures, where healthcare services are provided, play a leading role in the transition towards ESG-driven models. Through an exploration of the existing scientific and grey literature, and a case study investigation, the study aims at scrutinizing financing models involving the private sector, to support public healthcare organizations to enhance the ESG performance of hospital infrastructures. The analysis conducted has led to the identification of various financing schemes to improve ESG performance in hospital environments such as Public–Private Partnership, Energy Performance Contracts (EPC), Green and Social Impact Project Bonds. The pivotal role of ESG strategies enables the strengthening of hospital infrastructures for the improvement of societies and the reduction of environmental impacts. The chapter emphasizes the critical role of sustainable healthcare investments within the ESG framework, particularly in relation to hospital infrastructure. By exploring various financial models and instruments designed to enhance ESG performance, the analysis highlights the opportunity for healthcare institutions to collaborate with the private sector to renovate the hospital building stock while achieving meaningful environmental, social, and financial outcomes.

Keywords Sustainability · Hospital facilities · ESG · Healthcare infrastructure renovation · Financing schemes

---

The original version of the chapter has been revised. A correction to this chapter can be found at https://doi.org/10.1007/978-3-031-86079-9_23

M. Dolcini (✉) · A. Brambilla · S. Capolongo
Design & Health Lab, ABC Department, Politecnico Di Milano, Italy
e-mail: michele.dolcini@polimi.it

© The Author(s), under exclusive license to Springer Nature Switzerland AG 2025, corrected publication 2025

A. d. b. Machado et al. (eds.), *Environmental, Social, Governance and Digital Transformation in Organizations*, Information Systems Engineering and Management 35, https://doi.org/10.1007/978-3-031-86079-9_15

1 Introduction

The healthcare industry stands at a pivotal crossroads, facing the dual challenge of providing high-quality patient care while concurrently addressing the global imperative of sustainability [1, 2]. Hospitals, as primary healthcare infrastructure of the healthcare network, are nowadays confronted with the responsibility of delivering excellent clinical outcomes while reducing their environmental footprint, enhancing social impacts, and practicing responsible governance [3]. Even though these challenges are recognized, there's still a notable gap in both research and practical tools to ensure that sustainability improvements have a lasting effect in healthcare [4]. Hospital institutions play a central role in safeguarding public health and well-being. However, their vast operational footprint, resource consumption and human capital intensity, also make them significant contributors to environmental, social, and economic depletion [5].

1.1 Hospitals as Complex Infrastructure with Changing Needs

Hospital facilities have historically been interpreted as long-lifespan infrastructures, lasting over a century. Nowadays the lifespan shortened, due to the constant and unpredictable evolution of scientific knowledge, demographic patterns, and healthcare delivery models [6]. Therefore, new hospitals are typically design to last around 50 years [7]. Public hospitals in several European countries face significant challenges in financing the transformation of their building stock, delivering care in facilities that do not meet contemporary standards. Budget constraints, and competing healthcare priorities have hindered the modernization of these hospitals, which struggle to secure the substantial investment required for renovation and energy-efficient upgrades [5]. In many cases, hospitals are operating in outdated buildings that lack the capacity to integrate advanced medical technologies, meet current energy efficiency standards, or provide optimal environments for patient care, staff well-being and safety [8]. The consequences are multifaceted, including higher operational costs, suboptimal patient outcomes, and difficulty in meeting the rising demand for sustainable, high-quality healthcare services [9]. Without innovative financing models or increased public investment, hospitals risk falling further behind in their ability to provide care that aligns with modern healthcare standards and environmental goals.

1.2 Sustainability Assessment for Healthcare Facilities

While building sector is one of the largest contributors to energy and material consumption at the global level [6], hospital infrastructure represent one of the most impacting settings for sustainability challenges, due to its specific environmental damage and the large use of human capital. Sustainable solutions in healthcare are a critical prerequisite for ensuring health, as they encompass the social, economic, and environmental dimensions of sustainability. These solutions address various aspects, including the architectural design, building components, the impact on the surrounding environment, and management considerations. A crucial element in enhancing the sustainability of healthcare buildings lies in the ability to effectively assess and measure sustainability performance. In contemporary scientific discourse, numerous contributions have emphasized the significant challenges associated with evaluating sustainability within hospital or healthcare infrastructures, owing to the inherent complexity of these environments [10–12].

ESG has been increasingly adopted as a reference framework for sustainability measuring and assessment in several sectors. Initiatives like the European Green Deal and EU Sustainable Finance Taxonomy (Regulation (EU) 2020/852) have accelerated ESG integration. Several benchmark frameworks have been developed under the ESG umbrella, serving different objective in different industries. In real estate, GRESB has been developed for assessment of sustainability performances in real estate within the ESG framework while other traditional and widely applied sustainability protocols, as LEED, WELL, BREEAM, clearly refers to their alignment to ESG. Existing globally-applied sustainability evaluations system have progressively struggled to find application for the healthcare sector [13]. The evaluation of a hospital's sustainability cannot be separated from an in-depth assessment of the healthcare processes it provides, which are fundamentally driven by managerial practices and operational activities. In this context, the assessment of sustainability in healthcare requires a holistic approach that includes service efficiency, resource utilization, and management strategies. To overcome the limitations presented, recent attempts have been made to harmonize measurability for healthcare facilities application [14, 15] with tailored solutions for hospital settings. Indeed, the contemporary debate in healthcare emphasizes the importance of developing and utilizing a comprehensive set of indicators to evaluate the performance of healthcare service management. These indicators could range from patient outcomes, resource allocation, staff performance, to the environmental impact of hospital operations [16]. By relying on such metrics, healthcare institutions can monitor and improve not only their service delivery but also their alignment with broader sustainability goals, ensuring that operational excellence is achieved alongside environmental and social responsibility.

1.3 Financing Sustainable Renovations of Hospital Facilities

In most European countries, healthcare infrastructures are financed through public resources, as part of healthcare budget allocation, or with the adoption of dedicated financial scheme for infrastructure development and renovation [17]. However, as a result of public-debt pressures and economic and financial crisis since late 2000s, most countries have adopted a more austere approach in relations to GDP share allocation to healthcare, leading to the introduction of private capital [18, 19]. Successive waves of privatization have progressively diminished the role of the public sector in the infrastructure domain, prompting the rise of public–private partnerships (PPPs) as a means to support capital formation [18]. Furthermore, attention has increasingly been directed towards the inadvertent impact of these privatization efforts on public sector operations, as well as their influence on public spending and the pursuit of greater efficiency in comparison to traditional public infrastructure development processes [20].

The lack of consistent and timely renovation initiatives in healthcare infrastructure, particularly in some EU countries with limited exposure to private capitals for infrastructure development like Italy, has resulted in a progressive deterioration in the quality and functionality of healthcare facilities [7]. Without adequate investment and modernization efforts, these facilities struggle to meet contemporary healthcare demands, leading to a diminished capacity to provide effective and efficient services. The absence of strategic collaborations between the public and private sectors has exacerbated the challenges, hindering the ability to implement critical upgrades and innovations necessary for maintaining high standards of care.

Additionally, the transformation of healthcare infrastructures to align with sustainability goals is a pressing and intricate challenge that demands significant financial investment. A balanced approach involving both public resources and private solutions is crucial to successfully navigating the path to sustainable healthcare systems. By leveraging a mix of public funding and private investment, healthcare systems can modernize their infrastructure, reduce their environmental impacts, and enhance their social contributions in alignment with contemporary economic and environmental transformation requirements. Ultimately, this dual approach will enable healthcare facilities to operate sustainably, while continuing to provide high-quality care to an expanding, aging and diverse population.

2 Methods

For the purposes of the research, an investigation of the existing scientific and technical literature has been conducted, to ensure a comprehensive exploration of both academic and non-academic (grey) literature, providing a holistic view of the financing mechanisms available for ESG performance enhancements in healthcare infrastructures.

2.1 Scope

The investigation sought to answer the following key research questions:

i. *What financial mechanisms are available for supporting ESG improvements in healthcare infrastructure?*
ii. *What ESG benefits can be achieved by innovative financial mechanisms for healthcare infrastructures?*

The primary goal of the investigation conducted was to identify and evaluate the range of financial instruments available for supporting ESG-driven renovations in healthcare facilities. A broad set of search terms and keywords was developed to capture the relevant literature, including: *"ESG financing in healthcare"; "Sustainable healthcare infrastructure financing"; "Green solutions for hospital renovation"; "Private capital in healthcare infrastructure"; "Financing the decarbonization in healthcare facilities"; "Healthcare facility modernization and ESG".*

Due to the technical and innovative field of research and the novelty of the topic in the academic debate, a review of grey literature has also been conducted. This was done to capture contributions from industry, government bodies, and research centres that often provide practical insights into financing mechanisms that may not be covered in peer-reviewed publications. The review focused on reports, policy documents, and industry publications from a range of organizations, including: Government agencies (e.g., European Union, World Bank, national healthcare departments); International organizations (e.g., WHO, OECD). Industry reports from financial institutions (e.g., reports from investment banks, credit rating agencies, and sustainable finance organizations like the Climate Bonds Initiative). Grey literature was identified through direct searches on organizational websites, Google Scholar, and institutional repositories. Search terms similar to those used for the investigation of scientific contribution, were applied, such as: "sustainable finance in healthcare," "ESG finance for hospitals," and "ESG improvements in hospital infrastructure."

The inclusion criteria for grey literature focused on practical insights, case studies, and real-world examples of financing schemes used in healthcare infrastructure projects aimed at ESG improvements. At the same time, exclusion criteria have been established, to exclude documents with limited relevance to the healthcare sector, or those focused solely on theoretical policy recommendations.

3 Results

The documents examined in the narrative review and technical sector analysis have highlighted a range of financial instruments involving private sector engagement, which can be adopted by the public sector to enhance ESG (Environmental, Social, and Governance) performance. By leveraging these tools, public entities can mobilize private resources to address infrastructure needs, ensuring that healthcare facilities are renovated and modernized in alignment with sustainability and social responsibility objectives.

All the financing mechanisms represent forms of collaboration between the private and public sectors. Public and Private Partnerships (PPPs) are designed to foster innovation in the renovation of healthcare infrastructure, addressing both financial and operational challenges. By leveraging the expertise, resources, and investment capabilities of the private sector alongside the regulatory oversight and societal goals of the public sector, these models can enable the development of more sustainable, socially responsible, and efficiently governed healthcare facilities [21]. This collaborative approach facilitates the modernization of healthcare systems, driving innovation while addressing the evolving needs of healthcare delivery in a responsible and sustainable manner.

3.1 Private Public Partnership

The allocation of public resources to infrastructure capacity development has been hindered in many European countries [22]. Hence, PPP has been defined as a crucial component for infrastructure development in several sectors, including healthcare. An increasing number of public authorities contracting private companies to build or manage public hospital facilities has been observed over the last three decades [9, 23]. Private Public Partnership, in this context, can be defined as a form of long-term cooperation, through a contract, between public and private entities, with the aim of financing, building and managing infrastructures, or providing services of public interest. PPP schemes represent a solid opportunity for the public sector to leverage the resources, skills and capabilities of the private sector to address a variety of infrastructural, management system and healthcare delivery challenges. The contributions present different results as regards the efficiency of PPP in the healthcare sector, both regarding the infrastructure construction and operation management [24, 25]. Similarly, in the view of the private sector, PPP may be an opportunity to gain access to new markets compared to traditional investment, with lower risk profiles, while contributing to a public good. Although public healthcare markets usually come with lower potential returns, they offer interesting volumes and diversification for service delivery portfolio [26].

In the UK, the model has been largely applied over the last three decades from upgrading and renovating the healthcare building stock. The UK Department of Health reported in 2004 that 80 major PPP hospital projects have been carried out, resulting in significant transformation of the healthcare infrastructure system, with associated costs exceeding £60 billion [27]. From a project perspective, greater private sector involvement in public–private partnerships (PPPs) helps to ease the financial burden on the public sector while transferring specific risks to private partners. Studies show that applying PPPs to infrastructure renovation can lead to enhanced social, economic, and environmental outcomes [28, 29].

In the application of the model to the healthcare sector, the following aspects have been underline as significant role of the private sector to achieve public objectives, maximizing sustainability performance improvement: (i) Enhancing efficiency by either reducing service costs or elevating service quality and effectiveness; (ii) Expanding access to health services for the entire population through cost-effective service provision and utilizing the private sector's resources; (iii) Improving facilities and the quality of healthcare building stock.

3.1.1 Infrastructure-Based PPPs

Under this model, which has been widely adopted at a global scale, the infrastructure development is demanded to the private partner. Most contracts under this "infrastructure-model" follow a design-build-finance-maintain (DBFM) structure [25]. The capital investment is sustained with the support of capital lender [30]. The repayment to the private partner for construction costs is structured through annual amortized payments over the contract's duration, combined with regular payments for ongoing maintenance services. This approach enables the private operator to secure long-term debt financing, ensuring the project's financial viability. Costs are periodically reevaluated in relation to the value generated by the project. Since the 1990s, this infrastructure model has seen widespread global adoption, particularly in healthcare infrastructure projects in countries such as Australia, Canada, the UK, and Spain [25].

The adoption of PPPs provides advantages for improving Environmental, Social, and Governance (ESG) performance in healthcare, aligning financial and operational goals with sustainability and social responsibility objectives. PPPs can contribute substantially to the environmental sustainability of healthcare facilities by enabling the integration of energy-efficient technologies, renewable energy systems, and sustainable building designs. Private sector expertise and investment drive the adoption of cutting-edge environmental technologies that public funding alone might not accommodate. These environmental enhancements reduce the carbon footprint of healthcare operations, improve resource efficiency, and align with global efforts to mitigate climate change [25, 31]. From a social perspective, PPPs enhance the capacity of healthcare systems to provide equitable, high-quality care. By attracting private investment, PPPs help to develop healthcare infrastructure that meets the needs of underserved populations, expanding access to essential services, particularly in rural or economically disadvantaged areas [20]. These partnerships can also reduce the strain on public resources, allowing for the reallocation of government funds to other critical areas of healthcare delivery. Hence, one of the key social benefits of PPPs is the ability to accelerate the construction of hospitals and healthcare facilities, ensuring patient-centered care delivery [32]. In Australia and Spain, for instance, PPPs have been leveraged to construct hospitals in underdeveloped regions, significantly improving healthcare access and outcomes for local populations [30]. Indeed, the main benefits that can be attribute to PPP models for infrastructure development are on-time, on-budget and on-quality construction, gathered in the value for money measurement, against the traditional procurement option for the public sector [20]. Lastly, ESG advantages of PPPs models are driven also by the intrinsic ability to spread financial risks between public and private entities [33]. The private sector often assumes construction and operational risks, which reduces the financial burden on governments. By allowing private entities to secure financing and bring in expertise, PPPs help the creation of financially sustainable projects that deliver long-term benefits for healthcare systems. Moreover, the performance-based structure of PPPs ties returns to the achievement of specific project goals, ensuring that public resources are spent efficiently and effectively [24, 33].

3.2 Energy Performance Contract

In the field of public and private cooperation, Energy Performance Contracts (EPC) represent a trending solution for physical asset upgrading and energy efficiency measures. The scenario of significant constraints on public spending capacity in recent years has highlighted the urgent need for both technical and financial instruments to modernize existing infrastructure and improve energy efficiency, with the goal of reducing operational costs. This growing pressure underscores the necessity for innovative solutions to address the challenges associated with upgrading facilities while simultaneously ensuring sustainable management practices and long-term financial viability.

Rather than mere technical interventions, EPC is a financing technique that uses utility cost savings from reduced energy consumption to repay the cost of installing energy conservation measures. The EPC involves two key actors: (1) the provider of the energy efficiency measure, typically an Energy Service Company (ESCo), which is responsible for delivering all the technical, commercial, and financial services necessary for the implementation of the energy efficiency project, while also assuming the investment and savings risk; and (2) the beneficiary of the energy service, in this case, the hospital facility, which benefits from the intervention and compensates the service provider through payments based on the savings achieved. This contractual arrangement not only facilitates the implementation of energy-efficient solutions but also aligns financial incentives with environmental objectives, making it a viable option for sustainable infrastructure improvements [34].

Therefore, EPCs allow healthcare institutions to implement energy efficiency projects without the need for significant upfront capital investment, as the financial risk is transferred to an Energy Service Company (ESCo). This structure aligns with the environmental goals of reducing carbon emissions, optimizing energy use, and enhancing the sustainability of healthcare operations.

3.3 Green Bonds

Green bonds are financial instruments within the green capital market, which is specifically designed to incorporate sustainability considerations, promoting the recognition of long-term benefits, associated with transitioning to a sustainable economy [35]. These instruments are often regarded as an innovative approach that can help increase sustainable infrastructure investments from institutional investors by improving the liquidity of infrastructure assets [36]. The critical role of debt instruments in healthcare infrastructure projects, driven by the substantial upfront capital requirements and long-term financing needs, underscores the importance of green finance in facilitating the development of complex infrastructure, such as hospitals. Debt instruments, including green bonds, provide the financial structure to support large-scale investments in sustainable healthcare facilities, ensuring alignment with environmental goals while addressing the sector's significant capital demands [36].

The adoption of green bonds as a financing mechanism for hospital renovation has significant implications for the healthcare sector, particularly in terms of aligning infrastructure development with sustainability goals. Green bonds offer to healthcare facilities opportunity for raising capitals that are specifically earmarked for environmentally beneficial projects, such as energy-efficient building upgrades, the incorporation of renewable energy sources, and the implementation of sustainable water and waste management systems [31]. By leveraging green bonds, hospitals can secure capital to finance renovations that lower their environmental and carbon footprints, while also signaling a strong commitment to ESG principles. This alignment with sustainability goals enhances their appeal to investors and stakeholders, who are increasingly prioritizing ESG-focused investment opportunities in their portfolios. The first green bond was issued in 2008 by World Bank, together with Swedish financial institutions and pensions funds to foster green transition. Since then, it has been applied in multiple sectors, including healthcare.

Green bonds have been widely used in other infrastructure sectors, such as transportation and energy, to fund projects that contribute to climate change mitigation and environmental sustainability [37]. Their extension into the healthcare sector is seen as a step to enable hospitals to modernize their facilities in line with global environmental goals. For example, green bonds can finance the installation of energy-efficient lighting, HVAC systems, and advanced insulation technologies, which not only reduce energy consumption but also lower operational costs over time [38]. Indeed, over the last years, the review suggested the evidence of some infrastructure projects for hospital construction, even if primarily in the US market (Table 1).

Table 1 Latest case-study evidence of green bonds issuance for hospitals

Operation	Hospital organization	Country	Year	Value
New hospital	Lucerne cantonal hospital	Switzerland	2024	n/a
Renovation and expansion	Boston medical center	US	2023	\$232 M
New hospital	Ohio state—ohio state Wexner medical center's	US	2021	\$600 M
New hospital	Elazığ city hospital	Turkey	2016	\$288 M
New hospital	Kaiser foundation hospitals	US	2017	n/a
New hospital	Tandem health partners	US	2014	\$231.5 M

Moreover, the issuance of green bonds offers a compelling opportunity to attract an expanding pool of environmentally conscious investors, thereby ensuring a stable and reliable source of long-term funding for hospital infrastructure projects. The associated benefits—including enhanced energy efficiency, reduced greenhouse gas emissions, and an improved institutional reputation—position green bonds as a highly promising financial instrument for advancing sustainable transformation in healthcare infrastructure. [39, 36.

3.4 Social Impact Bonds

Social Bonds or Social Impact Bonds (SIBs) represent an innovative financial instrument within the realm of impact investing. Under the SIBs scheme, government and commissioners enter into agreements with social service providers, such as social enterprise and non-profit organizations and private investors to pay for the delivery of healthcare and social outcomes [40, 41]. SIBs involve the government to pay for an improvement in specific social outcomes, while investors provide the upfront capital to deliver the intervention, bearing the financial risk [20]. The measurability of achieved outcomes is a critical component of this financial model, as payments are directly tied to the attainment of the expected results.

In the healthcare sector, SIBs hold the potential to foster the development of novel health and social programs [40, 42]. Through collaboration between governments and private sector entities, new capital sources can be accessed to finance the initiatives. This partnership leverages the strengths of both actors in addressing social challenges, while simultaneously reducing the reliance on public debt for funding [41]. Thus, SIBs are aimed at overcoming the shortcomings of the current scenario, driven by the scarcity of financial resources. In conclusion, the literature reviewed in this investigation highlights an increasing shift toward financial support for healthcare services through the use of Social Impact Bonds (SIBs). In this context, the adoption of this financing mechanism has significant implications for social sustainability and ESG performance of the healthcare systems. However, its application in the development and renewal of healthcare infrastructure capital remains largely unexplored in the scientific debate, and marginal, when considering real applications. As noted by [20], this financing tool is primarily employed as a means for the public sector to progressively reduce its direct investment in capital. Furthermore, particularly with regard to channelling private resources into the construction or renewal of hospital infrastructure via SIBs, it is essential to strengthen the government's role through the implementation of risk-sharing mechanisms [20]. In this regard, Social Impact Bonds (SIBs) may serve as an effective strategy for attracting private investment to address the existing infrastructure gap between the current state and the optimal standards in healthcare construction to maximize public value.

4 Discussion

This research explores the critical role of innovative financing mechanisms in enhancing the Environmental, Social, and Governance (ESG) performance of healthcare infrastructures, particularly through the renovation and modernization of hospital facilities.

The analysis demonstrates that the adoption of public–private partnerships (PPPs) solutions, green bonds, social impact bonds (SIBs), and Energy Performance Contract (EPC), offers a viable solution to overcome the financial limitations of the public sector. The integration of these innovative financing mechanisms into healthcare infrastructure projects has significant implications for improving ESG performance. From an environmental perspective, green bonds and energy performance contracts (EPCs) contribute to the decarbonization of healthcare facilities, helping hospitals to meet sustainability targets and reduce their overall environmental impact [31]. Social impact bonds and PPPs addresses the social agenda by enabling healthcare systems to reduce disparities in healthcare access, ensuring that underserved populations benefit from improved facilities and services.

Finally, the analysis reveals that traditional public funding models, while foundational for healthcare development, have faced substantial limitations due to fiscal constraints, growing healthcare demands, and the increasing complexity of hospital projects. The stagnation in infrastructure development not only affects the quality of care but also escalates operational costs and undermines efforts to meet sustainability goals such as reducing greenhouse gas emissions. Consequently, there is a pressure for greater involvement of private resources to enhance the renovation process. This scenario aligns with the ongoing expansion of sustainable finance, as well as the increasing emphasis on ESG funds within the framework of the European Green Deal and the EU Taxonomy.

5 CRediT Authorship Contribution Statement

Michele Dolcini: Writing—review & editing, Writing—original draft, Visualization, Methodology, Investigation, Formal analysis, Conceptualization. **Stefano Capolongo**: Validation. **Andrea Brambilla**: Review & Editing, Validation, Supervision, Conceptualization,

6 Declaration of Competing Interest

The authors declare that they have no known competing financial interests or personal relationships that could have appeared to influence the work reported in this paper.

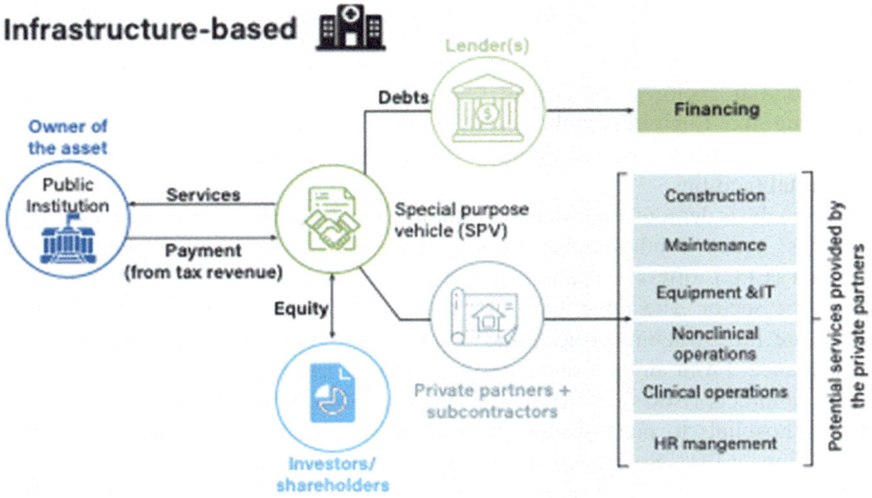

Fig. 1 The infrastructure-based PPP [43] Adapted from: Scamihorn [43]

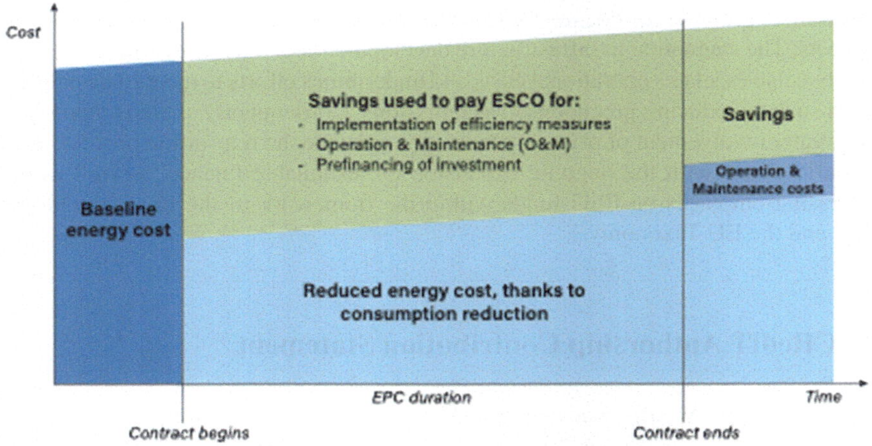

Fig. 2 Energy Performance contract models [43] Adapted from: Scamihorn [43]

Acknowledgements This article originates from the scientific and professional debate carried out during the short course *La Finanza nella Pubblica Amministrazione: Modelli Innovativi In Sanità*, held at the Politecnico di Milano, with Prof. Stefano Capolongo and Prof. Paolo Petinelli as chairs. The course was an enriching occasion for discussion among the academic, corporate, and public spheres regarding financing mechanisms to renovate and modernize the Italian healthcare building stock.

Figures
1 and

2 have been adapted from the MSc thesis entitled *Beyond Profit, Sustainable Healthcare Investments: Assessing Profitability and Stakeholder Perspectives in Hospital Infrastructure*, developed by Michael Ross Scamihorn in 2024 with Prof. Andrea Ciaramella, as supervisor. MD and AB co-supervised this brilliant master research, which investigated the relationship between the improvement of ESG performance in healthcare infrastructures and the increase in real estate asset value.

Data Availability Data will be made available on request.

References

1. McGain, F., Naylor, C.: Environmental sustainability in hospitals—a systematic review and research agenda. J. Health Serv. Res. Policy **19**(4), 245–252 (2014). https://doi.org/10.1177/1355819614534836
2. Romanello, M., Di Napoli, C., Green, C., Kennard, H., Lampard, P., Scamman, D., Walawender, M., et al.: The 2023 report of the lancet countdown on health and climate change: the imperative for a health-centred response in a world facing irreversible harms. The Lancet **402**(10419), 2346–2394 (2023). https://doi.org/10.1016/S0140-6736(23)01859-7
3. Quitmann, C., Sauerborn, R., Danquah, I., Herrmann, A.: 'Climate change mitigation is a hot topic, but not when it comes to hospitals': a qualitative study on hospital stakeholders' perception and sense of responsibility for greenhouse gas emissions. J. Med. Ethics **49**(3), 204–210 (2023). https://doi.org/10.1136/medethics-2021-107971
4. Kempeneer, S., Peeters, M., Compernolle, T.: Bringing the user back in the building: an analysis of ESG in real estate and a behavioral framework to guide future research. Sustainability **13**(6), 3239 (2021). https://doi.org/10.3390/su13063239
5. Capolongo, S., Gola, M., Brambilla, A., Morganti, A., Mosca, E. I., Barach, P.: COVID-19 and healthcare facilities: a decalogue of design strategies for resilient hospitals. Acta Bio Medica Atenei Parmensis **91**(9-S), 50–60. (2020) https://doi.org/10.23750/abm.v91i9-S.10117
6. Rechel, B.: European observatory on health systems and policies. Investing in Hospitals of the Future. Observatory Studies Series 16. Regional Office for Europe, World Health Organization, Copenhagen (2009)
7. Stefano, C.: "Annual Report 2023." Joint Research Partnership Healthcare Infrastructures. (2024) https://www.fondazionepolitecnico.it/en/initiatives/jrp-healthcare-infrastructures/
8. WHO.: "WHO guidance for climate-resilient and environmentally sustainable health care facilities. ISBN 978-92-4-001222-6. Geneva: World Health Organization. (2020) file:///Users/micheledolcini/Desktop/personale/9789240012226-eng.pdf.
9. McKee, M., Edwards, N., Atun, R.: Public-private partnerships for hospitals. Bull. World Health Organ. **84**, 890–896 (2006)
10. Brambilla, A., Capolongo, S.: Healthy and sustainable hospital evaluation—a review of POE tools for hospital assessment in an evidence-based design framework. Buildings **9**(4), 76 (2019). https://doi.org/10.3390/buildings9040076
11. Capolongo, S., Bottero, M.C., Buffoli, M., Lettieri, E (eds.).: Improving sustainability during hospital design and operation: a multidisciplinary evaluation tool. Green energy and technology. Springer International Publishing, Cham (2015) https://doi.org/10.1007/978-3-319-14036-0
12. Castro, M.D.F., Mateus, R., Bragança. L.: A Critical analysis of building sustainability assessment methods for healthcare buildings. Environ., Dev. Sustain. **17**(6), 1381–1412 (2015) https://doi.org/10.1007/s10668-014-9611-0
13. Buffoli, M., Capolongo, S., Bottero M.: Sustainable healthcare: how to assess and improve healthcare structures' sustainability. Annali Di Igiene: Medicina Preventiva e Di Comunità, no. 5 (Settembre), 411–18 (2013) https://doi.org/10.7416/AI.2013.1942

14. Brambilla, A., Apel, J.M., Schmidt-Ross, I., Buffoli, M., Capolongo, S.: Testing of a multiple criteria assessment tool for healthcare facilities quality and sustainability: the case of german hospitals. Sustainability **14**(24), 16742 (2022). https://doi.org/10.3390/su142416742
15. Castro, M.D., Fátima, R.M., Bragança, L.: Healthcare building sustainability assessment tool—sustainable effective design criteria in the Portuguese context. Environ. Impact Assess. Rev. **67**, 49–60 (2017). https://doi.org/10.1016/j.eiar.2017.08.005
16. Vainieri, M., Noto, G., Ferre, F., Rosella, L.C.: A performance management system in healthcare for all seasons? Int. J. Environ. Res. Public Health **17**(15), 5590 (2020). https://doi.org/10.3390/ijerph17155590
17. Dubas-Jakóbczyk, K., Kocot, E., Tambor, M., Szetela, P., Kostrzewska, O., Siegrist Jr R.B., Quentin, W.: The association between hospital financial performance and the quality of care – a scoping literature review. Int. J. Health Policy Manag. (2022). https://doi.org/10.34172/ijhpm.2022.6957
18. Barlow, J., Köberle-Gaiser, M.: Delivering innovation in hospital construction: contracts and collaboration in the UK's private finance initiative hospitals program. Calif. Manage. Rev. **51**(2), 126–143 (2009). https://doi.org/10.2307/41166483
19. Colla, P., Hellowell, M., Vecchi, V., Gatti, S.: Determinants of the cost of capital for privately financed hospital projects in the UK. Health Policy **119**(11), 1442–1449 (2015). https://doi.org/10.1016/j.healthpol.2015.08.018
20. Vecchi, V., Casalini, F.: Is a social empowerment of ppp for infrastructure delivery possible? Lessons from social impact boNDS. Ann. Public Coop. Econ. **90**(2), 353–369 (2019). https://doi.org/10.1111/apce.12230
21. Barlow, J., Roehrich, J., Wright, S.: Europe sees mixed results from public-private partnerships for building and managing health care facilities and services. Health Aff. **32**(1), 146–154 (2013). https://doi.org/10.1377/hlthaff.2011.1223
22. Moro Visconti, R., Morea, D.: Healthcare digitalization and pay-for-performance incentives in smart hospital project financing. Int. J. Environ. Res. Public Health **17**(7), 2318 (2020). https://doi.org/10.3390/ijerph17072318
23. Wright, S., Barlow, J., Roehrich, J.K.: Public-private partnerships for health services: construction, protection and rehabilitation of critical healthcare infrastructure in Europe. In: Competitive Government: Public Private Partnerships, pp 125–51 (2019) https://doi.org/10.1007/978-3-030-24600-6_7
24. Hellowell, M.: Public private partnerships and the quality and efficiency of healthcare services. In: Veronica, V., Hellowell, M. (eds.) Public-Private Partnerships in Health, pp 1–13. Springer International Publishing, Cham (2018) https://doi.org/10.1007/978-3-319-69563-1_1
25. UNECE.: A preliminary reflection on the best practice in PPP in healthcare sector: a review of different PPP case studies and experiences (2012) https://unece.org/ppp/icoe/health
26. Yamout, G., Jamali, D.: A critical assessment of a proposed public private partnership (PPP) for the management of water services in Lebanon. Water Resour. Manage. **21**(3), 611–634 (2007). https://doi.org/10.1007/s11269-006-9033-3
27. De Neufville, R., Lee, Y.S., Scholtes, S.: Using flexibility to improve value-for-money in hospital infrastructure investments." In: 2008 First International Conference on Infrastructure Systems and Services: Building Networks for a Brighter Future (INFRA), pp 1–6. IEEE, Rotterdam, Netherlands (2008) https://doi.org/10.1109/INFRA.2008.5439668
28. Hueskes, M., Verhoest, K., Block, T.: Governing public-private partnerships for sustainability. Int. J. Project Manage. **35**(6), 1184–1195 (2017). https://doi.org/10.1016/j.ijproman.2017.02.020
29. Meng, J., Ye, Z., Wang, Y.: Financing and investing in sustainable infrastructure: a review and research agenda. Sustain. Futur. **8**(December), 100312 (2024). https://doi.org/10.1016/j.sftr.2024.100312
30. Roehrich, J.K., Lewis, M.A., George, G.: Are public-private partnerships a healthy option? A systematic literature review. Soc. Sci. Med. **113**, 110–119 (2014). https://doi.org/10.1016/j.socscimed.2014.03.037

31. Climate Bond Initiative.: "Green Bonds Market Summary." (2021) https://www.climatebonds.net.
32. Vecchi, V., Hellowell, M. (eds.): Public-Private Partnerships in Health. Springer International Publishing, Cham (2018). https://doi.org/10.1007/978-3-319-69563-1
33. Rodrigues, N.J.P.: Public–private partnerships model applied to hospitals—a critical review. Healthcare **11**(12), 1723 (2023). https://doi.org/10.3390/healthcare11121723
34. Martiniello, L., Morea, D., Paolone, F., Tiscini, R.: Energy performance contracting and public-private partnership: how to share risks and balance benefits. Energies **13**(14), 3625 (2020). https://doi.org/10.3390/en13143625
35. Bracking, S.: Green bond market practices: exploring the moral 'balance' of environmental and financial values. J. Cult. Econ. **17**(3), 279–296 (2024). https://doi.org/10.1080/17530350.2024.2312864
36. Maltais, A., Nykvist, B.: Understanding the role of green bonds in advancing sustainability. J. Sustain. Financ. Invest. 1–20 2020. https://doi.org/10.1080/20430795.2020.1724864
37. Anh, T.U., Chuc, T.S., Rasoulinezhad, E.: Factors influencing the green bond market expansion: evidence from a multi-dimensional analysis. J. Risk Financ. Manag. **13**(6), 126 (2020). https://doi.org/10.3390/jrfm13060126
38. Shishlov, I., Morel, R.: Beyond transparency: unlocking the full potential of green bonds. Inst. Clim. Econ. n.d. http://www.i4ce.org/download/unlocking-the-potential-of-green-bonds/
39. Flammer, C.: Green bonds: effectiveness and implications for public policy. Environ. Energy Policy Econ. **1**:95–128. https://doi.org/10.1086/706794
40. Hulse, E.S., Grace, R.A., McPake, B., Lee, J.T.: Use of social impact bonds in financing health systems responses to non-communicable diseases: scoping review. BMJ Glob. Health **6**(3), e004127 (2021). https://doi.org/10.1136/bmjgh-2020-004127
41. Katz, A.S., Brisbois, B., Zerger, S., Hwang, S.W.: Social impact bonds as a funding method for health and social programs: potential areas of concern. Am. J. Public Health **108**(2), 210–215 (2018). https://doi.org/10.2105/AJPH.2017.304157
42. Crowley, D.M.: The role of social impact bonds in pediatric health care. Pediatrics **134**(2), e331–e333 (2014). https://doi.org/10.1542/peds.2013-4056
43. Scamihorn, M.R.: Beyond profit, sustainable healthcare investments: assessing profitability and stakeholder perspectives in hospital infrastructure. Milano: Politecnico di Milano (2024)

Strengthening Organizational Resilience in Response to Technostress Challenges

Serdar Yener, Helena Belchior Rocha, Aykut Arslan, and Stefan Chichevaliev

Abstract In this study, the relationship between environmental, social and governance dimensions, which are the sustainability dimensions of organizations, and techno-stress were investigated through a literature review method and the strategies to reveal the negative effects of technostress were tried to be presented. Technostress, which is defined as the negative reaction of employees who are exposed to the use of technological tools and methods intensively, threatens the sustainability of enterprises. In businesses where technostress is experienced intensely, productivity, effectiveness, performance, quality of life perception and satisfaction may be adversely affected. In the process of technostress, which is defined as a type or dimension of stress, individuals may experience anxiety, hopelessness and fatigue. Burnout can also be encountered as technostress becomes chronic. These situations, which are also defined as the dark side of technology, can also negatively affect the technological transformation of organizations. The transformation of organizations into a fully sustainable structure develops depending on the environmental interactions, social responsibilities and governance practices of the organizations. This concept, which is used as ESG, covers many issues such as morality, ethics, inclusivity, diversity, green deal, blue economy, and respect for the environment in organizations. In this study, within the framework of the dimensions that fall under the concept of ESG from a broad perspective, the relationship between technostress and the measures to be taken against technostress are discussed. In this way, it is possible to increase

S. Yener
Sinop University, Sinop, Türkiye
e-mail: serdar_yener@hotmail.com

H. B. Rocha (✉)
Iscte-University Institute of Lisbon, Lisbon, Portugal
e-mail: helena_rocha@iscte-iul.pt

A. Arslan
Piri Reis University, Sinop, Türkiye
e-mail: aarslan@pirireis.edu.tr

S. Chichevaliev
Vrije Universiteit Brussel, Brussels, Belgium
e-mail: Stefan.Chichevaliev@vub.be

© The Author(s), under exclusive license to Springer Nature Switzerland AG 2025
A. d. b. Machado et al. (eds.), *Environmental, Social, Governance and Digital Transformation in Organizations*, Information Systems Engineering and Management 35, https://doi.org/10.1007/978-3-031-86079-9_16

the resilience and sustainability of organizations in the changing and transforming world. Within the framework of the ESG concept, strategies such as social responsibility programs, education and training, awareness practices, detoxification, moral and ethical leadership, risk programs and green IT have been proposed to reduce technostress within the framework of ESG. These strategies can be diversified and increased. With these strategies, organizations can increase employees' perception of quality of life and satisfaction. In the study, constraints and suggestions for future studies are also presented. However, this study is intended to be used as a reference together with other studies in the technostress literature.

Keywords Technostress · Resilience · Quality of life

1 Introduction

Today, organizations use many tools of technology intensively to increase their performance, improve their effectiveness, and improve their creativity [11]. In fact, this intensive use of technology often brings with it many direct/indirect and positive/negative effects. Within the framework of occupational health and safety (OHS) criteria, organizations have responsibilities to eliminate psychological and physical threats and hazards that their employees may encounter. The negative effects of technology also affect the psychological and physical health of employees. Techno-stress has been added to the measures taken within the framework of OHS. With the definition of the concept of technostress, preventive measures are developed within the framework of OSH [2].

The fact that technology affects our personal lives as well as our professional life has expanded the definition of technostress. Technostress, which blurs the separation of family and work by changing the way we do business in professional life, has started to pose a danger to our family domain while increasing efficiency and effectiveness [10].

As stated in the abstract, technostress is defined as the psychological or physiological reaction of individuals to the technology they are exposed to. Technological tools, software, programs, social media tools, and sharing platforms, which were created especially by taking into account business needs and psychological needs, increased the interaction time of individuals with technology, eliminated the boundaries of work and private life, and made individuals constantly connected during their waking time [5].

While organizations want to gain sustainable competitive advantage within the framework of their own approaches, ignoring the negative effects of technology brings different problems [1, 8]. Sustainable competitive advantage organizations require having multidimensional strategies. In this process, which is also defined as sustainability today, environmental, social and governance dimensions affect the success of organizations [7]. However, ESG requires interactive relationships with stakeholders from an analytical perspective, open and transparent processes, moral and ethical responsibilities, management of diversity and differences, and respect for natural and other environmental elements.

Organizations should take into account the quality of life perceptions and satisfaction of employees within the framework of ESG principles, and take measures to ensure the sustainable use of technology in terms of technostress within the framework of OHS [6]. This approach makes it imperative for organizations to have an analytical technostress orientation strategy. In addition, a large-scale stakeholder analysis, that is, the classification of internal/external stakeholders, should also be defined. In a governance approach that takes into account all stakeholders, full participation of stakeholders in decision-making processes should be ensured along with moral and ethical behavior. Many stakeholder groups such as customers, employees, shareholders, public institutions, natural environment actors should take an active part in decision-making processes with governance strategies. On the other hand, the social dimension requires businesses to respect the dynamics of the social environment in which they operate. The social dimension requires respecting and taking into account the components of the society, its cultural codes, interpersonal and social relations. This can give businesses an analytical perspective. Otherwise, both the employees and the customers of organizations that cannot have and adapt to the codes of the society they are in will be adversely affected. The development of all these approaches based on technostress is another challenge. Social strategies to reduce or prevent technostress can make not only the business but also the society resilient.

When considered from the framework of technostress, businesses with sustainable technological capabilities can fulfill their responsibilities within the framework of ESG.

In this study, technostress strategies that will increase the sustainability of enterprises and improve their resilience are discussed within the framework of ESG dimensions. Measures and strategies to eliminate the negative effects of technostress are presented. Besides strategy suggestions on technostress, the limitations of the study and suggestions for future studies were also presented.

2 Technostress: Origins and Influence

The word "technostress" was first used in the literature by [3] and was used to describe the psychological response to the negative situations brought by technology. In terms of computers, the 1980s are also considered as the years when personal computers began to spread worldwide. Especially in the west, it has been seen that with the use of computers, employees can do more than their daily work and speed up the work, and this situation has been magically disseminated for organizations. In the process that has come to the present day, organizations have continued to use almost every tool of technology intensively in order to take advantage of the competitive advantage that technology has added to them. Although technostress started to be used in the literature in 1984, studies on technostress have started to become widespread after the 2000s with the effect of the internet. The widespread use of the internet has enabled individual users to enter technology outside of the business world, and this has had an irresistible effect with the development of social media tools. Since social media tools were later used as communication tools in the business environment, it has become impossible for users to distinguish between family and work. The inability of individuals to adapt to this situation and their reactions are considered as technostress symptoms. These reactions can be anxiety, distress, exhaustion, headache, insomnia, and physical ailments [11]. Although the concept of technostress first defined the inability of individuals to adapt to technology, other symptoms that occur as mentioned above show that the scope of the concept is greater than what is seen. Especially in the western world, which interacts with technology very intensely, the interaction of employees with technology is higher than known. In the Western world, scientists have increased the interest in the concept of technostress due to the emergence of different phenomena related to both the business and private lives of individuals [9]. In the studies, it is emphasized that the negative situations stated to be caused by technostress are based on the unpredictable progress of technology and increasing dependence [10]. The fact that technology is always accessible and attracts users with different dimensions as it reaches, and that it has many features that make people's lives easier increases dependency. While the increasing use of technology reduces the physical social interactions of individuals, it increases their virtual social interactions and creates a vicious circle. The increasing need for performance of businesses also contributes to this vicious circle and blurs the boundaries of work and private life [10]. According to the studies and as described above, technostress has many negative effects, both psychological and behavioral. Although the reasons for businesses to use technology are to increase performance, efficiency and efficiency, it has been observed that efficiency and effectiveness decrease and productivity decrease with the experience of technostress [4]. With the decrease in job satisfaction, it was observed that the cases of turnover also increased, while the cases of burnout started to increase [11]. While the increase in turnover rates has caused financial damage to businesses, it also negatively affects the organizational culture. In addition, the health costs of employees' technostress-induced ailments started to increase after a while [5]. When the literature is examined, it is observed that the determinants of technostress differ, but they are generally grouped under the same topics. As determinants of technostress [10, 11]:

- Techno-overload: It defines the increase in people's workload, the need to work harder, to work faster due to technology. For example, before the use of computers in businesses, the number of work units that an average employee can complete in a day is limited. Therefore, the mental activities of individuals for the work they do before using the computer are limited. However, the number of jobs done with the use of technology has increased several times and individuals have started to think about more work mentally.
- Techno-invasion: It defines the intense use of technology that disturbs the private life of individuals and eliminates the boundaries between work and private life. Individuals lose this distinction over time and feel that they are constantly working.

Especially with the invasion of private life with technology, the perception of private life also disappears. For example, the fact that individuals respond to a message about work during the family hours can be seen as the occupation of the individual's family space.

- Techno-complexity: With the intense occupation of business and private life by technology, it defines the need for individuals to learn many technologies in business and private life and the obligation to learn new technologies in order to adapt. For example, the fact that individuals are forced to use virtual tools to make bank transactions forces elderly individuals who have never used this technology.
- Techno-insecurity: It defines the fear of isolation of individuals who think that they cannot adapt to new technologies. In addition, the fear of losing their professions because they think that they will not adapt to new technologies creates a perception of insecurity.
- Techno-uncertainty: Changing and developing new technologies change previous habits or lifestyles. Since this change and development are continuous, it creates uncertainty. In addition, uncertainty is growing even more as changing and developing technology connects different applications in different areas of life. For example, making daily bill payments virtually with the use of technology, making flight or bus reservations virtually pushes individuals to this area.

The above-mentioned techno-overload, techno-uncertainty, techno-invasion, techno-complexity, and techno-insecurity are identified as the main determinants or factors of technostress. These determinants of technostress can interact with many different variables, both personal and corporate, resulting in different outcomes. While these outcomes are predominantly negative, they engender adverse circumstances within enterprises.

2.1 Technostress Determinants

When we look at the studies on technostress, the determinants are organizational-based, situational and individual-based. Individual determinants of technostress are often based on demographic characteristics. Determinants (factors) of technostress are presented below [9, 11]

Techno-overload: Technological workload refers to the increase in the responsibilities of employees with the use of technology and the effect of these responsibilities on other responsibilities. Individuals who are trying to fulfill increasing responsibilities are forced to work faster and harder. In addition, due to the interconnection of technological tools, employees are expected to do many jobs at the same time.

Techno-invasion: Techno-invasion describes the invasion of employees' lives by technology. With technology occupying the lives of employees, employees in the workplace are expected to be constantly accessible and in communication. This situation is experienced intensely in enterprises where continuous communication is provided with the technological infrastructures used. In fact, tools such as computers, smartphones and tablets that they can use outside of work are allocated free of charge to those who start working, ensuring that employees are constantly accessible. This means that employees use the time they allocate for themselves outside of work for work.

Techno-complexity: It has been mentioned above that one of the main determinants of technostress is constantly developing and changing technology. Individuals who try to adapt to the ever-changing and developing technology are also expected to develop their talents and skills. In this case, the constantly developing and changing technology creates a complex situation, and individuals experience the confusion of adapting in this complexity.

Techno-insecurity: The fear of losing one's jobs in case of not being able to adapt to the ever-changing and developing technology is defined as techno-insecurity. This anxiety, which will adversely affect the normal lives of individuals, can turn into a condition over time and lead to a sense of cynicism.

Techno-uncertainty: Techno-uncertainty is the uncertainty caused by ever-changing and evolving technology. In this case, individuals cannot predict what they may encounter in the future and cannot prepare for it. Therefore, individuals cannot prepare themselves and therefore cannot adapt to technology.

When the determinants of technostress are examined, it can be seen that it is caused by organizational, situational and individual factors. However, developing strategies that take these determinants into account can prevent technostress from occurring.

2.2 Strategies to Improve the Resilience of Businesses Against Technostress

There are several dimensions for businesses to improve their resilience to technostress. As mentioned above, the determinants of technostress can be organizational, personal and situational. In this context, business managers should increase resilience by improving organizational, personal and situational dimensions. Because, like all types of stress, the most important way to combat technostress is to develop strategies to increase resilience. In this sense, increasing individual awareness, developing situational factors in a way that increases resilience, and organizing organizational determinants in a supportive way can increase resilience. Developing emotional intelligence, practices or strategies to reduce stress, creating flexible working conditions and educating individuals in this direction can also be shown among these strategies. Resilience, both organizational and individual, is an important tool in coping with stress. Within the framework of ESG factors it can implement some of the following strategies to combat technostress:

- Techno-load and Time Pressure: Policies that encourage work-life balance and reduce excessive workload and time pressure can reduce technostress.

- Lack of Control: Policies and practices of businesses that increase employees' control over the work can make them feel more valuable and empowered. In particular, personnel empowerment practices can also put them in the decision-making mechanism.
- Information Overload: Policies and practices of businesses that will ensure that employees have the competencies to balance technological processes can make them more self-confident. Employees who are well-versed in technology do not experience fear and anxiety.
- Role Ambiguity: The transparent and open processes of businesses enable individuals to act in a controlled manner by reducing role confusion. With the reduction of role ambiguity, individuals can also be active in decision-making processes.

The described vicious circle businesses face underscores their pursuit of sustainable competitive advantage through technology use, even as they contend with the fact they are exposed to the dark side of technology. Nevertheless, businesses must work to maintain a balanced approach to their technology interactions in a balanced way in their struggle to gain sustainable competitive advantage. Measures to prevent or reduce the occurrence of technostress, which will adversely affect the quality of life of employees, can ensure a balanced increase in performance, productivity, and effectiveness in the medium and long term. These policies and practices are related to all dimensions of ESG. ESG principles can provide sustainability to businesses within the framework of ethical principles in all aspects.

In this part of the research, examples of strategies that will reduce or eliminate the effects of technostress and increase the resilience of businesses within the framework of ESG are given.

2.3 *Digital detox and Mindfulness*

Mindfulness, which is shown among the strategies for individuals, is used in many cultures with different tools. Mindfulness increases individuals' emotional control. In addition, it provides skills that will increase the self-awareness level of individuals. Mindfulness includes several tools, both physical and psychological. Thanks to mindfulness, physical and psychological endurance can increase. Thanks to awareness, individuals are enabled to predict the role of technology in daily life. The methods that can be used to raise awareness are presented below;

- Mindful Breathing: It refers to the motivation of employees by applying breathing techniques in a focused and concentrated manner. Deep and concentrated breathing can reduce employees' anxiety and increase their calmness. This can enable employees to make decisions by acting more coolly. There are many free platforms available on the internet to learn these techniques.
- Mindful Awareness: It refers to enabling individuals to be aware of their thoughts and feelings before judging other individuals in their interactions in public. Instant reactions with prejudices increase stress in society and cause stressors to interact

with each other. In this case, you cannot predict the source of the stressor that is stressing you. Emotional intelligence therapies are especially useful to make it easier to understand the feelings and thoughts of other individuals. At the same time, it can be considered to mentor experienced individuals in this field.

Digital Detox: Taking a break from the use of technological tools you use while carrying out daily activities can reduce stress. Getting away from technological gadgets, taking short walks, or doing physical activities during work breaks or vacations can make this easier. An appropriate measure may be to prohibit the flow of work-related information through technological-based communication tools outside of working hours.

2.4 Ethical Leadership

Ethics is a concept that refers to behaviors and attitudes that are considered correct in any context (reference). When considered in terms of businesses, the expectations of stakeholders shape ethical behavior.

The effects of the determinants of technostress can be mitigated through ethical leadership practices. By determining the expectations of employees, one of the most important stakeholder groups, in an empathetic way, measures can be taken to reconcile the interests of the business with the interests of the stakeholders. As can be seen from the above definition of the concept of ethics, ethics is related to human beings and human behavior. Therefore, for ethical leaders, the human factor has priority among other factors. Measures can be taken to ensure that people who are the target of technostress are least affected by technostress, and that people use balanced technology:

- Supportive Work Environment: Ethical leaders can create an organizational climate that is suitable for the natural characteristics and expectations of employees. The biggest expectation of employees is the work environment where they will be satisfied. The psychological and physical dimensions of satisfaction should be questioned. Organizational pairs that respect the honor and dignity of people, supportive, prioritize employee health, and value happiness and peace can be created. It is not possible to meet all the expectations of the employees. In this case, strategies that align the interests of the business with the interests of the employees should be developed. The fundamentals of these developed strategies should be presented transparently. Decision-making processes, quality control circles and meetings, stakeholder meetings can be held to include employees.
- Balancing the Use of Technology: Ethical leaders must identify the risks present in the use of technology. The use of technology that will endanger the health of individuals, intensive and tiring use of technology should be prevented. The boundaries between work and family life should be clear. The private lives of individuals should not be disturbed outside of working hours in work-related matters.

- Personnel Empowerment: One of the psychological needs of employees is development. Using employees' talents in challenging activities to make them feel valued, forming teams to guide each other, and matching understaffed with talented staff can reduce anxiety about technology. On the other hand, using them in tasks that will improve them can increase their perception of value. Environments can be created that will enable employees to share their views on work without any worries or concerns.

2.5 *Technostress Risk Assessment*

Occupational health and safety programs try to take into account each factor that may pose a risk to businesses. Every factor that threatens the psychological and physical health of employees is considered a potential source of risk. In this context, a risk assessment of the technostress risk in enterprises is also made. The level of impact and likelihood of technostress risk should be investigated. In this context, technostress emerges and all organizational practices or organizational culture components should be taken into account. In the event of technostress, it should be examined how the sustainability of the business may be affected. Technostress risk programs should be comprehensive in order to increase the resilience of businesses and employees. Within this framework, the following activities can be carried out:

- Survey and Feedback: Surveys can be applied to reveal the effect of technostress on individuals. Surveys and other measurement tools to measure the instant situation are used to see the risk levels of individuals. Surveys to be applied using one-to-one face-to-face methods can reveal concrete data.
- Statistics: Another data that can reveal the effects of technostress is historical statistics. For example, employees' work cycle rates, turnover rates, illness, etc. Statistics such as the statistics of the time they are away from work, hospital and other health problems can be indicators for problems in businesses. The details of these statistics can give clues about technostress. With the statistics used, projections for the future can be created and predictions and comparisons can be made.
- Risk Mitigation Strategies: Strategies to reduce the risk of technostress are determined according to the risk programs, questionnaires, feedbacks and statistics used. Focused strategies to reduce the risk of technostress can be useful. Focused strategies that can be used as part of the overall strategic plan of the business can be to organize technology usage criteria, communication criteria, criteria for relationships, and to provide training by creating policies related to them.
- Technostress Controls: Technostress risk programs, statistics and feedback, as well as the processes and results of the activities to be carried out in line with the strategic plan, should be observed. For this, responsible leaders at all levels should carry out routine checks in their area of responsibility. Routine controls can also demonstrate the effectiveness of strategic plans and practices.

2.6 Sustainability and Well-Being of Employees

The main target of the threat of technostress is the health of individuals. The fact that individuals are affected by technostress and that they experience burnout syndrome by experiencing chronic technostress affects the performance, efficiency and effectiveness of businesses. That's why ethical leaders should implement strategies that care about the health of individuals. In this context, leaders can take a number of measures:

- Healthy Living Programs: Healthy living programs can be created to help individuals maintain their physical and psychological health. In addition to taking measures that will not disrupt business processes in working environments, awareness trainings, facilitators, sports or exercise hours, ergonomic working conditions can be provided for individuals to have healthy living behaviors. A healthy life can be contributed to by agreeing with local sports clubs or halls and organizing joint programs.
- Flexible Working Conditions: Working conditions that will increase the work-life balance of employees can enable the individual to fulfill their responsibilities in their private life. Within the framework of family-friendly working strategies to be implemented by businesses, clarification of work-family boundaries, flexible working practices, and remote working practices can be made. The criteria for regulating them should be regulated.
- Resilience Training: It is recommended that business managers initiate practices that will improve the resilience of employees against technostress and provide mentoring or consultancy services. It is recommended to plan and implement resilience trainings, especially considering the determinants of technostress. In this context, if technological self-efficacy training, time management training, emotional intelligence and stress management skills can be given to the personnel, resilience against technostress can be increased.

2.7 Education and Training

Technostress management training can be given to business managers to increase the resilience of their personnel against technostress. Starting from the definition of the concept of technostress, the determinants of technostress, sample applications based on studies with technostress can be the subject of these trainings. In addition, the technostress experiences of the employees can be discussed in these trainings. In these discussions or meetings, the foundations of the technostress risk planning of the business can also be laid. Because employees are the target audience for technostress experiences. In particular, if it can be revealed which dimension of technostress is more prominent, strategies can be planned to prevent it.

- Technological Self-Efficacy Training: When the determinants of technostress are examined, the technological self-efficacy of the employees must be high in order to overcome the determinants such as techno-complexity, techno-uncertainty, techno-workload. Employees with high technological self-efficacy have high self-confidence and can make effective plans in the situations they encounter.

- Psychological Health Resources: It is known that employees use their psychological, physical and emotional resources while maintaining their normal lives. While fulfilling their responsibilities in different areas of life, they try to use these resources effectively and efficiently. Psychological resources are an important resource that will increase the resilience and fighting power of employees in technostress. Trainings and practices can be made to increase the psychological resources of the employees.

3 Discussion

The main objectives of businesses are to achieve sustainable competitive advantage. By gaining sustainable competitive advantage, businesses guarantee their existence. In this context, it becomes a necessity for organizations to use ESG approaches. Taking measures to reduce technostress with ethical leadership approaches can increase the resilience of organizations. Organizations need to meet the needs and expectations of stakeholders as they carry out their technological transformations. In this study, practical suggestions were presented to the enterprises. In this way, while the inclusive and diversity-supporting policies of businesses increase, performance, effectiveness and efficiency are positively affected. After all, all these efforts can affect the sustainability of the business. Sustainability requires businesses to have permanent, strong, environmentally and nature-friendly, social and participatory processes. The result of each strategy presented in the literature review is aimed at increasing the resilience of the enterprise.

The strategies or policies that businesses use to increase resilience can increase the happiness of employees and create a strong organizational culture. It can provide flexible working conditions. It can provide the integration of ESG dimensions.

Social responsibility practices implemented by businesses may consist of components such as awareness trainings for individuals threatened by technostress, digital detoxification, cognitive awareness, moral and ethical leadership, and technostress risk programs. These practices make businesses have character and increase their sustainability.

There is no standard in the implementation of all these strategies. The environments in which the businesses are located, organizational culture, individual characteristics and other contextual variables determine the strategy to be used.

Although the study presents many strategies based on a literature review in relation to ESG on technostress, it has some limitations. These mentioned constraints are strategies that need to be considered at the micro level and may not be generalized. Since it is at the micro level, contextual features are important in the implementation of these strategies.

The strategies or tactics presented in this study are suggestions prepared assuming that businesses have certain resources. For businesses that don't have these resources, these strategies may not work. In this case, businesses need to develop other minimums.

Another limitation of the study was that the negative consequences of the proposed strategies were not taken into account. For example, interrupting the use of technological tools can weaken the emotional bond by distracting individuals from their work. Such negative consequences can be observed.

On the other hand, the study generalized enterprises and presented standard tools. Individuals' differences affect their resilience and technological self-efficacy. This, in turn, also affects their level of exposure to technostress. Studies that about individual differences can increase the validity of the study.

4 Final Considerations

In the contemporary landscape of the business world, the looming threat of technostress poses a significant challenge to organizations striving for sustainable competitive advantage. However, proactive measures to mitigate technostress can effectively fortify this advantage in the dimensions of environmental, social and corporate governance reducing the impact of technostress. In this study, firstly, was introduced the concept of technostress, then the concept of ESG is explained and the relationship between them is revealed, examples of strategies of enterprises to resist technostress have been developed. Although based on a literature review, this study (essay) is important for all enterprises stakeholders. Increasing the resilience of individuals against technostress can make them resilient not only in business life but also in other areas of life. The different strategies mentioned above, have been presented as suggestions in studies conducted with different variables in different publications (references). While the appealing side of technology attracts individuals, the dark side pushes individuals into a vicious circle.

Like every element in life, there is a need for balance, this balance is also needed in the use of technology. Therefore, it is necessary to pay attention to this, when developing a strategy against technostress. Business managers need to show empathy and leadership based on ethical and moral criteria. There is a need for ethical leadership practices that value people for stakeholders affected by technostress. It should be ensured that employees work in environments where they feel comfortable in the enterprise. Establishing a balance in employees' work and family lives requires ethical leadership. Otherwise, the number of unhappy employees may increase and the performance, efficiency and effectiveness of businesses may decrease. Attention should be paid to the physical and emotional health, especially the psychological health of the employees. Businesses with happy employees can turn into strong businesses. Supportive training of employees on technostress can empower them. This, in turn, increases their job satisfaction and decreases their turnover levels. And therefore, it can increase the resilience and sustainability of businesses. The effects of technostress began to be discovered very late. Due to the consequences of technostress, increasing business cycle rates and expenditures, decreasing performances and efficiency of enterprises required urgent solutions. From an ESG perspective, there is an emphasis on technostress in all three dimensions of sustainability. There are measures to be taken in terms of environmental, social and corporate governance. This can be in the form of raising awareness by raising awareness, increasing resilience with education and training policies, and supporting it with organizational practices.

Regular meetings with employees can help to observe their experiences. In addition, stakeholder participatory decision-making environments can provide strong feedback. It should be part of the strategies in trainings that will increase awareness about technostress and strengthen resilience.

Each strategy mentioned above in the literature review can be used as a good strategy to reduce the levels of technostress. Practices that will increase Work-Family/Family-work balance, open communication channels, use of environmentally friendly technology, sustainable technology management can be given as examples.

References

1. Bencsik, A., Juhász, T.: The impact of technostress onorganizational functioning. Probl. Perspect. Manag. **21**(1), 230–241 (2023). https://doi.org/10.21511/ppm.21(1).2023.20
2. Brivio, E., Gaudioso, F., Vergine, I., Mirizzi, C.R., Reina, C., Stella-ri, A., Galimberti, C.: Preventing technostress through positive technology. Frontiers in Psychology **9**, 256 (2018). https://doi.org/10.3389/fpsyg.2018.02569
3. Brod, C.: Technostress: the human cost of the computer revolution. Addison Wesley (1984)
4. LaTorre, A., et al.: Technostress: an epidemiological study in Italy. Int. J. Environ. Res. Public Health **16**(24), 5151 (2019)
5. Nastjuk, I., Trang, S., Grummeck-Braamt, J.V., Adam, M.T.P., Tarafdar, M.: Integrating and synthesising technostress research: a meta-analysis on technostress creators, outcomes, and IS usage contexts. Eur. J. Inf. Syst. **33**, 361–382 (2023)
6. Nature.: Integration of environmental, social, and governance (ESG) criteria. Scientific Reports, **13**, 1–32 (2023)
7. Penn, R.L.: Mitigation strategies of technostress on supply chain management. Walden University Unpublished Doctoral Thesis (2016)
8. Salanova, M., Llorens, S., Cifre, E.: The dark side of technologies: technostress among users of information and communication technologies. Int. J. Psychol. **48**(3), 422–436 (2013). https://doi.org/10.1080/00207594.2012.680460
9. Tarafdar, M., Maier, C., Laumer, S., Weitzel, T.: Explaining the link between technostress and technology addiction for social networking sites: a study of distracion as a coping behavior. Inf. Syst. J. **30**(1), 1–214 (2020). https://doi.org/10.1111/isj.12253
10. Tarafdar, M., Tu, Q., Ragu-Nathan, B.S., Ragu-Nathan, T.S.: The impact of technostress on role stress and productivity. J. Manag. Inf. Syst. **24**(1), 301–328 (2007)
11. Yener, S., And, Arslan A., Kılınç, S.: Moderating roles of technological self-efficacy and time management in the technostress and employee performance relationship through burnout. J. Inf. Technol. People **34**(7), 1890–1919 (2021)

The Relationship Between Technostress and Environment, Social, and Governance (ESG)

Helena Belchior Rocha, Serdar Yener, Aykut Arslan, and Sebahattin Kilinç

Abstract This study was conducted to reveal the relationship between technostress and ESG (Environment, Social, and Governance) in the business world. Besides definition of technostress and ESG concepts, the impact of technostress, which is one of the growing dangers of today's world, on the sustainability of businesses and the measures that businesses can take against technostress are discussed. The main goal of businesses is to make a profit, and they try to make a profit by gaining sustainable competitive advantage. The ability of businesses to achieve sustainable competitive advantage depends on many factors. Business-related factors, situational factors, individual factors affect the sustainable competitive advantage of enterprises. Technostress has an intense relationship with these three key factors that affect sustainable competitive advantage. In today's world, where technology is increasingly integrated into all life, we encounter the negative effects of technology as well as its good aspects. While businesses integrate technology into processes to increase their performance, employees necessarily interact with technology intensively. Outside of work, individuals use e-commerce or other applications to meet their daily needs while trying to meet their psychological needs through technological tools such as social media. As can be seen, individuals have interacted intensively with technology both in the business environment and in their living spaces outside of work, and this interaction is increasing. Today, we can see some of the dark side of technology, but if the interaction continues to increase in this way, it is almost impossible to predict the consequences of this interaction in the future. The danger of technostress, which is seen as a threat to sustainable competitive advantage for businesses, is more complex.

H. B. Rocha (✉)
Portugal Iscte, University Institute of Lisbon, Lisbon, Portugal
e-mail: helena_rocha@iscte-iul.pt

S. Yener
Sinop University, Sinop, Türkiye

A. Arslan
Piri Reis University, Sinop, Türkiye

S. Kilinç
Ministry of National Defense of Türkiye, Ankara, Türkiye

© The Author(s), under exclusive license to Springer Nature Switzerland AG 2025
A. d. b. Machado et al. (eds.), *Environmental, Social, Governance and Digital Transformation in Organizations*, Information Systems Engineering and Management 35, https://doi.org/10.1007/978-3-031-86079-9_17

Because businesses both need technology and have to meet this need in a delicate balance. In this study, the relationship between technostress and sustainability was tried to be revealed based on the literature review. The study also suggested several strategies that businesses can use to combat technostress. Stakeholders of businesses can be involved in decision-making processes with participatory processes. It is also recommended that businesses make their technological transformations within the framework of ESG criteria. In addition, the need for a governance process in which all stakeholders participate is emphasized.

Keywords Technostress · Governance · Business environment

1 Introduction

In today's world, technology has become an indispensable part of our lives. Technology has changed our daily life routines and deeply affected our communication, work and social relations [1]. While technology enters our lives with its attractive aspects, it becomes indispensable after a while. While businesses accelerate technological transformations in order to automate their daily processes, increase their performance, improve efficiency and effectiveness, individuals have started to increase their interaction with technology to meet their social and psychological needs by getting caught up in the general trend of society [2]. As individuals and organizations increase their interaction with technology for innocent purposes, the dark side of technology also reveals itself [3]. The reactions of individuals as a result of the change in their normal lives due to the negative effects of technology and the inability of individuals to adapt to it are called technostress [4]. Technostress is defined as a modern-day disease in recent studies [5]. The physical and psychological reactions of individuals as a result of their inability to meet and adapt to the needs of changing technology are also considered as technostress. While technostress is defined as the emergence of the above-mentioned reactions as a result of intensive technology use, it negatively affects the well-being of individuals while harming the routines of individuals [6].

Sustainability has been used extensively in the literature as a measurement tool and target for businesses in recent years. The concept of sustainability emphasizes that businesses do not only exist to make a profit, but also have to have interactional relationships with different stakeholders and the natural environment in the environment in which they operate [7]. It is emphasized in many studies that businesses have responsibilities towards the natural environment and social environment and that they should fulfill these responsibilities with balanced relations [8]. These responsibilities of businesses are conceptualized within the framework of Environmental, Social and Governance dimensions of sustainability. Today's global businesses have long been criticized for ESG factors and are forced them to take more initiative in ESG areas.

The fact that technology increasingly affects both organizations and individuals makes it impossible for businesses to determine their sustainability strategies without considering the technostress factor [9]. The sustainability of businesses varies based on their intellectual capital, that is, their employees. The high quality and motivation of the employees creates a multiplier effect for businesses. Employees are the key to innovation, performance, effectiveness and efficiency. In the literature, it is emphasized in many studies that employees are the main determinants of the sustainability of businesses [10]. The target of the negative consequences of technostress is the employees. Employees' inability to adapt to technology may cause them to experience technostress, and if this becomes chronic, it may cause burnout [5]. While technostress negatively affects the motivation of individuals, it also reduces the performance, effectiveness and efficiency of businesses due to low satisfaction and well-being. Businesses with decreased satisfaction and well-being levels decrease their loyalty and loyalty to the business over time, and their turnover rates increase. Therefore, it is necessary to understand the negative aspects of technology and its relationship with environmental, social and governance dimensions, which are sustainability dimensions. Leaders who resolve this relationship can take precautions against technostress and ensure their sustainability with proactive measures.

In this study, the complex relationship between the ESG dimensions of technostress and sustainability, which is briefly mentioned above, has been tried to be revealed based on the literature review. In the introductory part of the study, while the concepts of technostress and sustainability are defined in general, clues about the relationship between them are revealed. In the literature review of the study, first of all, the concept of technostress has been put forward comprehensively with its sub-dimensions. According to the studies conducted in the literature, studies with different variables are mentioned. Technostress determinants, its effect on individuals and businesses, the effects of technostress in businesses with changing and transforming technology, the disappearance of boundaries between work and private life with the effect of technostress, the physical and psychological symptoms of technostress, the effects of technostress on the well-being and performance of individuals are presented.

After the definition of the concept of technostress in the literature review, the concept of sustainability has been defined comprehensively with its sub-dimensions (ESG). It has been emphasized that businesses cannot only focus on making a profit in today's world, but should take into account the social and natural environment in which they operate. It is presented with examples that it is impossible for businesses isolated from their social and natural environment to survive today. The responsibilities of enterprises towards the social and natural environment are exemplified. Exemplary strategies such as nature-friendly strategies for the natural environment, reduction of carbon footprint, waste management, and use of renewable energy sources are specified. Within the responsibilities towards the social environment, volunteering and charity activities towards disadvantaged groups and initiatives to increase participation were presented.

Then, the relationship between technostress and sustainability was tried to be presented comprehensively. Then, the conclusion and discussion part are presented. In addition, the limitations of the study and suggestions for future studies are presented.

2 Technostress

2.1 Conceptual Framework

For businesses, the use of technology means an increase in performance, effectiveness and efficiency [11]. While businesses increase productivity by improving their production processes with the use of technology, they also provide management and coordination by using it for communication and innovative purposes. At the same time, in terms of marketing, they use technology to ensure their integration into international markets. The negative effects of technology, which is used extensively to perform almost all functions of the enterprise, have also begun to be seen [12]. Some symptoms seen in individuals who are heavily exposed to technology have started to increase suspicion and criticism of technology. First of all, with the intense interaction of technology, the limits of human beings have started to be pushed, their living spaces have started to be threatened, and individual resources have started to be depleted. These problems, which occur with different symptoms, have begun to be defined by a disorder defined as technostress [13] in the literature. Technostress is defined as the negative reactions of individuals as a result of their inability to adapt to the changes that threaten their lives with the use of technology [4]. When we look at the studies, it is observed that technostress reduces the well-being and productivity of individuals, negatively affects job attendance and increases turnover.

Intensive studies on technostress, which was conceptualized in 1984 [13], took place after the 2000s, when internet technologies began to develop. Because the 2000s were the years when not only business employees but also other individuals felt the negative effects of technology due to technology addiction [2]. After the 2000s, studies on technostress have generally focused on revealing the determinants, determining the results and explaining their relationships with other variables.

When we look at the studies, it is observed that one of the main causes of technostress is the fast and constantly changing technology [14]. Employees are expected to adapt to the ever-changing and transforming technology. From the point of view of employees, adapting to constantly changing technology can create a perception of inadequacy [15]. Anxiety and worry may arise when s/he thinks that s/he will not be able to adapt. In addition, with the technology used, the separation of work and private life of employees is eliminated. It is observed that sometimes unwittingly and sometimes knowingly forced to fulfill work-related responsibilities in the private sphere of life. This causes the perception of employees to deteriorate in their perceptions of work and private life.

When we look at the studies, it is observed that the symptoms of technostress occur in many ways. Examples of physical symptoms include insomnia, constant headaches, and a feeling of fatigue. Psychological symptoms include anxiety, worry, and hopelessness [5]. Physical and psychological symptoms negatively affect the well-being of individuals and reduce their productivity, performance, effectiveness and efficiency.

2.2 Technostress Determinants

When we look at the studies on technostress, the determinants are organizational-based, situational and individual-based [16]. Individual determinants of technostress are often based on demographic characteristics. Determinants (factors) of technostress are presented below [2, 5].

Techno-overload: Technological workload refers to the increase in the responsibilities of employees with the use of technology and the effect of these responsibilities on other responsibilities. Individuals who are trying to fulfill increasing responsibilities are forced to work faster and harder. In addition, due to the interconnection of technological tools, employees are expected to do many jobs at the same time.

Techno-invasion: Techno-invasion describes the invasion of employees' lives by technology. With technology occupying the lives of employees, employees in the workplace are expected to be constantly accessible and in communication. This situation is experienced intensely in enterprises where continuous communication is provided with the technological infrastructures used. In fact, tools such as computers, smartphones and tablets that they can use outside of work are allocated free of charge to those who start working, ensuring that employees are constantly accessible. This means that employees use the time they allocate for themselves outside of work for work.

Techno-complexity: It has been mentioned above that one of the main determinants of technostress is constantly developing and changing technology. Individuals who try to adapt to the ever-changing and developing technology are also expected to develop their talents and skills. In this case, the constantly developing and changing technology creates a complex situation, and individuals experience the confusion of adapting in this complexity.

Techno-insecurity: The fear of losing one's jobs in case of not being able to adapt to the ever-changing and developing technology is defined as techno-insecurity. This anxiety, which will adversely affect the normal lives of individuals, can turn into a condition over time and lead to a sense of cynicism.

Techno-uncertainty: Techno-uncertainty is the uncertainty caused by ever-changing and evolving technology. In this case, individuals cannot predict what they may encounter in the future and cannot prepare for it. Therefore, individuals cannot prepare themselves and therefore cannot adapt to technology.

When the determinants of technostress are examined, it can be seen that it is caused by organizational, situational and individual factors. However, developing strategies that take these determinants into account can prevent technostress from occurring.

2.3 Technostress Effects

Technostress, which has been extensively introduced above, has significant destructive effects on individuals and organizations [17]. The decrease in job satisfaction and quality of life perceptions of individuals with the intellectual capital of organizations due to technostress leads to a decrease in their performance. At the same time, psychological and physical disorders or disorders appear over time. It is known that burnout can also be experienced when technostress becomes chronic [5]. In addition, the organizations in which individuals are involved are also affected by these negative results. The purpose of the establishment of organizations is to make a profit in general and to achieve this through sustainability. Today, the sustainability determinants of organizations consist of environmental, social and governance (ESG) dimensions [7]. The environmental, social and governance responsibilities of organizations interact intensely with the perception of technostress. For example, harmonizing and meeting the expectations of all stakeholders, especially employees, carbon footprint, energy use, and participation of stakeholders can be seen as responsibilities that are both technology-related and may cause technostress. Therefore, from the point of view of ESG, the interaction of organizations with technology creates challenging responsibilities and tasks for organizations. Organizations' failure to fulfill these responsibilities and duties can threaten their sustainability.

2.4 Technostress in the Workplace

Considering the determinants of technostress, many factors may be caused by the business. Factors such as Techno-overload, Techno-invasion, Techno-complexity, Techno-insecurity, Techno-uncertainty have both organizational and situational causes. The working conditions of businesses can reveal techno-overload. Increased workload with the use of technology in a way that exceeds the abilities and skills of employees can pose problems. The operational processes of enterprises can threaten the lives of employees through techno-invasion. Physical and psychological occupational health and safety measures in the working conditions of enterprises can lead to stress. Businesses' training and development programs for employees' skills and abilities can change the perceptions of techno-insecurity and techno-uncertainty in their abilities.

2.5 Mitigating Technostress

Organizations can adopt various strategies to mitigate the effects of technostress. These include implementing work-family/family-work balance policies, providing training and support to enhance individuals' technological skills, creating appropriate working environments that do not threaten the physical and psychological safety of employees, and adopting ethical leadership approaches that ensure transparency and inclusivity in decision-making processes.

3 Environment, Social and Governance

3.1 Environmental Sustainability

In terms of strategic management, environmental factors are categorized from the broader external environment (PEST) to the internal environment of the enterprise, with specific consideration given to the environment [18]. Enterprises exert various influences on their surroundings through their operations, impacting the culture, economy, livelihoods, and ecosystem of the locales in which they operate. Within the context of the environmental dimension of ESG, businesses typically focus on their ecological footprint and impact on the natural environment. The impact of an organizations on the environment is generally questioned in terms of how much carbon footprint it creates around it with its activities, how it processes its wastes if it produces waste, and how it affects the natural assets around it [19]. For instance, implementing measures such as installing filters in production organization chimneys, transitioning from solid fuels to renewable energy sources, and utilizing energy-efficient vehicles can mitigate the environmental impact of businesses. Other environmentally conscious practices include reducing waste through enhanced recycling efforts and devising strategies aimed at minimizing harm to natural resources [20]. Failure to adopt such measures may exacerbate climate change and deplete natural resources, further compounding environmental degradation caused by business activities. Examples of environmental sustainability initiatives include:

- Energy Efficiency Studies: The use of tools and equipment with high energy efficiency by businesses can reduce energy consumption. Switching from fossil fuel consumption to renewable energy sources as an energy source can contribute to environmental sustainability.
- Waste Management Studies: If there is potential to repurpose waste generated during organizational production processes, arranging for its utilization in other sectors or implementing recycling initiatives, including zero waste research and practices, can significantly contribute to environmental sustainability.
- Sustainable Supply Chain: Obtaining raw materials and semi-finished products used by the organizations from sources that are not harmful to the environment can ensure a sustainable supply chain. An example of this is that production organizations use materials that require less water and use raw materials produced naturally.

3.2 Social Sustainability

The relationships that organizations establish with customers, employees, suppliers, and other members of society within the communities in which they operate determine the degree of social sustainability [20]. Psychological and physical health of employees, working standards, customer satisfaction, suppliers' perceptions of the business, socio-cultural relations can be seen as social sustainability topics. For example, if the working hours or conditions of the employees threaten their family lives, it will be a threat to social sustainability.

Organizations should also contribute to issues such as: the protection and improvement of the social structure in the society in which they operate. From this point of view, socially responsible businesses can engage in activities that will increase the participation of disadvantaged groups within society in which they are located or fund these activities [21]. In addition, it is observed that societies have become more diverse with the increase of human movements and migrations, due to wars, regional crises, and climate change (reference). In these complex and diverse societies, fostering inclusion and diversity can serve as exemplary social sustainability initiatives for both organizations and the communities they inhabit. Recognizing every member of society as a stakeholder in organizational endeavors, efforts aimed at enhancing the quality of life and satisfaction of all stakeholder groups contribute to social sustainability. The following practices can be shown as examples of social sustainability activities:

- Employee Well-Being: It can be viewed as organizations implementing practices to meet the expectations and needs of their employees. For example, flexible working conditions, elimination of working conditions that threaten psychological and physical health, providing support for a healthy and fit life, providing sports or exercise opportunities, health programs can be given as examples.
- Diversity and Inclusion: Implementing practices and policies that foster inclusivity and diversity within business workplaces, while also working to reduce or eliminate prejudices, and providing essential mentoring services to marginalized groups, can serve as exemplary initiatives. All kinds of diversity and inclusion practices that will increase the participation of minorities, immigrants and other disadvantaged groups in society should be evaluated within this framework.
- Community Participation: Examples of these initiatives include donation and volunteering programs, as well as facilitating practices aimed at enhancing the socio-cultural development of the communities in which businesses operate. Organizations that encourage or provide funding for their employees to participate in such activities can make a positive impact in this regard.

3.3 Sustainability in Governance

Businesses have a multitude of stakeholders, both direct and indirect, internal and external. These include business owners or shareholders, employees, managers, customers, suppliers, and other members of society. Organizations adopting a corporate governance approach attempt to involve all these stakeholder groups in their decision-making processes. In particular, a corporate governance strategy that identifies the expectations of each stakeholder group from the business and works towards meeting them to the maximum strengthens sustainability [22]. Establishing stakeholder expectations, defining performance metrics to fulfill these expectations, outlining authorities, duties, and responsibilities, disclosing processes, and managing, auditing, and controlling them can foster corporate governance. This framework requires robust internal audit processes and independent audits. The guidelines that determine all these corporate governance processes and their control are strategies that increase the sustainability of businesses. With corporate governance, businesses end creating value from their various stakeholders and guarantee their sustainability in the long term. Some of the practices that can be shown as examples of corporate governance for businesses are:

- Diversity of the Board of Directors: The tool of decision-making mechanism of businesses is the board of directors. Having representatives of different stakeholders on the boards of directors of businesses is an important stage in corporate governance. It is an important indicator that representatives of all levels or diversity stakeholder groups are involved in the decision-making mechanism in the organizations. On boards of directors, gender, race, religion, age, etc. A structure that takes into account differences/diversity is a strong indicator of corporate governance.
- Managers' Compensation: In businesses, managers manage the business on behalf of the shareholders within the framework of the proxy theory [23]. Within context, the fact that the managers who work to protect the assets of the shareholders and increase them in accordance with their expectations benefit from the gains achieved, increases the loyalty of the managers. Some businesses draw long-term career plans for employees within this framework.
- Transparency: Transparency entails businesses conducting all their operations openly and transparently, ensuring that all stakeholders have access to the information they seek. This includes the preparation of regular reports such as balance sheets and income statements according to established criteria, demonstrating openness and honesty. It is very important that business shareholders or investors have access to the information about the business, in addition, it is also important for businesses to routinely compile sustainability reports and substantiate their contents. Some enterprises provide intensive information about their activities to stakeholders with routine reports.

4 ESG and Technostress

4.1 The Relationship Between ESG and Technostress

In today's world, technology permeates every aspect of individuals' lives, extending beyond computers to encompass smartphones, virtual networks, project teams services, and products. It has become an indispensable part of daily life, erasing boundaries between different spheres. The traditional constraints of work hours are no longer applicable, and the demarcation between work and family life is increasingly blurred. Virtually every industry incorporates technology, necessitating a fundamental proficiency in technology for individuals across various sectors [14]. The fact that technology occupies the lives of individuals so much, negatively affects healthy decision-making and the perception of quality of life. In these conditions, individuals find it difficult to adapt to both different areas of life and their work. The negative impact of technology on the normal living standards of individuals and the inability of individuals to adapt to this situation is considered as technostress [13]. Individuals may experience problems such as disappointment, pessimism, cynicism, headache, and insomnia during the technostress process [5]. The gradual increase in employee ailments caused by technostress in Western countries reduces the performance of businesses and increases the employee cycle and health expenses. The sustainability of organizations where technostress is felt so intensely is also endangered, because organizations have become unable to carry out their activities without using technology-based tools in their production or commercial processes. Therefore, while organizations are experiencing technological transformation, they inevitably experience the dark side of technology, in this sense, organizations must maintain a balanced approach to technology usage and shield themselves from potential adverse impacts while striving to achieve sustainable competitive advantage, which is one of their main objectives. It's essential to recognize that while technological tools can enhance effectiveness and efficiency, they can also serve as sources of technostress. Hence, organizations must consider the effect of technostress in their environmental, social and corporate governance processes, to achieve sustainability.

In terms of sustainability, ESG dimensions require businesses to be transparent, accountable, committed to ethical values, inclusive, respectful of diversity, minimizing their environmental impacts, and respecting the socio-cultural structure of the society. In terms of ESG, as explained above, it requires organizations to have an approach that advocates for the conservation of natural resources in their production and commercial processes, think about climate change and its future effects, and bases the supply chain on organic processes. At the same time, it should have an approach that will include it in the decision-making process in a way that meets the expectations of all its stakeholders, especially employees and customers.

In the corporate governance process, fostering the sustainability of organizations trying to meet employee expectations is fundamental. Incorporating psychological and physical safety measures within workspaces, along with implementing practices aimed at enhancing morale and motivation, profoundly impacts sustainability.

Furthermore, concerning technostress, one of the most significant factors affecting employees in the workplace is the unrestricted and unpredictable utilization of technological tools, encroaching upon employees living spaces. While employees are expected to adapt quickly to new technologies, they are also expected to work at a pace that will eliminate the boundaries of their work and family living spaces. In terms of ESG, it is foreseen that the positive contributions of employees to businesses in these processes will not be sustainable. It is essential for businesses to base their employee's relationship with technology on ethical foundations. Sustainable competitive advantage necessitates businesses to operate effectively and efficiently. In order for businesses to work effectively and efficiently, they need to use technologies that facilitate business processes intensively. Because technology contributes to efficiency while increasing the amount of work done per unit time and contributes to efficiency while facilitating the achievement of targets. Technologies that provide energy efficiency, technologies that enable optimization of resources, and technologies that enable remote working can be given as examples and its use contribute to reducing the environmental impact of businesses, at the same time, it can increase the effectiveness of businesses in fulfilling their social responsibilities. However, the intensive use of technology can also adversely affect the health of employees. Therefore, businesses today struggle in a vicious circle between sustainability and technology-related stress, this leads us to the fact that businesses should act in a controlled manner in the use of technology. There is a need for strategies that will positively affect the quality-of-life perceptions of stakeholders, especially employees, improve work/family life balance, and eliminate physical and psychological ailments. There are studies revealing that technostress can lead to burnout when it becomes chronic [5]. By taking these negative consequences into account, businesses can achieve better results when combining the use of technology with ESG criteria.

In the context of Environmental, Social, and Governance (ESG) factors, technostress can have significant implications. The unrestricted and unpredictable use of technological tools can encroach upon employees' living spaces, leading to technostress. This can threaten the social sustainability of organizations as it can negatively impact the psychological and physical health of employees, working standards, and work-life balance. Therefore, organizations must consider the effect of technostress in their ESG processes to achieve sustainability.

Considering the determinants of technostress, many factors can affect this process. While organizational factors come to the fore, personal and situational factors are also affected [4]. The interaction of organizational, situational and personal factors makes it difficult to reveal the effect of the determinants of technostress. For example, the techno-overload factor refers to the excessive interaction of individuals with technology and their intensive and long-term use of technology. Within the framework of ESG, measures can be taken to eliminate the intense and long-term interactions of individuals with technology and to increase the work/family balance.

The techno-invasion factor describes the infestation of the family living space with technology and the elimination of work-life boundaries with the intense exposure of individuals to technology [2]. Involving employees in decision-making processes to develop suggestions that will eliminate these situations lead to creativity and alternative solutions. Consequently, employees who are involved in decision-making processes feel empowered and valued.

Techno-complexity is defined as the transformation of user's perceptions by constantly evolving and changing technological tools [3]. The increasing integration of technology into various aspects of life poses a challenge for individuals to adapt to these developments. As technology permeates more spheres of existence, individuals find themselves compelled to enhance their technological knowledge and competence, thereby adding to the complexity of this effort. Within the ESG framework, exists an opportunity to provide guidance and training to individuals heavily reliant on technology. By offering targeted education and support, initiatives can be made to bolster individuals' technological skills and proficiency, ultimately empowering them to cross the digital landscape more effectively, ensuring that their skills in terms of technology can be increased.

Creating appropriate working environments based on an approach that balances the expectations of stakeholders with the use of technology, and creating environments that will not threaten the physical and psychological occupational safety of employees will eliminate techno-insecurity or reduce its impact. In this case, employees do not think that not being able to adapt to constant changing will jeopardize their job security or increase their career anxiety. On the other hand, ethical approaches in business leadership transform both the work areas and the behavior of individuals into positive ones. With ethical leadership, it can be ensured that processes are transparent and open and easier to include relevant stakeholders in decision-making processes.

For business managers, it is indispensable to consider all environmental, social, and governance factors impacting sustainability in their decision-making during technological transformations. In this way, the entire organization can embrace technological changes, fostering sustainability through participatory and inclusive decision-making processes.

4.2 ESG-Aligned Technostress Policies

While ESG strategies can be created to eliminate or reduce the effects of technostress in businesses, technostress policies that will support sustainability can also be created. These measures to reduce the effects of technostress in businesses may include:

- Work-Family/Family-Work Balance Policies: Imbalance or conflict is defined as when employees' responsibilities in business life threaten family life or when their responsibilities in family life threaten business life. Therefore, plans should be made to prevent the responsibilities of individuals in business life from threatening family life. Or policies can be created to support each other, including flexible working or remote working options.
- Transparent Communication Policies: Communication makes it easier to solve misunderstandings and other related problems in businesses. Therefore, problems

can be solved if measures are taken to create effective and transparent communication forms with all employees. It might make sense for this to start with senior management. Routine or instant information bulletins and routine meetings of senior management on issues of interest to stakeholders can strengthen communication.

- Sustainable Technology Use Policies: Sustainability can also be defined as the continuation of activities without threatening the expectations of all stakeholders and the environment. In this case, sustainable technology use policies can be created to align the expectations and needs of employees, who can be seen as one of the most important stakeholder groups, with the expectations of the business. These policies may be in a way that does not threaten or improve the above-mentioned individuals' living spaces outside of work.

5 Results and Discussion

Based on the literature review of this study, the relationship between technostress and sustainability can be examined from a broad framework. In this study, the relationship between technostress determinants and sustainability components is presented comprehensively. In addition, it is emphasized that if the phenomenon of technostress cannot be handled in a meaningful way, the sustainability of the business may be endangered.

- Environmental Sustainability: From the point of view of the impact of technostress on environmental sustainability, the excessive use of technology creates a techno-overload for individuals and increases energy consumption and energy waste for businesses. Energy waste can be reduced by using energy-efficient vehicles, but more efficient solutions for techno-overload should be considered.
- Social Sustainability: Another issue affected by technostress is social sustainability. Since technostress affects the well-being and life satisfaction of individuals, it also threatens individuals to establish healthy interactions in society. Burnout through chronic technostress can increase job turnover. In order to prevent these, it is recommended to use technology in a way that does not threaten the social and psychological lives of individuals. For example, work-family balance, diversity and diversity management policies can be recommended.
- Governance Sustainability: The use of technology can positively and negatively affect the corporate governance processes of businesses. In order to strengthen corporate governance, businesses involve their direct/indirect stakeholders in more decision-making processes with technological tools, while their intensive use can cause boredom and distancing. The accountability and transparency of businesses can also be increased with a balanced use of technology.
- ESG-Aligned Technostress Policies: Sustainability-oriented technostress policies may contain situational differences. The situational characteristics of each business are different from other businesses. Therefore, businesses can produce situational solutions suitable for their own characteristics. Work-family balance

policies, diversity and diversity management policies, and policies to strengthen decision-making processes by increasing participation can improve the sustainability of businesses.

- Technostress and ESG Sustainability: Although the study is based on a literature review, it contains many tips to increase the sustainability of businesses. First, technostress is a complex process that requires a very serious perspective. It negatively affects the sustainability of businesses by threatening the health of employees. Therefore, businesses can guarantee their sustainability by balancing the use of technology in line with the expectations and needs of stakeholders.

As comprehensively explained in the study, the relationship between technostress and sustainability is very complex. Technology can make the business processes of businesses efficient and effective. For example, by reducing remote working opportunities and flexible working conditions, it can positively affect both social and environmental sustainability (reducing carbon footprint and emissions). On the other hand, faster execution of work with excessive use of technological tools may cause an increase in the workload, and the constant change and self-renewal of technology may lead to an increase in the perception of uncertainty and confusion.

That's why businesses need to make their technology usage decisions very carefully. Businesses need to make technology use decisions by comprehensively examining the negative effects that may arise from the use of technology and taking into account all dimensions of sustainability. In addition, it is thought that businesses can reduce the negative effects of technology with the measures they can take. For example, training and guidance to be given to employees, continuous feedback and meetings can produce positive results in this regard.

6 Final Considerations

6.1 Limitations and Futher Studies

While the study provides comprehensive information between technostress and sustainability, it also has some limitations. It also contains suggestions for future studies on the subject.

- Scope of Study: The scope of the study is determined only to explain the theoretical definition of technostress and its effect on ESG. However, testing the same variables in field studies can reveal more concrete results.
- Technostress Measurement: Technostress measurement is carried out by a process that requires subjective measurement tools. In the literature, it is seen that tools related to technostress measurement have been developed. In addition, tools that

will reveal the technostress effects of enterprises and make objective measurements can provide an analytical approach to the subject. For example, the technological workload scale of enterprises and the techno-invade scale can be developed.

- Cultural Factors: The study discussed the relationship between technostress and sustainability in general terms. Cultural differences that would reveal the perceptual differences of individuals were not taken into account. Field studies that take into account cultural differences can produce interesting results. Knowing cultural differences can enable businesses to create more effective policies.
- Technological Changes: The study has been prepared by taking into account the current state of technology and its current effects. It is emphasized in different parts of the study that technology is constantly renewed and changing. Since the study was prepared based on current technology, its effects and results were evaluated within this framework. In accordance with the changing technology and tools, the work can be revised in the future.
- ESG Sustainability Practices: In this study, the general dimensions of sustainability were taken into consideration. However, the sustainability literature for businesses may include broader variables and determinants. Therefore, the study can be improved with different sustainability examples.
- Interdisciplinary Approach: The study dealt with the relationship between technostress and sustainability within the framework of business in general. However, technostress concerns many different disciplines. For example, in disciplines such as information technologies, psychology, sociology, organizational behavior, health, studies that will reveal the effect of technostress or interdisciplinary studies can be carried out in the future.
- Mitigation Strategies: Since the study is generally based on the relationship between technostress and sustainability, strategies to reduce the effect of technostress are not given comprehensively. Although a few general suggestions have been presented, new studies are needed to reveal strategies suitable for the situations and conditions of the enterprises.

This study is designed to present the relationship between technostress and sustainability from a logical framework. The results obtained in the study can create valuable recommendations for businesses. However, within the framework of the suggestions presented above, there is a need to carry out studies that will enrich this relationship in the future.

References

1. Tiwari, V.: Relationship among technostress productivity and role stress. Purushartha **2020**(13), 68–83 (2020)
2. Tarafdar, M., Maier, C., Laumer, S., Weitzel, T.: Explaining the link between technostress and technology addiction for social networking sites: a study of distracion as a coping behavior. Inf. Syst. J. **30**(1), 1–214 (2020). https://doi.org/10.1111/isj.12253

3. Ragu-Nathan, T.S., Tarafdar, M., Ragu-Nathan, B.S., Tu, Q.: The consequences of technostress for end users in organizations: conceptual development and empirical validation. Inf. Syst. Res. **19**, 417–433 (2008)
4. Tarafdar, M., Tu, Q., Ragu-Nathan, B.S., Ragu-Nathan, T.S.: The impact of technostress on role stress and productivity. J. Manag. Inf. Syst. **24**(1), 301–328 (2007)
5. Yener, S., Arslan A., Kılınç, S.: Moderating roles of technological self-efficacy and time management in the technostress and employee performance relationship through burnout. J. Inf. Technol. People **34**(7) (2021)
6. Upadhyaya, P.: Building peace through education and culture: evolving UN perspectives. Strateg. Anal. **44**(5), 429–437 (2020)
7. McKinsey (2022). ESG Framework|McKinsey. Available online: https://www.mckinsey.com/capabilities/strategy-and-corporate-finance/ourinsights/five-ways-that-esg-creates-value. Accessed on 5 May 2024
8. Slavković, M., Sretenović, S., Bugarčić, M.: Remote working for sustainability of organization during the COVID-19 pandemic: the mediator-moderator role of social support. Sustainability **14**(1), 70 (2021). https://doi.org/10.3390/su14010070
9. Christ-Brendemühl, S., Schaarschmidt, M.: The impact of service employees' technostress on customer satisfaction and delight: a dyadic analysis. J. Bus. Res. **117**, 378–388 (2020). https://doi.org/10.1016/j.jbusres.2020.06.021
10. Hill, J.: Environmental, social, and governance (esg) investing: a balanced analysis of the theory and practice of a sustainable portfolio. Academic Press, Cambridge, MA, USA (2020) ISBN 0128186933
11. Kim, D.G., Lee, C.W.: Exploring the roles of self-efficacy and technical support in the relationship between techno-stress and counter-productivity. Sustainability **13**(8), 4349 (2021). https://doi.org/10.3390/su13084349
12. Duarte, M., Caeiro, S.S., Farinha, C.S., Moreira, A., Santos-Reis, M., Rigueiro, C. and Simão, J.: Integration of sustainability in the curricula of public higher education institutions in Portugal: do strategic plans and self-report align? Int. J. Sustain. High. Educ. **24**(9), 299–317 (2023). https://doi.org/10.1108/IJSHE-01-2023-0001
13. Brod, C.: Technostress: the human cost of the computer revolution. Addison Wesley (1984)
14. Weinert, C., Maier, C., Laumer, S., Weitzel, T.: Technostress mitigation: an experimental study of social support during acomputer freeze. J. Bus. Econ. **90**(8), 1199–1249 (2020). https://doi.org/10.1007/s11573-020-00986-y
15. Yasin, E., Abdelmaboud, A., Saad, H., Qoura, O.: Side and the light side of technostress related to hotel innovations: transforming the hospitality industry or threat to human touch. Int. J. Tour. Archaeol. Hosp. **1**(1), 44–59 (2021). https://doi.org/10.21608/ijtah.2021.243721
16. Li, L., Wang, X.: Technostress inhibitors and creators and their impacts on university teachers' work performance in higher education. Cogn. Technol. Work **2020**(23), 315–330 (2020)
17. Park, H.J., Cho, J.S.: The influence of information security technostress on the job satisfaction of employees. J. Bus. Retail Manag. Res. **11**, 66–75 (2016)
18. Daugaard, D.: Emerging new themes in environmental, social and governance investing: a systematic literature review. Account. Financ. **2020**(60), 1501–1530 (2020)
19. Gupta, A., Sharma, U., Gupta, S.K.: The role of ESG in sustainable development: an analysis through the lens of machine learning. In: Proceedings of the 2021 IEEE international humanitarian technology conference (IHTC) Online, pp. 1–5. IEEE, New York, NY, USA (2021)
20. van Duuren, E., Plantinga, A., Scholtens, B.: ESG integration and the investment management process: fundamental investing reinvented. J. Bus. Ethics **2016**(138), 525–533 (2016)
21. Amel-Zadeh, A., Serafeim, G.: Why and how investors use ESG information: evidence from a global survey. Financ. Anal. J. **74**, 87–103 (2018)

22. Bhandari, K.R., Ranta, M., Salo, J.: The resource-based view, stakeholder capitalism, ESG, and sustainable competitive advantage: the firm's embeddedness into ecology, society, and governance. Bus Strategy Environ. **31**, 1525–1537 (2022)
23. Theodorou, A., Dignum, V.: Towards ethical and socio-legal governance in AI. Nat. Mach. Intell. **2**, 10–12 (2020). https://doi.org/10.1038/s42256-019-0136-y

Examining Successful ESG Strategies: A Collection of Case Studies from Various Sectors in Oman

Sangeeta Tripathi

Abstract This paper presents a collection of case studies highlighting successful environmental, social, and governance (ESG) strategies implemented across various sectors in Oman. These case studies examine the diverse approaches Omani organizations adopt to address ESG challenges and capitalize on opportunities in Oman. The research methodology employed in this study is based on in-depth case analysis, drawing on secondary data sources, company reports, and industry publications. The case studies span multiple sectors, including energy, manufacturing, tourism, and finance, providing a comprehensive overview of successful ESG strategies in Oman. Each case study delves into the specific challenges faced by the respective sector and examines the strategies and initiatives implemented to achieve positive ESG outcomes. Key focus areas include environmental conservation, community engagement, sustainable resource management, employee welfare, and ethical governance practices. Findings from the case studies demonstrate the importance of integrating ESG considerations into business operations, highlighting the positive impacts on organizational performance, reputation, and long-term sustainability. The case studies also shed light on the significance of stakeholder engagement and collaboration in successfully implementing ESG strategies, emphasizing the need for strong partnerships between businesses, government entities, local communities, and non-governmental organizations. The research findings contribute to the existing knowledge of ESG strategies and practices, showcasing real-world examples of successful implementation in Oman. The insights gained from these case studies can serve as valuable references for businesses and organizations seeking to enhance their ESG performance and align with global sustainability goals.

Keywords Environmental · Social · Governance · ESG strategies · Case studies · Oman · Sustainability · Sector-specific · Stakeholder engagement

---

S. Tripathi (✉)
University of Technology and Applied Sciences, Salalah, Oman
e-mail: sangeetakapil.234@gmail.com

© The Author(s), under exclusive license to Springer Nature Switzerland AG 2025

A. d. b. Machado et al. (eds.), *Environmental, Social, Governance and Digital Transformation in Organizations*, Information Systems Engineering and Management 35, https://doi.org/10.1007/978-3-031-86079-9_18

1 Introduction

Environmental, social, and governance strategies have gained significant attention in recent years as worldwide organizations recognize the importance of sustainable and responsible business practices. Due to this evolving global business landscape, the shift towards sustainable and responsible practices makes environmental, social, and governance principles indispensable for companies and governments. Embracing ESG is necessary for Oman to carefully balance economic growth and environmental sustainability while addressing financial constraints and managing finite natural resources. Furthermore, as Oman addresses challenges such as water scarcity, youth unemployment, and global economic uncertainties, it is of utmost importance to prioritize integrating ESG principles to promote resilience and ensure responsible development. To successfully achieve the transformative goals of Oman Vision 2040, Omani industries must adopt a comprehensive and sustainable approach to integrate ESG into their broader policy framework effectively. This will necessitate implementing adaptable strategies, innovative solutions, and collaborative efforts [13].

Oman Vision 2040 clearly explains the principles of ESG, encompassing long-term aspirations and focusing on four main pillars: People and Society, Economy and Development, Sustainable Environment, and governance and Institutional Performance. For Oman, integrating ESG is not just a strategic imperative but a pathway to adhere to international standards, attract more significant investment, and drive sustainable development in line with Oman Vision 2040 [13]. Compelling ESG message was highlighted during the recent climate and dialogue event organized by the Environment Society of Oman (ESO), Sohar International Bank, and ASYAD, with the participation of the Financial Services Authority (FSA) and the National Programme for Fiscal sustainability and capital market development [34].

With its diverse economic landscape, Oman has witnessed the implementation of various ESG initiatives across different sectors [15]. This research paper offers a comprehensive analysis of case studies that showcase Oman's advancements in this domain. Branch et al. [8] Oman has committed to reducing its greenhouse gas emissions emissions by 2% between 2020 and 2030 as part of its second NDC submitted in July 2021. The country has also outlined plans and initiatives aligned with its 2040 vision and national energy strategy to lower carbon emissions from the energy sector. The goal is to facilitate a gradual shift towards a low-carbon economy and achieve a 7% reduction in greenhouse gas emissions emissions by 2030 [20, 31]. Oman's leading organizations have adopted innovative and impactful strategies to address ESG challenges and capitalize on opportunities. Oman has set out five primary objectives: attaining environmental sustainability, lowering energy expenses, minimizing economic and social repercussions, and ensuring energy supply security. The country has presented a roadmap for emissions reduction and the necessary budget to achieve net-zero emissions by 2050. Data has been gathered and verified by Oman's Carbon Management lab to authenticate the nation's emissions baseline. The estimated carbon emissions for Oman in 2021 stand at 90 Mt CO2e, with 95% of these emissions attributed to five sectors: industry, oil and gas, power, transportation, and buildings [20].

Petroleum Development Oman (PDO) plays a pivotal role in Oman's economy, and its operations and results have a direct influence on the well-being and prospects of various stakeholders such as local communities, the government, shareholders, customers, suppliers, regulatory bodies, municipal authorities, academic institutions, non-governmental organizations, the media, as well as employees and contractors [17]. PDO is one of the companies at the forefront of environmental stewardship. Omantel, the country's leading telecommunications operator, has embraced sustainable digital transformation [4] to enhance operational efficiency and reduce its ecological footprint. Sohar Aluminium, a prominent player in the aluminium industry, has pioneered a circular economy model that minimizes waste and promotes resource efficiency. The Oman Tourism Development Company has spearheaded the development of sustainable tourism destinations. At the same time, the Oman Food Investment Holding Company has demonstrated a solid commitment to social responsibility and community empowerment [32]. Recognizing the significant Environmental impact of its operations, Petroleum Development Oman has taken proactive measures to address the challenges and bolster its environmental stewardship. PDO has implemented a comprehensive environmental Management System that aligns with international standards. Extensive ecological studies, including detailed environmental baseline analysis and in-depth sociological surveys, have informed the development of highly project-specific ecological impact assessments. One of the projects, the Salalah Liquefied Petroleum Gas extraction project, was designed with a strong focus on environmental protection, incorporating measures to minimize air emissions and water consumption and ensure proper waste disposal [12]. Implementing innovative mitigation measures has been crucial in addressing the unique environmental and sociological challenges posed by the Salalah Gas Pipeline project, which traverses sensitive ecosystems and communities in Oman.

2 Methodological Approach

This research aims to investigate the effective strategies and fundamental principles of environmental, social, and governance (ESG) practices across various sectors of Omani industries by analyzing data from various sources. The study proposes a disclosure-based approach to assess ESG performance, tackle challenges, and uncover prospects. This approach entails collecting information from companies' annual reports about their initiatives, programs, and resource allocation that impact stakeholders. The study recommends using content analysis to evaluate ESG performance within the disclosure-based approach.

The research relies on various secondary data sources to develop the index, including annual reports, sustainability reports, websites, and other relevant publications. This comprehensive approach yields a thorough understanding of ESG practices within Omani industries.

3 Petroleum Development Oman: Environmental Stewardship

As the world increasingly recognizes the need for sustainable practices, the oil and gas industry faces mounting pressure to address environmental, social, and governance concerns. Petroleum Development Oman, a major Middle Eastern energy sector player, is no exception. However, PDO's unwavering commitment to sustainability sets it apart. Petroleum Development Oman's objective and approach involve creating a sustainable and environmentally friendly future to optimize value for Oman [28]. That is why the company's environmental stewardship efforts have been instrumental in mitigating the industry's environmental impact. PDO has become the first company in the Middle East to attain ISO-14001 certification for environmental management.

Additionally, it is the inaugural company within the Shell network to receive the prestigious ISO-55001:2014 certifications for expertly managing assets to facilitate the production, treatment, storage, and transportation of oil and gas, as recognized by the Lloyd's Register Quality Assurance (pdo.co.om). PDO acknowledge the permanence of sustainability, the reality of climate change, and the crucial need to minimize its most severe effects. The company's focus is defined by three main strategic priorities: ensuring the safety and competitiveness of hydrocarbon operations in terms of cost and carbon, establishing a diverse hydrocarbon portfolio, and developing carbon capture, usage, and storage as well as low carbon fuels [28]. PDO's comprehensive environmental Management System aligns with international standards.

PDO has taken multi-pronged environmental initiatives to address its environmental impact. The company also focuses on sustainable water management practices, including water recycling and reuse, to minimize water consumption in its operations. Water produced during oil and gas extraction is treated and reused in its operations, reducing the overall water footprint. PDO is committed to protecting biodiversity in its operating areas, implementing conservation programs and environmental impact assessments to preserve local ecosystems [28].

PDO is committed to supporting the national initiative to achieve Net Zero greenhouse gas emissions emissions by 2050. The company has successfully reduced its carbon footprint by implementing flare reduction, renewable energy, production water management, and energy efficiency projects, resulting in a reduction of 0.1 million tonnes of CO2e and a total emission of 10.6 million tonnes of CO2e [28]. PDO has also conducted extensive ecological studies, including detailed environmental baseline assessments and in-depth sociological surveys, to inform the development of highly project-specific environmental impact assessments [31].

Furthermore, Petroleum Development Oman has updated its internal environmental engineering guidelines to reflect its growing commitment to environmental stewardship. PDO is a member of international and regional organizations such as the United Nations Global Compact, Regional Clean Sea Organization, Oman Energy Association, IPIECA and GCC Standardization, Gulf Corporation Council, Oman Society for Petroleum Services, National Oil Company Steering Committee [2]. The company has implemented innovative measures to reduce gas consumption for thermal-enhanced oil recovery operations, such as harnessing waste heat from power generation to provide steam for its EOR projects (pdo.co.in). This approach not only minimizes fuel gas consumption but also helps to reduce the company's carbon footprint [29]. PDO's risk management involves identifying potential issues that could be detrimental and assessing their potential impact. The next step is to evaluate the risks and seek out possible solutions. They also categorize the level of risk as high, medium, or low. Based on this risk assessment, they prioritize the safety of their stakeholders and review and revise the report if necessary [14].

Additionally, PDO has made strides in reducing greenhouse gas emissions emissions and phasing out harmful chemicals like halons and CFCs. The company's proactive approach to environmental monitoring, operational control, and corrective action, which is aimed at continuous improvement, has continuously enhanced its environmental performance. PDO's efforts garnered recognition, receiving numerous awards and certifications for its environmental performance [30]. PDO's Ras Al Hamra development project received the best sustainable project award at the Dossier Construction's Infrastructure Award and Summit 2019 (pdo.co.in). The company has also received the Oman Sustainability Week Platinum Award 2024 (mem.gov.om). The company's focus on knowledge management and organizational performance has also significantly influenced its environmental initiatives. Addressing knowledge management challenges, such as capturing and sharing best practices, establishing communities of practice, and continuously improving performance, PDO actively engages with several stakeholders, including government entities, industry associations, academic institutions, and local communities, to address their concerns gather feedback, collaborate on environmental initiatives and maintain positive relationships [28].

PDO's internal environmental engineering guidelines have been updated to reflect a growing awareness of the need to protect Oman's unique and internationally recognized ecosystems. Implementing ESG growth focuses on four points: baseline the existing business and operating model and stakeholders' expectations; define a target for ESG aspirations; quantify the ESG impact; policies; and integrate ESG into operations. Alhanshi et al. [3] and sustainability into the overall strategy.

PDO worked incredibly hard to achieve the challenging goals of reducing its greenhouse gas emissions emissions and increasing its renewable energy mix. One of PDO's flagship projects is the development of large-scale solar photovoltaic plants, which have significantly reduced the organization's reliance on fossil fuels and associated greenhouse gas emissions emissions. By transitioning to renewable energy sources, PDO has lowered its carbon footprint and contributed to Oman's broader sustainability goals.

PDO's commitment to its workforce is evident in its comprehensive employee welfare strategies. The company has also strengthened its ethical governance practices, ensuring compliance with regulations and fostering a culture of transparency and accountability. The company's comprehensive training and competence development programs have further solidified its position as a responsible and forward-thinking energy leader in the region [6].

Petroleum Development Oman's successful ESG initiatives have generated several benefits for the company and the broader community. By proactively addressing environmental challenges, PDO has demonstrated its commitment to responsible operations, which has enhanced its corporate reputation and strengthened stakeholder trust. The company's investments in eco-friendly technologies and processes have resulted in tangible cost savings, improved operational efficiency, and reduced environmental impact. Petroleum Development Oman's dedication to environmental protection extends beyond its operational activities. The company actively engages with local communities, regulatory authorities, and other stakeholders to foster a collaborative approach to environmental management [13].

4 ESG Strategy Initiatives by Sohar Aluminium

Sohar Aluminium, a leading player in Oman's industrial landscape, has proactively addressed environmental, social, and governance issues. Recognizing the need for sustainable development, the company has implemented various initiatives to mitigate its environmental impact, engage with local communities, and foster ethical governance practices [35]. Sohar Aluminium's commitment to sustainability is not just a local endeavour but a global one. By investing in holistic management systems, the company is dedicated to enhancing regional economic and social progress while ensuring transparent and responsible governance. In 2019, the company conducted a thorough review to align its business strategy with the UN's Sustainable Development Goals (SDGs). This alignment is a formality and a commitment to monitor and contribute to global sustainability objectives, making a significant impact worldwide.

Sohar Aluminium's environmental conservation efforts encompass energy efficiency, water management, and waste reduction. The company has invested in renewable energy sources, such as solar power, to reduce its carbon footprint. Siyabi and AHakro [35] Additionally, it has implemented water recycling and treatment systems to optimize water usage and minimize wastewater discharge. In line with Oman's "Vision 2020" for economic diversification and environmental sustainability, Sohar Aluminium's initiatives contribute to the country's broader goals.

The company's core pillars are environment, health, safety, people, community, innovation, growth and financial strength. Sohar Aluminium actively engages with the local community, supporting various social initiatives and collaborating with stakeholders to address their concerns. The company's employee welfare programs, which include occupational health and safety measures, skills development, and fair employment practices, are crucial for fostering a positive and inclusive work environment.

Sohar Aluminium's commitment to ethical governance is reflected in its robust policies, codes of conduct, and adherence to national and international standards. The company's governance structure ensures transparency, accountability, and responsible decision-making, which is essential for building trust with stakeholders and maintaining the company's social license to operate. Furthermore, Sohar

Aluminium's adoption of global sustainability reporting frameworks, such as the Global Reporting Initiative, demonstrates its dedication to transparent and comprehensive disclosure of its sustainability performance. The company received ISO 14001 and OHSAS 18,001 Certification for its environmental management effort. Sohar Aluminium has invested in renewable energy, water recycling, and waste management to reduce environmental impact [7]. There are some other certifications, such as ISO 9001 (Quality Management), ISO 14001 (Environment), OHSAS 18,001, and ISO45001 (Occupational Health & Safety), obtained by the company to ensure good governance and employee welfare [35, 37].

While Sohar Aluminium's sustainability initiatives have been commendable, the energy-intensive industry's production process can have significant environmental impacts, including high energy consumption, water usage, and greenhouse gas emissions. The company mitigates challenges through continuous innovation, investment, and stakeholder collaboration [37].

Sohar Aluminium is deeply committed to supporting the communities in which it operates. The company's corporate social responsibility (CSR) framework is designed to implement various projects that positively impact different segments of society. The company's community activities, which are a testament to its dedication to integrity, transparency, and pride, instil a sense of reassurance and confidence among stakeholders. Sohar Aluminium's mission is to drive wealth, prosperity, and growth in the Al Batinah region through excellence and ownership, which underlines the importance of pursuing sustainable business operations. (Sohar Aluminium (n.d.)

The CSR committee steers Sohar Aluminium's CSR strategy, which provides strategic guidance in identifying communities in need, allocating budgets, organizing targeted activities, and mobilizing community engagement. The company's focus on CSR and its unwavering commitment to high investment in CSR programs demonstrate its value to stakeholders, making them feel important and valued. The company's CSR projects include Jusoor Projects, Sustainable Projects, Education, social Contribution and Voluntary Projects, Environment, health, and Safety, and Gulf Aluminium Council and OQ Group Initiatives.

Additionally, ensuring ethical governance and regulation compliance can be an ongoing challenge, particularly in a dynamic regulatory environment. Sohar Aluminium's commitment to transparency and accountability helps to mitigate these risks [36].

Sohar Aluminium's sustainability initiatives demonstrate a comprehensive approach to addressing environmental, social, and governance challenges. By aligning its business strategy with global sustainability goals, investing in environmental conservation, fostering employee welfare, and maintaining transparent governance, Sohar Aluminium contributes to a more sustainable future for Oman and the global community.

5 ESG Strategies and Initiatives by Bank Muscat

Bank Muscat, a leading financial institution in Oman, has been at the forefront of integrating environmental, social, and governance principles into its business strategies and operations. The bank's distinctive ESG initiatives, which include environmental conservation, community engagement, sustainable resource management, employee welfare, and ethical governance practices, set it apart as a trailblazer in the industry [9].

Environmental conservation is a prime focus for Bank Muscat, as the bank recognizes the importance of sustainable practices in addressing pressing environmental challenges. Throughout the past year, Bank Muscat has made remarkable strides in bolstering awareness and education on environmental, social, and governance (ESG) matters, a pivotal aspect of ESG. Furthermore, Bank Muscat has made significant headway in developing an ESG framework, charting clear pathways and strategic actions to propel the organization toward fulfilling their ESG objectives Bank [10].

One of the bank's core priorities is to manage its environmental footprint. Bank Muscat has implemented robust measures to conserve natural resources, reduce carbon emissions, and promote renewable energy. For instance, the bank has retrofitted its facilities with energy-efficient technologies to lower electricity consumption. In 2022, Bank Muscat achieved a significant milestone by slashing its electricity consumption by 6% compared to the previous year, bringing it down to 11,636 MWH in 2022 from 12,356 MWH in 2021, a feat we are particularly proud of Bank Muscat Annual [9]. Additionally, Bank Muscat has been actively integrating green financing practices into its operations, providing funding to support environmentally friendly initiatives and projects within the Omani community [16].

The bank's efforts have included upgrading branches with energy-efficient lighting, installing solar panels, and launching paperless initiatives to minimize paper consumption Bank Muscat Annual [9]. Due to these efforts, Bank Muscat has reduced significant electricity consumption from 2020 (11,747) to 2021. Bank Muscat has also reduced its fuel consumption through fleet optimization and employee awareness programs, achieving 68,052 compared to 93,658 L of petrol from 2022 to 2021 and diesel by 1103 from 2022 to 1126 L in 2021, demonstrating its commitment to environmental sustainability Bank Muscat Annual [9]. In 2022, the amount of paper sent for recycling decreased by 90% from 138,400 kg to 14,528 kg, while the amount of e-waste sent for recycling decreased by 46% from 21,950 kg to 11,780 kg. Additionally, there was a 7% increase in general waste sent to landfills, rising from 5,726 m3 to 6,125 m3 Bank [10]. Bank Muscat achieved a 27% decrease in emissions compared to 2021 in their carbon footprint and environmental impact. This demonstrates the company's dedication to reducing its environmental footprint, which has been accomplished by implementing various measures, such as enhancing energy efficiency Bank [10].

Bank Muscat's ESG initiatives extend beyond environmental conservation, emphasizing community engagement and social responsibility. The bank has steadfastly supported Oman's community development, education, and healthcare initiatives. Bank Muscat has also implemented employee-centric programs to foster a nurturing and inclusive work environment, focusing on employee well-being, training, and professional development [11, 35].

Regarding governance, Bank Muscat has instituted robust ethical practices and transparent reporting mechanisms to ensure accountability and ethical decision-making. The bank has established a sustainability committee that oversees the implementation of its ESG strategy and ensures compliance with relevant regulations and best practices Bank Muscat Annual [9]. The committee's main objective is to ingrain exemplary practices into the bank's core business operations and corporate culture. It meticulously evaluates the bank's sustainability initiatives against industry benchmarks. It shares its valuable insights and recommendations with the Executive Management, which holds the highest authority in ESG matters within the bank. This ensures the seamless integration of sustainability into Bank Muscat's overarching business strategy. The committee convenes biannually to deliberate on policies and strategies influencing economic performance, internal development, community empowerment, and responsible banking practices [11]. The Materiality Assessment process started by examining operations, stakeholders, and existing ESG performance to determine the impact of past ESG issues. After analyzing the context, the bank assessed the actual and potential impacts of identified issues and compared them with best practices, global frameworks, and trends. Bank Muscat consulted with internal stakeholders such as the Finance, Risk, Procurement, Human Resources, and Corporate Communications Departments to evaluate the significance of the identified impacts through online surveys and consultations. The updated list of material issues is categorized based on the level of impact. This comprehensive approach has enabled Bank Muscat to establish a robust ESG framework that aligns with global sustainability standards and practices. The Bank Muscat's Board of Directors plays a crucial role in strengthening governance and representing the concerns of shareholders and stakeholders. The bank promotes transparency, integrity, and ethical behaviour by clearly defining the responsibilities of the Chairman and the Chief Executive Office. Through its governance practices, Bank Muscat has positioned itself as a responsible and trustworthy financial institution prioritizing sustainable value creation for all its stakeholders [11].

Bank Muscat has been acknowledged by both national and international organizations for its sustainable efforts and initiatives. The bank has received several distinctions, including the Best Corporate Digital Bank award from Global Finance, the Best Bank for ESG in Oman by Euromoney, and the Exceptional Priority Banking Services Award by the Signature 100 Luxury Awards for its Asalah Priority Banking. Additionally, EMEA Finance recognized Bank Muscat as the Best Private Bank. It was honoured with the Gold Award for Corporate Banking innovation—Dynamic Forex Service for Corporate Clients at the Infosys Finacle innovation [6]. Meethaq, a subsidiary of Bank Muscat, also achieved significant milestones by securing three prestigious awards in 2023. These include the Market Leadership Bank in Oman by

Global Islamic Finance Awards (GIFA) for the second consecutive year, the Leading Islamic Banking Brand in Oman 2023 by Global Brands, and the Most Innovative Islamic Window for SMEs by Global Business Outlook. Additionally, Meethaq was awarded the Best Digital Islamic Bank in Oman 2023 by Global Finance [11].

Bank Muscat takes a proactive approach to managing cybersecurity. By conducting continuous vulnerability scanning, annual internal penetration testing, and external testing, the organization assesses and maintains its IT infrastructure's cyber security and safety.

Bank Muscat's initiatives, such as workshops and activities related to green and sustainable finance, are positively impacting. Their efforts towards developing a Sustainable Finance Framework (SFF) are aligned with Oman's vision, 2040. Through the SFF, the bank offers financial instruments such as green, social, and sustainability bonds, loans, and sukuk to support renewable energy projects, thus promoting a greener economy. Bank Muscat first launched green products in the sultanate [11].

The bank introduced a green financing Loan solution in 2019. It encouraged customers to install rooftop solar panels to save money and contribute to a more sustainable future. The bank's unwavering strategic focus on digitalization, including (specific digital solutions or platforms), enhances operational efficiency and significantly improves the customer experience.

6 ESG Strategies and Initiatives by Omantel

As a leading telecommunications provider in Oman, Omantel has recognized the importance of integrating environmental, social, and governance principles into its business operations. The company's ESG strategies and initiatives span various areas, including environmental conservation, community engagement, sustainable resource management, employee welfare, and ethical governance practices. Omantel's sustainability framework identifies, prioritizes, and manages sustainability risks and opportunities. Moreover, it facilitates the alignment to operating responsibly by maintaining international best practices and standards. Omantel's sustainability endeavours create enduring value for company stakeholders, encompassing customers, employees, shareholders, and society [24].

The growing emphasis on corporate social responsibility and sustainability has prompted companies to adopt ESG strategies to enhance their competitive advantage and appeal to socially conscious consumers AL-Saleh [1] and investors. In this context, ESG investing has become increasingly prevalent, particularly in emerging and frontier markets, where companies must navigate complex challenges such as poverty, urbanization, pollution, and corruption.

Omantel's ESG strategies and initiatives are multi-faceted, addressing many focus areas to drive positive impact. The company has implemented initiatives to promote energy efficiency, renewable energy usage, and water conservation in the environmental domain. Omantel has succeeded by combining its operations, procedures, and deep knowledge in the communications and digital technology sector [22].

Regarding social impact, Omantel is actively engaged in community development programmes, supporting initiatives that improve the well-being of the people in Oman. The company also prioritizes employee welfare, offering training and development opportunities, fair compensation, and a safe and inclusive work environment. Siyabi and AHakro [5, 35]. Supporting Tropical Cyclone Shaheen Relief Activities, Participation in the National Campaign for Immunization, IT & Coding Curriculum, Upgrade to transform graduating students, Renewable Energy lab, Edlal platforms, Colour their world, ICT internship, Omantel Olympics of Coding, innovation Corner by Omantel, Iot of Things Lab, Mehnati, Bolld Donation campaign, Reversal Vending Machines, Activation of innovation Programmes, Ultrasound device for liver surgery at Royal Hospital, Scholarships for a bachelor's degree for students of social security families, Pediatric Unit for Sohar Hospital, Omantel research funds in collaboration with Sultan Qaboos University to promote research in the field of ICT, Mubadera award to support and strengthen Omani women's associations, Summer academy programmes in partnership with GUtech, Outward Bound Oman Programme and Mekunu Cyclone relief campaign are some of the initiatives undertaken by Omantel (omantel.com).

Omantel's governance practices are guided by strong ethical principles, emphasizing transparency, accountability, and compliance with relevant laws and regulations. The company has established robust risk management and internal control systems to mitigate potential governance risks. The company's pioneering strategies have propelled it to deliver cutting-edge solutions across various consumer and business sectors. With a commitment to exceeding subscriber expectations, Omantel provides an unparalleled experience. By investing in emerging technologies and offering state-of-the-art solutions in cloud computing, ICT, AI, cybersecurity, and more, the company is actively contributing to the realization of Oman Vision 2040. Its technological prowess drives innovation and positions it as a leader in the ever-evolving landscape of advanced technologies Times of [23]. Omantel has set up a Sustainability Committee to oversee sustainability efforts and provide regular updates to the board. The committee also supports the Board of Directors in fulfilling its oversight responsibilities for Omantel's sustainability strategy and initiatives. This includes environmental, social, and governance aspects, as well as integrating sustainability practices into Omantel's core business operations. The committee comprises representatives from relevant departments, senior management, and external experts as needed [24].

7 ESG Strategies and Initiatives by OMRAN Group

The Oman Tourism Development Company has been at the forefront of integrating environmental, social, and governance principles into its business operations and strategic initiatives. OMRAN Group's ESG framework and ESG Baseline reaffirm the group's commitment to sustainability in the tourism sector. This framework aligns directly with Oman Vision 2040, international standards and the United Nations Sustainable Development Goals. It aims to promote responsible development. OMRAN Group's initiative and efforts aim to enhance regional and international investments, contributing to the growth of Oman's tourism sector [26].

The Omran group significantly influences local communities, preserves nature, and safeguards cultural heritage. Omran Group Company guides ESG principles based on environmental stewardship, social-economic prosperity, and responsible governance. Its strategic themes under environmental management focus on climate change, environment, and biodiversity. Under social and economic prosperity, the company focuses on human capital, social well-being, and education. Responsible governance focuses on ESG governance, corporate compliance, accreditations, standards, and synergies [27].

The tourism industry has seen a growing emphasis on sustainable business forms, such as ecotourism, alternative tourism, and environmentally friendly travel. Oman's tourism sector, represented by the Oman Tourism Development Company, has recognized the need to sustain the resources that attract visitors, ensuring the industry's long-term viability.

The Oman Tourism Development Company has implemented a comprehensive set of ESG strategies and initiatives, focusing on key areas such as environmental conservation, community engagement, sustainable resource management, employee welfare, and ethical governance practices. OMRAN Group's operations embrace proactive measures to confront climate change risks and champion the integration of clean technologies. The company's efforts are dedicated to minimizing the ecological impact by responsibly overseeing natural resources and biodiversity and embracing circular practices that ensure sustainability in every aspect. Along with this, OMRAN Group focuses on Human-centric operations, nurturing a secure and thriving work environment and prioritizing the well-being of its public and employees. The company services focus on local communities, individuals, cultural diversity, and heritage. To promote and maintain ethical business standards, the company encourages transparent communication and responsible governance, emphasizing the crucial role of shaping sustainable, long-term strategies to drive economic growth that benefits the company and society [27].

The company's impactful initiatives are making waves across various digital platforms. With a steadfast commitment to elevating Oman's status as a world-class tourist destination, "Visit Oman" has made remarkable strides in 2023, driving an influx of high-quality inbound leisure traffic to the country. Additionally, the Omaran Company has garnered prestigious accolades, including the SUSTAINABLE CITY

YITI GCC Housing Award 2023, the awe-inspiring MUSANDAM ZIPLINE, officially recognized by Guinness World Records as the world's longest zipline over water, Oman SAIL's historic silver medal at the Asian Games, and the esteemed title of "Best Venue" at the Middle East Event [6] for OCEC. These remarkable achievements underscore the company's unwavering dedication to excellence and innovation in the travel and tourism industry [25].

As part of its ongoing commitment to creating a positive social impact, the OMRAN group has extended its partnership with the Oman Food Bank for a second year in 2023, focusing on collecting surplus food from its hotels and resorts to support those in need. The group has distributed 7,000 meals in the past year, collected 4,200 kg of excess food, and recycled 2,025 papers. Additionally, the OMRAN Group has collaborated with the Dhofar Municipality to develop various destinations, such as the Jarzeez public facilities, Itlalat Darbat Waterfalls, Maghsail Beach, and Itlalat Hamreer, enriching the tourism experiences through public services and recreational facilities. Furthermore, the group organized two successful carnivals with the Ministry of Heritage and Tourism and the Governor's Offices in Al Buraimi and Musandam, attracting over 25,000 and 13,000 visitors, respectively. This reflects the group's dedication to promoting tourism and cultural exchange. Moreover, the OMRAN Group signed two Memorandums of Understanding at the World Tourism Market (WTM—London), aiming to drive foreign direct investment, establish a real estate development fund, and enhance cross-destination tourism between the Musandam Governorate and the Emirate of Ras Al Khaimah. The group also participated in various industry events, such as the ITB Berlin, to promote tourism in Oman and attract investment [25].

The OMRAN Group is committed to empowering its employees through comprehensive development programs and initiatives, fostering a culture of inclusivity and professional growth. The group has launched several initiatives and programmes to enhance In-Country Value (ICV) and human resource development in line with its long-term strategic goal to create job opportunities in the tourism and hospitality sectors. These initiatives include Midhyaf, Masarat, Tadreeb, and the Graduates Employment Initiative, all contributing to developing a thriving and sustainable tourism sector [25].

Regarding environmental sustainability, the OMRAN Group has implemented various measures to reduce its carbon footprint and promote eco-friendly practices. The group has undertaken initiatives to conserve water, energy, and natural resources and manage waste effectively. The Oman Tourism Development Company (OMRAN Group) has reached a significant milestone in its journey toward excellence. The company has accomplished an impressive feat by completing 58 million safe person-hours without any lost time, injuries, or accidents, further solidifying its exceptional performance in tourism, hospitality, and real estate development [27].

The Sultanate of Oman has seen significant economic growth and development in recent years, driven by its vast natural resources and strategic location. However, as the country looks to diversify its economy and transition towards a more sustainable future, implementing environmental, social, and governance practices has become a critical challenge for many of its leading organizations [18].

8 Insights from the Case Studies

Omantel, the country's leading telecommunications provider, needs help incorporating ESG principles into its operations. It concerns energy consumption, e-waste management, and the company's social impact on local communities. Siyabi and AHakro [35] Petroleum Development Oman, the state-owned oil and gas company, has faced pressure to reduce its carbon footprint and address environmental concerns while balancing economic development and sustainable practices.

Similarly, Sohar Aluminium, one of the most significant industrial players in Oman, has grappled with the environmental impact of its operations, particularly in terms of water usage and waste management. The OMRAN Group, Oman's premier tourism development company, has encountered challenges in integrating sustainability into its hotel and resort projects and addressing social concerns such as labour practices and community engagement [19, 33, 35]. One of the challenges could be related to Corporate social Responsibility (CSR). The idea of businesses contributes to society's and the community's overall welfare. It is widely acknowledged that every business is responsible for addressing societal issues and taking proactive steps to resolve them. Therefore, businesses have a significant onus to operate as good corporate citizens and be more socially responsible. While traditional CSR mainly involved donations and charity-oriented activities, the modern understanding of CSR encompasses a broader range of activities, including both internal and external CSR initiatives. Internal CSR focuses on good governance, meeting business objectives, fulfilling targets, and employee compensation.

On the other hand, external CSR encompasses a broader spectrum of business activities that benefit all stakeholders. However, measuring the impact of these CSR activities can pose several challenges. For instance, if a business adequately discloses its CSR performance, it becomes easier to quantify the impact of these activities. Additionally, comparing the CSR performance of different companies can be challenging, mainly when one business focuses on a specific aspect of CSR with many contributions. At the same time, another business has a smaller but more diversified CSR performance. When examining the concept of CSR, it is essential to consider these challenges. Therefore, a proposed conceptual model aims to comprehensively cover various CSR activities and address these challenges within the proposed framework.

The financial sector, represented by Bank Muscat, is crucial in Oman's adoption of ESG principles. The bank has tried incorporating sustainability into its lending practices and investment decisions but faces obstacles in ensuring the wider business community aligns with ESG standards.

After carefully examining the case studies from different Omani Industries, it has been found that Omani industries significantly focus on the following ESG criteria. (See Fig. 1).

However, the country's corporate governance and disclosure frameworks have yet to fully mature to effectively support and enforce ESG compliance across the private sector [32]. As Oman continues its economic diversification and transition towards

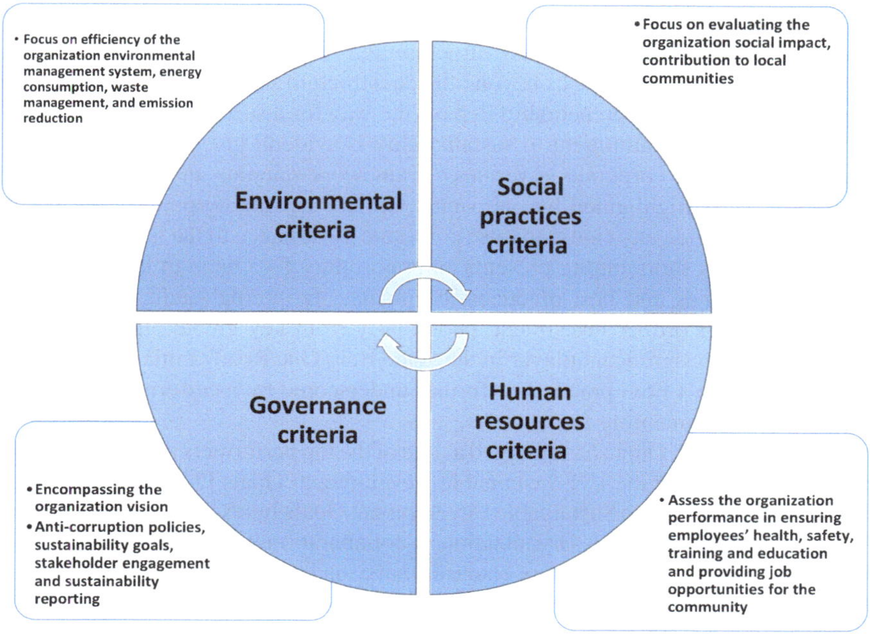

Fig. 1 ESG criteria focused prominently by Omani Industries. *Sources* Designed by the Author based on case studies

a more sustainable future, it will be crucial for the public, private sector, and civil society to work collaboratively to address the challenges of ESG implementation.

9 The Strength of ESG Practices in Oman

The Sultanate's commitment to a green economy and sustainable development has been well-documented. However, Oman's corporate governance landscape is not as mature as other developed nations, with current codes falling short of providing a clear path towards enhancing sustainability [32]. This gap represents a significant opportunity for Oman to strengthen its ESG practices and position itself as a regional leader in sustainable development.

Achieving and maintaining sustainable development has become increasingly prevalent for many countries over the last decades, and Oman is no exception [37]. The country has made plans for economic diversification to reduce its dependency on the fast-depleting oil reserves through building competencies across other sectors of the economy [37]. This economic diversification has led to a substantial boom in several industries, creating a pressing need to prioritize the efficient use of natural resources [37].

The Omani government has recognized the importance of this shift and has allocated significant investments in infrastructure and industrialization. This emphasis on sustainable practices extends beyond the construction sector, as evidenced by the country's focus on entrepreneurship to pave the way for a sustainable future [18].

While Oman's commitment to sustainability is evident, integrating robust ESG practices within its corporate governance framework remains an area of ongoing development. As highlighted by one study, Oman's current corporate governance codes do not provide any clause to raise corporate governance to the level of sustainability [32]. This shortcoming presents an opportunity for Oman to leverage international guidelines and best practices to enhance its internal audit function and ultimately attract further investment, particularly from key global players such as China, which is a significant player in the "One Belt, One Road" initiative [33]. The strength of Oman's ESG practices is further underscored by its efforts to align with international sustainability frameworks.

Due to all efforts, Oman has emerged as one of the top performers in the Arab world in terms of achieving the UN Sustainable Development Goals [21]. According to a report titled Arab Region Sustainable Development Goals Index 2022, released by the World Government Summit Organization in collaboration with the Mohammed bin Rashid School of Government, six countries have made significant progress towards the SDGs. These countries include Jordan, Tunisia, the United Arab Emirates, Algeria, Morocco, and Oman, with Oman achieving a total index score of 66 or more. However, the report also highlights the substantial variations in the progress of Arab countries towards the SDGs. The average regional score for the Arab region in terms of achieving sustainable development goals stands at 58.2 out of 100. The report emphasizes the need for more regional effort and collaboration to meet the sustainable development goals in the next eight years (Muscat Daily).

10 Conclusion

A thoughtful and multifaceted strategy is essential to seamlessly weave ESG principles into Oman's economic and social landscape. The first crucial step involves crafting a detailed ESG Framework that articulates precise guidelines and standards for corporate entities and government agencies. By aligning this framework with international best practices while carefully tailoring it to resonate with Oman's distinct socio-economic context, we can establish a comprehensive roadmap for organizations. This roadmap will empower them to report on key ESG metrics transparently, enhancing their accountability and fostering a culture of trust.

Building on this foundation, it is equally vital to cultivate a rich understanding of ESG principles among diverse stakeholders. Strategic investments in educational and training programs can significantly uplift awareness among business executives, policymakers, and the wider community. Academic institutions, including universities, colleges, and various professional organizations, have a significant role to play by integrating ESG topics into their curricula. This will ensure that the next generation of leaders is equipped with the knowledge and skills to champion sustainable practices. The initiatives already undertaken by the Financial Services Authority (FSA) to enhance capacity-building and awareness represent significant progress.

Effective ESG implementation hinges on meaningful stakeholder engagement. Organizations can better understand the varied needs and priorities by fostering dialogue among regulators, community members, civil society, and other key players. This approach not only raises awareness but also strengthens capacity. By integrating a wide range of viewpoints into the decision-making process, Oman can ensure that its ESG initiatives are sensitive to the concerns of its populace and effectively address their needs.

To further attract investments that adhere to ESG standards, Oman could explore a range of appealing incentives such as tax breaks, grants, and favourable financing options for firms that exhibit exemplary ESG performance. Additionally, the establishment of green financing mechanisms, like green bonds, would mobilize capital for projects dedicated to environmental and social enhancement. This approach broadens investment opportunities and connects with a vast pool of socially responsible capital. The recent regulations introduced by the FSA regarding bonds and sukuk have set a solid foundation for organizations to access these essential resources, opening the door to innovative financing solutions. Furthermore, the Ministry of Finance's announcement of the National Framework for Sustainable Finance positions Oman on a promising trajectory towards sustainable development, guiding the nation toward a brighter, more responsible future.

References

1. AL-Saleh, S.M., Khdeer, J.M.: The role of sustainability reporting in attracting foreign investment in light of the global reporting initiative (GRI). (2023). https://doi.org/10.21271/zjhs.27.4.18
2. Al-Toubi, S., Malik, H.: Knowledge management and organisational performance with a case study from PDO. Springer Nature, pp. 437–469. (2018). https://doi.org/10.1007/978-3-319-71434-9_18
3. Alhanshi, M., Ojiako, U., Williams, T.: Managing strategic resources in petroleum industry projects. Taylor Francis **33**(11), 1043–1060 (2020)
4. Alraja, M.N., Hussein, M.A., Ahmed, H.M.S.: What affects the digitalization process in developing economies? Evidence from the SME sector in Oman. Inst Adv Eng Sci (IAES) **10**(1), 441–448 (2020). https://doi.org/10.11591/eei.v10i1.2033
5. Asmyatullin, R.R., Тыркба, X., Ruzina, E.I.: Smart cities in GCC: comparative study of economic dimension. IOP Publishing, 459(6), 062045–062045. (2020). https://doi.org/10.1088/1755-1315/459/6/062045
6. Awards.: (2023). Retrieved October 8, 2024, from https://www.pdo.co.om/en/about/awards/Pages/default.aspx
7. Benameur, K., Mostafa, M.M., Hassanein, A., Shariff, M.Z., Al-Shattarat, W.: Sustainability reporting scholarly research: a bibliometric review and a future research agenda. Springer Sci. Bus. Med. (2023). https://doi.org/10.1007/s11301-023-00319-7
8. Branch, M., Goldberg, L.R., Hand, P.: A guide to ESG portfolio construction. Euromoney Inst. Investor. **45**(4), 61–66 (2019). https://doi.org/10.3905/jpm.2019.45.4.061

9. Bank Muscat Annual Report.: (2021). https://www.annualreports.com/HostedData/AnnualReports/PDF/bank-muscat_2023.pdf
10. Bank Muscat.: (2022). https://www.bankmuscat.com/en/about/SiteAssets/SustainabilityReport_2022_Eng.pdf
11. BM Sustainability Report.: (2023). https://www.bankmuscat.com/en/about/SiteAssets/Sustainability_Report_2023_ENG.pdf
12. Company, O.G.: (n.d). Salalah liquefied petroleum gas project
13. ESG Country Report.: (2023). https://astra.grandviewresearch.com/country-esg-outlook/oman-vision
14. Executive, H.A.S.: (2024). PDO Risk Management. https://pdo.co.om/hseforcontractors/Pages/DocDescription.aspx?docid=15928
15. Fulton, M.R., Kahn, B.M., Sharples, C.: Sustainable investing: establishing long-term value and performance. RELX Group Neth. (2012). https://doi.org/10.2139/ssrn.2222740
16. Ghosh, S.K., Ghosh, P.K., Chowdhury, S.: Essential of central bank's regulatory policy to strengthen green banking practice and reporting in a country. Asian J. Financ. Acc. **10**(2), 133–133 (2018). https://doi.org/10.5296/ajfa.v10i2.13599
17. Lamki, M.S.A., Terken, J.J.: Hydrogeology in petroleum development Oman. (1996)
18. Magd, H., McCoy, M.P.: Entrepreneurship in Oman: paving the way for a sustainable future. Elsevier BV **15**, 1632–1640 (2014). https://doi.org/10.1016/s2212-5671(14)00634-0
19. National Net Zero Carbon Program Report.: (2023). https://www.ea.gov.om/media/gqgdr52d/net-zero-carbon-report-2023-final.pdf
20. Oman Sustainability Week Platinum Award.: (n.d). Retrieved from https://mem.gov.om/en-us/Media-Center/News/ArtMID/608/ArticleID/1389/Oman-Sustainability-Week-Awards-2024-Celebrates-Achievements-in-Sustainability).
21. Oman Has Emerged as One of the Top Performers.: (n.d.). Retrieved October 16, 2024, from https://www.muscatdaily.com/2022/09/12/oman-among-6-arab-countries-to-achieve-60-of-unsdgs/
22. Omantel CSR initiatives Retrieved October 13, 2024, from https://www.omantel.om/csr/key-initiatives
23. Oman, T.T.O.: Omantel as digital leader. (2024) https://cdn-3.timesofoman.com/article/145048-omantel-highlights-its-sustainability-role-as-digital-leader
24. Omantel Sustainability Report.: (2023). https://www.omantel.om/wcm/connect/b24599a5-3699-49f6-805e-365a441a1bd2/Omantel+Sustainability+Report+%28Final+version-GRI%29+%281%29.pdf?MOD=AJPERES&CVID=o-gLFTl
25. OMRAN Annual Report.: (2023). https://www.omran.om/wp-content/uploads/2024/09/3009_Annual-Report-_EN.pdf
26. OMRAN ESG Framework.: (n.d). https://www.omran.om/archives/45826
27. OMRAN ESG Report.: (2024). https://www.omran.om/wp-content/uploads/2024/09/OMRAN-Group_ESG-Report.pdf
28. PDO Who We Are.: Can be retrieved from (https://pdo.co.om/en/Pages/WhoWeAre/Our-Memberships-And-Awards.aspx)
29. PDO Approach to Sustainability.: Can be retrieved from https://pdo.co.om/en/Pages/Sustainability/OurApproach.aspx
30. PDO, O.: PDO-2022-Sustainability-Report. (2022). https://edoman.om/wp-content/uploads/2024/03/PDO-2022-Sustainability-Report.pdf
31. Phimister S (2023) Oman Industry ESG Whitepaper 2023
32. Rehman, A.: Can sustainable corporate governance enhance internal audit function? Evidence from Omani public listed companies. Multi. Digit. Publ. Inst. **14**(11), 537–537 (2021). https://doi.org/10.3390/jrfm14110537
33. Sarker, M.N.I., Hossin, M.A., Yin, X., Sarkar, M.K.: One belt one road initiative of China: Implication for future global development. Mod. Econ. **9**(4), 623–638 (2018)
34. Siyabi, A.S.A., AHakro, A.N.: Green economy and sustainable development in the sultanate of Oman. rScroll. (2020). https://doi.org/10.47611/jsr.vi.915

35. Sohar Aluminium.: (n.d.). Contributing to the community. Sohar-Aluminium.com. Retrieved October 9, 2024, from https://www.sohar-aluminium.com/Contributing-to-the-community
36. SA Sustainability Report.: (2021). https://www.sohar-aluminium.com/sites/default/files/2022-06/SA%20_Sustainability%20Report%20_%202021%20_%20Eng.pdf
37. Saleh, M.S., Alalouch, C.: Sustainable construction in Oman: the potential role of the industrialized building systems. Sultan Qaboos Univ. **17**(1), 1–1 (2020). https://doi.org/10.24200/tjer.vol17iss1pp1-10

Toward a Sustainable Digital Future: A Comprehensive Study on Digital Transformation and Environmental Sustainability

Khizar Hayat, Qingyu Zhang, and Mudassir Husnain

Abstract The digital advancements have brought about unprecedented disruptions, reshaping traditional practices and opening new avenues for innovation. Digital transformation (DT) signifies a monumental shift across organization and society, propelled by the evolution of technological digitalization like big data analysis, artificial-intelligence, the internet of things, and cloud services provider. However, within this digital revolution, there remains a significant gap in understanding its implications for environmental sustainability. This research will endeavor to bridge this gap by conducting a comprehensive analysis of DT's impact on environmental sustainability. Through a meticulous review of existing literature, the study will expose the various disruptions that DT has set off within this domain. The findings will culminate in developing a robust framework, delineating transformations across four pivotal areas: urban sustainability, waste management, pollution control and sustainable production. Within these key domains, a nuanced exploration will disclose a multitude of subcategories, each representing a distinct aspect of transformation. For instance, under pollution control, sensor technologies and data analytics advancements have revolutionized monitoring and mitigation efforts, enabling real-time interventions and more precise control measures. Similarly, innovative solutions powered by digital technologies have facilitated efficient recycling processes in waste management, minimizing waste generation and promoting circular economy practices. Furthermore, the research will go beyond simple classification and provide a guideline for subsequent investigations. This emphasizes the critical nature of evaluating performance indicators, organizational capabilities, and DT strategies for environmental sustainability.

K. Hayat · Q. Zhang (✉)
Institute of Business Analytics and Supply Chain Management, College of Management, Shenzhen University, Shenzhen, China
e-mail: qzhang@szu.edu.cn

K. Hayat
e-mail: khizarhayat@szu.edu.cn

M. Husnain
HNU-ASU International College, Hainan University, Shenzhen, China
e-mail: mudassir.husnain@szu.edu.cn

© The Author(s), under exclusive license to Springer Nature Switzerland AG 2025
A. d. b. Machado et al. (eds.), *Environmental, Social, Governance and Digital Transformation in Organizations*, Information Systems Engineering and Management 35, https://doi.org/10.1007/978-3-031-86079-9_19

Keywords Systematic literature review · Environmental sustainability · DT and internet of things

1 Introduction

Advancements in digital technology have sparked a digital revolution that has profoundly disrupted businesses, impacting their strategic frameworks, structural dynamics, and methods for creating value. It has caused significant disruption in international marketplaces and sectors and opened the door for new business models [40]. The rapid rise of technologically advanced corporations has left traditional enterprises especially exposed, drastically altering the competitive environment. Globally, notable instances include the significant transformation of the hotel industry by platforms such as Booking.com and Airbnb, the upheaval of the music business by Spotify, and the considerable digitization of well-known companies like Starbucks [22]. Cloud-computing, mobile technologies, AI, big data analytics, and the Internet of Things (IoT) have all contributed to these upheavals. In recent years, the rapid acceleration of DT has redefined industries, economies, and social structures, promising unprecedented opportunities for efficiency, innovation, and global connectivity [10]. From AI to block-chain technology, the digital revolution is shaping the future of business and influencing global sustainability efforts. As nations and corporations works to reduce their carbon footprint and align with the UN Sustainable Development Goals, the role of digital technologies in promoting environmental sustainability has become a central focus of contemporary research [37, 76]. However, while technological advancements offer potential solutions to environmental challenges, they also introduce new complexities, particularly regarding resource consumption, waste generation, and energy demands. This duality highlights the need for a deeper exploration of the DT and environmental sustainability relationships [20, 55].

The emergence of sustainable digital practices is driven by the realization that technology, when strategically employed, can significantly reduce environmental impacts. For instance, adopting cloud computing has reduced energy consumption for data storage, while AI and big data analytics enable more efficient supply chain management, minimizing waste and optimizing resource use [3, 64]. Furthermore, digital platforms have created opportunities for circular economy models, where waste is minimized and products are reused or recycled. Despite these advancements, the environmental cost of DT, such as the rising demand for electricity to power data hubs and the e-waste created by short device life cycles, cannot be ignored. The balance between leveraging digital technologies for sustainability and mitigating their negative externalities remains an area for further investigation. Previous studies have explored various facets of this relationship. For instance, research has shown that block-chain can enhance supply chain transparency, thereby encouraging sustainable business practices [33]. Studies have illustrated the IoT's potential to optimize energy consumption in smart cities and reduce greenhouse gas emissions [33]. Additionally, cloud computing has been examined for its ability to decrease the environmental impact of IT infrastructure [38]. However, many of these studies are fragmented, focusing on isolated technologies or specific sectors without addressing DT's broader, systemic implications on environmental sustainability. Moreover, few have delved into the paradoxical nature of this relationship, where digital advancements may both support and hinder sustainability efforts [11].

The existing literature, while insightful, has left several critical gaps unaddressed. First, there is limited research that integrates the various digital technologies into a cohesive framework that comprehensively examines their collective impact on environmental sustainability. Second, much of the prior work focuses on the prospective benefits or the challenges of DT without providing a balanced analysis that considers both aspects in tandem. Lastly, there is a dearth of empirical indication assessing the longstanding sustainability outcomes of DT at both macro (industry-wide) and micro (firm-level) scales. These gaps hinder the development of actionable strategies for leveraging digital technologies to achieve sustainable development goals. digital technology is the leading DT driver, although other aspects have contributed. These include changing customer expectations, greater digital competition, and more data. It affects healthcare, social dynamics, consumer behavior, and organizational structures [29]. The digital revolution is also predicted to impact the sustainability triangle, notably environmental sustainability. Technology improves health, the environment, and the food supply chain. Further study is vital to discover the broader consequences of DT on corporate social responsibility, societal implications, organizational performance, and environmental sustainability, which have been disregarded in the literature. Addressing pollution and resource depletion requires focusing on how digital change may help the environment.

The study intentions to discourse these gaps by comprehensively analyzing the intersection between DT and environmental sustainability. The study will investigate how digital innovations such as, IoT, block-chain, Cloud-computing and AI can contribute to environmental goals while also considering the unintended consequences that may arise from their deployment. Furthermore, this research will employ empirical data to assess the long-term sustainability outcomes of DT, offering a holistic perspective that can guide policymakers, business leaders, and researchers in fostering a more sustainable digital future. This study shows the uncertainties around DT's implications on environmental sustainability, which motivated our investigation. Our research seeks to clarify how the digital revolution affects environmental sustainability and influences policy formulation and goal-setting across regions, a significant gap. The digital revolution disrupts environmental sustainability, according to the report. Through a thorough literature analysis, we identify and categorize the effects of the impact of DT on environmental sustainability is divided into four primary areas: energy efficiency, waste management, sustainable manufacturing, and pollution control, each with subcategories. We also suggest studying DT strategy, organizational competencies, and efficiency in environmental sustainability research.

The implications of this study are manifold. Practically, it will provide businesses with insights into integrating digital technologies into their sustainability strategies without exacerbating environmental degradation. For policymakers, the findings will offer guidance on formulating regulations that encourage using digital tools to support sustainability while mitigating potential risks. Finally, this study will contribute to the academic discourse by providing a more nuanced understanding of the complex relationship between DT and environmental sustainability, serving as a foundation for future research in this evolving field.

2 Theoretical Background and Literature Review

2.1 Digital Transformation (DT)

DT involves integrating digital technology to change company models, improve consumer experiences, and boost operational efficiency. Beyond adopting technology, DT is a strategy transformation that affects all areas of an organization, from value generation to decision-making [40]. Various digital technologies have driven DT, each with different consequences on industry and sustainability. AI, IoT, block-chain, cloud computing, and big data analytics are DT subcategories that enhance sustainability goals. AI has transformed various sectors by improving information processing and decision-making. Research shows that applications of AI in supply chain management and energy systems may minimize waste and optimize resource use [8]. AI's high energy consumption, particularly in large-scale machine learning models, may outweigh its environmental advantages [10]. Comparative studies like [39] demonstrate that AI enhances operational efficiency but requires further research to establish its sustainability impact. Smart cities and environmental management benefit from real-time monitoring provided by IoT, which connects common things to the Internet. Smart grids and IoT systems decrease emission levels, improve governance of water resources, and optimize domestic and commercial energy consumption [40]. IoT's environmental advantages are more immediate and apparent than AI's, but data security, privacy, and the massive electronic waste of IoT devices are significant problems [76].

Block-chain's decentralized and transparent design might transform banking, logistical issues, and energy. Khalid et al. [33] Found that block-chain improves supply chain transparency, simplifying manufacturing environmental effect monitoring. The high energy needs of proof-of-work consensus procedures have been criticized for damaging sustainability [66]. Block-chain has unique transparency and security but more significant energy costs than IoT and AI, making it a double-edged sword for sustainability. big data analytics provides data-driven insights for sustainable decision-making, while cloud computing conveniently stores and processes massive volumes of data [58]. Say cloud adoption saves energy usage and material waste by reducing physical infrastructure. The increased demand for cloud services strains data centers, which need plenty of energy for cooling and maintenance. big data analytics improves supply chains, environmental estimations, and sustainability.

However, it also raises data privacy and environmental issues regarding data storage [16, 28].

2.2 Environmental Sustainability

Environmental sustainability focuses on meeting the needs of the present without compromising the ability of future generations to meet their own needs, with particular emphasis on conserving natural resources, reducing emissions, and protecting ecosystems [12]. As global challenges such as climate change, resource depletion, and biodiversity loss become more pressing, there is a growing interest in leveraging digital technologies to support sustainability efforts. Environmental sustainability has been elevated to the center due to the convergence of the circular economy and Industry 4.0, both dedicated to enhancing resource efficiency [34]. In current period, the sustainability-based sharing economy is redefining stakeholder relationships in a world that is becoming more interconnected. Ecological sustainability, a fundamental principle of sustainability, guarantees that our endeavors to satisfy current requirements do not harm the environment, thereby protecting the environment for future generations [2, 38]. Integrating environmental sustainability principles into operations can increase an organization's value, making digitalization efforts more significant. Environmental protection has become a critical priority that requires the concentrated attention of businesses and governments due to the increasing levels of pollution and resource depletion [5].

Consequently, organizations are increasingly under pressure from stakeholders and markets to adopt environmentally sustainable practices. By prioritizing sustainability as a fundamental component of their operations, businesses can improve their brand image and provide consumers additional value [47]. Consequently, companies must integrate environmental sustainability into their evolving business models to achieve a more substantial impact as they implement DT strategies.

2.3 DT and Environmental Sustainability

The combination of environmental sustainability and DT has become a critical area of research in the direction of sustainable development goals and net-zero emissions. Research has demonstrated that the integration of digital technologies into environmental strategies can expedite the accomplishment of these objectives. Significant advancements in society and industry are driven by digital technologies, such as artificial intelligence (AI), big data analytics, mobile technologies, the Internet of Things (IoT), and social platforms [7]. Additionally, these technologies are being implemented more frequently to advance environmental sustainability [68]. Numerous organizations are introducing new products and platforms powered by digital technologies to strengthen sustainability initiatives. Plant Village, Smart Water Management Systems, and Peter Ma's AI-driven solution for detecting waterborne diseases have been the subject of research that has highlighted AI-based innovations. In an investigation conducted by [39], the potential for digitalization to improve environmental sustainability was investigated. The study examined nine situations in multiple nations where big data and IoT were implemented to address sustainability challenges and enhance environmental performance. Furthermore, the investigation underscored that digital workplaces have the potential to contribute to the advancement of environmental sustainability.

Businesses increasingly employ AI, IoT, and big data analytics to mitigate environmental pollution and reduce carbon emissions to establish sustainable business practices. big data analytics revolutionizes measuring and mapping environmental impacts, providing innovative solutions to monitor and improve sustainability [41, 60]. For instance, big data can be employed to create methods that enhance traceability in the food system and certify products based on their environmental performance, such as their carbon footprint or the sustainability practices used in their production processes. In the same vein, block chain technology is a powerful instrument that has the potential to enhance sustainability in the business and industrial sectors significantly. Block chain advances sustainable practices by facilitating the extension of product life cycles, optimizing resource utilization, and reducing carbon emissions.

The research of [59] emphasizes the potential of DT to mitigate environmental risks, monitor compliance with environmental regulations, and engage in proactive sustainability efforts. This is in contrast to older studies on sustainability, which were primarily concerned with resource management. This indicates a transition from a reactive to a proactive approach, in which businesses incorporate sustainability into their primary operations and adhere to regulations through digital channels.

3 Research Method

To achieve the research objectives, this study implements a structured and systematic methodology for evaluating extant literature regarding the intersection of environmental sustainability and DT. The paper will derive key research trends and identify potential areas for future exploration using the Systematic Literature Review (SLR) method. SLR has been widely acknowledged as a highly effective method for synthesizing research findings and identifying voids in the literature [46, 53]. Following established protocols from prior research, we adopted a meticulous, step-by-step procedure for selecting and analyzing relevant articles. Based on the previous and relevant studies, the following steps were implemented for systematic literature review.

3.1 Selection of Sources

In addition to the Web of Science, this study utilized Scopus, another highly regarded academic database, to ensure a comprehensive literature search. Scopus is known for its extensive collection of peer-reviewed journals and robust citation analysis tools, making it an ideal source for identifying key research articles. By incorporating multiple databases, we aimed to capture a broader spectrum of relevant literature and provide a more exhaustive overview of research trends in DT and environmental sustainability.

3.2 Setting Research Criteria

To ensure the identification of relevant articles, we established clear search criteria and divided the literature search into two key phases. In the first phase, we concentrated on exploring research related to DT, focusing on its fundamental concepts and technological advancements. In the second phase, we expanded the search to target environmental sustainability initiatives driven by digital technologies specifically. Our focus included cloud computing, artificial intelligence (AI), block-chain, the Internet of Things (IoT), big data, and intelligent systems. These technologies have been pivotal in reshaping environmental sustainability practices across industries and organizations, driving innovation and disruption in sustainability efforts. The initial search phase began with broad keywords such as DT, sustainable DT, environmental sustainability, sustainability, waste management, air pollution control, sustainable manufacturing, circular economy, sustainable cities, and urban sustainability. These keywords generated a large number of results due to their cross-disciplinary nature. To narrow our focus, we applied reference search techniques by examining the bibliographies of the retained papers, which helped us gain deeper insights into the literature trends within our research domain. However, as our primary interest was understanding how digital technologies like AI, big data, IoT, social media, and cloud computing drive transformations in environmental sustainability, we refined our keywords in the second phase to capture these technologies' role in sustainability efforts precisely.

3.3 Selection Criteria

A significant number of articles were identified during the comprehensive literature search. We exclusively focused on those that satisfied the predetermined selection criteria that were previously discussed. The final sample was subject to the following criteria: (a) the article must be a research study published in a reputable international journal, (b) it must provide practical applications for minimizing environmental impact, and (c) the article must focus on environmental sustainability and incorporate at least one of the specified digital technologies. Initially, 140 articles were selected from Scopus and Science Direct based on these parameters. Subsequently, 140 articles were refined to 70 after a more rigorous filtering procedure. We thoroughly screened the full-length articles to expedite the review process. We compiled a table that includes the paper title, author(s), publication year, journal name, and category (e.g., technology and environmental concentration). To facilitate a more comprehensive evaluation, each article was meticulously summarized to provide a thorough understanding of its scope, methodology, and findings. The last group of studies was selected based on the relevance of each study to our research concentration.

3.4 Content Analysis and Synthesis

Once the final set of articles was selected, a comprehensive content analysis was performed following a systematic approach. Like the literature search process, we categorized each article in a detailed table, including publication details, scope, technological focus, and specific environmental impacts. The selected articles showcased the use of digital technologies to mitigate environmental impacts by enhancing various organizational functions and processes. We identified key patterns and trends through an in-depth analysis, allowing us to group the articles into distinct themes. We developed a framework from these themes that serves as a foundation for future research. The following section provides an in-depth overview of the key findings from this analysis (Fig. 1).

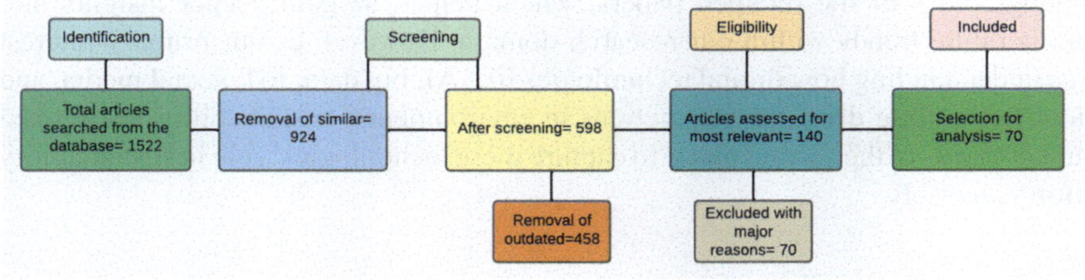

Fig. 1 Steps for selection and finalizing the target articles (PRISMA standard)

4 Results

Upon conducting a comprehensive examination of the selected articles, it was discovered that various digital technologies, including artificial intelligence, block-chain, big data analytics, IoT, cloud computing, and digital platforms, significantly influence environmental sustainability. Our analysis demonstrated that these technologies enable transformations in various critical areas, such as energy efficiency, sustainable logistics, resource optimization, and eco-friendly supply chain management. The studies consistently illustrated how digital innovations transform various aspects of sustainability practices, offering essential insights into how businesses and industries can utilize these technologies to accomplish environmental objectives.

The selected articles are grouped into themes due to their alignment with specific environmental areas, even though they span a variety of digital technologies. For example, energy efficiency, pollution control, waste management, and sustainable manufacturing research frequently include big data, IoT, AI, social media, and cloud computing. Consequently, we categorized these studies into four primary themes, subdivided into distinct subcategories delineated in the proposed framework. Figure 2 summarizes the total quantity of documents reviewed in each category. It is worth noting that pollution control has been the subject of most research on digital disruptions, with waste management following closely behind. Critical societal challenges, including carbon emissions, climate change, and natural calamities, were the focus of research in the pollution control theme. Similarly, many studies concentrated on waste management, significantly contributing to our overall sample.

We have observed a consistent increase in publications in the categories we have identified over the years. It is anticipated that this upward progress will continue as

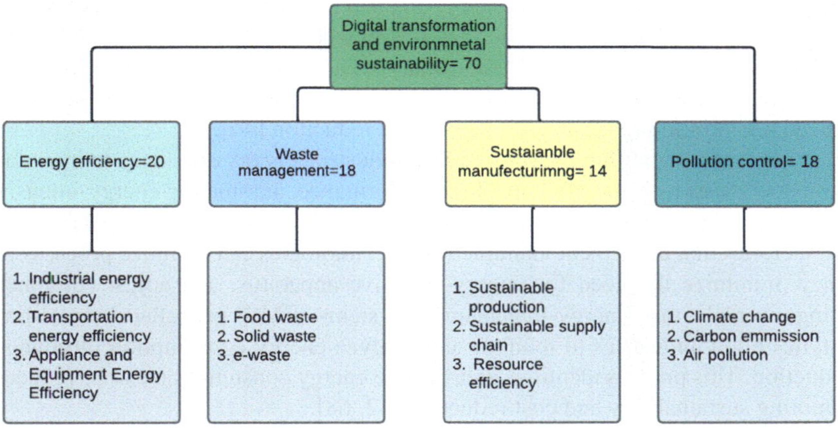

Fig. 2 Division of articles on DTs in environmental sustainability

further studies are integrated into the body of research in the future. Despite undertaking a thorough literature search without imposing time restrictions, no relevant publications predating 2015 met our inclusion criteria, as previously stated. The forthcoming literature is expected to contain additional research consistent with our search parameters as digital technologies continue to develop.

4.1 Energy Efficiency

DT and environmental sustainability are significantly influenced by energy efficiency. The potential to optimize energy use is increased dramatically as industries, organizations, and governments adopt digital technologies [57]. Integrating intelligent systems, data analytics, and automated controls is made possible by DT, streamlining energy management processes, reducing waste, and improving operational efficiency. For example, implementing advanced sensors and Internet of Things (IoT) devices in transportation networks, factories, and structures enables real-time energy consumption monitoring, resulting in improved resource management and reduced carbon footprints [10]. Additionally, the optimization of energy consumption is furthered by digital solutions, including AI and machine learning (ML), which forecast demand patterns, identify inefficiencies, and suggest energy-saving measures. This DT enhances sustainability initiatives and substantially contributes to attaining global energy efficiency objectives [33].

Digital technologies have transformed manufacturing operations in industrial energy efficiency by automating energy-intensive processes, facilitating predictive maintenance, and optimizing machine performance [39]. To reduce energy consumption, factories implementing manufacturing technologies can maximize process efficiency, reduce the necessity for energy-intensive machinery, and modify equipment settings in real time. Real-time analytics-powered energy management systems (EMS) enable industries to monitor and analyze energy consumption throughout the production process, thereby identifying areas where energy consumption can be reduced and promoting sustainability and cost reduction [40].

Digital technologies have revolutionized industrial energy efficiency in manufacturing operations by optimizing machine performance, automating energy-intensive processes, and enabling predictive maintenance [53]. To decrease energy consumption, factories that implement manufacturing technologies can optimize process efficiency, minimize the need for energy-intensive apparatus, and adjust equipment settings in real-time. Energy management systems (EMS) propelled by real-time analytics allow industries to monitor and analyze energy consumption throughout production. This process identifies areas where energy consumption can be reduced, promoting sustainability and cost reduction [42, 68].

In the same vein, digital innovations have significantly enhanced the energy efficacy of appliances and equipment. IoT-powered intelligent appliances allow consumers to monitor and regulate their energy consumption more effectively, frequently from remote locations. AI algorithms are now integrated into energy-efficient devices, which enable them to modify their operation in response to environmental conditions and real-time energy consumption [2, 44, 68]. For example, smart thermostats and lighting systems automatically adapt to minimize energy consumption in unoccupied rooms. Conversely, intelligent meters provide homes and businesses with comprehensive information about energy consumption patterns. These developments decrease energy consumption and motivate consumers to adopt more sustainable energy habits [3, 20, 27] (Table 1).

Table 1 DT in energy efficiency

Category	Findings	Relevant studies
Industrial energy efficiency	The substantial contribution of DT to the promotion of environmental sustainability. By facilitating real-time monitoring, predictive maintenance, and automation of energy-intensive activities, the integration of digital technologies such as IoT, AI, and machine learning optimizes energy use in manufacturing processes. In addition, efficient energy management systems enable industries to monitor energy consumption patterns, identify inefficiencies, and implement corrective actions, resulting in substantial energy savings	[10, 33, 39, 46, 66, 70, 72]
Transportation energy efficiency	Real-time data systems, autonomous vehicles, and smart transportation systems enhance logistical operations, thereby decreasing energy consumption in supply chains. Implementing DT in the transportation sector contributes to sustainability by reducing energy consumption and emissions and promoting more efficient and environmentally friendly transportation solutions	[3, 8, 20, 27, 37, 39, 40]
Appliance and equipment energy efficiency	Smart appliances fueled by IoT, AI, and machine learning enable real-time monitoring and control of energy usage, resulting in optimized performance and reduced consumption. These technologies allow digital thermostats, lighting systems, and energy-efficient appliances to automatically adapt to environmental conditions and utilization patterns, thereby substantially reducing energy waste. Integrating AI-driven energy management systems offers consumers actionable energy consumption insights, promoting long-term sustainability	[2, 42, 44, 68]

In the same vein, digital innovations have significantly enhanced the energy efficacy of appliances and equipment. IoT-powered intelligent appliances allow consumers to monitor and regulate their energy consumption more effectively, frequently from remote locations [68]. AI algorithms are now integrated into energy-efficient devices, which enable them to modify their operation in response to environmental conditions and real-time energy consumption. For example, smart thermostats and lighting systems automatically adapt to minimize energy consumption in unoccupied rooms. Conversely, intelligent meters provide homes and businesses with comprehensive information about energy consumption patterns. These developments decrease energy consumption and motivate consumers to adopt more sustainable energy habits [2].

4.2 Waste Management

The rapid pace of urbanization and the increasing population levels have resulted in solid waste becoming a global environmental concern. In the same vein, food waste has significant environmental and social consequences in both urban and rural regions across the globe [26]. Consequently, digital waste management solutions are gaining momentum, encouraging innovation and opening new business opportunities. Various materials, including construction trash, metals, wood, and polymers, are sorted by color using advanced technology in AI-powered systems like Zen Robotics' heavy picker [63]. As previously mentioned, we organized the literature on solid, electronic, and food waste to facilitate a more comprehensive comprehension of these concerns (Table 2).

Digital technologies have become an essential component of numerous environmental sustainability initiatives. AI has been implemented in multiple domains, such as energy efficiency, biodiversity conservation, refuse management, geographical mapping, environmental risk assessment, and climate change mitigation [26, 49]. For example, AI has been employed in e-waste management to optimize collection services by user preferences. Furthermore, recycling and waste management initiatives are progressively employing big data. Ismail et al. [29] Employed big data to monitor and identify unlawful waste dumping in Hong Kong from 2011 to 2017, analyzing more than two million disposal records from approximately 5,700 projects. This is a notable example. Similarly, implemented big data to assess and contrast the waste management strategies of public and private organizations in the construction sector [29].

Various digital technologies, including IoT, cloud computing, and social media, are revolutionizing waste management practices. For instance, an IoT-driven solution was implemented at the Asian Institute of technology (AIT) to manage food waste, substantially reducing the waste generated by the campus community. Additionally, research by [13] utilized the theory of planned behavior to explore how social media can influence recycling behaviors, highlighting its potential to promote

Table 2 DT in waste management

Category	Findings	Relevant studies
Solid waste	Acknowledging the critical role of digitalization in improving traditional waste management practices in developing countries, the research also emphasizes the innovative potential of digital solutions in enhancing sustainability through waste recycling. The waste management sector is experiencing a substantial transformation driven by digital technologies. This transition to digitalization has created new opportunities for promoting a sustainable approach that emphasizes preventing, reducing, and recycling refuse from specific waste streams	[9, 34, 43, 45, 71]
Food waste	Digital technologies such as IoT and big data are significantly improving industrial and household food waste disposal. IoT-enabled systems enable organizations to monitor and manage food waste in real time, which allows them to identify inefficiencies, optimize disposal processes, and trace waste generation. In the same way, big data analytics can analyze large volumes of waste-related data, providing valuable insights into food waste patterns and assisting in developing more efficient waste reduction strategies	[1, 6, 13, 15, 30, 48]
e-waste	Implementing innovative solutions to optimize waste management strategies and improve recycling processes as e-waste increases is imperative. Organizations can enhance their capacity to monitor e-waste, improve collection methods, and fortify classification procedures by employing cutting-edge technologies like artificial intelligence, machine learning, and IoT	[11, 26, 29, 32, 49, 51, 63]

more sustainable waste management practices. These technologies provide innovative ways to reduce waste, improve recycling, and encourage environmentally conscious behaviors [6].

4.3 Sustainable Manufacturing

A manufacturing system that is more sustainable and environmentally friendly can result in a reduced ecological footprint, improved worker safety, increased profitability, and decreased operational costs [1]. Significant interest in academic discussions is being generated by innovative and sustainable manufacturing. This method prioritizes using renewable energy sources, sustainable production methods, energy efficiency, and environmentally beneficial practices [21] (Table 3).

In the field of smart manufacturing, digital innovations such as AI, IoT, blockchain, and machine learning are essential for promoting sustainable practices across

Table 3 DT in sustainable manufacturing

Category	Findings	Relevant studies
Sustainable production	With the emergence of the circular economy, there is an increasing interest in implementing digital technology-driven manufacturing processes prioritizing environmental sustainability. The objective of these methods is to reduce ecological damage while simultaneously fostering cost savings and resource efficiency. Industries can reduce waste, optimize production cycles, and reuse materials by incorporating digital tools, all of which contribute to reducing the environmental impact	[5, 7, 21, 47, 67]
Sustainable supply chain	Digital advancements enable businesses to optimize their value chains by substantially reducing waste and enhancing resource efficiency. These technologies promote more sustainable consumption by optimizing production processes, reducing irrelevant material consumption, and reducing environmentally detrimental pollution. Companies can reduce their environmental impact and encourage more responsible operations throughout the supply chain by identifying inefficiencies, enhancing recycling efforts, and promoting circular practices through data analytics, AI, and IoT	[31, 35, 54, 59, 65]
Resource efficiency	DT is essential for the promotion of environmental sustainability and the improvement of resource efficiency. AI, block-chain-driven optimization, big data analytics, and the Internet of Things (IoT) are technologies that optimize the use of energy, water, and essential materials across industries. These digital tools facilitate automated control, predictive analytics, and real-time monitoring, enhancing resource management, reducing waste, and lowering emissions. For instance, AI-driven systems can anticipate energy demand and optimize usage to reduce waste, while IoT sensors monitor resource consumption to improve efficiency	[52, 57, 72]

various industries. Numerous studies have demonstrated that these technologies facilitate sustainable supply chain management while enhancing energy efficiency, waste reduction, and eco-friendly operations [7, 21, 67]. For example, [5] Developed a data-driven optimization framework for sustainable production that reduces waste and energy consumption. Similarly, [47] Investigated the function of AI in predictive maintenance, resource optimization, and environmental footprint reduction. These technologies improve the environmental performance of businesses, as well as the sustainability of their operations and green manufacturing processes.

4.4 Pollution Control

In light of the substantial risks to public health and ecosystems posed by accelerated urban development, environmental protection necessitates concentrated efforts to address the pollutants discharged into surface water. The chemical industry, energy sectors such as coal mining and oil refinement, and massive industries like steel and iron fabrication are major contributors to this pollution [41]. These sectors have been identified as significant sources of air and water contamination. Numerous nations have implemented stringent regulations over the years to mitigate the detrimental effects of these pollution-intensive industries to regulate emissions and enhance environmental quality [60].

Innovative solutions to environmental challenges are being introduced by digital technologies, revolutionizing pollution monitoring, control, and management. As evidenced by studies, these technologies are reshaping efforts to combat air pollution, reduce carbon emissions, enhance effluent treatment, and manage climate-related disasters [19, 61]. For example, artificial intelligence has emerged as a potent instrument for resolving intricate, dynamic environmental concerns by optimizing pollution control procedures. In the same way, big data analytics facilitates the advancement of low-carbon transportation systems, including green vehicles, which in turn reduces CO2 emissions and promotes sustainability [28, 62]. Ma et al. [41] Have shown that big data can improve decision-making by establishing cost-effective carbon emission limits, minimizing risks and expenses, and reducing environmental damage (Table 4).

Numerous investigations have investigated the utilization of IoT technology to supervise and regulate air pollution. IoT sensors offer valuable insights into the enhancement of air quality by providing real-time data that enables the precise monitoring of pollution levels, including PM2.5, across various locations and time frames [14, 38]. Furthermore, social media has been acknowledged as a highly effective instrument for the evaluation of the effects of air pollution and disaster management. It is a critical resource for the management of environmental crises and the collection of public insights due to its perceived advantages and simplicity of use. pollution control has also been implemented through the use of mobile technologies [75]. For instance the smart sensors and data-driven models to measure real-time mobile air quality, which allowed for more precise predictions of air quality. These developments underscore the potential of digital instruments to resolve environmental challenges [60].

Table 4 DT in pollution control

Category	Findings	Relevant studies
Climate change	Digital technologies are increasingly crucial in the mitigation of environmental impacts, particularly in the reduction of carbon footprints, the management of natural disasters, and the mitigation of climate change. Companies are progressively employing these technological advancements to establish sustainable practices. AI, IoT, and data analytics are being implemented to optimize energy consumption, monitor emissions, and improve disaster response systems	[17, 19, 50, 56, 61, 69]
Carbon emission	The incorporation of digital technologies to support environmentally sustainable practices is becoming more prevalent as businesses and governments worldwide strive to reduce carbon emissions. The transition to digitalization is facilitating the more efficient monitoring and management of resource efficiency, energy consumption, and emissions. The increasing emphasis on digital-driven sustainability is contributing to the acceleration of efforts to address climate change and promote long-term environmental stewardship	[16, 28, 41, 62, 74]
Air pollution	Digital technologies offer a wealth of new opportunities to analyze, manage, and mitigate air pollution in urban areas worldwide. Through the use of IoT sensors, big data analytics, and AI, cities can better understand pollution patterns, forecast air quality issues, and implement timely interventions	[4, 14, 38, 60, 75]

5 Discussion

This research emphasized the areas of environmental sustainability in which digital technologies are revolutionizing the management of resources, emissions, production processes, and urban development. The results indicated that digital technology enables businesses to implement sustainable practices or develop new environmentally-focused business models. DT is a catalyst for improving environmental sustainability in both scenarios. The study underscored the fact that the environmental advantages of these technologies were novel and inconceivable a decade ago. For example, utilizing and optimizing patterns is a revolutionary sustainability strategy. These innovations have the potential to significantly impact a variety of sectors, such as government, industry, and non-profit organization DT, these entities can improve policymaking, environmental sustainability, and quality of life.

The DT in this sector provides a shared platform to address environmental degradation through sustainable practices. The ongoing digitalization of sustainability presents many opportunities for companies to enhance existing business models or develop innovative ones. This change suggests that conventional businesses will compete in new sectors to establish a competitive edge and distinguish themselves.

Nevertheless, it is imperative to further our comprehension of how these digital sustainability trends can be exploited to establish a competitive advantage. The results of this study indicate that, despite the extensive literature on the broad relationship between digitalization and environmental sustainability, there is a deficiency in the comprehension of the precise capabilities that businesses require to endure a sustainable DT. Nike, [18] for example, has implemented strategies to reduce its ecological footprint. Additionally, specific organizations endeavor to integrate sustainability principles into their operations to mitigate environmental harm.

In contrast, organizations are oriented around sustainability, providing AI-driven waste management solutions. The capabilities required for each sustainable DT approach will differ substantially. Consequently, we suggest a research agenda to investigate these domains further.

5.1 Theoretical Contribution

Digital technologies are altering the measurement and management of ecological sustainability issues. However, there is still a discrepancy in academic comprehension regarding the necessary capabilities to integrate environmental sustainability into DT efforts and the manner in which organizations should adjust to these disruptions. The objective of this investigation is to inspect the necessity for more thorough research to examine the effects of DT on sustainability. The safety of the atmosphere from pollution and resource depletion remains a significant challenge, thus it is imperative to prioritize this objective. Our objective is to enhance the current body of knowledge by creating a framework that demonstrates the manner and location of DTs in the environmental sustainability sector. The framework emphasizes substantial disruptions in sectors, including sustainable production, pollution control, waste management, and resource efficiency. As previously mentioned, future research can further explore the capabilities required for these transformations, their impact on organizational enactment, and the role of digitalization, broadening our comprehension of sustainability. Additionally, our review will encourage association among academics in the environmental sustainability arena, thereby fostering collaborative endeavors to address shared challenges.

5.2 Practical Implications

The results of our investigation will be beneficial to policy-makers in both community and private enterprises, as they will enable them to classify and order critical areas for investment in environmental sustainability. These insights will allow companies to enhance and develop their sustainability strategies, thereby establishing a mutually beneficial for all stakeholders. For example, our research indicates that digital technologies offer distinctive opportunities to either develop new business models that prioritize environmental sustainability or incorporate sustainable practices into their current models through digital adoption. These examples can be used as a starting point for strategic discussions in a broad way, providing decision-makers with guidance on plans and implementation strategies. Managers can initiate the process by concentrating on incremental variations, such as the digitalization of specific processes or sub-processes. This approach will enhance the overall business models and facilitate the advancement of sustainability objectives.

5.3 Future Research Agendas

Future studies on DT and environmental sustainability should first examine how firms understand and implement sustainability into their DT initiatives. This would entail studying strategic incentives, how sustainability drives digital innovations, and long-term competitiveness and operational effects. Further study is needed to determine what competencies firms need to integrate environmental sustainability into DT effectively. In addition to leadership and organizational cultural changes, this may need technical, management, and operational abilities. Additionally, firms that want to transform their business models toward sustainability entirely must grasp infrastructure, value propositions, and supply chain management adjustments. Research should distinguish between companies with fully sustainable business strategies and those that incorporate sustainability into their CSR without changing their fundamental operations [23–25]. Organizations with varying sustainability objectives require different degrees of commitment, resource demands, and strategic methods.

Future studies could also examine how environmentally friendly DT skills improve organizational performance. This involves identifying the technology infrastructure, human capital, and financial expenditures needed to digitalize environmental sustainability activities. Research should also examine if digitalizing these processes gives organizations a competitive edge, highlighting the strategic benefits of combining sustainability with digital activities. These studies will reveal the immediate advantages and difficulties of environmentally friendly digital strategies for organizations.

Future studies should focus on efficiently integrating environmental sustainability objectives with digital activities across company functions. It is crucial to identify the digital business strategies businesses require for environmentally sustainable DTs and how technology may support sustainability while preserving business success. Research should also examine the basic drivers of sustainable DT, such as regulatory constraints, market demand, and company values. Understanding these factors can help organizations integrate digitization with environmental responsibility.

References

1. Abbate, S., Centobelli, P., Cerchione, R.: The digital and sustainable transition of the agri-food sector. Technol. Forecast. Soc. Chang. **187**, 122222 (2023). https://doi.org/10.1016/J.TECHFORE.2022.122222
2. Aheleroff, S., Xu, X., Lu, Y., Aristizabal, M., Pablo Velásquez, J., Joa, B., Valencia, Y.: IoT-enabled smart appliances under industry 4.0: a case study. Adv. Eng. Inform. **43**, 101043 (2020). https://doi.org/10.1016/J.AEI.2020.101043
3. Alanazi, F., Alenezi, M.: Driving the future: leveraging digital transformation for sustainable transportation. J. Infrastruct. Policy Dev. **8**(3), 3085 (2024). https://doi.org/10.24294/JIPD.V8I3.3085
4. Amann, M., Kiesewetter, G., Schöpp, W., Klimont, Z., Winiwarter, W., Cofala, J., Pavarini, C.: Reducing global air pollution: the scope for further policy interventions. Philos. Trans. R. Soc. A **378**(2183), 20190331 (2020). https://doi.org/10.1098/RSTA.2019.0331
5. Andriushchenko, K., Buriachenko, A., Rozhko, O., Lavruk, O., Skok, P., Hlushchenko, Y., Muzychka, Y., Slavina, N., Buchynska, O., Kondarevych, V.: Peculiarities of sustainable development of enterprises in the context of digital transformation. Entrepreneurship Sustain. Issues. **7**(3), 2255 (2020). https://doi.org/10.9770/jesi.2020.7.3(53)
6. Annosi, M.C., Brunetta, F., Bimbo, F., Kostoula, M.: Digitalization within food supply chains to prevent food waste. Drivers, barriers and collaboration practices. Ind. Mark. Manage. **93**, 208–220 (2021). https://doi.org/10.1016/J.INDMARMAN.2021.01.005
7. Bai, C., Zhou, H., Sarkis, J.: Evaluating Industry 4.0 technology and sustainable development goals–a social perspective. Int. J. Prod. Res. **61**(23), 8094–8114 (2023). https://doi.org/10.1080/00207543.2022.2164375
8. Baldini, G., Barboni, M., Bono, F., Delipetrev, B., Duch Brown, N., Fernandez Macias, E., ... Nepelski, D. (2019). Digital Transformation in Transport, Construction, Energy, Government and Public Administration. Retrieved from https://publications.jrc.ec.europa.eu/repository/handle/JRC116179
9. Batista, M., Caiado, R.G., Quelhas, O.L., Lima, G.B., Leal Filho, W., Yparraguirre, I.T.: A framework for sustainable and integrated municipal solid waste management: Barriers and critical factors to developing countries. J. Clean. Prod. **20**(312), 127516 (2021). https://doi.org/10.1016/J.JCLEPRO.2021.127516
10. Benzerga, S., Hauf, D., Pretz, M., Bounfour, A.: When energy revolution meets digital transformation. Smart Innov. Syst. Technol. **73**, 65–81 (2018). https://doi.org/10.1007/978-3-319-59424-8_7/FIGURES/11
11. Bhaskar, K., Kumar, B.: Electronic waste management and sustainable development goals: Is there a business case for linking the two? J. Indian Bus. Res. **11**(2), 120–137 (2019). https://doi.org/10.1108/JIBR-01-2018-0051/FULL/XML
12. Biswas, D., Jalali, H., Ansaripoor, A.H., De Giovanni, P.: Traceability versus sustainability in supply chains: the implications of blockchain. Eur. J. Oper. Res. **305**(1), 128–147 (2023). https://doi.org/10.1016/J.EJOR.2022.05.034
13. Bolwig, S., Tanner, A.N., Riemann, P., Redlingshöfer, B., Zhang, Y.: Reducing consumer food waste using green and digital technologies. (2021). Retrieved from https://forskning.ruc.dk/en/publications/reducing-consumer-food-waste-using-green-and-digital-technologies
14. Chien, F., Sadiq, M., Nawaz, M.A., Hussain, M.S., Tran, T.D., Le Thanh, T.: A step toward reducing air pollution in top Asian economies: the role of green energy, eco-innovation, and environmental taxes. J. Environ. Manage. **297**, 113420 (2021). https://doi.org/10.1016/J.JENVMAN.2021.113420
15. de Souza, M., Pereira, G.M., de Sousa, L., Jabbour, A.B., Chiappetta Jabbour, C.J., Trento, L.R., Borchardt, M., Zvirtes, L.: A digitally enabled circular economy for mitigating food waste: understanding innovative marketing strategies in the context of an emerging economy. Technol. Forecast. Soc. Chang. **173**, 121062 (2021). https://doi.org/10.1016/J.TECHFORE.2021.121062

16. Oliveira, K.K., De Souza, R.A.: Digital transformation towards education 4.0. Inform. Educ. Int. J. **21**(2), 283–309 (2022)
17. Dwivedi, Y.K., Hughes, L., Kar, A.K., Baabdullah, A.M., Grover, P., Abbas, R., Wade, M.: Climate change and COP26: are digital technologies and information management part of the problem or the solution? An editorial reflection and call to action. Int. J. Inf. Manage. **63**, 102456 (2022). https://doi.org/10.1016/J.IJINFOMGT.2021.102456
18. Frei, R., Jack, L., Krzyzaniak, S.A.: Sustainable reverse supply chains and circular economy in multichannel retail returns. Bus. Strategy Environ. 1–16. (2020) https://doi.org/10.1002/bse.2479
19. Freitag, C., Berners-Lee, M., Widdicks, K., Knowles, B., Blair, G.S., Friday, A.: The real climate and transformative impact of ICT: A critique of estimates, trends, and regulations. Patterns (2021). https://doi.org/10.1016/J.PATTER.2021.100340/ATTACHMENT/BFBCC1A8-8DC6-4A94-AF57-843E8F882A25/MMC2.PDF
20. Gomes, J.M.M., Caiado, R.G.G., Mareth, T., Santos, R.S., Scavarda, L.F.: Lean transportation tools towards waste reduction and digital transformation in the offshore industry an action-research. Int. J. Lean Six Sigma ahead-of-print(ahead-of-print). (2024) https://doi.org/10.1108/IJLSS-12-2023-0234/FULL/XML
21. Guan, L., Li, W., Guo, C., Huang, J.: Environmental strategy for sustainable development: role of digital transformation in China's natural resource exploitation. Resour. Policy **87**, 104304 (2023). https://doi.org/10.1016/J.RESOURPOL.2023.104304
22. Harnrungchalotorn, S., Phayonlerd, Y.: Starbucks with corporate social responsibility: how starbucks succeeds in a business world with CSR (Master's thesis, 42. (2018). https://doi.org/10.1177/1096348017727057
23. Hayat, K., Jianjun, Z., Ali, S., Ageli, M.M.: Eco-advertising and ban-on-plastic: the influence of CSR green practices on green impulse behavior. J. Knowl. Econ. (2022). https://doi.org/10.1007/s13132-022-01014-w
24. Hayat, K., Jianjun, Z., Ullah, S.: Relationship of Corporate Social Responsibility and Impulse Buying: Role of Corporate Reputation and Consumer Company Identification. International Journal of Management Research and Emerging Sciences **13**(1), 123–154 (2023)
25. Hayat, K., Qingyu, Z.: The synergistic effects of green innovation strategies on sustainable innovative performance with the mediation of green innovative competitive advantage. (March 2023), pp. 1–18. (2024) https://doi.org/10.1002/csr.2770
26. Heeks, R., Subramanian, L., Jones, C.: Understanding e-waste management in developing countries: strategies, determinants, and policy Implications in the Indian ICT Sector. Inf. Technol. Dev. **21**(4), 653–667 (2015). https://doi.org/10.1080/02681102.2014.886547
27. Huang, Y., Hu, M., Xu, J., Jin, Z.: Digital transformation and carbon intensity reduction in transportation industry: empirical evidence from a global perspective. J. Environ. Manag. **344**, 118541 (2023). https://doi.org/10.1016/J.JENVMAN.2023.118541
28. Hussain, M., Mir, G.M., Usman, M., Ye, C., Mansoor, S.: Analysing the role of environment-related technologies and carbon emissions in emerging economies: a step towards sustainable development. Environ. Technol. (2022). https://doi.org/10.1080/09593330.2020.1788171
29. Ismail, H., Hanafiah, M.M.: A review of sustainable e-waste generation and management: present and future perspectives. J. Environ. Manag. **264**, 110495 (2020). https://doi.org/10.1016/J.JENVMAN.2020.110495
30. Joubert, R.L., Jokonya, O.: A systematic literature review of factors affecting the adoption of technologies in food waste management. Proced. Comput. Sci. **181**, 1034–1040 (2021). https://doi.org/10.1016/J.PROCS.2021.01.298
31. Junge, A.L.: Digital transformation technologies as an enabler for sustainable logistics and supply chain processes—an exploratory framework. Braz. J. Oper. Prod. Manag. **16**(3), 462–472 (2019). https://doi.org/10.14488/BJOPM.2019.V16.N3.A9
32. Kazancoglu, Y., Ozkan-Ozen, Y.D., Mangla, S.K., Ram, M.: Risk assessment for sustainability in e-waste recycling in circular economy. Clean Technol. Environ. Policy **24**(4), 1145–1157 (2022). https://doi.org/10.1007/S10098-020-01901-3/TABLES/8

33. Khalid, M.: Energy 4.0: AI-enabled digital transformation for sustainable power networks. Comput. Ind. Eng. **193**, 110253 (2024). https://doi.org/10.1016/J.CIE.2024.110253
34. Kurniawan, T.A., Meidiana, C., Dzarfan Othman, M.H., Goh, H.H., Chew, K.W.: Strengthening waste recycling industry in Malang (Indonesia): Lessons from waste management in the era of Industry 4.0. J. Clean. Prod. **382**, 135296 (2023). https://doi.org/10.1016/J.JCLEPRO.2022.135296
35. Lerman, L.V., Benitez, G.B., Müller, J.M., de Sousa, P.R., Frank, A.G.: Smart green supply chain management: a configurational approach to enhance green performance through digital transformation. Supply Chain Manag. **27**(7), 147–176 (2022). https://doi.org/10.1108/SCM-02-2022-0059/FULL/PDF
36. Leveraging Digitalization for Improving Energy Efficiency.: (n.d.). https://doi.org/10.1007/978-981-19-6778-8_22
37. Li, W.: Research on the impact of digital economy on industrial energy efficiency. Front. Bus. Econ. Manag. **7**(1), 115–119 (2023). https://doi.org/10.54097/FBEM.V7I1.3911
38. Li, Y., Dai, J., Cui, L.: The impact of digital technologies on economic and environmental performance in the context of industry 4.0: a moderated mediation model. Int. J. Product. Econ. **229**, 107777 (2020). https://doi.org/10.1016/J.IJPE.2020.107777
39. Llopis-Albert, C., Rubio, F., Valero, F.: Impact of digital transformation on the automotive industry. Technol. Forecast. Soc. Chang. **162**, 120343 (2021). https://doi.org/10.1016/J.TECHFORE.2020.120343
40. Lv, Z., Shang, W.: Impacts of intelligent transportation systems on energy conservation and emission reduction of transport systems: a comprehensive review. Green Technol. Sustain. **1**(1), 100002 (2023). https://doi.org/10.1016/J.GRETS.2022.100002
41. Ma, Q., Tariq, M., Mahmood, H., Khan, Z.: The nexus between digital economy and carbon dioxide emissions in China: the moderating role of investments in research and development. Technol. Soc. **68**, 101910 (2022). https://doi.org/10.1016/J.TECHSOC.2022.101910
42. Maciej, S., Becker, F.G., Cleary, M., Team, R.M., Holtermann, H., The, D., فاطمی, ح.: The effect of digital technologies on energy efficiency policy. Int. J. Electr. Eng. and Sustain., 7(1), 1–8. (2023). 10.2/JQUERY.MIN.JS
43. Madureira, C., Amorim, M., Ferreira Dias, M., de Sousa, M., Silva, C., Travassos Rosário, A., Carmo Dias, J.: Sustainability and the digital transition: a literature review. Sustainability **14**, 4072 (2022). https://doi.org/10.3390/SU14074072
44. Mao, C., Chang, D.: Review of cross-device interaction for facilitating digital transformation in smart home context: a user-centric perspective. Adv. Eng. Inform. **57**, 102087 (2023). https://doi.org/10.1016/J.AEI.2023.102087
45. Marchesi, L., Lunesu, M.I., Tonelli, R.: Blockchain technology for certifying waste management within the digital transformation for industry and SME. UMAP 2024—Adjunct Proceedings of the 32nd ACM Conference on User Modeling, Adaptation and Personalization, pp. 415–418. (2024). https://doi.org/10.1145/3631700.3665588
46. Maroufkhani, P., Desouza, K.C., Perrons, R.K., Iranmanesh, M.: Digital transformation in the resource and energy sectors: a systematic review. Resour. Policy **76**, 102622 (2022). https://doi.org/10.1016/J.RESOURPOL.2022.102622
47. Martínez-Peláez, R., Ochoa-Brust, A., Rivera, S., Félix, V.G., Ostos, R., Brito, H., Mena, L.J.: Role of digital transformation for achieving sustainability: mediated role of stakeholders, key capabilities, and technology. Sustainability **15**(14), 11221 (2023). https://doi.org/10.3390/SU151411221
48. Mazzucchelli, A., Gurioli, M., Graziano, D., Quacquarelli, B., Aouina-Mejri, C.: How to fight against food waste in the digital era: key factors for a successful food sharing platform. J. Bus. Res. **124**, 47–58 (2021). https://doi.org/10.1016/J.JBUSRES.2020.11.055
49. Modarress Fathi, B., Ansari, A., Ansari, A.: Threats of internet-of-thing on environmental sustainability by E-waste. Sustainability **14**(16), 10161 (2022). https://doi.org/10.3390/SU141610161
50. Mondejar, M.E., Avtar, R., Diaz, H.L.B., Dubey, R.K., Esteban, J., Gómez-Morales, A., Garcia-Segura, S.: Digitalization to achieve sustainable development goals: steps towards a smart green

planet. Sci. Total. Environ. **794**, 148539 (2021). https://doi.org/10.1016/J.SCITOTENV.2021.148539
51. Moradi, R., Yazdi, M., Haghighi, A., Nedjati, A.: Sustainable resilient E-waste management in London: a circular economy perspective. Heliyon, 10(13). (2024) https://doi.org/10.1016/J.HELIYON.2024.E34071/ASSET/FD6ED1F7-7C4F-4C47-BF6C-2C3C221E5A7E/MAIN.ASSETS/GR14_LRG.JPG
52. Mosconi, E.M., Marconi, M., Morea, D., Pellegrini, M.M., Ruan, T., Gu, Y., Qu, R.: Research on the practical path of resource-based enterprises to improve environmental efficiency in digital transformation. Sustainability **14**(21), 13974 (2022). https://doi.org/10.3390/SU142113974
53. Negri, M., Cagno, E., Colicchia, C., Sarkis, J.: Integrating sustainability and resilience in the supply chain: a systematic literature review and a research agenda. Bus. Strategy Environ. 1–29. (2021). https://doi.org/10.1002/bse.2776
54. Oubrahim, I., Sefiani, N., Happonen, A.: The influence of digital transformation and supply chain integration on overall sustainable supply chain performance: an empirical analysis from manufacturing companies in morocco. Energies **16**(2), 1004 (2023). https://doi.org/10.3390/EN16021004
55. Pauliuk, S., Koslowski, M., Madhu, K., Schulte, S., Kilchert, S.: Co-design of digital transformation and sustainable development strategies—what socio-metabolic and industrial ecology research can contribute. J. Clean. Prod. **343**, 130997 (2022). https://doi.org/10.1016/J.JCLEPRO.2022.130997
56. Plekhanov, D., Franke, H., Netland, T.H.: Digital transformation: a review and research agenda. Eur. Manag. J. **41**(6), 821–844 (2023). https://doi.org/10.1016/J.EMJ.2022.09.007
57. Quttainah, M.A., Ayadi, I.: The impact of digital integration on corporate sustainability: emissions reduction, environmental innovation, and resource efficiency in the European. J. Innov. Knowl. **9**(3), 100525 (2024). https://doi.org/10.1016/J.JIK.2024.100525
58. Raihan, A.: (n.d.). The Influence of Digital Transformation on Environmental Sustainability The 2023 PhD International Conference on Digital Transformation and Sustainable Development The Influence of Digital Transformation on Environmental Sustainability. Retrieved from https://www.researchgate.net/publication/377529823
59. Rehman Khan, S.A., Ahmad, Z., Sheikh, A.A., Yu, Z.: Digital transformation, smart technologies, and eco-innovation are paving the way toward sustainable supply chain performance. Sci. Prog. 105(4). (2022). https://doi.org/10.1177/00368504221145648/ASSET/IMAGES/LARGE/10.1177_00368504221145648-FIG4.JPEG
60. Ren, S., Hao, Y., Wu, H.: Digitalization and environment governance: does internet development reduce environmental pollution? J. Environ. Plan. Manag. **66**(7), 1533–1562 (2023). https://doi.org/10.1080/09640568.2022.2033959
61. Salam, A.: Internet of things for environmental sustainability and climate change. pp. 33–69. (2024). https://doi.org/10.1007/978-3-031-62162-8_2
62. Shang, Y., Raza, S.A., Huo, Z., Shahzad, U., Zhao, X.: Does enterprise digital transformation contribute to the carbon emission reduction? Micro-level evidence from China. Int. Rev. Econ. Financ. **86**, 1–13 (2023). https://doi.org/10.1016/J.IREF.2023.02.019
63. Sharma, H., Kumar, H., Mangla, S.K.: Enablers to computer vision technology for sustainable E-waste management. J. Clean. Prod. **412**, 137396 (2023). https://doi.org/10.1016/J.JCLEPRO.2023.137396
64. Song, J., He, G., Wang, J., Zhang, P.: Shaping future low-carbon energy and transportation systems: digital technologies and applications. IEnergy **1**(3), 285–305 (2022). https://doi.org/10.23919/IEN.2022.0040
65. Stroumpoulis, A., Kopanaki, E.: Theoretical perspectives on sustainable supply chain management and digital transformation: a literature review and a conceptual framework. Sustainability **14**(8), 4862 (2022). https://doi.org/10.3390/SU14084862
66. Tiefenbeck, V.: Bring behaviour into the digital transformation. Nat. Energy **2**(6), 1–3 (2017). https://doi.org/10.1038/nenergy.2017.85

67. Varriale, V., Camilleri, M.A., Cammarano, A., Michelino, F., Müller, J., Strazzullo, S.: Unleashing digital transformation to achieve the sustainable development goals across multiple sectors. Sustain. Dev. (2024). https://doi.org/10.1002/SD.3139
68. Verhoef, P.C., Broekhuizen, T., Bart, Y., Bhattacharya, A., Qi Dong, J., Fabian, N., Haenlein, M.: Digital transformation: a multidisciplinary reflection and research agenda. J. Bus. Res. **122**, 889–901 (2021). https://doi.org/10.1016/J.JBUSRES.2019.09.022
69. Vial, G.: Understanding digital transformation: a review and a research agenda. Manag. Digit. Transform. (2021). https://doi.org/10.4324/9781003008637-4
70. Wang, S., Zhang, R., Yang, Y., Chen, J., Yang, S.: Has enterprise digital transformation facilitated the carbon performance in Industry 4.0 era? Evidence from Chinese industrial enterprises. Comput. Ind. Eng. **184**, 109576 (2023). https://doi.org/10.1016/J.CIE.2023.109576
71. Wu, D., Xie, Y., Lyu, S.: Disentangling the complex impacts of urban digital transformation and environmental pollution: evidence from smart city pilots in China. Sustain. Cities Soc. **88**, 104266 (2023). https://doi.org/10.1016/J.SCS.2022.104266
72. Xu, Q., Li, X., Guo, F.: Digital transformation and environmental performance: evidence from Chinese resource-based enterprises. Corp. Soc. Responsib. Environ. Manag. **30**(4), 1816–1840 (2023). https://doi.org/10.1002/CSR.2457
73. Zhang, J., Lyu, Y., Li, Y., Geng, Y.: Digital economy: an innovation driving factor for low-carbon development. Environ. Impact Assess. Rev. **96**, 106821 (2022). https://doi.org/10.1016/J.EIAR.2022.106821
74. Zhang, L., Gu, F., He, M.: The influence of digital transformation on the reconfigurability and performance of supply chains: a study of the electronic, machinery, and home appliance manufacturing industries in China. Sustainability **16**(7), 2689 (2024). https://doi.org/10.3390/SU16072689
75. Zhang, X., Han, L., Wei, H., Tan, X., Zhou, W., Li, W., Qian, Y.: Linking urbanization and air quality together: a review and a perspective on the future sustainable urban development. J. Clean. Prod. **346**, 130988 (2022). https://doi.org/10.1016/J.JCLEPRO.2022.130988
76. Zhang, Y., Zhang, P., Li, Z., Sun, D.: Impact of enterprise digital transformation on the efficiency of total factor energy utilization—from the perspective of differences in enterprise energy efficiency. Trans. Econ. Bus. Manag. Res. **6**, 276–291 (2024). https://doi.org/10.62051/9BKKDK89

Implementing Digital ESG Strategies in Waste Management Companies

Paolo Biancone, Davide Calandra, and Roberto Marseglia

Abstract Environmental, Social, and Governance (ESG) criteria are becoming increasingly central to the operational strategies of organizations, especially in sectors with significant environmental impacts, such as waste management. This chapter explores how digital technologies can be integrated with ESG initiatives within waste management companies to enhance sustainability practices. The study focuses on companies in four Italian regions—Valle d'Aosta, Piemonte, Liguria, and Toscana—mapping their adoption of ESG reporting and digital tools. The research identifies key gaps in ESG reporting where digital technologies, such as IoT sensors, blockchain, and advanced analytics, can improve sustainability outcomes. These tools are analyzed for their role in optimizing waste collection routes, reducing operational costs, improving recycling rates, and enhancing the accuracy of ESG metrics. The chapter also examines the disparities in digital adoption across regions, highlighting the challenges faced by companies that have not yet fully embraced digital transformation. Through multiple case studies, the chapter demonstrates the tangible benefits of digital technologies in enhancing transparency, stakeholder engagement, and operational efficiency. It also outlines how these tools can support more reliable and accessible ESG reporting, fostering greater accountability and compliance with regulatory requirements. Ultimately, the chapter argues that digitalization is essential for achieving long-term sustainability goals in the waste management sector. By adopting digital tools, waste management companies can meet regulatory demands and lead the transition toward a more sustainable future. This chapter provides a roadmap for companies looking to improve their sustainability performance, offering insights into best practices and identifying areas where further digital adoption is needed.

P. Biancone · D. Calandra (✉)
Department of Management "Valter Cantino", University of Turin, Turin, Italy
e-mail: davide.calandra@unito.it

P. Biancone
e-mail: paolo.biancone@unito.it

R. Marseglia
Alia Servizi Ambientali S.P.A, Florence, Italy
e-mail: r.marseglia@aliaserviziambientali.it

© The Author(s), under exclusive license to Springer Nature Switzerland AG 2025

A. d. b. Machado et al. (eds.), *Environmental, Social, Governance and Digital Transformation in Organizations*, Information Systems Engineering and Management 35, https://doi.org/10.1007/978-3-031-86079-9_20

Keywords ESG reporting · Digital technologies · Waste management · Transparency · Sustainability

1 Introduction

In recent years, there has been a significant shift towards companies adopting environmental, social, and governance (ESG) criteria across various sectors. ESG criteria serve as a framework Through which organizations can evaluate and communicate their sustainability efforts, social responsibilities, and governance practices. This shift is primarily driven by increasing regulatory pressures, consumer demand for corporate accountability, and the growing recognition that sustainable business practices are critical for long-term success and resilience [4, 5]. The role of ESG is particularly important in industries with substantial environmental impacts, such as waste management. waste management companies are at the forefront of environmental sustainability, tasked with reducing waste, enhancing recycling efforts, and mitigating pollution while also adhering to ethical and transparent governance practices [24].

Despite the growing emphasis on ESG reporting, there are significant challenges regarding the integration of these criteria within the operations of waste management companies. The most pressing challenge is the lack of standardized frameworks and tools to ensure accurate, transparent, and comprehensive reporting. Without these frameworks, companies struggle to effectively communicate their sustainability practices and often fail to meet the transparency expectations of stakeholder. digital technologies have emerged as powerful enablers of ESG reporting, offering solutions such as sensors, blockchain, and big data analytics to improve the precision, accessibility, and accountability of sustainability data [1, 11]. These technologies not only enhance operational efficiency but also enable companies to meet the increasing demand for transparency from regulators, consumers, and investors [21].

However, there is a clear gap in the literature concerning how waste management companies leverage digital technologies to enhance their external ESG reporting. While existing studies highlight the benefits of digital tools, they often focus on broader sectors or need to consider the regional disparities in digital adoption within specific industries like waste management. Additionally, there is limited understanding of how these technologies are integrated at different levels within organizations and across diverse geographical contexts. This raises an important question about how digital tools can address the specific challenges of ESG reporting in waste management. Therefore, our research question (RQ) is: How are digital technologies enhancing the quality and transparency of ESG reporting within the waste management sector, and how do regional disparities influence their adoption and effectiveness?

Our chapter aims to map the current state of ESG integration within the waste management sector, focusing on companies in four Italian regions: Valle d'Aosta, Piemonte, Liguria, and Toscana. The study explores how companies in this sector incorporate digital tools into their sustainability efforts, highlighting regional disparities in digital adoption and ESG reporting practices. The objective is to provide a comprehensive overview of how digitalization can improve sustainability and transparency, focusing on its potential to enhance reporting accuracy, operational efficiency, and stakeholder engagement. By identifying best practices and gaps in digital adoption, this research seeks to offer insights into how waste management companies can leverage technology to meet regulatory demands and lead the way towards a more sustainable future.

The results of this study highlight the disparities in the adoption of ESG reporting and the use of digital technologies among waste management companies in four Italian regions: Valle d'Aosta, Piemonte, Liguria, and Toscana. The findings suggest that companies with higher levels of external digital adoption demonstrate improved transparency, interactivity, and accessibility in their ESG reports. These companies are better equipped to meet sustainability targets and engage stakeholder Through real-time data sharing and interactive platforms. However, there remains a significant gap in the use of these technologies across regions, with many companies needing more digital tools to optimize their sustainability reporting. Therefore, our study makes three main contributions: (1) Mapping the current state of ESG integration within the waste management sector, particularly in Italy, providing insights into regional differences in ESG reporting practices; and (2) Identifying the role of digital technologies in enhancing sustainability practices, improving operational efficiency, and increasing the transparency of ESG reports.

The rest of the chapter is structured as follows. Paragraph 2 examines the current state of research on ESG reporting and digital technologies in waste management. It discusses key theoretical variables such as transparency, interactivity, readability, and accessibility, and how digital tools like IoT, blockchain, and big data analytics enhance these. Paragraph 3 describes the qualitative research design, including the literature review and case study methods used to analyze waste management companies across four Italian regions. Paragraph 4 presents the study's findings, focusing on the disparities in ESG reporting and digital adoption among waste management companies in Valle d'Aosta, Piemonte, Liguria, and Toscana. Paragraph 5 concludes the chapter by addressing theoretical and practical implications, limitations, and future research scope.

2 Literature Review

The growing pressure on companies to adopt sustainable business practices has significantly increased the importance of Environmental, Social, and Governance (ESG) reporting [12]. ESG criteria provide a framework for companies to assess and communicate their impact on environmental sustainability, social responsibility, and governance practices [4, 5]. This shift towards greater transparency and accountability has been especially crucial in sectors like waste management, where environmental concerns are a key priority. digital technologies have become vital in enhancing the efficiency and precision of ESG reporting, particularly in industries like waste management that deal with complex operations [21].

2.1 The Role of ESG in the Waste Management Sector

Waste management is closely tied to environmental sustainability, making it an essential industry for ESG integration. Properly handling, treating, and disposing waste have significant environmental and societal implications [24]. ESG reporting in this sector typically emphasizes waste reduction, recycling rates, energy efficiency, and pollution control metrics. However, governance issues, such as compliance with regulations and ethical business practices, are also vital [17].

The literature underscores the importance of transparency in building trust with key stakeholder, including consumers, investors, and regulators [10]. sustainability reports incorporating ESG metrics allow waste management companies to demonstrate their commitment to mitigating environmental impacts while addressing social and governance concerns. Despite this, the adoption of ESG practices in the waste management industry varies considerably across regions, driven by local regulations, market demand, and the availability of digital tools to support the tracking and reporting of sustainability metrics.

2.2 Digital Technologies in Enhancing ESG Reporting

Digital transformation in waste management has introduced several technologies that improve both operational efficiency and ESG reporting capabilities. These technologies include sensors, data analytics platforms, and blockchain for ensuring transparency and data security [1]. Based on the work [20], several theoretical variables, such as transparency, interactivity, and readability, are crucial for evaluating how digital tools can improve ESG reporting.

Transparency is one of the most important variables that digital technologies can enhance. For example, IoT sensors monitor environmental metrics like waste collection routes, recycling rates, and pollution control in real time, significantly improving the transparency of ESG reporting [28]. Companies that leverage IoT can provide their stakeholder with more accurate and up-to-date information, thus fulfilling their transparency obligations.

Blockchain technology further supports transparency by creating immutable records of waste handling and disposal, ensuring accountability, and reducing the risk of data manipulation or fraud in sustainability reports [13]. This is particularly beneficial in regions with complex regulatory frameworks or where multiple stakeholder are involved in waste management processes.

Another critical technology is big data analytics, which enables companies to process large volumes of operational data and optimize processes for sustainability

[7]. By using predictive analytics, companies can forecast waste generation and adjust their operations to reduce energy consumption and emissions. By integrating these digital tools into their ESG frameworks, companies can generate more precise and transparent real-time, data-driven sustainability reports.

Linked closely to transparency is the interactivity of ESG reports. Data visualization tools and interactive web-based platforms allow stakeholder to explore sustainability metrics dynamically, deepening their understanding of ESG data [20]. For example, waste management companies that provide interactive dashboards can offer stakeholder real-time updates on environmental performance, making their ESG reports more engaging and informative [27].

Readability is equally important, ensuring that ESG reports are accessible to a broader audience, including non-experts. Digital platforms enable companies to present complex data in more user-friendly formats, such as graphs and charts [8]. This is especially relevant for waste management companies, where technical data on waste processing and environmental metrics can be difficult for the public to understand. Improving readability helps companies communicate their sustainability efforts more effectively.

Lastly, accessibility refers to how easily stakeholders can access ESG information. Digital tools like cloud-based platforms and online portals allow companies to offer stakeholder 24/7 access to their sustainability reports. This ensures that ESG data is always readily available and easy to navigate, promoting greater accountability and encouraging stakeholder engagement [9].

3 Methodology

3.1 Research Design

This study employs a qualitative research design to explore the integration of Environmental, Social, and Governance (ESG) criteria within waste management companies, focusing on digitalization's role in enhancing sustainability practices. According to [16], qualitative methods are particularly effective in allowing researchers to understand and develop an analytical framework from primary or secondary data. This approach is well-suited for investigating phenomena inductively, providing real-time observation and the ability to track "in vivo" processes, thereby facilitating the development of new theoretical insights from real-world practices [2, 20]. Qualitative methodologies are also invaluable in the social sciences for their interpretative approaches, crucial for understanding how phenomena evolve and for establishing connections and epistemological virtuosity [3]. Compared to quantitative methods, qualitative research allows for contextual negotiations, accommodating unforeseen changes in the research design that can add value to the initial contextual assumptions [14, 19]. In this study, we adopted a dual approach combining theoretical analysis with empirical validation Through an exploratory case study, as suggested by [6,

29]. This approach is particularly suitable for exploring how digital tools and ESG reporting interact within the waste management sector. Exploratory case studies guide research toward addressing specific research questions, even when explicit propositions still need to be fully developed [29].

3.2 Data Analysis

Data collection for this study was conducted in two phases. First, a thorough literature review was performed to identify critical variables and theories related to ESG reporting and digitalization in the waste management sector. The literature review in Sect. 2 was conducted using a Scopus analysis, with keywords such as "ESG reporting," "waste management," and "digitalization," focusing on peer-reviewed articles published in English [15]. This initial phase provided the study's theoretical foundation and helped identify the main themes and variables for further investigation. The second phase involved an exploratory case study to empirically test the theories and variables identified in the literature review. Italy was chosen as the case study location due to its diverse regional practices and the significant role of local municipalities in waste management. Italy's complex governance structure, which includes 7,904 municipalities providing a range of public services, makes it an ideal context for exploring the implementation of ESG practices and digitalization [18]. The sample included waste management companies operating in four Italian regions: Valle d'Aosta, Piemonte, Liguria, and Toscana. These regions were selected based on their varying levels of ESG reporting and digital adoption, providing a diverse set of cases for analysis. The data was collected from the companies' sustainability reports, websites, "Transparent Administration" sections, and social media pages for the year 2023 to comprehensively understand their ESG practices and digitalization efforts. Content analysis was conducted independently by two researchers, who coded the data using a binary system: "0" if the variable was not present and "1" if it was present. In cases where discrepancies arose, particularly for subjective variables such as "ease of reading" and "ease of access," the research team met to standardize the final coding decision [22, 23]. This triangulation of data sources and collaborative analysis ensured the reliability and validity of the findings [26].

For the qualitative analysis, Atlas.ti Cloud was utilized as the primary software tool. Atlas.ti is a robust qualitative data analysis software that facilitates the systematic coding and analysis of sustainability reports and other relevant documents. The cloud-based platform enabled collaborative work among researchers and ensured secure access to data from multiple locations, enhancing the study's rigor and depth [25].

4 Results

4.1 An Overview of ESG Reporting in Selected Waste Management Companies

The study involved a detailed mapping of waste management companies operating in four Italian regions: Valle d'Aosta, Piemonte, Liguria, and Toscana (Table 1). These regions were selected due to their diverse approaches to environmental management, digital adoption, and sustainability reporting. The mapping provided a comprehensive view of the current state of ESG integration within the waste management sector.

Valle d'Aosta is the smallest region among the four, with a limited number of waste management companies. The study identified one company operating in this region that notably needs to engage in sustainability reporting. This absence of reporting indicates a potential gap in the region's approach to integrating ESG principles into waste management practices. The lack of formalized sustainability initiatives could be attributed to the region's smaller scale of operations, which might result in fewer resources dedicated to developing and reporting ESG strategies.

Piemonte is a more industrialized region with many waste management companies. The study identified 17 companies in this region, of which 6 are subsidiaries of more giant corporations that consolidate their sustainability reporting within the parent company's disclosures. Out of these, six companies produce independent sustainability reports, indicating a moderate commitment to ESG transparency. The companies in Piemonte that engage in sustainability reporting tend to incorporate a range of ESG metrics, including environmental impact assessments, waste reduction targets, and community engagement initiatives. These companies also adopt digital tools such as IoT sensors to monitor waste processes and big data analytics to optimise operations. This digital adoption is closely linked to their ability to track and report on ESG metrics more effectively. However, ten companies in Piemonte do not engage in sustainability reporting, highlighting a significant regional disparity. This gap suggests that while some companies lead the way in ESG practices, others may lag due to a lack of resources, awareness, or regulatory pressure.

Table 1 Regional distribution of sustainability reporting practices among waste management companies in 4 Italian regions

Region	Companies with sustainability reporting	Companies without sustainability reporting	Total companies
Valle d'Aosta	0	1	1
Piemonte	6	11	17
Liguria	3	4	7
Toscana	5	6	11

Source Authors' elaboration

In Liguria, the study identified 7 companies involved in waste management (excluding parent companies' disclosure). Like Piemonte, there is a noticeable split in ESG reporting practices. Of these, three companies are subsidiaries whose ESG efforts are reported at the parent company level, and three companies publish their sustainability reports. Liguria's companies that report on ESG practices tend to focus on environmental stewardship, particularly in marine waste management and pollution control areas, which are critical issues given the region's extensive coastline. These companies also show some integration of digital technologies, although the level of digital adoption varies significantly across the region. The five companies in Liguria that do not engage in sustainability reporting could benefit from increased awareness and incentives to adopt ESG practices and digital tools. The region's environmental challenges, particularly in coastal management, underscore the importance of such practices for long-term sustainability.

Toscana presents a contrasting scenario. Five of the 11 companies identified in the region engage in sustainability reporting. The absence of ESG reporting in six companies in Toscana suggests a significant opportunity for improvement, mainly Through adopting digital tools that could facilitate better environmental management and transparency. The lack of reporting in these companies may reflect several underlying factors, such as insufficient regulatory enforcement, limited access to digital technologies, or a lack of incentives for companies to pursue sustainability goals. However, the region's potential for adopting ESG practices remains high, particularly if targeted initiatives encourage digitalization and waste management transparency.

4.2 *Internet and Sustainability Reporting in Selected Waste Management Companies*

The analysis of internet-based sustainability reporting among waste management companies in the four Italian regions reveals a wide disparity in how transparency, interactivity, readability, accessibility, integration with financial reporting, and social media usage are applied (Table 2). These factors are critical in determining how effectively companies communicate their ESG initiatives to stakeholders.

In Valle d'Aosta, the only company, Quendoz, demonstrates moderate transparency by providing some information online but lacks interactive reports, readability features, and integration with widespread financial reporting. The company does, however, ensure a certain level of accessibility, and it actively engages on social networks.

In Piemonte, several companies show varying levels of commitment to online ESG reporting. Amiat, Società Canavesana Servizi, Seta, Cidiu, Egea, and Amag Ambiente lead the way, providing not only transparent and accessible reports but also offering features like interactive reports and enhanced readability. Notably, Seta and Egea fully integrate their ESG data with financial reporting and use social networks to reach a wider audience. However, other companies, such as ConSer VCO and

Table 2 Internet reporting in waste management companies among four Italian regions

Region	Transparency	Interactive reports	Readability	Accessibility	Integrate popular financial reporting	Social networks
Valle d'Aosta						
Quendoz	1	0	0	1	0	1
Piemonte						
ConSer VCO	1	0	0	0	0	0
Medio novarese ambiente	1	0	0	0	0	0
Assa novara	1	1	0	0	0	0
ASM vercelli	1	1	0	0	0	0
Amiat	1	1	1	1	0	1
SEAB	1	0	0	0	0	0
Società canavesana servizi	1	1	1	1	0	1
Seta	1	1	1	1	1	1
Tekno service	1	0	0	0	0	0
Cidiu	1	1	1	1	0	1
Etambiente	1	0	0	0	0	0
Dockslanterna	1	0	0	0	0	0
Egea	1	1	1	1	1	1
ASP	1	0	0	0	0	0
Cosmo	1	0	0	0	0	0
Amag ambiente	1	1	1	1	0	0
Ideal service	1	0	0	0	0	0
Liguria						
Amaie energia e servizi	1	0	0	0	0	0
De Vizia	1	0	0	0	0	0
SAT	1	0	0	0	0	0
ATA	1	0	0	0	0	0
AMIU genova	1	1	1	1	1	1
Aprica	1	1	1	1	0	1
ACAM Ambiente	1	1	1	1	0	1
Toscana						

(continued)

Table 2 (continued)

Region	Transparency	Interactive reports	Readability	Accessibility	Integrate popular financial reporting	Social networks
Nausicaa	1	0	0	0	0	0
Asmiu	1	0	0	0	0	0
Ersu	1	0	0	0	0	0
SEA Ambiente	1	0	0	0	0	0
AAMPS	1	1	1	1	0	1
REA	1	1	1	1	0	1
SEI toscana	1	1	1	1	0	1
Alia servizi ambientali	1	1	1	1	0	1
Sistema ambiente	1	1	1	1	0	1
Ascit	1	0	0	0	0	0
Geofor	1	0	0	0	0	0

Source Authors' elaboration

Tekno Service, display minimal digital engagement, offering essential transparency but lacking interactivity, readability, and social media presence.

In Liguria, AMIU Genova, Aprica, and ACAM Ambiente excel in their online reporting by offering comprehensive, interactive, and readable reports. These companies integrate ESG data with financial reporting and actively utilize social media. Conversely, companies like Amaie Energia e Servizi and De Vizia offer only essential transparency with limited digital engagement, lacking interactivity and social media presence.

A similar pattern emerges in Toscana, with companies such as AAMPS, REA, SEI Toscana, Alia Servizi Ambientali, and Sistema Ambiente showcasing high levels of transparency and digital engagement. These companies provide accessible, readable, and interactive reports, although none fully integrate ESG metrics with popular financial reporting. Their use of social networks also highlights a broader attempt to engage stakeholder online. On the other hand, companies like Ascit and Geofor demonstrate minimal digital presence, offering only fundamental transparency without the features needed for effective online ESG reporting. Therefore, the analysis underscores that companies with higher levels of digital adoption and online engagement tend to provide more transparent and comprehensive ESG reports. However, many companies still need digital tools to optimize their sustainability reporting and communication, indicating substantial room for improvement across the regions.

5 Discussion and Conclusion

This study has explored the integration of Environmental, Social, and Governance (ESG) criteria within the waste management sector, focusing specifically on how digital technologies can enhance the reporting and transparency of sustainability practices. In industries like waste management, where environmental impact is substantial, accurate and transparent sustainability reporting is crucial [24]. Digital tools like sensors, blockchain, and big data analytics have improved operational efficiency and enhanced ESG reporting [1, 21]. This research has identified significant disparities in the adoption of digital technologies for ESG reporting by conducting an exploratory case study of four Italian regions. Companies with integrated digital tools demonstrate higher levels of transparency and interactivity in their reports, while those lagging in technological adoption face challenges in meeting these standards [28].

5.1 Theoretical Implications

The findings of this study contribute to the broader theoretical discourse on ESG reporting and digital transformation, particularly within environmentally sensitive sectors like waste management. Theoretical variables such as transparency, interactivity, readability, and accessibility—drawn from existing literature—were key to evaluating how digital tools improve ESG reporting [8, 20]. This study reinforces the notion that digitalization is not just a tool for operational enhancement but a fundamental driver for increased transparency and stakeholder engagement [27]. Moreover, this research aligns with theoretical perspectives that link technology adoption to corporate sustainability, emphasizing that companies able to incorporate advanced technologies into their ESG strategies tend to achieve greater trust and competitive advantage [5].

5.2 Practical Implications

From a practical perspective, this study offers valuable insights for waste management companies, policymakers, and stakeholder involved in sustainability reporting. Companies that integrate digital tools such as IoT sensors and blockchain can provide real-time data on environmental performance, improving both accuracy and timeliness in ESG reporting [13, 28]. The use of interactive and accessible digital platforms is also highlighted as a method for improving stakeholder engagement and compliance with regulatory frameworks [1]. Policymakers can use these findings to create incentives for the broader adoption of digital tools across the sector, offering guidance on better integrating ESG metrics into digital reporting frameworks [18]. For companies that have not yet adopted digital tools, the research underscores the necessity of investing in these technologies to enhance both operations and sustainability reporting.

5.3 Limitations

This research is not without limitations. The most significant limitation lies in the desk-based nature of the analysis, which primarily relies on publicly available data such as sustainability reports and websites. This approach excludes companies' internal processes, where technology could be applied to optimize ESG reporting and sustainability tracking. For example, while the study offers insights into external communication, it needs to delve into how digital tools are employed internally to manage and process ESG data. Additionally, the focus on four regions in Italy may limit the generalizability of the findings to other geographical contexts, suggesting that the insights derived here may only partially apply to waste management sectors in different regulatory or market environments.

5.4 Future Scope

Future research should address the internal processes of waste management companies, investigating how digital tools are used to manage ESG data at the operational level. A more comprehensive analysis of the internal processes—including adopting technologies such as IoT, blockchain, and big data—could provide insights into the real potential of these tools to enhance ESG reporting. Furthermore, future studies could conduct longitudinal analyses to assess the long-term impact of digital technology on ESG performance, exploring whether digital adoption leads to measurable improvements in sustainability outcomes over time. Expanding the research to include waste management companies in other countries or regions would also provide comparative insights and improve the generalizability of the findings.

References

1. Asif, M., Searcy, C., Castka, P.: ESG and industry 5.0: the role of technologies in enhancing ESG disclosure. Technol. Forecast. Soc. Change **195**, 122806 (2023). https://doi.org/10.1016/j.techfore.2023.122806
2. Cardoni, A., Kiseleva, E., Terzani, S.: Evaluating the intra-industry comparability of sustainability reports: the case of the oil and gas industry. Sustainability **11**(4), 4 (2019). https://doi.org/10.3390/su11041093
3. Dal Mas, F., Valentina, N., Massaro, M., Bagnoli, C.: Combining deep and digital technologies as a path towards twin transition: the "future farming" case study. IEEE Transactions on

Engineering Management, IEEE Transactions on Engineering Management. pp. 1–12. (2024). https://doi.org/10.1109/TEM.2024.3444592
4. Darnall, N., Ji, H., Iwata, K., Arimura, T.H.: Do ESG reporting guidelines and verifications enhance firms' information disclosure? Corp. Soc. Responsib. Environ. Manag. **29**(5), 1214–1230 (2022). https://doi.org/10.1002/csr.2265
5. Eccles, R.G., Serafeim, G.: The performance frontier: innovating for a sustainable strategy. Harvard Business Review. (2013). https://hbr.org/2013/05/the-performance-frontier-innovating-for-a-sustainable-strategy
6. Fløttum, K., Gjesdal, A.M., Gjerstad, Ø., Koteyko, N., Salway, A.: Representations of the future in English language blogs on climate change. Glob. Environ. Chang. **29**, 213–222 (2014). https://doi.org/10.1016/j.gloenvcha.2014.10.005
7. Goloshchapova, I., Poon, S.-H., Pritchard, M., Reed, P.: Corporate social responsibility reports: topic analysis and big data approach. Eur. J. Finance **25**(17), 1637–1654 (2019). https://doi.org/10.1080/1351847X.2019.1572637
8. Grossi, G., Biancone, P.P., Secinaro, S., Brescia, V.: Dialogic accounting through popular reporting and digital platforms. Med. Acc. Res. **29**(7), 75–93 (2021). https://doi.org/10.1108/MEDAR-01-2021-1163
9. Gu, Y., Katz, S., Wang, X., Vasarhelyi, M., Dai, J.: Government ESG reporting in smart cities. Int. J. Acc. Inf. Syst. **54**, 100701 (2024). https://doi.org/10.1016/j.accinf.2024.100701
10. Hoang, T.H.V., Pham, L., Nguyen, T.T.P.: Does country sustainability improve firm ESG reporting transparency? The moderating role of firm industry and CSR engagement. Econ. Model. **125**, 106351 (2023). https://doi.org/10.1016/j.econmod.2023.106351
11. Lanzalonga, F., Marseglia, R., Irace, A., Biancone, P.P.: The application of artificial intelligence in waste management: Understanding the potential of data-driven approaches for the circular economy paradigm. Management Decision, ahead-of-print (ahead-of-print). (2024). https://doi.org/10.1108/MD-10-2023-1733
12. Lanzalonga, F., Oppioli, M., Calandra, D., Secinaro, S.: The impact of ESG performance on intangible assets and intellectual capital in the food and beverage industry. Management Decision, ahead-of-print (ahead-of-print). (2024). https://doi.org/10.1108/MD-09-2023-1664
13. Liu, X., Wu, H., Wu, W., Fu, Y., Huang, G.Q.: Blockchain-enabled ESG reporting framework for sustainable supply chain. In Scholz, S.G., Howlett, R.J., Setchi R. (AC. Di), Sustainable Design and Manufacturing 2020 pp. 403–413. Springer. (2021). https://doi.org/10.1007/978-981-15-8131-1_36
14. Massaro, M., Dumay, J., Bagnoli, C.: Transparency and the rhetorical use of citations to Robert Yin in case study research. Med. Acc. Res. (2019). https://doi.org/10.1108/MEDAR-08-2017-0202
15. Massaro, M., Dumay, J., Guthrie, J.: On the shoulders of giants: undertaking a structured literature review in accounting. Acc. Auditing Acc. J. **29**(5), 767–801 (2016). https://doi.org/10.1108/AAAJ-01-2015-1939
16. Miles, M.B., Huberman, A.M., Saldana, J.: Qualitative data analysis. SAGE. (2014)
17. Morioka, S.N., de Carvalho, M.M.: A systematic literature review towards a conceptual framework for integrating sustainability performance into business. J. Clean. Prod. **136**, 134–146 (2016). https://doi.org/10.1016/j.jclepro.2016.01.104
18. Padovani, E., Iacuzzi, S., Jorge, S., Pimentel, L.: Municipal financial vulnerability in pandemic crises: a framework for analysis. J. Public Budg. Acc. Financ. Manag. **33**(4), 387–408 (2021). https://doi.org/10.1108/JPBAFM-07-2020-0129
19. Reinecke, J., Arnold, D.G., Palazzo, G.: Qualitative methods in business ethics, corporate responsibility, and sustainability research. Bus. Ethics Q. **26**(4), xiii–xxii (2016). https://doi.org/10.1017/beq.2016.67
20. Secinaro, S., Brescia, V., Calandra, D.: Internet reporting and digital governance: evidence from multiple Italian municipalities. In Handbook of Accounting and Public Governance pp. 283–301. Edward Elgar Publishing. (2024). https://www.elgaronline.com/edcollchap/book/9781800888456/book-part-9781800888456-28.xml

21. Secinaro, S., Calandra, D., Degregori, G.: New technologies in supporting ESG criteria and the implementation in the new normal: Mapping the field and proving future research paths Corp. Gov. Res. Dev. Stud. Open Access, 1, 1. (2023). https://doi.org/10.3280/cgrds1-2023oa15788
22. Secinaro, S., Dal Mas, F., Brescia, V., Calandra, D.: Blockchain in the accounting, auditing and accountability fields: a bibliometric and coding analysis. Acc. Auditing Acc. J. **35**(9), 168–203 (2021). https://doi.org/10.1108/AAAJ-10-2020-4987
23. St. Pierre, E.A., Jackson, A.Y.: Qualitative data analysis after coding. Q. Inq. **20**(6), 715–719 (2014). https://doi.org/10.1177/1077800414532435
24. Tashmanov, G.D., Samieva, Z.T., Tolmachev, A.V., Palesheva, N.V.: ESG management and environmental reporting of small and medium business in the service sphere in support of the green economy in Russia and central Asia. In Popkova, E.G., Sergi, B.S. (Di, A.C.), ESG Management of the Development of the Green Economy in Central Asia (pp. 155–162). Springer International Publishing. (2023). https://doi.org/10.1007/978-3-031-46525-3_17
25. Taylor, R.P., Cordeiro, C.J.D., Giordano, D., Hover, J., Kouba, T., Love, P., McNab, A., Schovancova, J., Sobie, R., Collaboration, on behalf of the A.: Consolidation of cloud computing in ATLAS. Journal of Physics: Conference Series, 898(5), 052008. (2017). https://doi.org/10.1088/1742-6596/898/5/052008
26. Thurmond, V.A.: The point of triangulation. J. Nurs. Scholarsh. **33**(3), 253–258 (2001). https://doi.org/10.1111/j.1547-5069.2001.00253.x
27. Tran, Y.T., Nguyen, N.P., Hoang, T.C.: The role of accountability in determining the relationship between financial reporting quality and the performance of public organizations: evidence from Vietnam. J. Acc. Publ. Policy **40**(1), 106801 (2021). https://doi.org/10.1016/j.jaccpubpol.2020.106801
28. Wu, W., Fu, Y., Wang, Z., Liu, X., Niu, Y., Li, B., Huang, G.Q.: Consortium blockchain-enabled smart ESG reporting platform with token-based incentives for corporate crowdsensing. Comput. Ind. Eng. **172**, 108456 (2022). https://doi.org/10.1016/j.cie.2022.108456
29. Yin, R.K.: Review of case study research. Design and methods 4th ed. Zeitschrift für Personalforschung / German Journal of Research in Human Resource Management, 26(1), 93–95. (2012)

ESG in an Industry 4.0 Context: A Promising Encounter

João Alvarez Peixoto, Andreia de bem Machado, Jaqueline Morbach, Ana Carolina Tramontina, and Marc François Richter

Abstract The intersection between ESG principles (environmental, social and governance) and Industry 4.0 reflects the convergence of sustainability and technological innovation. Industry 4.0 is characterized by automation, digitalization and the use of advanced technologies such as the Internet of Things (IoT), big data and artificial intelligence, and has the potential to transform production processes, making them more efficient and sustainable. The research problem was therefore central: what are the links between Industry 4.0 and ESG? To answer this research problem, a bibliometric review was carried out examining scientific production on the relationship between ESG and Industry 4.0, with the aim of mapping the trends and challenges encountered by academics and professionals in this area. The integration between ESG and Industry 4.0 is promising, since digital technologies can help optimize production processes, improve environmental and social management and increase transparency and corporate governance. However, the survey pointed out that the application of ESG principles in Industry 4.0 is still at an early stage, with few studies focusing on early stages, with few studies focusing on practical implementation. It concludes that there is a growing trend towards the adoption of

J. A. Peixoto
Graduate Program in Science, Technology, Engineering and Mathematics (PPGSTEM), State University of Rio Grande do Sul, Porto Alegre, Brazil
e-mail: joao-peixoto@uergs.edu.br

A. de bem Machado
Engineering and Knowledge Management Department, Federal University of Santa Catarina (UFSC), Florianópolis, Brazil
e-mail: andreia.bem@ufsc.br

J. Morbach · A. C. Tramontina · M. F. Richter (✉)
Graduate Programme in Environment and Sustainability (PPGAS), State University of Rio Grande do Sul (Uergs), Porto Alegre, Brazil
e-mail: marc-richter@uergs.edu.br

J. Morbach
e-mail: jaquemorbach@gmail.com

A. C. Tramontina
e-mail: ana-tramontina@uergs.edu.br

© The Author(s), under exclusive license to Springer Nature Switzerland AG 2025
A. d. b. Machado et al. (eds.), *Environmental, Social, Governance and Digital Transformation in Organizations*, Information Systems Engineering and Management 35, https://doi.org/10.1007/978-3-031-86079-9_21

these practices, driven by pressure for greater sustainability and social responsibility in the industrial context.

Keywords Sustainability · Industry 5.0 · Bibliometry environment · Social · Governance

1 ESG Approach

ESG refers to the three main groups of criteria used to assess a company's performance in terms of environmental, social and governance responsibility. In Industry 4.0, the ESG concept becomes even more relevant, as the digital transformation brings with it new challenges and opportunities in relation to these criteria: (1) Environmental: In Industry 4.0, companies seek to minimize their environmental impact by adopting cleaner and more efficient technologies. This can include implementing more sustainable production processes, using renewable energies, reducing the waste of natural resources and effectively managing waste [16], (2) Social: In the context of Industry 4.0, companies need to consider the social impact of their operations, both internally and externally. This includes aspects such as employee welfare, workplace safety, diversity and inclusion, and the impact of operations on the development of local communities. In addition, automation and artificial intelligence can have significant implications for employment and income distribution, making it even more important for companies to address these issues responsibly [16], and (3) governance: corporate governance in Industry 4.0 concerns management and decision-making practices within companies. This includes aspects such as transparency, business ethics, corporate responsibility and the quality of relations with stakeholders, including shareholders, customers, suppliers and local communities. With the increasing complexity of industrial operations enabled by technology, good governance becomes fundamental to guaranteeing the sustainability and long-term success of companies [16]. Figure 1 shows these three ESG indicators.

The ESG is particularly important today for a number of reasons, such as: (a) Sustainability and environmental awareness: many countries are implementing stricter environmental laws and regulations, and companies need to comply with them to avoid penalties and litigation. In addition, reputation is fundamental for any company. In this sense, a strong environmental awareness improves a company's image and brand perception. This can lead to greater customer loyalty and satisfaction [9], (b) Social responsibility: employee satisfaction: Companies that set high social standards attract and retain talented employees and thus take employee satisfaction seriously. This includes, for example, fair working conditions, equal rights and also valuing the topic of diversity. In terms of customer expectations, consumers increasingly value social responsibility and favour companies that adopt ethical business practices [19]. (c) governance: with regard to transparency and accountability, good governance practices ensure that companies operate transparently and that their managers are held accountable. This fosters the confidence of investors and other

Fig. 1 The three pillars of ESG segmented

stakeholders. In addition, effective governance structures help companies to better identify and manage risks, which leads to a more stable and sustainable business [16], (d) Financial performance: nowadays, with the importance of sustainability, many investors take ESG criteria into account when making investment decisions, as companies with solid ESG practices are seen as less risky and more sustainable. In addition, studies have shown that companies with solid ESG practices tend to perform better in the long term, as they are better prepared for future challenges and changes in the market [7], (e) Competitive advantage: ESG initiatives generally encourage innovation, as companies look for new and more sustainable methods and technologies to improve their business processes. This also brings advantages to companies that integrate ESG practices, as they can gain access to new markets and business opportunities that remain closed to less sustainable competitors. In short, ESG is not only an ethical imperative, but also provides a strategic advantage for the company [3]. To summarize, ESG influences the market and the economy by shaping investment decisions, financial performance, regulatory compliance, market access, innovation and corporate reputation. ESG practices help companies to act in a more sustainable and responsible way, which promotes their long-term success and the prosperity of society as a whole. In this way, companies that integrate ESG practices into their operations also contribute to a more sustainable world, as well as reaping significant economic and competitive benefits. Figure 2 illustrates ESG and its relationship with the market.

Some practical examples of the impacts on industry promoted by the ESG are as follows: (a) Clean Manufacturing: companies must adopt advanced technologies to reduce pollution and waste, promoting cleaner and more efficient manufacturing processes; (b) Circular Economy: the implementation of circular economy models,

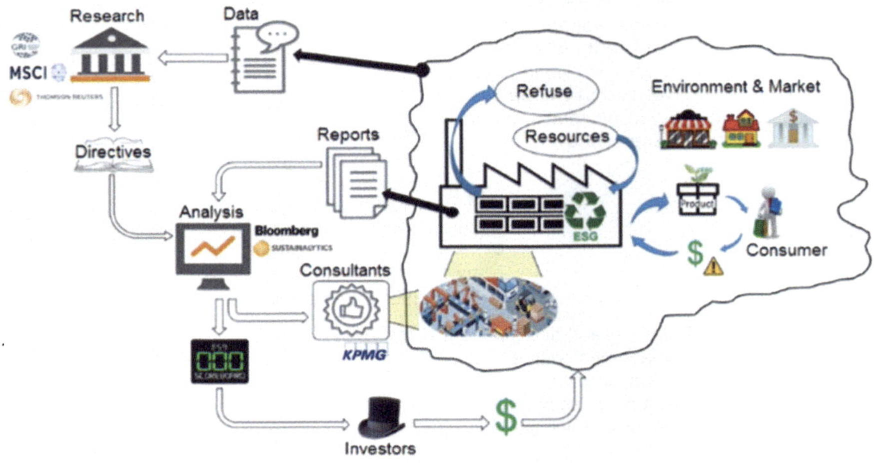

Fig. 2 ESG and its relationship with the market

where waste from one production phase is reused as inputs for new ones, minimizing waste and maximizing resource efficiency; (c) Environmental certifications: obtaining certifications such as ISO 14001 (environmental management), which demonstrate the company's commitment to responsible environmental management; (d) Community Partnerships: collaboration with Non-Governmental Organizations (NGOs) and other organizations in order to promote community development, education and health, benefiting the areas where the company operates; (e) Green Technologies: investment in research and development of technologies that reduce the environmental impact of industrial operations, such as carbon capture and storage (CCS) and bioenergy.

2 ESG and its Connection to Sustainability and the SDGs

The connection between ESG and sustainability is deep and intrinsic, with both concepts sharing the goal of promoting practices that are beneficial to the environment, society and the economy in the long term. In addition, the United Nations' Sustainable Development Goals (SDGs) provide a comprehensive framework to guide these efforts. The relationship between ESG and sustainability can be demonstrated in each of the 3 pillars of sustainability.

On the Environmental side, both the management of natural resources and the reduction of emissions and pollution are issues linked to sustainability, and strongly related to the SDGs. ESG practices focus on the conservation and sustainable use of natural resources, including water, energy and materials. This is in line with sustainability, which aims to preserve the environment for future generations. ESG promotes the reduction of greenhouse gas emissions and pollution, contributing to the mitigation of climate change and the protection of biodiversity.

In the Social section, issues such as social equity and justice, as well as the health and well-being of employees are linked to sustainability and the SDGs. ESG emphasizes social justice, including human rights, fair working conditions and community development. Social sustainability seeks to create equitable and just societies where all individuals can thrive. ESG also promotes practices that improve the health and well-being of communities, such as safe and healthy working environments, which are essential for social sustainability.

With regard to the governance pillar, ESG relates to transparency, ethics and stakeholder engagement, also linked to the SDGs. ESG requires governance practices that promote transparency, responsibility and ethics. Corporate sustainability depends on a robust structure to ensure that companies operate in a responsible and resilient manner over the long term. In addition, ESG also encourages engagement with all interested parties (stakeholders) to ensure that business decisions consider social and environmental impacts, reflecting a sustainable and holistic approach.

It makes sense that ESG is linked to the SDGs. These are 17 global goals set by the UN to be achieved by 2030, covering a wide range of sustainability issues. They are part of the 2030 Agenda for Sustainable Development, a global action plan adopted by all UN member states in September 2015 [20]. Figure 3 shows the 17 SDGs.

The 2030 Agenda includes 17 overarching goals and 169 specific targets aimed at ending poverty, protecting the planet and ensuring that all people enjoy peace and prosperity by 2030. ESG can be an effective way for companies to contribute to these goals. ESG aligns with some of the key SDGs, presented in Table 1.

Fig. 3 The 17 Sustainable development goals (SDGs)

Table 1 The relationship between ESG and the SDGs

SDGs	Pillars of ESG	Relationship between ESG and SDG
SDG 1: No poverty	Social	ESG promotes fair wages and decent working conditions, helping to reduce poverty
SDG 3: Good health and well-being	Social	ESG policies support safe and healthy work environments, contributing to employee well-being
SDG 5: Gender equality	Social and governance	ESG promotes gender equality through inclusive practices and diversity policies
SDG 6: Clean water and sanitation	Environmental	ESG encourages efficient water use and effluent management, contributing to the availability of clean water and sanitation
SDG 7: Affordable and clean energy	Environmental	Companies that adopt ESG practices often invest in renewable energy and energy efficiency technologies
SDG 8: Decent work and economic growth	Social and governance	ESG supports fair working conditions and sustainable economic growth
SDG 9: Industry, innovation and infrastructure	Environmental and governance	ESG promotes sustainable innovation and the development of resilient infrastructure
SDG 12: Responsible consumption and production	Environmental and governance	ESG encourages sustainable consumption and production practices, such as circular economy and waste management
SDG 13: Climate action	Environmental	ESG focuses on reducing carbon emissions and developing climate mitigation and adaptation strategies
SDG 15: Life on land	Environmental	ESG promotes biodiversity conservation and sustainable management of ecosystems
SDG 16: Peace, justice and Strong institutions	Governance	ESG promotes governance practices that are transparent, responsible, and ethical
SDG 17: Partnerships for the goals	Governance	ESG encourages collaboration between businesses, governments, and other organizations to achieve sustainable development goals

3 Industry 4.0 and its Value Offer

According to the evolutionary perspective, industrial revolutions are triggered by the introduction of new technologies or concepts, promoting qualitative changes in the economic system. And every change has its degree of drama, based on the consequences it causes, the need to let go of the old technology and the necessary commitment to the new.

The past shows the occurrence of four moments in which the industrial structures underwent significant changes, with regard to the use of technologies or segments of concepts, which radically changed the way consumer goods were manufactured. In their time, this had a positive impact on production performance, but also on the workforce, which had to adapt to this new reality, modified by real revolutions [8]. Figure 4 shows the historical organization of the industrial revolutions.

And it would be no different in the Fourth Industrial Revolution, in which goods and consumer markets have been transformed by the advance of digital technologies and the search for diversified products. Access to information has given consumers greater choice, whether in the functionalities they are looking for in a product or in the manufacturers available. Big brands are no longer exclusive and consolidated products are being favored over other similar products with identical functionalities. Abundant production continues to be in demand, but there is a trend towards mass production of highly personalized products [10].

In an Industry 4.0 scenario, small production batches and the need to produce diversified products are requirements to cater for a market that uses digital technologies to shape its product desires and choose where, how and when to buy. The manufacturing environment also needs to evolve to new concepts, because there is no going back to Process Planning and Control (PCP) to redo the plan for every peculiarity of the product, every insertion of a new resource or every lack of a resource in the production system [8].

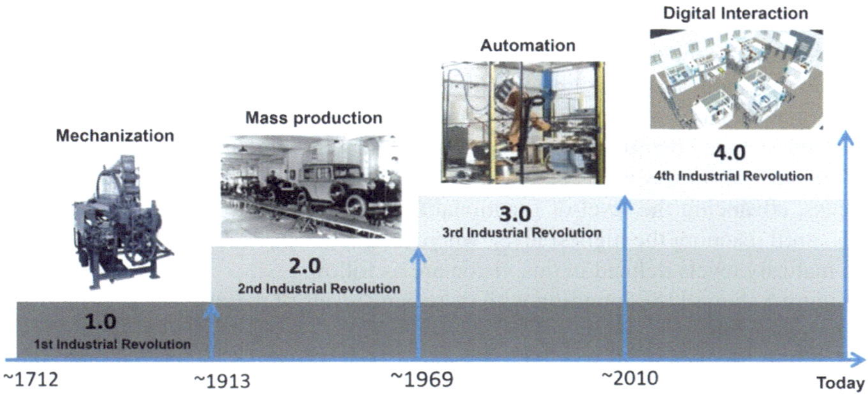

Fig. 4 The industrial revolutions

An approach proposed by [21] categorizes the industrial revolutions as eras: (a) 1st Industrial Revolution: the age of steam, in which manufacturing mechanisms and systems were introduced that used water and steam power, (b) 2nd Industrial Revolution: the age of electricity, in which mass production systems were introduced, using electricity to power these systems; (c) 3rd Industrial Revolution: the information age, in which electronics and automated systems were introduced; (d) 4th Industrial Revolution: the era of cyber-physical systems, in which machines are made up partly of physical–mechanical elements and partly of digital systems.

But disruptive change only happens when there is an understanding among all the players involved, from management to the factory workforce. The 4th Industrial Revolution draws attention because the signs of its existence occurred in the second decade of this century [2]. At the end of the first decade there were still discussions about the need to use Industry 4.0-enabling tools in Brazil's manufacturing environments (Confederação Nacional [5]), something that in the second decade became an essential requirement for building digital and interactive production systems, the premises of Industry 4.0.

The vision of [8] already pointed to the principles of Industry 4.0 as a reference for its fulfilment, and it is up to the company to use the premises that provide the most results.

These are: (a) Interoperability: allowing all equipment and machines in a factory or industrial environment, even if they come from different suppliers, to communicate via networks; (b) Virtualization: enabling data obtained from physical products and equipment to be transmitted to virtual models and simulations, mirroring real behaviour in the virtual environment; (c) Decentralization of production process controls: with products that will manage decision-making in manufacturing and production processes in real time. (d) Adapting production in real time: where data will be analyzed the moment it is collected, allowing production to be altered or transferred to other silos in the event of failures or in the production of customized goods; (e) Service-orientation: in which data and services will be made available on an open network, in order to make their functionalities available for allocation; (f) Modular equipment and production line systems: which make factories more flexible and adaptable to the necessary changes.

With companies mobilizing to implement Industry 4.0 enabling tools, meeting the premises that characterize it and achieving the desired results, new approaches to the subject are emerging. The vision of [18], based on the [2], proposes a path to be followed, starting with a level of technology sufficient to digitize a manufacturing process, advancing the level of functionalities, starting with Industry 4.0 enabling tools, until reaching the highest level, where adaptability is contemplated (Table 2). The maturity levels defined in this vision are as follows:

Figure 5 shows [18] definition with increasing levels of maturity to meet Industry 4.0 requirements.

Figure 6 Shows the approach by Hermann et al. [8] and Schuh et al. [18], with their assumptions.

Table 2 Maturity levels

Level	Maturity	Description
1	Automation	With the ability of machines to be managed using digital logic controllers
2	Connectivity	In which machines can connect to each other, sharing digital data
3	Visualization	In which manufacturing data can be visualized and displayed through display panels or dashboards
4	Transparency	In which data is presented in context, so that its meaning is clearly understood
5	Predictability	In which the machine can analyze the behavior of its manufacturing variables and predict process failures
6	Adaptability	In which the machine adjusts its process flow to adapt to a new manufacturing condition in real-time

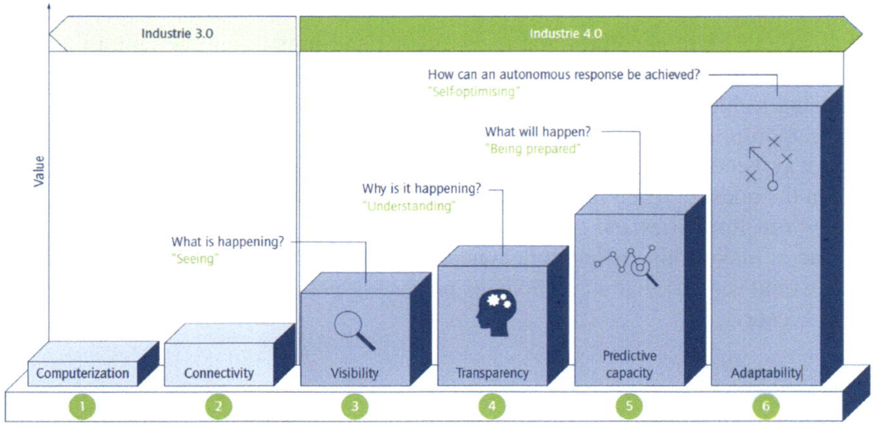

Fig. 5 Schuh's approach based on the Acatech guide

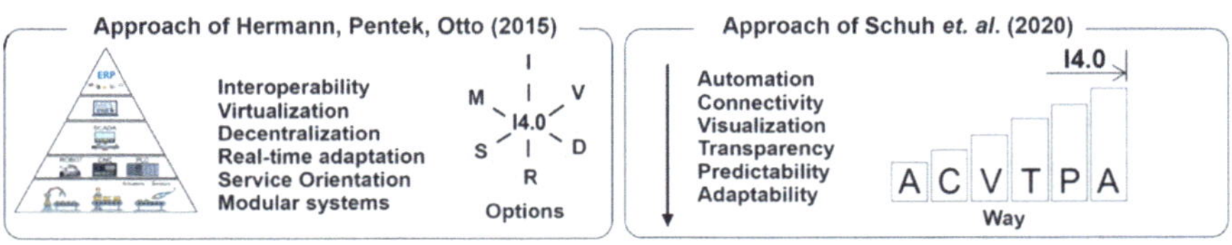

Fig. 6 Hermann and Schuh's approach to the premises of Industry 4.0

Both premises eventually arrive at the same result, showing themselves to be different means to the same end. But for a more skeptical and cautious industrial market, the path defined by the premises presented by Schuh ends up being more adherent, as it presents the difficulty gradually, leading the entrepreneur to

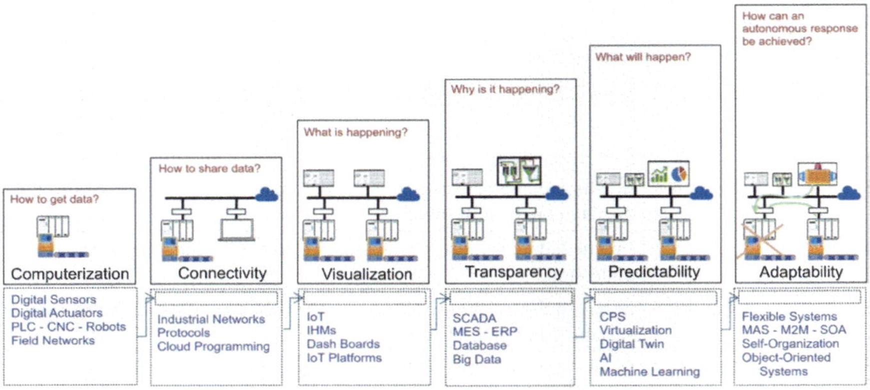

Fig. 7 The relationship between maturity level and enabling tools in the context of Industry 4.0

invest step by step, absorbing the technologies and implanting the culture among his employees. Figure 7 shows an illustration of the Acatech guide, with the most promising technologies representing each level of the path.

The appropriation of one technology motivates the search for another, once it is on the way, becoming a motivating factor for the entrepreneur and his manufacturing team in the quest to reach more technologically advanced stages. They benefit from the new functionalities and gains that these stages bring to the production process. This factor makes this vision more coherent from a business perspective. Which, in itself, would suggest that all companies producing goods and services are adhering to this scenario.

4 ESG and Industry 4.0: A Bibliometric Analysis

The integration of practices that define Industry 4.0 has not only redefined production processes, but has also revolutionized the way companies operate and interact with the world, creating an intersection with ESG principles, which are increasingly relevant to sustainability and corporate responsibility. It is therefore important to evaluate other research on the subject in order to justify its relevance. This chapter presents a bibliometric analysis of scientific publications that relate Industry 4.0 to the ESG approach, on methodological procedures.

A systematic online database search was used as the literature search method, followed by a bibliometric analysis of the results. Bibliometrics is an information science methodology that uses mathematical and statistical methods to map documents and publication patterns from bibliographic records stored in databases [6]. Bibliometrics allows for relevant counts such as: production by region, temporality of publications; research by area of knowledge; count of literature related to the citation of the study; and the impact factor of a scientific publication. Mathematical and statistical data that helps to systematize the results of research and minimized the occurrence of biases when looking at a particular topic [14]. For the bibliometric analysis of the present study, the process was organized into three distinct stages: planning, collection and results. These stages converged to answer the guiding research question: what are the relationships between Industry 4.0 and ESG?

The planning was carried out in June 2024. At this stage, certain criteria were defined, such as limiting the search to electronic databases without considering physical catalogues in libraries, given the large number of documents in web search databases. Within the scope of the planning, the Web of Science (WOS) database was stipulated as relevant to the research field, due to its relevance in academia and its interdisciplinary nature.

Considering the research problem, search terms were delimited during the planning phase, namely: 'Industry 4.0' and 'ESG', concomitantly to be present in the 'title', 'abstract' and 'keyword' fields, without delimiting time, language or any other restriction that could limit the result. Data collection retrieved a total of 25 indexed papers, with records from 2021 (first publication) to 2024 (June). As a result of this collection, it was identified that these 25 papers were written by 87 authors, linked to 75 institutions from 30 different countries. These papers were used to identify and index the publications, which are distributed across 11 areas of knowledge and 4 types of publication. Table 3 shows the results of this data collection in a general bibliometric analysis.

In order to analyze the results in greater depth, these results were exported to bibliographic management software called EndNoteWeb. This data enabled the organization of relevant information in a bibliometric analysis, such as: time distribution; main authors, institutions and countries; type of publication in the area; main keywords and the most referenced works. Firstly, the time distribution of the papers was analyzed, showing that publications were rather timid in 2021, with only 2 papers in the area. In 2022 there was an increase in publications, totaling 5 publications in the area. 2023 was the year with the highest number of publications (13), with a reduction in 2024. It is important to remember that the bibliometric study was carried out in

Table 3 General bibliometric data on the relationship between Industry 4.0 and ESG

Parameters	Values
Database	Scopus
Search terms	"Industry 4.0" and "ESG"
Search fields	"title", "abstract" and "keywords"
Total number of retrieved works	25
Authors	87
Institutions	75
Countries	30
Fields of knowledge	11
Type of publication	4

June 2024 and therefore only included articles published in the first six months of the year.

Two pioneering papers published in 2021 were identified. The first of these, entitled 'Corporate transformation towards Industry 4.0 and financial performance: The influence of environmental, social, and governance (ESG)', by the authors Fadi Alkaraan, Khaldoon Albitar, Khaled Hussainey and V.G. Venkatesh. The authors discuss the importance of corporate transformation towards Industry 4.0 (CTTI4.0), but there is almost no research into how companies communicate CTTI4.0 information in their annual reports and how this affects financial performance. The second article, entitled 'A fuzzy rule-based Industry 4.0 maturity model for operations and supply chain management', by the authors Gusmão Caiado and collaborators, discusses the link between disruptive technologies and Industry 4.0 (I4.0). Of the 25 papers, there is a varied list of authors, institutions and countries that feature prominently in research on circular economy, green recovery and sustainability. When analyzing the countries with most publications in the area, it can be seen that India and France stand out with 5 and 4 publications, respectively, with an average of 20% and 16% of the total number of publications.

Figure 8 shows the time distribution of papers correlating Industry 4.0 and ESG, and shows the main countries involved.

Brazil has only one publication in the area, which is equal to 5% of the total number of publications, implying that the discussion of this subject is still innovative in the country in the academic sphere. Another analysis is related to the identification of prominent authors in the area, where it was found that there is no reference author in the area in the WOS database, in terms of the number of publications in independent journals.

From the general survey, it was also possible to analyze the type of document in which research in the area of Industry 4.0 and ESG fits. It can be seen that publications are concentrated in journal articles, with 70% of the total number of documents. In total, there are 4 categories of indexations marked, and a group called undefined aggregates the other possible indexations, as shown in Fig. 9.

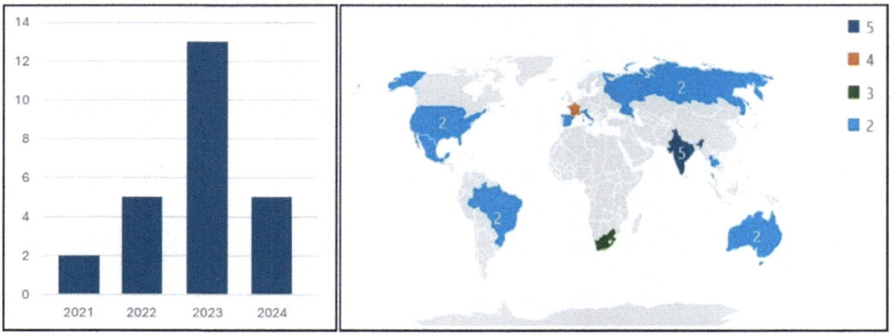

Fig. 8 The time distribution of papers and the countries with most publications papers correlating Industry 4.0 and ESG

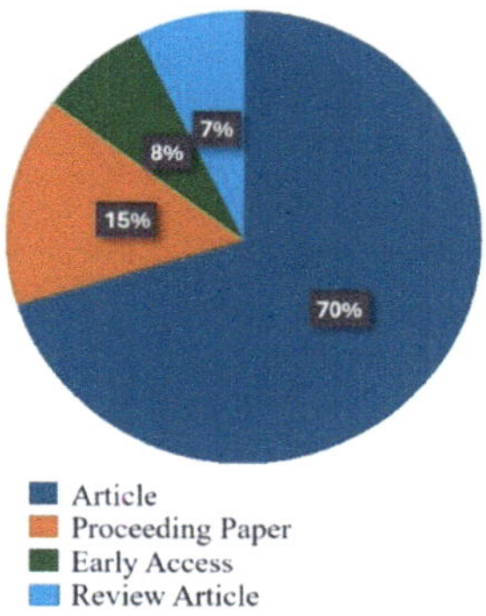

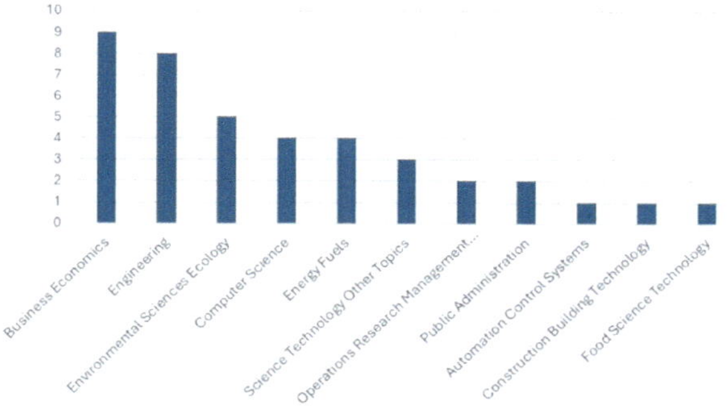

Fig. 9 Distribution of publications by type of journal, and field of knowledge that relates Industry 4.0 to ESG

From the bibliometric analysis, based on the group of papers retrieved from the WOS database, it was possible to identify the knowledge areas of the publications. Business economics stands out with an average of 22% of publications, followed by engineering with 20%, as shown too in Fig. 9.

Of the 25 resulting documents, 19 documents were excluded, being articles with no open access, books, book chapters, and open access articles that did not adhere to the research question: "what are the links between Industry 4.0 and ESG?" Thus, a total of 6 scientific articles were used to answer the research question, according to the systematic summary in Table 4.

The relationship between Industry 4.0 and ESG is multifaceted and complementary, with three main pillars: environmental, social and governance. Environmentally, Industry 4.0 promotes the adoption of advanced technologies that improve energy efficiency and reduce emissions, in line with sustainability practices. Socially, digitalization and automation boost job security, inclusion and community development, while creating challenges for the workforce that need to be managed responsibly. In governance, the growing complexity of operations demands greater transparency, ethics and corporate responsibility, facilitated by the integration of cyber-physical systems and digital technologies that improve management and communication with stakeholders. These interactions not only help companies achieve a more sustainable and ethical performance, but also make them more competitive and resilient in the long term.

5 Industry 4.0 Meets ESG

Industry 4.0 is enabled by technological tools, concepts and methods, with the aim of guaranteeing functionalities that lead to the interaction of data from a manufacturing process, from its production order to its delivery to the user. The 4th Industrial

Table 4 Systematic table of selected articles

Author(s)	Title	What are the relationships between Industry 4.0 and ESG?
[15]	The impact factors of Industry 4.0 on ESG in the energy sector	The article identifies a factor of integration in Industry 4.0, taking into account the ESG effect
[4]	Digitalization of supply chain management with Industry 4.0 enabling technologies: a sustainable perspective	The article outlines Industry 4.0 and its integration with technologies in analyzing sustainability performance in any organization using environmental, social, and governance (ESG) issues and metrics
[1]	Sustainable strategic investment decision-making practices in UK companies: The influence of governance mechanisms on synergy between industry 4.0 and circular economy	The article details the relationship between Industry 4.0 technologies and governance mechanisms (ESG, board composition, risk management, internal audit) across these two domains and their impact on company performance
[17]	Technologies empowered environmental, social, and governance (ESG): an Industry 4.0 landscape	The article explains the assimilation of Industry 4.0 technologies, including internet of Things (IoT), artificial intelligence (AI), blockchain, and big data for achieving ESG goals, and the organization's data are presented in detail to study the progress of these technologies toward ESG
[22]	Industry 4.0-enabled environment, social, and governance reporting: a case from a Chinese energy company	The article proposes using Industry 4.0 and its technologies to improve ESG processes
[11]	Moderating ESG compliance between Industry 4.0 and green practices with green servitization: examining its impact on green supply chain performance	The article reveals that green servitization mediates and ESG compliance moderates the relationship between Industry 4.0 and green practices

Revolution is not seen as an invention, but rather as an innovation, especially since the tools used were created for other purposes and used to implement a new way of doing business. One way of characterizing a company's level of maturity is its ability to appropriate technological tools, which is classified in a continuous and growing way, so that the tools are added to the production system as the level of maturity grows [18].

As with Industry 4.0, the ESG approach can also be seen from the digital tools used to generate functionalities that prove their actions and enable their environmental, social and governmental objectives, based on digital transformation [13]. In this way, it is possible to correlate the Industry 4.0 maturity index with the ESG approach, based on digital tools that the company uses. It is worth remembering that having a digital tool does not mean that it is being used for environmental, social and governmental purposes, but that there is a possibility of doing so. Which would be difficult to do if the company didn't have such tools. Figure 10 shows an illustration, inspired by the

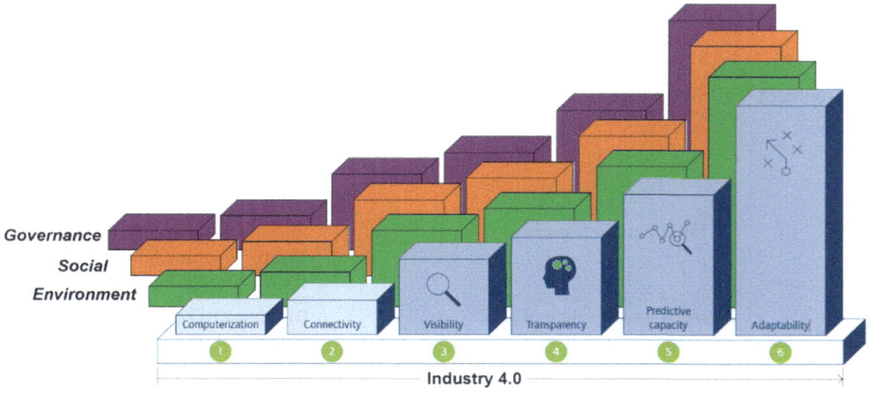

Fig. 10 Industry 4.0 Maturity Scale and its relationship with ESG Pillars

Industry 4.0 maturity scale, with a look at the ESG pillars, seeking to answer what maturity index a company needs to have for each topic in the Industry 4.0 pillars.

For Industry 4.0 Maturity, each topic is related to one or more enabling technologies, based on digital transformation. Table 5 details each of the digital tools for each ESG pillar topic, by [16].

By relating the Industry 4.0 enabling tools to the same ESG enabling tools, it is possible to relate the maturity level to the ESG pillar topic, assuming that they are the same technology. Figures 11, 12 and 13 shows these relationships.

In this relationship, some tools are not exactly the same. But have the same functionalities and can be related. This is the case with digitization (Dig) in ESG, which at the Industry 4.0 maturity level was considered to be IoT platforms. Or in the case of digital platforms (DP), which at the Industry 4.0 maturity level were considered dashboards. And traceability (RF), which was considered to be the use of protocols at the I4.0 maturity level. The others have an identity that allows direct correlation.

In an analysis of the degree of maturity required in Industry 4.0 for each topic in each ESG pillar, considering the highest level required, the graph in Fig. 14 was edited, with all topics itemized.

A first analysis shows that for a company to fully meet the environmental pillar (E) of the ESG, it will need to have a maximum Industry 4.0 maturity level, reaching adaptability (6th level). However, 3 of its 6 topics can already be covered with a visualization maturity level (3rd level). The social pillar (S) of the ESG can be fully met with a maturity level of predictability (5th level). And the government pillar (G) of the ESG can be achieved with the transparency maturity level (4th level). The analysis also shows that for all the topics in the ESG pillars, the minimum level the company must have is the connectivity maturity level (2nd level).

Within this analysis, the correlation that exists between the maturity levels of Industry 4.0 and the topics of the ESG pillars makes sense. However, for ESG, there is no aggregation between the topics. It is possible to meet one and not meet the

Table 5 ESG elements related to transformation tools digital

ESG pillar	Topic of each element	Digital tool that drives
Environment	Waste management	Big data (BD); Internet of things (IoT); Virtualization (Vir); Machine-to-machine (M2M); 3D printing (3D)
	Full energy efficiency	Internet of things (IoT); Machine-to-machine (M2M); Cloud computing (Cloud)
	Impact on climate change	Big data (BD); Internet of things (IoT); Digital platforms (DP)
	Rational use of natural resources	Internet of things (IoT); Digital platforms (DP)
	Environmental prevention	Cloud computing (Cloud) Digital platforms (DP)
	Reverse logistics	Process digitization (Dig); Internet of things (IoT); Traceability (TR)
Social	Working conditions	Cloud computing (Cloud); Internet of things (IoT); Traceability (TR); Virtualization (Vir)
	Employee engagement	Cloud computing (Cloud); Big data (BD)
	Positive impact on the community	Cloud computing (Cloud); Process digitization (Dig); Digital platforms (DP)
	Human and labor rights	Digital platforms (DP)
	Adequate and continuous training	Digital platforms (DP); Virtualization (Vir); Process digitization (Dig)
	Customer relations	Digital platforms (DP); Virtualization (Vir); Internet of things (IoT)
	Respect for employees	Digital platforms (DP)
	Inclusion policies	Virtualization (Vir); Digital platforms (DP)
Governance	Financial and accounting transparency	Cloud computing (Cloud); Digital platforms (DP)
	Equitable management	Cloud computing (Cloud)
	Ethical conduct	Digital platforms (DP)
	People management	Big data (BD) Process digitization (Dig)
	Reporting channels	Big data (BD)

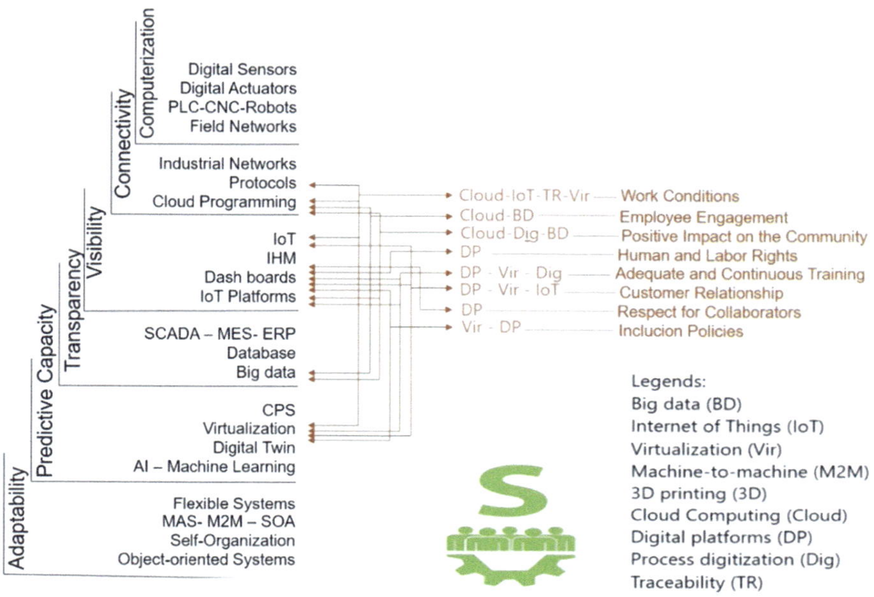

Fig. 11 Industry 4.0 Maturity Levels and their Relevance to the Social pillar of ESG

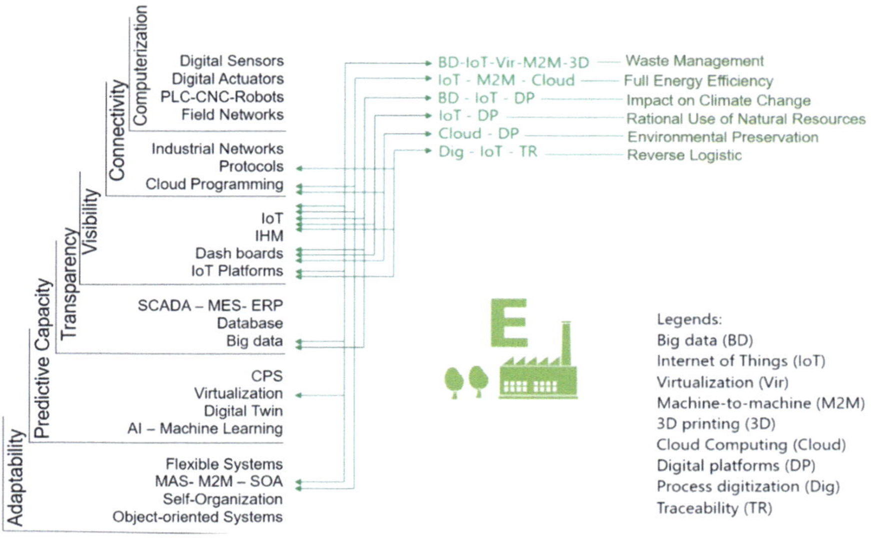

Fig. 12 Industry 4.0 Maturity Levels and their Relevance to the Environmental pillar of ESG

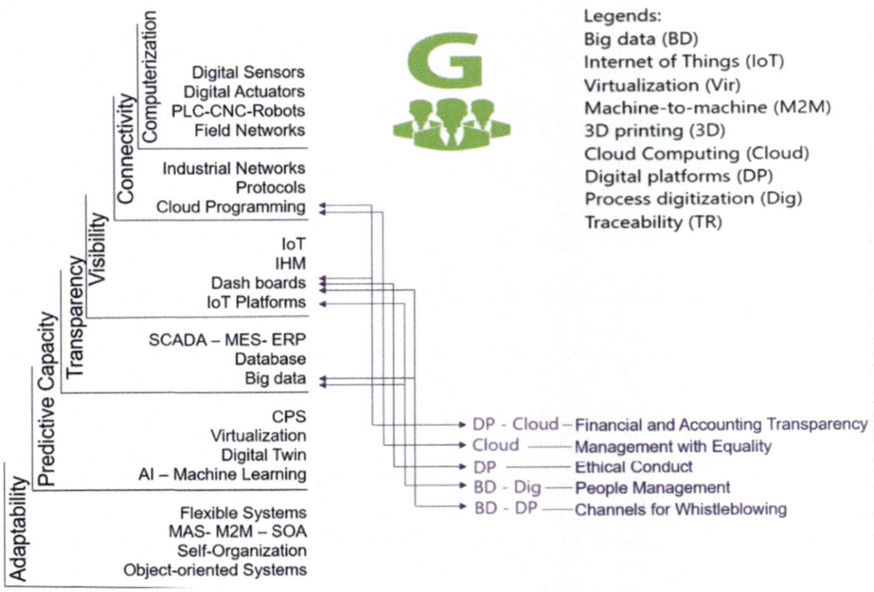

Fig. 13 Industry 4.0 Maturity Levels and their Relevance to the governance pillar of ESG

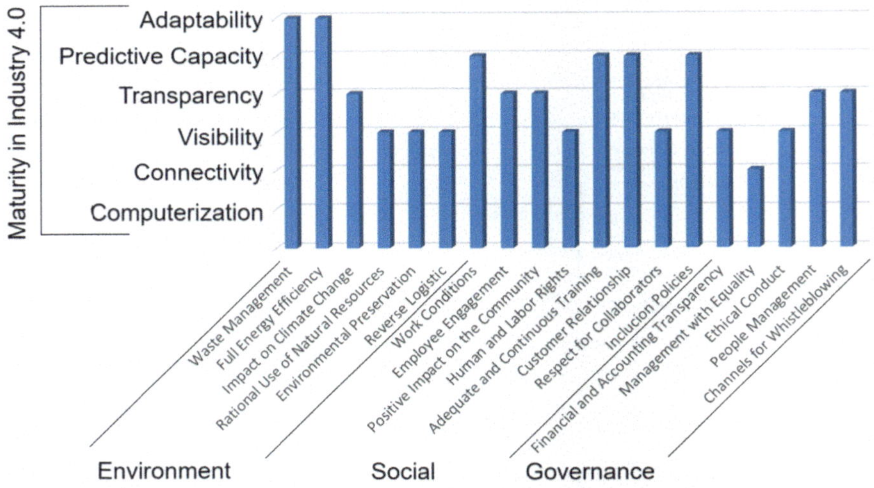

Fig. 14 ESG topics in relation to the Industry 4.0 maturity level

other, regardless of the technological level that the company has. Furthermore, the distribution of the different maturity levels of Industry 4.0 within the ESG pillars makes this clear. On the other hand, a company that is interested in meeting the specific topics of the ESG pillars can check the necessary technologies and their relationship with Industry 4.0, and can meet both concepts. Meeting ESG criteria will bring market importance, with the act of taking care of environmental, social and governmental issues being a value offer. Likewise, meeting the requirements of Industry 4.0 provides a more agile, diversified and informative production process, offering value to the consumer.

6 A Step Towards Industry 5.0

In a scientific scenario, there are research lines that point out that Industry 4.0 has brought connectivity, interaction and the massive use of digital transformation in machines, processes and the entire production chain of goods and services. There is a concern about machines, which must be properly autonomous or conveniently collaborative. The next step is the importance of the environment in which this manufacturing process takes place, but, above all, concern for the people who are on the other side of the collaboration, interacting with the machines and the importance of this manufacturing process. Thus, Industry 5.0 is seen as an evolution of Industry 4.0, by adding concern for the human being, the environment in which they operate, their relationship with work and society. In fact, scientific publications point in this direction.

As Maddikunta et al. [12] cite in their article 'Industry 5.0: A survey on enabling technologies and potential applications', which, in their definitions, refer to human, social and environmental factors, indicating that Industry 5.0 is focused on the human being, based on the principles of industrial recycling, which include the 6Rs: recognize, reconsider, realize, reduce, reuse and recycle. The article also mentions the return of human labor to the factory, where humans and machines are combined to increase process efficiency by harnessing human brain power and creativity; the integration of the strengths of cyber-physical production systems (CPS) and human intelligence to create synergistic factories, seeking innovative, ethical and human-centered designs; the consideration of human factors alongside technologies in industrial systems; and the era of the Social Smart Factory, where cobots communicate with humans using corporate social networks to enable continuous communication between humans and machines.

In this scenario, as the findings of science are confirmed, there is a proposal that envisages Industry 5.0 with the union of Industry 4.0 and ESG actions. Since the digital tools are the same, as is the information pertinent to the two proposals (Industry 4.0 and ESG), this interpretation makes sense. Figure 15 shows the ESGI (Environmental, Social, governance and Industrial) model with its digital tools that enable each topic, which consists of the union of Industry 4.0 topics with the ESG Pillars.

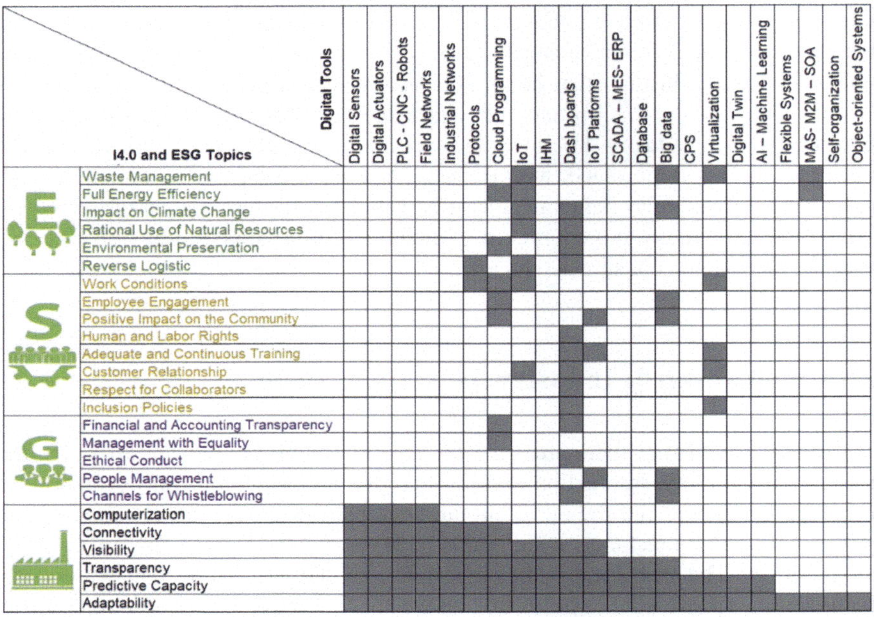

Fig. 15 ESGI and the digital tools that drive

ESGI is a proposal from this work. It doesn't represent Industry 5.0, but it could be a good proposal. More studies and tests are needed in the industrial market itself, in order to detect its adherence to the subject. However, the purpose here is not to coin a new term or propose a new concept. The important contribution of this work is to report scientifically on the importance of using the enabling tools of Industry 4.0 and the enabling tools of the ESG Pillars, as a way of having a complete, efficient, effective production environment that has environmental, social and governmental concerns within its corporate culture, as a company policy.

7 Conclusion

As progress is made towards a scenario in which sustainability is a central theme, it is essential that industrial processes are transformed with a view to greater transparency, better use of resources and greater environmental and social responsibility. In the advance of Industry 4.0 and the emergence of Industry 5.0, ESG is finding new perspectives for applying its principles effectively, with a view to creating a solid basis for sustainability and greater responsibility on the part of corporations.

The digital technologies available, which characterize the era of Industry 4.0, make it possible to monitor and optimize industrial processes, leading to improvements in environmental management, reducing waste of resources and materials, and minimizing the impacts of industrial activities. It is important to note that the new technologies also offer methodologies that favour an elaborate form of governance and greater alignment with ESG principles.

Digital transformation can bring important benefits to the workplace and the well-being of employees, reducing working hours and improving working conditions. However, this integration, which places human beings at the center of technological innovation, is one of the characteristics of Industry 5.0. The new phase aims not only to optimize processes, but also to value the human contribution and guarantee social inclusion and employee well-being through a more integrated vision of the three ESG pillars.

From the intersection of Industry 4.0 and 5.0, it is possible to generate business models that strive for productivity, but without renouncing the sustainable vision of the future. Companies that act responsibly, in line with ESG principles, better meet the expectations of their employees, have better tools to overcome adversity and create a more positive impact on society and the environment, as well as generating more conscious and responsible industrial development.

From the bibliometric analysis carried out, it was possible to identify that the correlation is still incipient, given the small number of studies found in the period evaluated, which indicates that ESG principles are not yet applied to their full potential in terms of industrial activity. Although recognized and valued, ESG principles still face challenges in becoming an integral part of industrial activities, which would often require procedural and structural changes on the part of corporations.

The trend is for digital tools to be used more and more in order to apply ESG principles in industry, and for corporations to priorities them. In this way, sustainability can be approached in a more integrated way, generating positive impacts on corporations, society and the environment. It is also important that the relationship between Industry 4.0 and 5.0 and ESG is disseminated and popularized, highlighting the importance of the issue and encouraging industries to adopt more sustainable practices.

References

1. Alkaraan, F., Elmarzouky, M., Hussainey, K., Venkatesh, V.G.: Sustainable strategic investment decision-making practices in UK companies: the influence of governance mechanisms on synergy between Industry 4.0 and circular economy. Technol. Forecast. Soc. Change, 187(2), 122187. (2023) https://www.sciencedirect.com/science/article/pii/S0040162522007089
2. Acatech.: Recommendations for implementing the strategic initiative industrie 4.0: final report of the industrie 4.0 working group. Berlim: Ministry of Education and Research (BMBF). (2013). https://en.acatech.de/publication/recommendations-for-implementing-the-strategic-initiative-industrie-4-0-final-report-of-the-industrie-4-0-working-group/

3. Bhandari, K.R., Ranta, M., Salo, J.: The resource-based view, stakeholder capitalism, ESG, and sustainable competitive advantage: the firm's embeddedness into ecology, society, and governance. Bus. Strateg. Environ. **31**(4), 1525–1537 (2022)
4. Chauhan, S., Singh, R., Gehlot, A., Akram, S.V., Twala, B., Priyadarshi, N.: Digitalization of supply chain management with industry 4.0 enabling technologies: a sustainable perspective. Processes **11**(1), 96 (2022). https://doi.org/10.3390/pr11010096
5. Confederação Nacional Da Indústria.: Oportunidades para a indústria 4.0: aspectos da demanda e oferta no Brasil. CNI. (2017). https://www.portaldaindustria.com.br/publicacoes/2018/2/oportunidades-para-industria-40-aspectos-da-demanda-e-oferta-no-brasil/
6. De Bem Machado, A., Souza, M.J., Catapan, A.H.: Systematic review: intersection between communication and knowledge. J. Inform. Syst. Eng. Manag. **4**(1), 86 (2019). https://doi.org/10.29333/jisem/5765
7. Gasperini, A.: Principles for responsible investment (PRI) and ESG factors. Climate Action (2020). https://doi.org/10.1007/978-3-319-95885-9_19
8. Hermann, M., Pentek, T., Otto, B.: Design principles for industrie 4.0 scenarios: a literature review. (2015). http://www.snom.mb.tu-dortmund.de/cms/de/forschung/Arbeitsberichte/Design-Principles-for-Industrie-4_0-Scenarios.pdf.
9. Koh, H.K., Burnasheva, R., Suh, Y.G.: Perceived ESG (environmental, social, governance) and consumers' responses: the mediating role of brand credibility, Brand Image, and perceived quality. Sustainability, 14(8), 4515. (2022). https://www.mdpi.com/2071-1050/14/8/4515
10. Kotler, P., Kartajaya, H., Setiawan, I.: Marketing 4.0: do tradicional ao digital (1st ed., Vol. 1). Sextante. (2017).
11. Kumar, M., Raut, R.D., Mangla, S.K., Chowdhury, S., Choubey, V.K.: Moderating ESG compliance between industry 4.0 and green practices with green servitization: examining its impact on green supply chain performance. Technovation **129**, 102898–102898 (2024). https://doi.org/10.1016/j.technovation.2023.102898
12. Maddikunta, P.K., Pham, Q.V., Prabadevi, B., Deepa, N., Dev, K., Gadekallu, T.R., Ruby, R., Liyanage, M.: Industry 5.0: a survey on enabling technologies and potential applications. J. Ind. Inform. Integr. **1**(26), 100257 (2022). https://doi.org/10.1016/j.jii.2021.100257
13. Machado, A.D.B., Richter, M.F., Peixoto, J.A.: Relations between digital transformation and sustainability post COVID-19. Adv. Logistics Oper. Manag. Sci. Book Series (2022). https://doi.org/10.4018/978-1-7998-8856-7.ch021
14. Machado, A.D.B., Santos, A.S., Sousa, M.J.: Digital transformation: management of smart cities. Emerald Publ. Limited EBooks (2023). https://doi.org/10.1108/978-1-80455-994-920231004
15. Nitlarp, T., Kiattisin, S.: The impact factors of Industry 4.0 on ESG in the energy sector. Sustainability **14**(15), 9198 (2022). https://doi.org/10.3390/su14159198
16. Peixoto, J.A., Machado, A.B., Richter, M.F.: ESG indicator metrics used by organisations to assess the degree of sustainability in companies. Int. J. Adv. Eng. Res. Sci. **9**(9), 589–600 (2022). https://doi.org/10.22161/ijaers.99.64
17. Saxena, A., Singh, R., Gehlot, A., Akram, S.V., Twala, B., Singh, A., Montero, E.C., Priyadarshi, N.: Technologies empowered environmental, social, and governance (ESG): an Industry 4.0 landscape. Sustainability **15**(1), 309 (2022). https://doi.org/10.3390/su15010309
18. Schuh, G., Anderl, R., Dumitrescu, R., Krüger, A., Hompel. M.: Industrie 4.0 maturity index: managing the digital transformation of companies—UPDATE 2020. (2020). https://en.acatech.de/publication/industrie-4-0-maturity-index-update-2020/
19. Tunio, R.A., Jamali, R.H., Mirani, A.A., Das, G., Laghari, M.A., Xiao, J.: The relationship between corporate social responsibility disclosures and financial performance: a mediating role of employee productivity. Environ. Sci. Pollut. Res. **28**, 10661–10677 (2021). https://doi.org/10.1007/s11356-020-11247-4
20. United Nations.: The 17 sustainable development goals. United Nations. (2015). https://sdgs.un.org/goals

21. Xu, L.D., Xu, E.L., Li, L.: Industry 4.0: state of the art and future trends. Int. J. Prod. Res. **56**(8), 2941–2962 (2018)
22. Yu, W., Gu, Y., Dai, J.: Industry 4.0-enabled ESG reporting: a case from a chinese energy company. J. Emerg. Technol. Acc. (2022). https://doi.org/10.2308/jeta-2022-014

Ethical Practices in Digital Transformation Over the Last Half Decade: A Framework for Sustainability and Integrity in Business

Albérico Travassos Rosário and Rui Nunes Cruz

Abstract Digital transformation, characterized by integrating digital technologies into all aspects of business operations, has profoundly reshaped the economic and social landscape. While it offers unprecedented opportunities for growth and innovation, it also raises significant questions regarding ethics and corporate responsibility. This study explores the ethical practices essential to effective digital transformation management, focusing on data privacy protection, cybersecurity, and equity of access to technologies. Reviewing the literature identified in the SCOPUS database over the past decade, we identify fundamental ethical principles and propose a normative framework to guide organizations. The research highlights the importance of robust and transparent governance, a commitment to digital inclusion, and the need for clear policies that ensure respect for human rights in the digital age. In the end, we discuss how the adoption of ethical practices not only mitigates legal and reputational risks, strengthens consumer confidence and promotes a sustainable business environment. This work contributes to the academic and practical discourse, offering valuable insights for business leaders and policymakers in navigating the complexities of digital transformation ethically and responsibly.

Keywords Ethical practices · Digital transformations

A. T. Rosário
GOVCOPP, Universidade Europeia, Lisbon, Portugal
e-mail: alberico@ua.pt

R. N. Cruz (✉)
UNIDCOM/IADE, Universidade Europeia, Lisbon, Portugal
e-mail: rui.cruz@universidadeeuropeia.pt

© The Author(s), under exclusive license to Springer Nature Switzerland AG 2025
A. d. b. Machado et al. (eds.), *Environmental, Social, Governance and Digital Transformation in Organizations*, Information Systems Engineering and Management 35, https://doi.org/10.1007/978-3-031-86079-9_22

1 Introduction

Digital transformation, defined as integrating digital technologies into various aspects of business, has fundamentally reshaped economic and social landscapes over the past decade. It allows organizations to quickly adapt to changing consumer expectations, increase operational efficiency, and foster innovation. However, as businesses increasingly adopt digital processes, ethical concerns surrounding data privacy, cybersecurity, and equitable access have become more urgent. These concerns demand deliberate governance and careful management to ensure integrity and sustainability in this digital era [23].

The strategic shift towards digitalization is primarily driven by the need to remain competitive and relevant in a rapidly evolving market. As [23] note, the role of senior leadership has become critical in steering organizational change and managing talent in a digital context. Leaders are now tasked with overseeing the seamless integration of technology, upholding ethical standards, and fostering an environment that is adaptive to technological shifts. This aligns with [66], who discusses the "Stages of Organizational Conscience," focusing on how digitalization introduces ethical complexities that affect corporate responsibility and governance. Rojas highlights the ethical dilemmas that surfaced during the pandemic, where increased digital reliance posed risks for ethical breaches, underscoring organizations' need to develop frameworks to address these challenges.

Fernández-Rovira et al. [24] emphasize how Big Data and AI are transforming customer relationships, raising ethical concerns regarding the use of personal data, particularly in industries such as tourism. They stress the importance of transforming the sector to align with digitalization while maintaining ethical standards [35].

One prominent trend within digital transformation is the rise of "coopetition," a strategy that blends cooperation with competition among businesses. Meena et al. [52, 53] delve into this dynamic, showing how companies can leverage shared resources and mutual growth while remaining competitive. Their research suggests that when governed ethically, coopetition can lead to significant improvements in business performance, fostering innovation and resilience in digital transformation. Additionally, coopetition has ethical implications, as businesses are expected to act transparently and fairly in collaborative ventures.

Central to digital transformation is innovation empowerment, particularly for firms aiming to disrupt traditional business models. Lin and Yi [47] explore how digital technologies enable businesses to overcome operational constraints through improved connectivity, automation, and data-driven decision-making. They emphasize that digitalization is not merely a technological upgrade but a transformative approach requiring rethinking strategic priorities. This transformation can enhance a firm's innovative capacity and long-term sustainability when managed ethically.

Digital transformation is not confined to specific regions or industries but is a global phenomenon. Voronenko et al. [83, 84] studied digital transformation trends, highlighting regions like Ukraine, where such efforts are essential for maintaining competitiveness and supporting national development. Their findings underline the international scope of digital transformation and the importance of ethical frameworks that account for regional differences, ensuring inclusive access to digital opportunities.

Finally, Paul et al. [62, 63] offer a multidisciplinary view on digital transformation, examining its impact across fields such as artificial intelligence, the Internet of Things, and augmented reality. They stress the importance of developing a research agenda that includes ethical considerations, particularly in crafting governance models that can respond to the challenges posed by emerging technologies. Paul and colleagues argue that while digital transformation creates a wealth of opportunities, it also raises ethical concerns that require continuous reflection, especially as new technologies evolve and become integrated into everyday operations.

The advancement of technologies like artificial intelligence and big data analytics drive digital transformation. Kim et al. [40, 41] explore the influence of AI on consumer behavior and marketing strategies, demonstrating how these technologies are revolutionizing the marketing landscape. Puntoni and Wertenbroch [64, 65] reflect on the role of consumer automation in creating value for both companies and consumers, highlighting the need for ongoing adaptation to technological innovations. Singh and Kaunert [78, 79] emphasize the importance of AI personalization in enhancing customer experiences, enabling brands to anticipate consumer needs before they are even expressed.

In conclusion, digital transformation presents both unprecedented opportunities and significant ethical challenges. While it drives growth and innovation, it also requires a solid commitment to ethical principles such as transparency, inclusivity, and data protection. To successfully navigate this complex landscape, organizations must adopt frameworks that enhance their digital capabilities and ensure that these transformations contribute to a sustainable and responsible business environment. This study seeks to contribute to this ongoing discourse by exploring the ethical practices necessary for responsible digital transformation, focusing on establishing a normative framework that aligns technological advancements with ethical corporate governance.

2 Methodological Approach

The researchers conducted a systematic literature review using the PRISMA (Preferred Reporting Items for Systematic Reviews and Meta-Analyses) framework. According to Page et al. [58], the PRISMA 2020 model is an effective tool for systematic reviews. It enables researchers to transparently report the rationale behind the research, the methods used, and the findings obtained.

Additionally, the PRISMA framework provides a structured approach that guides researchers through the critical stages of study identification, screening, eligibility assessment, and inclusion. It encourages thorough documentation of the research protocol, allowing others to replicate and validate the methods. This rigorous approach enhances the overall quality and reliability of academic discourse, as noted by Rosário and Dias [67], and contributes to the refinement of future research methodologies in the field.

Moreover, the insights gained from this comprehensive literature review can offer valuable practical guidance for industry professionals and policymakers. Unlike traditional literature reviews, the LRSB model takes a more detailed approach to examining the existing literature on the topic [67]. This method condenses the body of research, focusing on the most relevant studies. It also provides an audit trail, enabling readers to assess the quality of the research methodologies, conclusions, and the studies included in the review.

Consequently, the LRSB model ensures the accuracy and validity of the data by meticulously screening and selecting information sources. This process unfolds through three phases and six steps, as shown in Table 1.

The investigators utilized the Scopus database to identify and select reputable sources widely recognized in academic and scientific communities. However, this study has certain limitations, as it focused exclusively on Scopus, thereby excluding other significant academic and scientific databases.

The literature search process began by selecting the Scopus database. This was followed by defining specific inclusion and exclusion criteria. These criteria included: (i) sources had to address the keywords and their related secondary components primarily; (ii) priority was given to peer-reviewed sources, such as journal articles and academic books, to ensure the credibility and rigor of the information; (iii) only sources available in English were considered to maintain consistency and accessibility; and (iv) selected sources needed to have a clear and detailed methodology.

Once the inclusion and exclusion criteria were established, the researchers searched for relevant studies in the Scopus database. Initially, the keywords "digital transformation," "business integrity," "ethics," and "sustainability" were used, yielding 109 documents from the period 2014 to 2024. The search was then refined by limiting the subject area to "Business, Management, and Accounting" and applying additional filters, such as "full text" availability and "English" language.

These filters narrowed the search results to 96 documents (N = 96), as illustrated in Fig. 1.

Table 1 Process of systematic LRSB

Fase	Step	Description
Exploration	Step 1	Formulating the research problem
	Step 2	Searching for appropriate literature
	Step 3	Critical appraisal of the selected studies
	Step 4	Data synthesis from individual sources
Interpretation	Step 5	Reporting findings and recommendations
Communication	Step 6	Presentation of the LRSB report

Source adapted from Rosário and Dias [67]

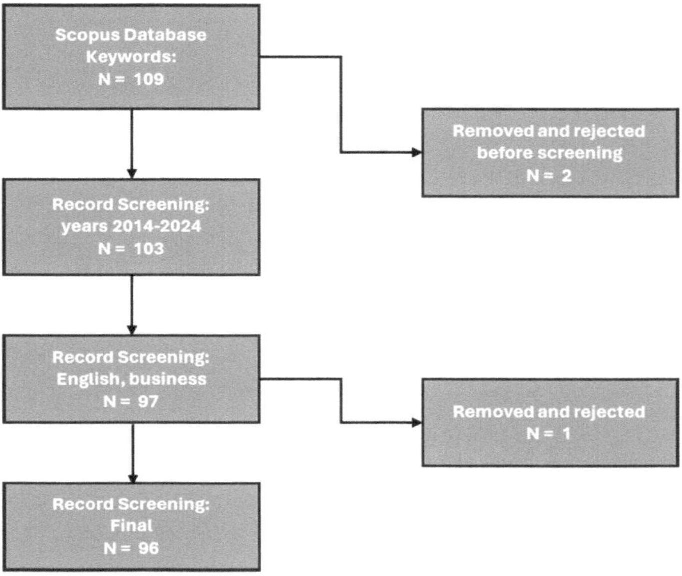

Fig. 1 PRISMA 2020 flow diagram for the literature search process. *Source* Own elaboration

Content and thematic analysis techniques were used to identify, examine, and present various documents. Subsequently, the 96 scientific and academic documents cataloged in Scopus were subjected to narrative and bibliometric analysis.

This approach aimed to explore the content and uncover recurring themes directly addressing the research questions. This process resulted in 96 documents, classified as 89 articles and 9 Book chapters.

3 Publication Distribution

Although our segmentation includes a 10-year-old screen, the topic is very new and has only begun to have editorial significance since 2019, but it is strongly increasing (Fig. 2).

Figure 3 shows the countries with the most active scientific output in research areas, focusing on India and China. Based on recent IGI Global publication indicators, these countries had the most publications in the period and are still very active in the editorial area for this subject.

This analysis aims to understand whether these countries prioritize utilizing artificial intelligence in business processes or if they emphasize boosting their productivity.

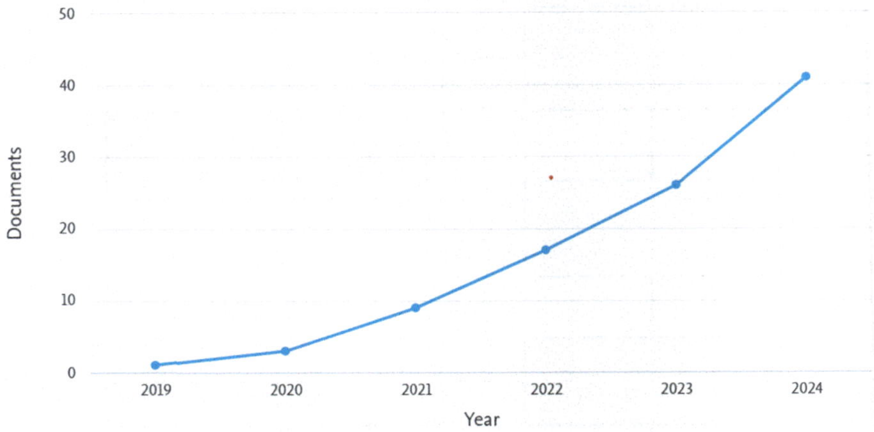

Fig. 2 Documents by year. *Source* Scopus analytics

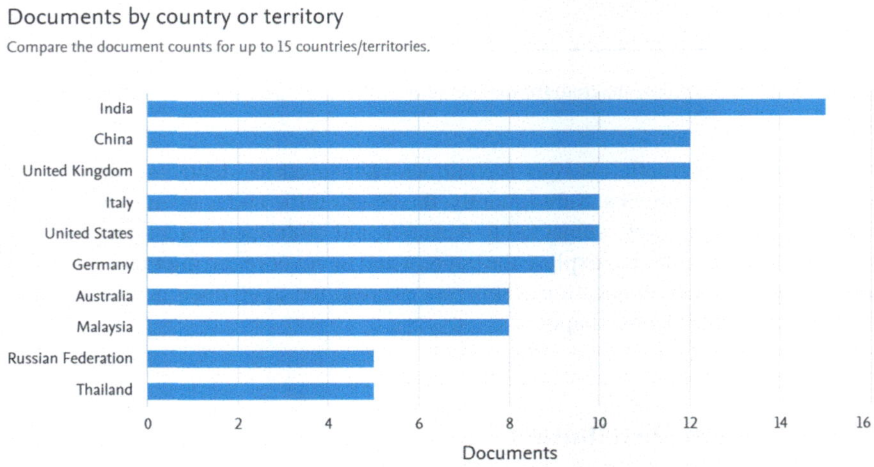

Fig. 3 Scientific production by country. *Source* Scopus analytics

Figure 4 highlights the area of Business and Management as the primary source of publications followed at some distance by the areas of social sciences, economics, and engineering.

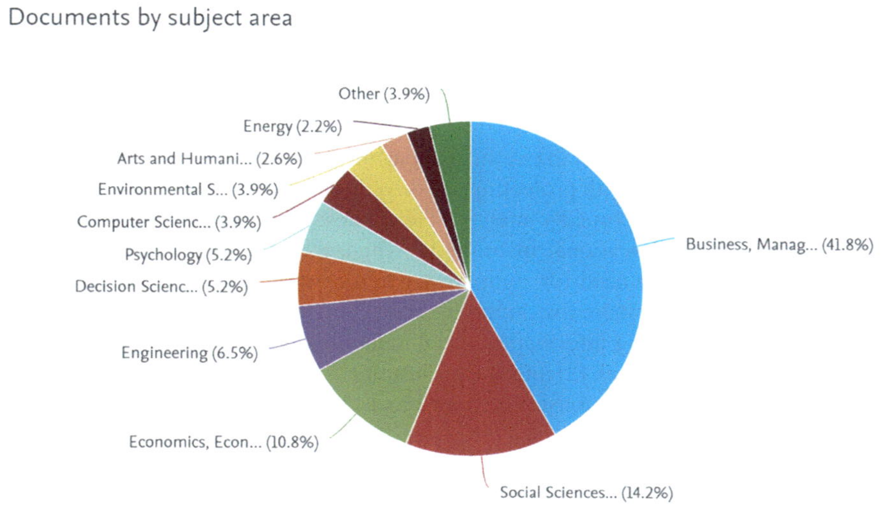

Fig. 4 Subject area. *Source* Scopus analytics

4 Theoretical Perspectives: Ethical Foundations in Digital Transformation

Ethical considerations in digital transformation encompass several key areas, including data privacy, cybersecurity, and equitable access to technology. As digital solutions increasingly influence organizational practices, it is essential to address these ethical foundations to foster a responsible and sustainable business environment [98].

Jain et al. [38, 39] examine the evolution of generative AI in marketing, highlighting ethical concerns related to personalization and data privacy. Similarly, Sahin and Soylemez [68, 69] explore the role of AI ethics and brand activism in digital consumption, emphasizing the need for companies to incorporate ethical considerations into their digital marketing strategies. Hermann [32] delve into AI-driven personalization, particularly the ethical implications of overusing consumer data for segmentation. Naz and Kashif [56, 57] further discuss issues of customer prioritization and the manipulation of consumer behavior through predictive marketing techniques.

4.1 Data Privacy and Protection

4.1.1 Cybersecurity

With the ongoing rise of digitalization, cybersecurity has become a critical concern for organizations tasked with protecting sensitive information from increasing cyber threats. Ivanov [36, 37] critically analyzes the cybersecurity risks posed by artificial intelligence (AI) in educational institutions, where data breaches and algorithmic biases signal the urgent need for comprehensive measures to protect personal data. These challenges underscore the importance of robust security protocols, which are essential for safeguarding information and maintaining fairness in AI systems.

Building on this, Ali et al. [5] introduce a sustainable blockchain framework for the halal food supply chain, demonstrating how blockchain's transparent and immutable features can address cybersecurity concerns. Their study reveals that blockchain technology can enhance data security, providing an equitable solution for managing information within digital ecosystems. By improving the integrity of supply chains, this approach offers a secure and sustainable method for data protection across various industries.

The environmental impact of digitalization is another critical factor, as Samuel [70, 71] discusses the connection between data-intensive research and cybersecurity. He emphasizes that protecting data safeguards privacy and minimizes unnecessary data duplication and storage, reducing the carbon footprint of digital operations. This perspective broadens the scope of cybersecurity, linking it to sustainability and the responsible use of digital resources.

Cybersecurity's importance extends beyond traditional sectors, as highlighted by Tettamanzi et al. [80, 81], who focus on the football industry within sustainable sports management. Their research shows that reliance on digital technologies to manage sensitive player and operational data demands robust cybersecurity protocols. Protecting this information is essential to ensuring the integrity and transparency of operations as the industry continues to digitalize.

Finally, Wang et al. [87, 88] contribute to the discourse by examining how cybersecurity intersects with gender equality in digital employment. While digitalization can promote inclusion and diversity, cybersecurity strategies must be gender-sensitive to avoid reinforcing biases or enabling breaches that disproportionately affect women and other vulnerable groups. Their findings remind us that the ethical implementation of cybersecurity measures plays a crucial role in ensuring equity and inclusivity across digital landscapes [77].

4.1.2 Equitable Access to Technology

Digital transformation holds the potential to either bridge or deepen digital divides, making equitable access to technology essential for fostering a fair and inclusive digital environment. Pandya and Kumar [60, 61] emphasize the role of Industry 4.0

technologies in enhancing the sustainability of micro, small, and medium-sized enterprises (MSMEs). With equitable access to these technologies, MSMEs may remain competitive. This highlights the need for supportive policies that ensure technology is accessible to all significantly smaller businesses that lack the resources of larger enterprises.

Building on the theme of inclusivity, Walkowiak [85, 86] explores neurodiversity initiatives within digital transformation—his research advocates for developing inclusive technologies that accommodate the diverse needs of neurodivergent individuals. By prioritizing digital accessibility, organizations can remove barriers that neurodiverse individuals may face when interacting with standard digital platforms, thus promoting a more inclusive digital landscape.

Similarly, Anwar et al. [7] investigate how blockchain technology can contribute to more secure and equitable access to digital resources. Their study of the Anon Chain system reveals how blockchain can reduce reliance on centralized authorities and democratize data access, particularly benefiting underserved regions where trust in traditional institutions may be lacking. This decentralized data-sharing approach is critical in expanding access to digital technologies.

The conversation on equitable access extends to supply chain management as well. Demir et al. [20] propose an intelligent and sustainable supply chain model integrating equitable access as a core component. They underscore the need for smaller entities to have access to advanced supply chain technologies, which more giant corporations often dominate. This approach not only supports digital inclusivity but also enhances supply chains' overall sustainability and efficiency.

Lastly, Ali et al. [5] highlight blockchain's capacity to promote digital equity within supply chains by decentralizing access to technology. Their work illustrates how blockchain can empower smaller suppliers, ensuring all stakeholders have equal opportunities to leverage technological advancements. This contributes to a more balanced and inclusive digital ecosystem.

This literature review underlines the ethical imperatives of data privacy, cybersecurity, and equitable access. Each study offers valuable insights into how these areas intersect with digital transformation. Organizations can foster more responsible, inclusive, and sustainable digital progress by addressing these foundational ethical concerns.

4.2 *Vital Ethical Principles in Digital Transformation*

The principles surrounding emotional and cognitive trust in conversational agents underscore transparency's critical role in establishing trust in AI-mediated interactions. As digital transformation progresses, it introduces ethical principles for promoting responsible practices. Key among these principles is transparency, commitment to inclusion, and respect for human rights. These foundational values are crucial for fostering trust and ensuring equitable access as organizations increasingly adopt digital solutions [1, 91].

Brüns and Meißner [13, 14] further explore the theme of trust by addressing the perceived authenticity of brands. They argue that overusing specific strategies can undermine the authenticity consumers associate with a brand. This connects back to the broader ethical concern of transparency in digital transformation, as overreliance on specific tactics can erode consumer trust. Similarly, Cloarec et al. [19] introduce the concept of the 'transformative privacy calculus,' which deals with the tension between personalization and privacy in social media. Their research highlights how transparency and trust are crucial to maintaining an ethical relationship with consumers, especially when balancing the benefits of personalization with the need for privacy.

In line with these studies, Aggarwal and Commuri [2] suggest that brand strategies must also incorporate ethical considerations to sustain authenticity and strengthen consumer relationships. This perspective reinforces the idea that maintaining trust and transparency is essential for long-term success in both AI interactions and branding. Ethical branding enhances consumer loyalty and aligns with the broader ethical principles guiding digital transformation, such as respect for human rights and commitment to inclusivity.

4.2.1 Transparency and Governance

Transparency in digital practices and governance structures is crucial for building and maintaining stakeholder trust. Samuel [70, 71] explores the ethical implications of data usage and its environmental impact, underscoring the need for transparency in data collection, management, and utilization. He argues that ethical governance should go beyond data privacy to include the environmental consequences of data-intensive processes, highlighting transparency as a critical factor in responsible digital operations and sustainability.

This focus on transparency is echoed by Eustachio et al. [22], who investigate sustainability leadership and emphasize that effective digital governance must be grounded in transparent practices. Their study identifies transparency as a core component of sustainable leadership, illustrating how transparent and open governance can steer digital initiatives toward more ethical outcomes. By embedding transparency in organizational strategies, leaders can set a precedent for responsible digital transformation.

Expanding on this, Al-Faouri et al. [4] examine governance frameworks in the context of intelligent strategies for technology application, emphasizing the importance of ethical data usage and transparency in innovation. Their research suggests that transparency is vital for ethical governance and serves as a catalyst for innovation. When organizations implement transparent and ethical data practices, they enhance stakeholder trust, crucial for successfully adopting and applying digital technologies.

Gaponenko et al. [28, 29] contribute to the discussion by focusing on transparency within value management strategies in Russian companies undergoing digital transformation. They argue that transparent governance aligns digital practices with corporate values, ensuring that ethical considerations are fully integrated into digital initiatives. Similarly, Lentz [45, 46] addresses the issue of ethical authorship in AI, asserting that transparency in data handling and AI development can reduce biases and foster moral responsibility. Lentz highlights that clear authorship and ethical accountability are central elements of transparent governance, essential for guiding digital transformation responsibly and ethically.

4.2.2 Commitment to Digital Inclusion

Digital inclusion is a critical ethical principle, ensuring that all individuals, regardless of background, have access to digital resources. Lin and Yi [47] explore how firm innovation through digital empowerment can foster inclusivity, arguing that by creating accessible digital platforms, organizations drive innovation and expand access to technology for a wider audience. This emphasis on inclusion reflects a growing recognition of digital empowerment's role in leveling the technological playing field.

Similarly, Voronenko et al. [83, 84] examine global trends in digital transformation and stress that digital inclusion is essential for achieving equitable technological progress. As digital transformation continues to accelerate globally, they argue that organizations must prioritize inclusion to prevent the deepening of digital divides, particularly in regions where access to technology remains limited. This perspective aligns with the broader need for organizations to consider the social impact of their digital initiatives [3].

In education, Ivanov [36, 37] underscores the importance of digital inclusion within AI ethics, particularly in higher education. He argues that AI systems must be designed inclusively to avoid discriminatory practices that could marginalize certain groups. His work highlights the ethical necessity of ensuring that digital transformation efforts, especially in sensitive fields like education, prioritize inclusivity to promote fairness and accessibility, as also stated by Sim et al. [77].

At the educational level, Machado et al. [51] conclude that generative AI tools can help create more efficient exercises. Teachers and educators can use technology to create instructional materials and present summaries of concepts more efficiently.

It is concluded that generative AI tools can also help create more efficient exercises. Teachers and educators can use technology to create instructional materials and present summaries of concepts.

Ali et al. [5] highlight the role of blockchain in promoting transparency and inclusion within the halal food supply chain. Their research demonstrates that blockchain technology can support digital inclusion by creating an accessible, decentralized data-sharing framework. This ensures that all stakeholders in the supply chain have equal access to information, further underscoring the potential of digital tools to drive inclusive practices.

Finally, Bar-Gil et al. [11, 12] focus on AI ethics in human resources (HR) analytics, emphasizing the importance of inclusive practices when integrating AI into HR processes. They warn of the ethical risks of AI systems that lack inclusive considerations, particularly in hiring and promotion decisions. Their study suggests that a commitment to digital inclusion is crucial for ensuring fair and ethical AI usage in corporate environments, where decisions made by AI can significantly impact individuals' careers.

4.2.3 Respect for Human Rights in Digitalization

Respect for human rights is a foundational element of ethical digital transformation, as digital systems increasingly shape various aspects of life. Gaponenko et al. [28, 29] argue that value-based management in digital transformation plays a crucial role in upholding human rights. Their research suggests that by embedding human-centric values into digital strategies, organizations can ensure that their efforts respect individual rights and dignity, highlighting the importance of aligning digital innovation with ethical principles.

This perspective is further expanded by Samuel [70, 71], who addresses health researchers' ethical responsibilities in considering their work's environmental impacts and links them to human rights. Samuel's study highlights how the ethical handling of environmental data in health research protects broader societal rights, reinforcing that digital practices, particularly those affecting public health, must prioritize human rights to maintain ethical integrity.

Similarly, Ali et al. [5] present a blockchain framework for the halal food supply chain, emphasizing transparency and respect for human rights within digitalized supply chain operations. They argue that blockchain can uphold ethical standards by ensuring accountability and fair treatment for all stakeholders, illustrating how digital solutions can reinforce human rights through transparency and decentralized governance.

Lin-Hi et al. [49, 50] focus on adopting health technologies, emphasizing that these technologies must align with ethical principles that safeguard individual rights. They argue that transparency and ethical governance are essential in protecting patient rights, ensuring that digital transformation in healthcare is conducted responsibly and with respect for human dignity.

Demir et al. [20] build on this by examining the intersection of supply chain sustainability and human rights. They stress that digital supply chains should prioritize sustainable practices that respect workers' and consumers' rights. By integrating human rights considerations into digital supply chain models, organizations can align their digital transformations with broader ethical and social values, ensuring that technological advancements do not come at the cost of fundamental rights.

This literature review on ethical principles in digital transformation identifies transparency, inclusion, and respect for human rights as essential pillars. These principles contribute to a responsible approach to digitalization, ensuring that technological progress benefits society while adhering to ethical standards. By committing to these values, organizations can foster trust and accountability in their digital initiatives, creating a more equitable and responsible digital future.

5 Proposed Normative Framework for Ethical Digital Transformation

A normative framework for ethical digital transformation provides structured guidance for organizations to systematically incorporate ethical principles. This framework covers crucial areas such as governance, compliance, cybersecurity risk management, and inclusive access policies. By establishing clear standards in these domains, organizations can ensure that their digital transformation efforts are efficient and aligned with ethical values and societal responsibilities.

Pahari et al. [59] contribute to this discussion by exploring the ethical effects of digital advertising on social networks. Their research offers insights into creating adaptable normative frameworks, which can help organizations navigate the complexities of digital advertising while maintaining ethical standards. This aligns with Angmo and Mahajan's [6] study on virtual influencers, an emerging trend that impacts consumer expectations and authenticity in digital spaces. Their work underscores the need for normative frameworks to address such evolving phenomena, ensuring that digital engagement remains authentic and ethical [44].

Building on this, Dwivedi et al. [21] highlight the importance of ethical frameworks within social networks and digital marketing. They emphasize that these frameworks are essential to prevent privacy intrusions and reinforce ethical standards in digital communication. This perspective complements the work of Fiore et al. [25], who propose using machine learning algorithms to predict healthy consumer choices. Their study demonstrates how AI can promote ethical and sustainable purchasing behavior, further illustrating the positive impact of ethical frameworks when applied to AI-driven systems.

5.1 Governance and Compliance

Effective governance and compliance are fundamental to achieving ethical digital transformation. Rojas [66] explores the concept of ethical governance within the post-pandemic digital landscape, emphasizing the necessity for responsive governance frameworks that can adapt to the rapid evolution of technology. He argues that the realities of the post-pandemic world demand governance models prioritizing ethical considerations, enabling organizations to address new societal and technological challenges with integrity.

Building on this idea, Samuel [70, 71] proposes governance frameworks tailored to data-intensive health research. His study focuses on safeguarding patient data and ensuring compliance with ethical standards in health-related digital transformations. Compliance with data protection regulations and the ethical use of health data are critical for maintaining patient trust and protecting their rights. Samuel's work highlights the essential role structured governance frameworks play in managing sensitive data, particularly in areas like healthcare.

Expanding this discussion to the supply chain sector, Al-Faouri et al. [4] examine innovative digitalized supply chain management strategies. They advocate for governance structures prioritizing ethical practices, suggesting that transparent governance policies can build stakeholder trust and align business operations with ethical standards. Their research emphasizes that ethical governance is critical to the success of digital transformations within supply chains.

Finally, Paul et al. [62, 63] take a broader view by addressing the future research needs in multidisciplinary digital transformation. They stress the importance of governance frameworks that can be adapted across various industries. They argue that the need for flexible, multidisciplinary governance models becomes more urgent as digital transformation spreads across sectors. These models are crucial for effectively addressing the diverse ethical challenges in different fields.

5.2 Cybersecurity Risk Management

Cybersecurity risk management is a critical component of any ethical digital transformation framework, as the increasing digitalization of processes and data makes systems more vulnerable to cyber threats. Ivanov [36, 37] delves into the cybersecurity risks associated with educational AI applications, emphasizing the potential for data breaches and the misuse of personal information. His work underscores the importance of implementing robust cybersecurity policies to protect user privacy, particularly in sectors like education that handle sensitive data.

Extending this discussion, Tettamanzi et al. [80, 81] highlight the sustainable management of cybersecurity in the sports business sector. As sports organizations adopt more digital technologies, they face unique challenges in protecting sensitive data, making industry-specific cybersecurity frameworks essential. Their study stresses that tailored cybersecurity strategies are vital to ensuring the secure and ethical digital transformation of industries with specialized needs.

Further broadening the scope, Meena et al. [52, 53] explore cybersecurity within the context of cooperation strategies—where organizations collaborate and compete simultaneously. They emphasize that in environments where resources and data are shared across organizational boundaries, robust cybersecurity measures are crucial to protecting shared information. Their research demonstrates that cybersecurity is an indispensable part of ethical risk management, particularly in collaborative settings that pose complex security challenges.

Ali et al. [5] propose a practical solution by proposing blockchain technology for improving cybersecurity in halal food supply chains. Their study shows how blockchain's immutable ledger can enhance data security, preventing tampering and unauthorized access. By incorporating blockchain into cybersecurity frameworks, organizations can mitigate risks while increasing transparency—a fundamental element of any ethical digital transformation.

5.3 *Policies for Inclusive Access*

Inclusive access policies are essential for ensuring that digital transformation benefits diverse stakeholders, creating a fair and equitable digital landscape. Pandya and Kumar [60, 61] explore the application of Industry 4.0 technologies in supporting the sustainability of micro, small, and medium-sized enterprises (MSMEs), advocating for policies that provide smaller businesses access to digital tools. Their research highlights that equitable access to digital resources is critical for promoting sustainability in MSMEs, enabling these enterprises to compete effectively in a digitalized market.

Expanding on the importance of inclusivity, Wang et al. [87, 88] examine how digital transformation can promote gender equality in employment settings. They find that policies supporting equitable digital access can help reduce gender disparities by creating more opportunities for underrepresented groups. Their study suggests that inclusive access policies drive ethical digital transformation and contribute to broader social objectives, such as advancing gender equality in the workplace.

Similarly, Lin and Yi [47] emphasize the role of policies that foster inclusive innovation, underscoring the importance of creating digital systems that are accessible to all. They propose that prioritizing inclusive design and development enables organizations to create digital tools that serve a wide range of users, including those from marginalized communities. Lin and Yi's study advocates for policies that prioritize inclusivity at every stage of digital transformation, ensuring that access to digital resources is equitable across all sectors of society.

This proposed normative framework provides a solid foundation for ethical digital transformation, integrating structured governance, risk management, and inclusive access policies. By implementing these principles, organizations can align digital advancements with ethical standards, ensuring that transformation efforts are both responsible and sustainable. Each element of the framework contributes to a cohesive strategy, ensuring that digital transformation serves the greater societal good alongside organizational goals.

6 Case Studies and Practical Examples

Case studies and practical applications offer valuable real-world insights into the ethical implementation of digital transformation across various sectors. These examples highlight how organizations confront ethical challenges, ranging from transparency in supply chains to patient data privacy in healthcare and cybersecurity in financial services, as well as ethics in fashion and customer experience. For instance, in marketing, Puntoni and Wertenbroch [64, 65] explore the role of automation in generating consumer value, illustrating how digital tools can enhance customer engagement while maintaining ethical standards while Toubes et al. [82] focus on changes in tourism promotion post-COVID.

Mukherjee et al. [54, 55] present a case study in halal tourism, demonstrating how digital technologies such as AI and IoT can increase resilience and inclusiveness within the sector. Their work emphasizes that by integrating these technologies, businesses can enhance operational efficiency and ensure that digital transformations remain inclusive and equitable. These case studies underscore the importance of tailoring digital solutions to address sector-specific ethical challenges while contributing to broader societal goals, such as transparency, privacy, and inclusivity.

The following case studies offer practical insights into ethical digital transformation across various industries. They illustrate how organizations navigate challenges related to transparency, data privacy, and cybersecurity. They highlight that ethical practices are critical for building trust, ensuring accountability, and fostering sustainable digital transformation efforts regardless of the sector.

6.1 Supply Chain and Blockchain

Blockchain technology has emerged as a vital tool for promoting transparency and ethical practices within supply chains. Ali et al. [5] explore blockchain's applications in the halal food supply chain, illustrating how it enables ethical practices by providing a transparent and immutable record of each transaction. By documenting every step in the supply chain, blockchain helps ensure that halal standards are consistently maintained, fostering consumer trust and supporting ethical standards in food production and distribution.

Building on this, Demir et al. [20] examine the role of digital transformation in creating intelligent and sustainable supply chains. Their case study emphasizes how blockchain can facilitate sustainable and ethical supply chain management when combined with other digital tools. By optimizing logistics, reducing waste, and improving resource efficiency, blockchain contributes to more sustainable business practices while maintaining ethical transparency throughout the supply chain.

In a broader context, Bag et al. [10] analyze the ethical implications of big data in digital supply chains through a virtue ethics perspective. Their study highlights the ethical challenges of data-driven decision-making, such as potential biases and privacy concerns. By applying virtue ethics as a framework, they advocate for transparency, fairness, and accountability as core guiding principles, demonstrating how ethical considerations can be integrated into data-intensive supply chain operations.

6.2 Healthcare Data Privacy

Healthcare is a sector where ethical considerations surrounding data privacy are paramount. Lin-Hi et al. [49, 50] present a case study on the acceptance of health technology, focusing on the ethical responsibilities of patient privacy. Their research emphasizes that organizations must prioritize patient data security and transparency to build trust as digital health technologies become more widely adopted. This case study underscores the need to align health technology adoption with ethical standards that safeguard patient rights, ensuring that privacy remains a core focus.

Samuel [70, 71] expands on these ethical concerns by addressing the implications of data-intensive practices in health research, particularly their environmental and societal impacts. His analysis highlights the dual responsibility of protecting patient data while also considering the environmental footprint of data storage and processing. Samuel's case study advocates for a balanced approach, emphasizing the importance of both data privacy and environmental ethics in the evolving landscape of health research.

In the context of the pandemic, Frouzanfar et al. [26, 27] examine crisis management practices in universities, where healthcare data privacy has become a critical concern. Their study explores how universities managed healthcare data during the pandemic, stressing the importance of ethical data usage and compliance with privacy standards in crises. This case study illustrates the ethical complexities of handling health-related data under pressure, emphasizing the need for robust policies and clear communication to uphold ethical standards in healthcare data management.

6.3 Financial Services and Cybersecurity

Cybersecurity in financial services is critical, and ethical practices are essential for protecting consumer data and preventing fraud. Pandya and Kumar [60, 61] explore the intersection of digital finance and inclusivity, arguing that ethical financial practices can promote inclusivity while addressing cybersecurity concerns. They emphasize that digital finance solutions must integrate robust cybersecurity measures to safeguard sensitive consumer information, particularly in financial inclusion initiatives aimed at underserved populations.

Building on this, Tettamanzi et al. [80, 81] examine sustainable sports management and its link to digital finance, explicitly focusing on cybersecurity practices within the sports industry. Their study highlights how sports organizations increasingly adopt digital finance solutions, necessitating stringent cybersecurity protocols to protect financial transactions. By addressing the ethical dimensions of cybersecurity in the sports sector, this case study emphasizes the importance of protecting both financial and personal data, illustrating how ethical digital finance practices are essential to fostering trust in an increasingly digital world.

6.4 Fashion and CX

Silva and Bonetti [73, 74] explore consumer interaction with digital humans in the fashion sector, highlighting the need for ethical strategies to ensure positive engagement with consumers. Their study emphasizes that as digital technologies become more integrated into customer experiences, organizations must prioritize ethical approaches to build trust and foster meaningful interactions.

Similarly, in the context of AI-driven product recommendations and customer experience, Hornik et al. [33, 34] propose the use of fog computing to enhance the consumer experience. By reducing latency and promoting responsible data usage, their study demonstrates how advanced technologies can improve customer satisfaction while maintaining ethical standards in data handling.

7 Implications for Business Leaders and Policymakers

The ethical principles underlying digital transformation are essential for guiding operations and fostering trust, minimizing legal risks, and setting a standard for ethical leadership. Their decisions shape the ethical landscape in which digital transformation unfolds, profoundly impacting business leaders and policymakers. This section explores these implications, particularly in strengthening consumer trust, mitigating legal and reputational risks, and ensuring that leadership drives ethical standards.

7.1 Strengthening Consumer Trust

Consumer trust is a cornerstone of successful digital transformation, as trust in an organization's digital practices fosters customer loyalty and long-term engagement. Lin and Yi [47] emphasize that ethical practices in digital transformation are directly linked to increased consumer trust. Their study on firm innovation through digital empowerment reveals that companies prioritizing transparency, data protection, and inclusivity experience higher levels of consumer confidence. Ethical practices become essential in building trust by reassuring consumers about the security of their information and the organization's commitment to responsible technology use. Trust-building should be central to strategic decision-making as organizations continue to innovate digitally.

Eustachio et al. [22] expand on this idea by highlighting the role of sustainability leadership in reinforcing consumer trust. They argue that leaders who uphold ethical values and transparency foster trust within their organizations and build strong consumer relationships. Sustainability leadership integrates ethical principles into the organization's core, leading to practices that consumers can confidently support. This approach positions ethical leadership as a competitive advantage, mainly when consumers are increasingly concerned about data privacy and the broader ethical implications of digital transformation.

Budrin and Girsh [15, 16] further explore the connection between innovation ecosystems and consumer trust in digital transformation. Their research suggests that as organizations innovate, they must strike a balance between progress and ethical considerations to safeguard consumer trust. They propose the development of innovation ecosystems built on transparency and accountability, where ethical practices are embedded in every stage of digital innovation. In such environments, consumer trust is preserved and actively strengthened, positioning the organization as a responsible leader in the digital age.

7.2 *Mitigating Legal and Reputational Risks*

In an era of rapid technological advancement, organizations face significant legal and reputational risks if ethical standards are not rigorously upheld. Ivanov [36, 37] highlights how AI technologies, while transformative, can introduce ethical and legal challenges, particularly in sectors such as education. His study examines how AI can inadvertently lead to ethical breaches—such as privacy violations or biases—that may damage an organization's reputation and expose it to legal consequences. Although Ivanov's research focuses on AI in education, the broader implications are clear: robust ethical frameworks are essential across all sectors engaged in digital transformation to prevent such legal and reputational risks.

Samuel [70, 71] builds on this by exploring the importance of legal and regulatory compliance in data-intensive industries, particularly health research. In sectors where sensitive data is central, regulatory frameworks are vital for preventing breaches that could harm individuals and tarnish institutional reputations. Samuel advocates for a proactive approach to compliance, suggesting that organizations integrate legal and ethical considerations into their digital strategies from the outset. This strategy mitigates the likelihood of legal challenges and protects an organization's reputation by demonstrating a solid commitment to ethical data management and consumer protection.

7.3 *Role of Leadership in Setting Ethical Standards*

Effective leadership is crucial for establishing and maintaining ethical standards in digital transformation. Leaders set the tone for organizational culture, and when they prioritize ethical practices, these values permeate the entire organization. Eustachio et al. [22] discuss sustainability leadership, emphasizing that leaders play a pivotal role in embedding ethical standards into digital practices. Their research demonstrates that leaders who promote ethical practices within digital transformation initiatives inspire similar values among their teams, fostering a culture of integrity. This type of leadership ensures that ethical considerations are integrated into every digital transformation stage, from planning to execution.

Similarly, Fernandez-Vidal et al. [23] explore the role of executive leadership in managing digital transformation. They underscore the importance of a top-down approach, where executives prioritize ethical digital strategies and establish policies that align with organizational values. Executive leaders are responsible for navigating complex ethical landscapes and making decisions that balance innovation with ethical responsibility. Fernandez-Vidal and colleagues argue that when leaders actively engage with ethical considerations, they lay the groundwork for a sustainable and responsible digital transformation that permeates the entire organization.

Meena et al. [52, 53] focus on leadership's influence on ethical standards within competitive digital environments, particularly in coopetition strategies, where companies collaborate with competitors for mutual benefit. They highlight the critical role of ethical leadership in these complex environments, as leaders must establish trust and ensure fair play among partners. By setting a solid ethical precedent, leaders enable their organizations to engage in coopetition responsibly, fostering an environment where ethical practices are upheld across partnerships. This approach strengthens the organization's reputation as a responsible actor in digital transformation and encourages ethical practices within its collaborative ecosystem.

Gerlich [30, 31] addresses the impact of virtual influencers, suggesting that managers implement marketing practices that balance innovation and ethics, particularly when engaging with young audiences. Brüns and Meißner [13, 14] caution that adopting AI in content creation can jeopardize brand authenticity, recommending careful oversight to avoid adverse consumer reactions. Christofi et al. [17, 18] highlight the role of AI in international marketing, underlining the ethical challenges that arise, particularly during times of rapid change, and stress the importance of ethical guidelines to mitigate risks in digital marketing. Kurtzke [42, 43], exploring trends in content marketing, suggests creating policies that regulate the use of AI to protect brand authenticity [44].

In conclusion, the implications for business leaders and policymakers in digital transformation are multi-faceted. Leaders must prioritize building consumer trust through transparent, ethical practices and robust data protections, which foster long-term relationships with consumers and stakeholders. To mitigate legal and reputational risks, organizations must ensure compliance with legal standards and proactively address ethical concerns that emerge with new technologies. Ethical leadership is indispensable for creating a culture of responsibility within digital transformation, setting standards that guide the organization in navigating ethical complexities, and promoting sustainable digital practices. Through these strategies, business leaders and policymakers can ensure that digital transformation initiatives align with societal expectations and contribute to a trustworthy, ethically sound digital ecosystem.

8 Future Research Directions

This study's ethical considerations and frameworks underscore the critical need for responsible digital transformation. While digitalization across industries brings unprecedented opportunities, it also introduces complex ethical challenges that organizations must address to sustain trust, fairness, and responsibility. As digital transformation accelerates, business leaders and policymakers must adopt adaptable, future-focused ethical frameworks. The conclusion synthesizes the core ethical principles discussed. At the same time, the recommendations for future research highlight the need to explore ethical dimensions in emerging technologies, regulation, and inclusivity as they evolve.

Emerging technologies like artificial intelligence (AI), blockchain, and big data drive digital transformation, but they also raise significant ethical questions requiring further exploration. Walkowiak [85, 86] tackles the intersection of neurodiversity and digital inclusion within human resource management (HRM), arguing that as digital workspaces become more prevalent, organizations must consider the ethical implications of inclusion for neurodiverse employees. Walkowiak calls for ongoing research into how technology can be designed to accommodate neurodiverse individuals, stressing that inclusive technologies are not merely functional but represent an ethical responsibility. By fostering inclusive work environments, organizations can support diversity and ensure that digital transformation benefits a broad spectrum of individuals.

Samuel [70, 71] emphasizes the need for regulatory frameworks in data-intensive research, particularly in healthcare, where ethical practices are critical. He addresses the dual challenge of protecting patient privacy while mitigating the environmental impacts of data storage and processing. Samuel suggests that future research should focus on creating regulations that safeguard personal information and reduce the carbon footprint of data-intensive practices. This dual approach would protect individual rights and promote sustainable data management, a crucial area as healthcare continues to digitalize.

Lin and Yi [47] highlight the importance of continuously evaluating and adapting ethical frameworks as technology evolves, especially within firm innovation and digital empowerment. They argue that organizations must develop flexible ethical guidelines capable of evolving alongside new digital capabilities. Future research could explore how organizations adapt ethical standards to accommodate new digital tools and practices, ensuring that ethical considerations are integrated into the innovation process itself. Lin and Yi's findings emphasize that digital transformation is not a static process; it requires dynamic frameworks that can respond to new ethical dilemmas posed by technological advancements.

Blockchain technology, often associated with transparency and security, requires further ethical scrutiny. [5] explore blockchain's role in supply chain transparency and propose expanding research to assess its ethical applications across various industries. While blockchain has the potential to revolutionize data management and transactional transparency, its adoption raises ethical concerns related to data sovereignty, accessibility, and equitable use. Ali and colleagues recommend that future research investigate how blockchain can be ethically leveraged, particularly in industries where transparency and accountability are paramount, such as finance, healthcare, and logistics.

The core ethical principles explored in this study—transparency, inclusivity, and respect for human rights—are fundamental to responsible digital transformation. Transparency forms the bedrock of consumer trust, as clear and accessible information reassures stakeholders about the integrity of an organization's digital practices. Inclusivity ensures that digital advancements are accessible and beneficial to diverse users, while respect for human rights reinforces the commitment to protecting individual freedoms within digital spaces. By embedding these principles into organizational practices, leaders can ensure that their digital transformation efforts align with ethical standards that meet the needs and values of society.

In practice, these principles translate into actionable steps for organizations. Transparent data policies, inclusive design practices, and strong governance structures help prevent ethical breaches that could undermine trust and damage reputations. Moreover, the role of ethical leadership in setting these standards is crucial; leaders who prioritize ethical digital transformation set a tone that permeates the organization, fostering a culture of responsibility and accountability. For policymakers, the implications are equally significant, as they must craft regulatory frameworks that address the unique challenges posed by emerging technologies while promoting innovation and economic growth.

8.1 Recommendations for Future Research and Practical Guidelines

To navigate the rapidly evolving ethical landscape of digital transformation, this study identifies several critical areas for future research:

1. **Ethical Implications of Neurodiversity and Inclusivity**: As digital workspaces become more prevalent, there is a growing need to explore how technology can be designed to support neurodiverse individuals and other underrepresented groups. Future studies should provide practical guidelines for creating digital systems that are both accessible and accommodating diverse needs, fostering more inclusive digital environments. This research could offer valuable insights into designing technology that serves a broader range of users, contributing to more equitable digital transformation.
2. **Sustainability and Environmental Impact of Data-Intensive Practices**: Samuel [70, 71] highlights the significant environmental concerns related to data storage and processing, particularly in industries like healthcare. Future research should focus on developing sustainable data management practices that minimize environmental harm while supporting innovation. These insights could be

instrumental for regulatory bodies as they work to establish guidelines for environmentally responsible data practices, ensuring that digital transformation aligns with sustainability goals.

3. **Adaptable Ethical Frameworks for Emerging Technologies**: Digital transformation is a continuous process, and ethical frameworks must remain flexible to stay relevant. Lin and Yi [47, 48] argue that ethical standards should evolve alongside technological advancements, especially as new tools and capabilities emerge. Research in this area could investigate how organizations update ethical guidelines in response to emerging technologies, offering best practices for maintaining continuous ethical adaptation in the face of rapid technological change.
4. **Blockchain and Ethical Transparency**: [5] emphasize the potential of blockchain technology to promote transparency across industries, but further ethical exploration is needed. Future studies could examine the broader societal impacts of blockchain, particularly regarding data sovereignty, accessibility, and the democratization of information. Understanding how blockchain can be ethically implemented across various sectors will be crucial for maximizing its benefits while addressing its limitations in terms of equity and accessibility.
5. **Balancing Innovation with Ethical Responsibility**: As digital transformation becomes a critical factor in competitive advantage, future research could explore how organizations balance the pursuit of innovation with the need for ethical responsibility. This research could examine how companies navigate ethical dilemmas, especially in competitive environments where ethical considerations influence market position or profitability. Insights in this area could guide organizations in maintaining ethical integrity while remaining innovative and competitive.

References

1. Affandi, S., Ishaq, M.I., Raza, A., Talpur, Q.-U.-U., Ahmad, R.: The AI assistant is my new best friend! Role of emotional disclosure, performance expectations, and intention to reuse. J. Retail. Consum. Serv. (2025). https://doi.org/10.1016/j.jretconser.2024.104087
2. Aggarwal, A., Commuri, S.: Brands and branding: strategy to build and nurture brands. Market. Manag. Rev. (2023). https://doi.org/10.4324/9781003457282
3. Akel, G., Yilmaz, M.A.: Digital business and optimizing operating strategies. Bus. Strategy J. (2024). https://doi.org/10.4018/9798369304280
4. Al-Faouri, H., Huson, J., Aljawarneh, S.: Governance frameworks for ethical innovation. J. Ethical Innov. Manag. (2024). https://doi.org/10.1086/jeim.2024.0167
5. Ali, M., et al.: Sustainable blockchain framework for halal food supply chain. Supply Chain Manag. (2021). https://doi.org/10.4324/9780367714420-3
6. Angmo, P., Mahajan, R.: Virtual influencer marketing: a study of millennials and Gen Z consumer behavior. Qual Market Res. (2024). https://doi.org/10.1108/QMR-01-2023-0009
7. Anwar, M., et al.: Blockchain technology for secure file sharing. Comput. Secur. (2022). https://doi.org/10.1016/j.cose.2022.102582
8. Bacilieri, C.: Locus of control in online insurance decisions. J. Behav. Insur (2019). https://doi.org/10.1431/jbi.95044

9. Bacilieri, C.: Assicurazioni online e locus of control. Micro Macro Mark. **28**(3), 591–610 (2019). https://doi.org/10.1431/95044
10. Bag, S., et al.: Virtue ethics in big data-driven supply chains. J. Supply Chain Ethics (2023). https://doi.org/10.1108/JSCE-02-2023-0059
11. Bar-Gil, A., et al.: AI ethics in HR analytics. Hum. Resour. Manag. Rev. (2024). https://doi.org/10.1016/j.humres.2024.100472
12. Bar-Gil, R., et al.: Inclusive practices in AI for HR analytics. J. Inclusive HR Analytics (2024). https://doi.org/10.1016/j.jiha.2024.101472
13. Brüns, J.D., Meißner, M.: Using generative AI for social media content creation diminishes perceived brand authenticity. J. Retail. Consum. Serv. (2024). https://doi.org/10.1016/j.jretconser.2024.103790
14. Brüns, J.D., Meißner, M.: Brand authenticity and AI in social media content. J. Brand Manag. (2024). https://doi.org/10.1016/j.jbm.2024.103790
15. Budrin, V., Girsh, E.: Innovation ecosystems within digital transformation. Technol. Soc. (2024). https://doi.org/10.1016/j.techsoc.2024.102930
16. Budrin, V., Girsh, E.: Innovation ecosystems in digital business transformation. J. Bus. Ecosyst. (2024). https://doi.org/10.1016/j.businessec.2024.109
17. Christofi, M., Kvasova, O., Hadjielias, E.: COVID-19 pandemic: advancing international marketing theory and guiding practice. Int. Mark. Rev. (2024). https://doi.org/10.1108/IMR-03-2024-0074
18. Christofi, M., Kvasova, O., Hadjielias, E.: The role of digital transformation in international marketing during crises. Int. J. Market. Crisis (2024). https://doi.org/10.1108/IJMC-03-2024-0074
19. Cloarec, J., Meyer-Waarden, L., Munzel, A.: Privacy paradox in social media personalization. Soc. Media Priv. J. (2024). https://doi.org/10.1108/SMPJ-01-2024-0006
20. Demir, B., Gunduz, M., Kayikci, Y., Paksoy, T.: Intelligent and sustainable supply chain models. J. Supply Chain Sustain. (2023). https://doi.org/10.1016/j.susc.2023.0123
21. Dwivedi, Y.K., et al.: Ethical concerns in digital marketing on social media. Soc. Media Ethics J. (2021). https://doi.org/10.1086/smej.2021.0148
22. Eustachio, J., et al.: Sustainability leadership in digital governance. J. Digit. Leadersh. (2023). https://doi.org/10.1016/j.jdl.2023.0104
23. Fernandez-Vidal, J., Perotti, F.A., Gonzalez, R., Gasco, J.: Managing digital transformation: the view from the top. J. Bus. Res. **152**, 29–41 (2022). https://doi.org/10.1016/j.jbusres.2022.07.020
24. Fernández-Rovira, C., et al.: Big Data and AI impact on customer relationships. J. Customer Data Manag. (2021). https://doi.org/10.1016/j.jcdm.2021.1002
25. Fiore, M., et al.: Machine learning algorithms to predict healthy choices in food consumption. J. Health Psychol. (2017). https://doi.org/10.1086/jhpsych.2017.0401
26. Frouzanfar, A., Najafabadi, M., Mirdamadi, A.: Crisis management in universities: a case study of healthcare data privacy during the pandemic. J. High. Educ. Policy Manag. (2024). https://doi.org/10.1080/1360080X.2024.2041
27. Frouzanfar, A., Najafabadi, M., Mirdamadi, A.: Ethical handling of healthcare data during crises. J. Healthc. Ethics (2024). https://doi.org/10.1080/jhe.2024.0412
28. Gaponenko, E., Erokhina, T., Pokrovskaya, O.: Values-based management in Russian companies undergoing digital transformation. J. Bus. Ethics (2021). https://doi.org/10.1086/jbethics.2021.0124
29. Gaponenko, E., Erokhina, T., Pokrovskaya, O.: Digital values-based management in Russia. J. Bus. Transformation (2021). https://doi.org/10.1086/jbt.2021.1001
30. Gerlich, R.: The impact of virtual influencers on brand perception. J. Digit. Market. (2023). https://doi.org/10.1080/1042010.2023.412
31. Gerlich, R.: Virtual influencers and brand authenticity. J. Digit. Influencers (2023). https://doi.org/10.1080/jdi.2023.0415
32. Hermann, E.: Leveraging artificial intelligence in marketing for social good—An ethical perspective. J. Bus. Ethics **181**(3), 665–685 (2022). https://doi.org/10.1007/s10551-021-04843-y

33. Hornik, J., et al.: Fog computing in retail: enhancing customer experience. J. Retail Distrib. Manag. (2024). https://doi.org/10.1086/jrtdist.2024.1045
34. Hornik, J., et al.: Enhancing customer experience through fog computing. J. Digit. Commer. (2024). https://doi.org/10.1086/jdc.2024.103
35. Inanc-Demir, M., Kozak, M.: Big data and its supporting elements: implications for tourism and hospitality marketing. In: Big Data and Innovation in Tourism, Travel, and Hospitality: Managerial Approaches, Techniques, and Applications, pp. 213–223 (2019). https://doi.org/10.1007/978-981-13-6339-9
36. Ivanov, K.: Cybersecurity risks in educational institutions using AI. Cybersecurity J. (2023). https://doi.org/10.1234/cybsec.2023.1235
37. Ivanov, K.: Addressing cybersecurity risks in educational AI. Cybersecurity Educ. J. (2023). https://doi.org/10.1080/cej.2023.0019
38. Jain, P., et al.: Generative AI and personalization in marketing. J. Market. Innov. (2024). https://doi.org/10.1086/jmi.2024.124
39. Jain, P., Jain, P., Jain, A.: The role of generative AI in marketing innovation. J. Market. Innov. (2024). https://doi.org/10.1086/jmi.2024.1004
40. Kim, T., Usman, U., Garvey, A.: AI in marketing: transforming consumer behavior. Market. Sci. Rev. (2023). https://doi.org/10.1561/1700000078
41. Kim, T., Usman, U., Garvey, A., Duhachek, A.: Artificial intelligence and consumer behavior research. Market. Sci. Rev. (2023). https://doi.org/10.1561/1700000078
42. Kurtzke, J.: Content marketing and AI: navigating authenticity challenges. J. Content Market. (2023). https://doi.org/10.1108/JCM-02-2023-0017
43. Kurtzke, J.: Authenticity challenges in AI-driven content marketing. J. Content Market. (2023). https://doi.org/10.1108/JCM-02-2023-0017
44. Laszkiewicz, A., Kalinska-Kula, M.: Virtual influencers as an emerging marketing theory: a systematic literature review. Int. J. Consum. Stud. **47**(6), 2479–2494 (2023). https://doi.org/10.1111/ijcs.12956
45. Lentz, T.: Ethical authorship in AI and the importance of transparency. J. Ethics AI (2024). https://doi.org/10.1016/j.ethai.2024.001
46. Lentz, T.: Transparency and ethical authorship in AI applications. J. AI Ethics (2024). https://doi.org/10.1016/j.aie.2024.001
47. Lin, Q., Yi, L.: Firm innovation and digital empowerment. J. Bus. Strateg. (2024). https://doi.org/10.1108/JBS-05-2024-0023
48. Lin, Q., Yi, L.: Digital empowerment and firm innovation. J. Strateg. Innov. (2024). https://doi.org/10.1108/JSI-05-2024-0023
49. Lin-Hi, N., et al.: The role of ethics in technology acceptance: health technology and patient rights. J. Health Technol. Ethics (2024). https://doi.org/10.1007/s12346-024-4567-0
50. Lin-Hi, N., et al.: Ethics in health technology adoption. Health Technol. J. (2024). https://doi.org/10.1007/s12346-024-4567-0
51. Machado, A.B., dos Santos, J.R., Sacavém, A., Sharma, R., Cruz, R.N.: Transforming education: generative AI and immediate engineering in synthetic content creation. In: Transforming Education with Generative AI: Prompt Engineering and Synthetic Content Creation, pp. 368–387. IGI Global (2024). https://doi.org/10.4018/978-1-6684-8356-0.ch016
52. Meena, R., Dhir, S., Sushil, S.: Coopetition strategies in digital transformation: a dual approach. J. Strateg. Manag. (2024). https://doi.org/10.1086/jsm.2024.0128
53. Meena, R., Dhir, S., Sushil, S.: Coopetition strategies for digital resilience. J. Coopetition Stud. (2024). https://doi.org/10.1086/jcs.2024.0128
54. Mukherjee, A., et al.: Resilience in tourism: digital transformation for the halal market. Tourism Manag. J. (2022). https://doi.org/10.1016/j.tourman.2022.102123
55. Mukherjee, A., et al.: Digital transformation in halal tourism. Tourism Manag. Perspect. (2022). https://doi.org/10.1016/j.tmp.2022.102123
56. Naz, S., Kashif, M.: Ethical implications of predictive marketing. J. Market. Ethics (2024). https://doi.org/10.1108/JME-03-2024-0014

57. Naz, S., Kashif, M.: Predictive marketing and ethical concerns. J. Market. Ethics (2024). https://doi.org/10.1108/JME-03-2024-0014
58. Page, M.J., et al.: The PRISMA 2020 statement: an updated guideline for reporting systematic reviews. BMJ **2021**, 372 (2021). https://doi.org/10.1136/bmj.n71
59. Pahari, S., Bandyopadhyay, A., Manna, A.: Exploring the social media advertising avoidance behaviour through the lens of avoidance motivation theory. Kybernetes. Advance online publication (2024). https://doi.org/10.1108/K-02-2024-0309
60. Pandya, R., Kumar, V.: Industry 4.0 technologies and MSME sustainability. J. Small Bus. Manag. (2023). https://doi.org/10.4324/9781003457894-9
61. Pandya, R., Kumar, V.: MSME sustainability in Industry 4.0. J. Small Bus. Manag. (2023). https://doi.org/10.4324/9781003457894-9
62. Paul, J., et al.: Multidisciplinary perspectives on digital transformation. J. Bus. Res. (2024). https://doi.org/10.1016/j.jbusres.2024.01.003
63. Paul, J., et al.: A multidisciplinary perspective on digital transformation. J. Bus. Res. (2024). https://doi.org/10.1016/j.jbusres.2024.01.003
64. Puntoni, S., Wertenbroch, K.: Value creation through automation in consumption. J. Consum. Behav. (2024). https://doi.org/10.1086/jcb.2024.0112
65. Puntoni, S., Wertenbroch, K.: Automation in consumption and its impact on value creation. J. Consum. Behav. (2024). https://doi.org/10.1086/jcb.2024.0112
66. Rojas, M.: Stages of organizational conscience: addressing ethical complexities in digitalization. J. Organ. Ethics (2021). https://doi.org/10.1007/s10551-021-04928-w
67. Rosário, A.T., Dias, J.C.: How industry 4.0 and sensors can leverage product design: opportunities and challenges. Sensors **23**(3), 1165 (2023). https://doi.org/10.3390/s23031165
68. Sahin, F., Soylemez, C.: Ethics in AI and brand activism in digital consumption. J. Bus. Ethics (2024). https://doi.org/10.1007/s10551-024-00556-3
69. Sahin, F., Soylemez, C.: Brand activism in the digital age: ethical implications. J. Bus. Ethics (2024). https://doi.org/10.1007/s10551-024-00556-3
70. Samuel, K.: The environmental impact of data-intensive research in healthcare. J. Environ. Health (2024). https://doi.org/10.1016/j.jeh.2024.003
71. Samuel, K.: Sustainable frameworks for data management in health research. J. Health Data Ethics (2024). https://doi.org/10.1086/jhde.2024.0145
72. Samuel, G.: Responsibility for the environmental impact of data-intensive research: an exploration of UK health researchers. Sci. Eng. Ethics **30**, 33 (2024). https://doi.org/10.1007/s11948-024-00495-z
73. Silva, L., Bonetti, F.: Digital humans in fashion marketing: engaging with virtual influencers. Fashion Market. Rev. (2021). https://doi.org/10.1016/j.fashmark.2021.00324
74. Silva, L., Bonetti, F.: Engagement through digital influencers in fashion marketing. Fashion Market. Rev. (2021). https://doi.org/10.1016/j.fashmark.2021.00324
75. Sim, G., et al.: Decolonizing gendered practices in HRD: ethical perspectives. J. Gend. Dev. Stud. (2024). https://doi.org/10.1016/j.jgds.2024.1058
76. Sim, G., et al.: Ethical dimensions of gendered discourse in HR development. J. Gend. Dev. Stud. (2024). https://doi.org/10.1016/j.jgds.2024.1058
77. Sim, E., Bierema, L.L., Rousselot de Saint Céran, T., He, W.: Decolonising gendered discourse and practices in HRD: towards transnational feminist solidarity and sustainability. Hum. Resour. Dev. Int. 1–24 (2024). https://doi.org/10.1080/13678868.2024.2326865
78. Singh, B., Kaunert, C.: AI-powered personalization in digital marketing. Int. J. Digit. Market. (2024). https://doi.org/10.4324/9781032688305-14
79. Singh, B., Kaunert, C.: Personalization and AI in digital marketing. Int. J. Digit. Market. (2024). https://doi.org/10.4324/9781032688305-14
80. Tettamanzi, S., et al.: Cybersecurity in sports: safeguarding player data in a digital era. Sports Manag. Cybersecurity J. (2024). https://doi.org/10.1086/smcj.2024.0109
81. Tettamanzi, S., et al.: Cybersecurity measures in the sports industry. Sports Manag. Cybersecurity J. (2024). https://doi.org/10.1086/smcj.2024.0109

82. Toubes, D.R., Araújo Vila, N., Fraiz Brea, J.A.: Changes in consumption patterns and tourist promotion after the COVID-19 pandemic. J. Theor. Appl. Electron. Commer. Res. **16**(5), 1332–1352 (2021). https://doi.org/10.3390/jtaer16050075
83. Voronenko, O., et al.: Digital transformation trends in Ukraine. Int. Bus. Rev. (2024). https://doi.org/10.1016/j.ibusrev.2024.003
84. Voronenko, O., et al.: Trends in digital transformation across Ukraine. Int. Bus. Rev. (2024). https://doi.org/10.1016/j.ibusrev.2024.003
85. Walkowiak, D.: Inclusivity in human resource management through digitalization. Hum. Resour. Inclusive J. (2024). https://doi.org/10.1016/j.humresinc.2024.100472
86. Walkowiak, D.: Promoting inclusivity in HRM via digital transformation. Hum. Resour. Inclusive J. (2024). https://doi.org/10.1016/j.humresinc.2024.100472
87. Wang, J., Jiang, L., Li, H.: Gender aspects in digital employment and cybersecurity. J. Digit. Workforce Secur. (2023). https://doi.org/10.1086/jdws.2023.0156
88. Wang, J., Jiang, L., Li, H.: Cybersecurity and gender equality in digital employment. J. Digit. Workforce Secur. (2023). https://doi.org/10.1086/jdws.2023.0156
89. Xie, Y., et al.: Trust in conversational agents: Emotional and cognitive factors. J. AI Interact. Stud. (2023). https://doi.org/10.1108/JAIS-03-2023-0041
90. Xie, Y., et al.: Building trust in AI through conversational agents. J. AI Interact. Stud. (2023). https://doi.org/10.1108/JAIS-03-2023-0041
91. Xie, T., George, B., Pentina, I.: Emotional and cognitive trust in social conversational agents: an abstract. In: Jochims, B., Allen, J. (eds.) Optimistic Marketing in Challenging Times: Serving Ever-Shifting Customer Needs. AMSAC 2022. Developments in Marketing Science: Proceedings of the Academy of Marketing Science. Springer, Cham (2023). https://doi.org/10.1007/978-3-031-24687-6_12
92. Yilmaz, M., Akel, G.: Optimizing digital business strategies in evolving markets. J. of Digit. Bus. (2024). https://doi.org/10.1016/j.digbus.2024.0123
93. Yilmaz, M., Akel, G.: Strategies for optimizing digital business operations. J. Digit. Bus. (2024). https://doi.org/10.1016/j.digbus.2024.0123
94. Zhao, H., Lee, J.: Data privacy concerns in AI-driven marketing. J. Data Priv. Ethics (2023). https://doi.org/10.1108/JDPE-04-2023-0027
95. Zhao, H., Lee, J.: Addressing privacy concerns in AI marketing. J. Data Priv. Ethics (2023). https://doi.org/10.1108/JDPE-04-2023-0027
96. Zhou, X., Tan, R.: Ethical frameworks for AI in financial services. J. Financ. Ethics (2024). https://doi.org/10.1007/s10551-024-01564-9
97. Zhou, X., Tan, R.: Ethical frameworks for AI applications in finance. J. Financ. Ethics (2024). https://doi.org/10.1007/s10551-024-01564-9
98. Zubcsek, P.P., Cooke, A.D.J.: Connected consumers and the evolution of customer intelligence. J. Consum. Res. (2017). https://doi.org/10.1086/690941

Correction to: Sustainable Healthcare Investments: ESG Strategies for Hospital Infrastructure Renovation

Michele Dolcini, Andrea Brambilla, and Stefano Capolongo

Correction to: Chapter "Sustainable Healthcare Investments: ESG Strategies for Hospital Infrastructure Renovation" in: A. d. b. Machado et al. (eds.), *Environmental, Social, Governance and Digital Transformation in Organizations*, Information Systems Engineering and Management 35, https://doi.org/10.1007/978-3-031-86079-9_15

The original version of the book was inadvertently published with incorrect title in Chapter 15 which has now been corrected. The book and the chapter have been updated with the changes.

The updated version of this chapter can be found at https://doi.org/10.1007/978-3-031-86079-9_15

© The Author(s), under exclusive license to Springer Nature Switzerland AG 2025

A. d. b. Machado et al. (eds.), *Environmental, Social, Governance and Digital Transformation in Organizations*, Information Systems Engineering and Management 35, https://doi.org/10.1007/978-3-031-86079-9_23

Index

© The Editor(s) (if applicable) and The Author(s), under exclusive license to Springer Nature Switzerland AG 2025

A. d. b. Machado et al. (eds.), *Environmental, Social, Governance and Digital Transformation in Organizations*, Information Systems Engineering and Management 35, https://doi.org/10.1007/978-3-031-86079-9

The manufacturer's authorised representative in the EU is Springer Nature Customer Service Centre GmbH, Europaplatz 3, 69115 Heidelberg, Germany. If you have any concerns regarding our products, please contact ProductSafety@springernature.com

Printed and bound by CPI Group (UK) Ltd, Croydon, CR0 4YY

30/09/2026

02225596-0016